ORIGIN AND EVOLUTION
OF
INTERPLANETARY
DUST

ASTROPHYSICS AND SPACE SCIENCE LIBRARY

A SERIES OF BOOKS ON THE RECENT DEVELOPMENTS
OF SPACE SCIENCE AND OF GENERAL GEOPHYSICS AND ASTROPHYSICS
PUBLISHED IN CONNECTION WITH THE JOURNAL
SPACE SCIENCE REVIEWS

PROCEEDINGS

VOLUME 173

ORIGIN AND EVOLUTION OF INTERPLANETARY DUST

PROCEEDINGS OF THE 126TH COLLOQUIUM OF THE
INTERNATIONAL ASTRONOMICAL UNION,
HELD IN KYOTO, JAPAN, AUGUST 27–30, 1990

Edited by

A. C. LEVASSEUR-REGOURD
Université Paris VI, Aeronomie CNRS, Verrières-le-Buisson, France

H. HASEGAWA
Osaka Sangyo University, Osaka, Japan

SPRINGER SCIENCE+BUSINESS MEDIA, B.V.

Library of Congress Cataloging-in-Publication Data

International Astronomical Union. Colloquium (126th : 1990 : Kyoto, Japan)
    Origin and evolution of interplanetary dust : proceedings of the 126th Colloquium of the International Astronomical Union, held in Kyoto, Japan, August 27-30, 1990 / edited by A.C. Levasseur-Regourd, H. Hasegawa.
        p. cm. -- (Astrophysics and space science library ; v. 173)
    Includes index.
    ISBN 978-0-7923-1365-6   ISBN 978-94-011-3640-2 (eBook)
    DOI 10.1007/978-94-011-3640-2
    1. Cosmic dust--Congresses. I. Levasseur-Regourd, Anny-Chantal.
II. Hasegawa, Hiroichi, 1926- . III. Title. IV. Series.
QB791.I663 1990
523.1'125--dc20   91-24311

TABLE OF CONTENTS

PREFACE
GROUPE PICTURE
LIST OF PARTICIPANTS
ORGANIZING COMMITTEES
NECROLOGY

I - INTERPLANETARY DUST :
  SPACE AND EARTH ENVIRONMENT STUDIES

3   PARTICULATE DETECTION IN THE NEAR EARTH SPACE ENVIRONMENT ABOARD
THE LONG DURATION EXPOSURE FACILITY LDEF : COSMIC OR TERRESTRIAL ?
(Invited review).
     J.A.M. McDonnell, K Sullivan, T.J. Stevenson and D.H. Niblett

11   STUDY OF COSMIC DUST PARTICLES ON BOARD LDEF AND MIR SPACE STATION
     J.C. Mandeville

15   THE PRESENT STATUS OF THE MUNICH DUST COUNTER EXPERIMENT ON BOARD
OF THE HITEN SPACECRAFT (Invited contribution)
     E. Igenbergs, A. Hüdephol, K. Uesugi, T. Hayashi, H. Svedhem, H. Iglseder,
     G. Koller, A. Glasmachers, E. Grün, G. Schwehm, H. Mizutani,
     T. Yamamoto, A. Fugimura, N. Ishii, H. Araki, K. Yamakoshi and K. Nogami

21   IN-SITU EXPLORATION OF DUST IN THE SOLAR SYSTEM AND INITIAL RESULTS FROM
THE GALILEO DUST DETECTOR (Invited review)
     E. Grün, H. Fechtig, M.S. Hanner, J. Kissel, B.A. Lindblad, D. Linkert,
     G. Morfill and H.A. Zook.

29   THE NASA SOLAR PROBE MISSION : IN SITU DETERMINATION OF INTERPLANETARY
OUT-OF-THE ECLIPTIC AND NEAR-SOLAR DUST ENVIRONMENTS
     B. T. Tsurutani and J. E. Randolph

33   COLLECTION OF COSMIC DUST FROM THE STRATOSPHERE
     W. Gucun, O. Ziyuan, X. Yiwen and W. Xiguang

37   DYNAMIC MODELLING TRANSFORMATIONS FOR THE LOW EARTH ORBIT SATELLITE
PARTICULATE ENVIRONMENT
     J.A.M. Mc Donnell, K. Sullivan, S.F. Green, T.J. Stevenson and
     D.H. Niblett

41   FACE-DEPENDENT IMPACT PROBABILITIES UPON LDEF FOR HELIOCENTRIC
PARTICLE ORBITS
     D. Steel

45 THE MUNICH DUST COUNTER : A COSMIC DUST EXPERIMENT ON BOARD OF THE
MUSES-A MISSION OF JAPAN
E. Igenbergs, A. Hüdepohl, K. Uesugi, T. Hayashi, H. Svedhem, H. Iglseder,
G. Koller, A. Glasmachers, E. Grün, G. Schwehm, H. Mizutani,
T. Yamamoto, A. Fujimura, N. Ishii, H. Araki, K. Yamakoshi and K. Nogami

49 COLLECTION OF STRATOSPHERIC MICROPARTICLES ABOVE THE SULFATE LAYER
USING BALLOON-BORNE COLLECTORS
J.R. Stephens, Y. Nakada, T. Onaka and F.J.M. Rietmeijer

53 A COSMIC MATTER ACCRETION EVENT AROUND 660 000 YEARS BEFORE PRESENT
FOUND IN TWO DATED, CENTRAL PACIFIC CORES
K. Yamakoshi, K. Nogami and R. Omori

57 PRELIMINARY STUDY ON NEOGENE MICROTEKTITES IN THE CORE COLLECTED
FROM NORTH PACIFIC
P. Hanchang, L. Zhenkun, Z. Shijie, M. Xueying and C. Zhifang

II - INTERPLANETARY DUST :
PHYSICAL AND CHEMICAL ANALYSIS

63 PHYSICAL AND MINERALOGICAL PROPERTIES OF ANHYDROUS INTERPLANETARY
DUST PARTICLES IN THE ANALYTICAL ELECTRON MICROSCOPE (Invited review)
J.P. Bradley

71 AQUEOUS ALTERATION IN HYDRATED INTERPLANETARY DUST PARTICLES
(Invited review)
K. Tomeoka

79 THE EFFECT OF TOTAL PRESSURE ON VAPORIZATION OF ALKALIS FROM
PARTIALLY MOLTEN CHONDRITIC MATERIEL
T. Shimaoka and N. Nakamura

83 CONDENTATION EXPERIMENTS OF MG-SILICATE MINERALS
A. Tsuchiyama

87 ULTRAVIOLET-INDUCED AMORPHIZATION OF CUBIC ICE AND ITS IMPLICATION FOR
THE EVOLUTION OF ICE GRAINS
A. Kouchi and T. Kuroda

91 SIMULATION IN LABORATORY OF SOLID GRAINS PRESENT IN SPACE
L. Colangeli, E. Bussoletti and V. Mennella

95 THE INFRARED SPECTRA OF SYNTHESIZED AMORPHOUS SILICATES WITH
COMPOSITIONS OF OLIVINE AND PYROXENE
C. Koike and A. Tsuchiyama

99   OPTICAL CONSTANTS OF KEROGEN FROM 0.15 TO 40 μm : COMPARISON WITH
METEORITIC ORGANICS
B.N. Khare, W.R. Thompson, C. Sagan, E.T. Arakawa, C. Meisse and
I. Gilmour

102   OPTICAL CONSTANTS OF BASALTIC GLASS FROM 0.0173 TO 50 μm
E.T. Arakawa, D.W. Young, J.M. Zhang, P.C. Eklund, B.N. Khare,
W.R. Thompson and C. Sagan

105   NOBLE METAL ENRICHMENTS IN COSMIC SPHERULES
K. Nogami, K. Misawa, R. Omori, M. Jianguo and K. Yamakoshi

109   STUDIES ON ISOTOPIC RATIOS OF OSMIUM AND IRIDIUM IN COSMIC SPHERULES
USING INSTRUMENTAL NEUTRON ACTIVATION ANALYSIS
K. Yamakoshi and K. Nogami

113   STRUCTURES OF AMORPHOUS SILICATE DUSTS SIMULATED BY MOLECULAR
DYNAMICS METHOD
A. Tsuchiyama and K. Kawamura

117   ASTROPHYSICAL INTERESTING COMPOUND GRAINS PRODUCED BY A GAS
EVAPORATION METHOD
C. Kaito and Y. Saito

121   MEASUREMENT OF FAR-INFRARED ABSORPTION FOR AMORPHOUS SILICATES
BETWEEN 27 AND 400 μm
C. Koike and H. Shibai

125   LABORATORY SPECTRA OF AMORPHOUS AND CRYSTALLINE OLIVINE : AN
APPLICATION TO COMET HALLEY IR SPECTRUM
A. Blanco, V. Orofino, E. Bussoletti, S. Fonti, L. Colangeli and J.R. Stephens

III - INTERPLANETARY DUST : ZODIACAL LIGHT AND OPTICAL STUDIES

131   THE ZODIACAL CLOUD COMPLEX (Invited review)
A.C. Levasseur-Regourd, J.B. Renard and R. Dumont

139   SPATIAL DISTRIBUTION AND ORBITAL PROPERTIES OF ZODIACAL DUST
(Invited review)
B. Kneissel and I. Mann

147   ON THE GEGENSCHEIN AND THE SYMMETRY PLANE
S.S. Hong and S.M. Kwon

151   ULTRAVIOLET OBSERVATIONS OF THE ZODIACAL LIGHT AND THE ORIGIN OF
INTERPLANETARY DUST GRAINS
C.F. Lillie

155   LIGHT SCATTERING BY DUST PARTICLES IN THE OUTER SOLAR SYSTEM
J.W. Hovenier and P.B. Bosma

159   LIGHT SCATTERING BY SOLAR SYSTEM DUST : THE OPPOSITION EFFECT AND
THE REVERSAL OF POLARIZATION
K. Muinonen and K. Lumme

163   THE OPTICAL PROPERTIES OF INTERPLANETARY DUST (Invited review)
P.L. Lamy and J.M. Perrin

171   THE INFRARED ZODIACAL LIGHT (Invited review)
M.S. Hanner

179   TEMPORAL AND SPATIAL VARIATIONS OF THE ATMOSPHERIC DIFFUSE LIGHT
S.M. Kwon, S.S. Hong, J.L. Weinberg

183   FINE RESOLUTION BRIGHTNESS DISTRIBUTION OF THE VISIBLE ZODIACAL LIGHT
S.M. Kwon, S.S. Hong, J.L. Weinberg and N.Y. Misconi

187   INTERPLANETARY DUST CLOSE TO THE SUN
I. Mann

191   FUTURE OBSERVATIONS OF THE F-CORONA WITH THE LASCO CORONAGRAPH
SPACE EXPERIMENT
P.L. Lamy, A. Llebaria, A. Maucherat, S. Koutchmy and F. Giovane

195   SYNTHETIC MAPS OF THE BRIGHTNESS AND POLARIZATION OF THE F-CORONA
Y. Fang, P.L. Lamy and A. Llebaria

199   OPTICAL PROPERTIES OF INTERPLANETARY DUST IN THE TANGENTIAL PLANE
J.B. Renard, A.C. Levasseur-Regourd and R. Dumont

203   SCATTERING CALCULATIONS ON THE BASIS OF FREDHOLM INTEGRAL EQUATION
METHOD
M. Matsumura and M. Seki

207   THE SCATTERING MATRIX OF RANDOMLY ORIENTED INFINITE CYLINDERS
P. Stammes

211   ASTEROIDAL DUST AND THE ZODIACAL EMISSION
W. T. Reach

IV - COMETARY DUST :
OBSERVATIONS AND EVOLUTION

217 SPECTROSCOPIC EVIDENCE OF ORGANIC MOLECULES RELEASED BY THE DUST
OF HALLEY'S INNER COMA
J. Clairemidi, P. Rousselot and G. Moreels

221 MODELING DUST FRAGMENTATION IN COMETS
I. Konno and W.F. Huebner

225 THE CONTRIBUTION OF LONG PERIOD COMETS TO THE INTERPLANETARY DUST
CLOUD
M. Fulle and G. Cremonese

229 LONG DUST TRAILS OF SHORT PERIOD COMETS (Invited contribution)
H.U. Keller and K. Richter

235 COMETS AS A SOURCE OF INTERPLANETARY AND INTERSTELLAR GRAINS
(Invited contribution)
F. Hoyle and N.C. Wickramasinghe

241 CHEMICAL COMPOSITION OF AN EMANATION FROM COMETS : IDENTIFICATION
OF THE 3 MICRON COMET FEATURE
A. Sakata, S. Wada and A.T. Tokunaga

245 SPECTROPOLARIMETRY OF COMET HALLEY
N. Visvanathan, Z. Meglick and D.T. Wickramasinghe

249 SCATTERING PROPERTIES OF COMETARY DUST BASED ON POLARIMETRIC
DATA
S. Mukai, T. Mukai and S.Kikuchi

253 SYNCHRONIC BAND AND ITS IMPLICATION IN THE COMETARY DUST
J.I. Watanabe and K. Nishioka

257 ICE PARTICLE EMISSION FROM COMETARY ANALOGUE SAMPLES
H. Kohl and E. Grün

261 THE DUST IN THE COMA OF COMET HALLEY
J.I. Hage and J.M Greenberg

265 POLYOXYMETHYLENE IN COMETARY DUST : LABORATORY TESTS
D.C. Boice, D.W. Naegeli and W.F. Huebner

269 ON THE ANTI-TAIL OF COMET BRADFIELD 1987XXIX
H. Akisawa, T. Oka and K. Sugawara

273 FORMATION MECHANISMS OF THE SPLIT TAIL OF COMET BRADFIELD 1987XXIX
K. Sugawara and J.I. Watanabe

277 SPATIAL DISTRIBUTION AND COLOR OF DUST IN HALLEY'S INNER COMA
J. Clairemidi, E. Brandon, P. Rousselot and G. Moreels

281 PENETRATION OF HYPERVELOCITY PROJECTILES INTO LOW DENSITY MATERIALS
A. Fujiwara, T. Kadono, A. Nakamura, T. Ishibashi and N. Fujii

285 POLARIMETRIC PROPERTIES OF HALLEY'S DUST
A.K. Sen, M.R. Deshpande and U.C. Joshi

V - METEOROIDS AND METEOR STREAMS

291 THE ORBITAL DISTRIBUTION AND ORIGIN OF METEOROIDS (Invited review)
D. Steel

299 A STUDY OF METEOR ORBITS OBTAINED IN JAPAN
B.A. Lindblad

303 THE MICROMETEOROID IN THE UPPER ATMOSPHERE
F. Kamijo

307 AN ESTIMATION OF METEOROID FLUX AT OUTER MARTIAN SPACE FOR STEADY
METEOR STREAMS
K. Nagasawa

311 THE IAU METEOR DATA CENTER IN LUND
B.A. Lindblad

315 η LYRID METEOR STREAM ASSOCIATED WITH COMET IRAS-ARAKI-ALCOCK,
1983VII
K. Ohtsuka

319 THE ANNUAL VARIATION OF RADIO METEOR ECHOES OBSERVED FROM 1981 TO
1985
K. Suzuki

323 LIFETIME OF METEOR STREAMS ASSOCIATED WITH COMET HALLEY
M. Hajdukova and A. Hajduk

327 THE TAURID COMPLEX : GIANT COMET ORIGIN ?
D.I. Steel, D.J. Asher and S.V.M. Clube

331 MASS DISTRIBUTION AND BULK DENSITY DISTRIBUTION OF INTERPLANETARY
DUST
A. Hajduk

335 THE INTERNATIONAL METEOR ORGANIZATION
M. Gyssens, A. Knöfel J. Rendtel and P. Roggemans

VI - CIRCUMPLANETARY DUST : COLLISIONAL AND ELECTROSTATIC PROCESSES

341 PHYSICAL PROCESSES ON CIRCUMPLANETARY DUST (Invited review)
J. A. Burns

349 DUST IN PLANETARY RING SYSTEMS (Invited review)
M. R. Showalter

357 READING SATURN'S RING SPOKES
L.R. Doyle and E. Grün

361 CATASTROPHIC DISRUPTION OF SOLID BODIES BY COLLISION EXPERIMENTAL
APPROACH (Invited review)
A. Fujiwara

367 METHODS, DIFFICULTIES AND FIRST RESULTS IN LABORATORY SIMULATION OF
COSMIC DUST ELECTRIC CHARGING
J. Sveska and E. Grün

371 ELECTROSTATIC FRAGMENTATION OF IRREGULARLY SHAPED PARTICLES
T. Mukai

375 PLASMA EMISSION FROM HIGH VELOCITY IMPACTS OF MICROPARTICLES
ONTO WATER ICE
T. Timmermann and E. Grün

379 VELOCITY DISTRIBUTION OF FRAGMENTS IN COLLISIONAL BREAKUP
A. Nakamura and A. Fujiwara

383 JETS OF FRAGMENTS FROM CATASTROPHIC BREAK-UP AND THEIR
ASTROPHYSICAL IMPLICATIONS
G. Martelli, P. Rothwell, P.N. Smith, I. Giblin, J. Martinsson, E. Ducrocq,
M. Wettstein, M. Di Martino and P. Farinella

VII - ORIGIN OF INTERPLANETARY DUST : FROM COMETS AND ASTEROIDS, BACK TO INTERSTELLAR DUST

389 COMETARY AND ASTEROIDAL SOURCES OF INTERPLANETARY DUST
(Invited review)
M.V. Sykes

397 CONSTRAINTS ON THE PARENT BODIES OF COLLECTED INTERPLANETARY DUST
PARTICLES (Invited review)
S.A. Sandford

405   CHARACTERISTICS OF INTERSTELLAR AND CIRCUMSTELLAR DUST (Invited
      review)
          A.G.G.M. Tielens

413   CHEMICAL COMPOSITION OF DUST EXPECTED FROM CONDENSATION MODELS
      (Invited review)
          T. Yamamoto

421   DISTRIBUTION OF DUST IN THE DISK AROUND BETA PICTORIS
          T. Nakano

425   OFF-DISK IMPLANTATION OF EARLY SOLAR WIND INTO A PLANETESIMAL-DUST
      CLOUD
          S. Sasaki

429   COMPARISON OF 3 MICRON FEATURES OF TRAPPED H_2O AND H_2O FROST IN SiO
      CONDENSATE WITH OBSERVED DUST FEATURES
          S. Wada, A. Sakata and A.T. Tokunaga

433   THE SOURCE COMPOSITION OF GALACTIC COSMIC RAYS AS POSSIBLY
      ORIGINATED FROM THE DUST IN THE CIRCUMSTELLAR AND INTERSTELLAR
      SPACE
          K. Sakurai

437   LABORATORY STUDIES OF GRAIN MANTLES IN INTERSTELLAR SPACE
          C.X. Mendoza-Gomez and J.M. Greenberg

COLLOQUIUM SUMMARY

443   THE INTERPLANETARY MEDIUM IS THRIVING (Invited review)
          J.M. Greenberg

AUTHOR INDEX

PREFACE

THE KYOTO COLLOQUIUM

It has almost become a tradition to periodically review the progress of our knowledge of interplanetary dust at an interdisciplinary level. After the Honolulu (1967), Heidelberg (1975), Ottawa (1979) and Marseilles (1984) meetings, it was decided to hold a meeting specially devoted to the Origin and Evolution of Interplanetary Dust in Kyoto, Japan on 27-30 August 1990. This colloquium was certainly appropriate in location, timing and objective.

The choice of the location was most appropriate, not only because of the charm of the beautiful city of Kyoto, but also because of the important involvement of our Japanese colleagues, both in observations of interplanetary, cometary, meteoritic, circumplanetary or circumstellar dust, and in physico-chemical analysis or theoretical developments. We owe many thanks to the Local Organizing Committee and to the Japanese supporting organizations (ICRR, NAO, ISAS, KIT) for the efficient and most pleasant way in which colloquium was organized.

The timing of the meeting was equally appropriate. During the eighties, new knowledge had emerged from comprehensive studies of cometary flybys or remote observations, while infrared space observations of asteroidal dust bands, cometary debris trails, and thermal emission from interplanetary dust cloud. More recently, new in-situ data were provided by the Long Duration Exposure Facility, the space station Mir, the Hiten spacecraft and the Galileo space probe; simultaneously the question of the interrelation between interstellar or circumstellar dust and interplanetary dust received much attention.

The colloquium was sponsored by commission 21 (Light of the Night Sky) from the International Astronomical Union, and cosponsored by commission 15 (Physical Study of Comets, Minor Planets and Meteorites) and commission 22 (Meteors and Interplanetary Dust) of the IAU, together with commission B of COSPAR. The Science Organizing Committee represented a broad range of nationalities and expertise. The programme, with 20 invited papers, 30 contributed talks and 60 poster presentations, attracted 122 registered participants from 15 different nations.

THE PROCEEDINGS

The order of presentation of these proceedings virtually follows the order in the meeting: Interplanetary dust by space and Earth environment studies, with special emphasis on new results (Part I); interplanetary dust by physical and chemical analysis, along with laboratory simulations on relevant ices and mineral (Part II); interplanetary dust by zodiacal light and optical studies with interpretation of the observations in the infrared range (Part III); Cometary dust, observations and evolution (Part IV); Meteoroids and meteor streams (Part V); Circumplanetary dust, collisional and electrostatic processes (Part VI); Origin of interplanetary dust, from comets as asteroids, back to interstellar and circumstellar dust (Part VII). The colloquium summary, prepared by Mayo Greenberg, shows - from its multifaceted approach - just how much the interplanetary medium is thrieving; it suggests that the question of the dual source of the interplanetary dust has now been answered, the final stage being to trace asteroids and comets back to the interstellar cloud out of which the solar system was born.

The great interest in the colloquium was obvious to all the participants and the editors have attempted to produce conference proceedings reflecting the quality of so many fine presentations on a wide range of interrelated topics. The discussion remarks were handed out at the end of the meeting to the authors, who were recommended to incorporate them in their manuscript. All articles were scrutinized by various experts, and the revised manuscripts were later edited, either in France or in Japan. It is indeed a great pleasure to thank the following referees for their help: J.A. Burns, N. Fujii, A. Fujiwara, T. Fukuoka, E. Grün, J.M. Greenberg, M.S. Hanner, I. Hasegawa, S.S. Hong, D. Jewitt, H. U. Keller, C. Koike, P. Lamy, A.C Levasseur-Regourd, B.A. Lindbad, K. Lumme, T. Maihara, J.A.M. McDonnell, H. Mizutani, S. Mukai, T. Mukai, Y. Nakagawa, T. Nakano, K. Nogami, A. Sakata, H. Tanabe, N.C. Wickramasinghe, K. Yamakoshi, T. Yamamoto. As is likely in a colloquium, not all the papers represent the points of view of the majority of the participants. It is nevertheless to be emphasized that the efforts of the referees are appreciated and their comments have been taken into account by the authors, to the benefit of the reader.

Even with many published papers, such a book has to remain a convenient tool for future daily work. It was therefore decided that the proceedings should not exceed 500 pages and that the emphasis would be put on conciseness: eight pages for invited reviews, six pages for invited contributions and four pages for contributions and posters. Besides, all the participants wanted the proceedings to be published in less than a year after the colloquium. Strict deadlines were imposed for manuscript submission, refereeing procedures, final submission and editing. It is indeed a pleasure to thank all the authors for their kind and efficient cooperation.

The editorial was mainly carried out at Université Pierre et Marie Curie (Paris VI)/Service d'Aéronomie du CNRS. I especially thank Pierrette Montagné, Edwige Regnault, Jean-Baptisie Renard and Emmanuel Villame for their assistance. The book has been produced from camera ready manuscripts prepared in 15 different countries all over the world. It nevertheless contains very few non-uniformities, since some papers, together with all the headings, were appropriately retyped by the publisher.

Finally, I wish to express my gratitude to Tadashi Mukai, secretary of Science oganizing Committee, and contact person with the publisher in Japan for his invaluable help both at the meeting and in the preparation of the proceedings.

A. Chantal Levasseur-Regourd
Pairs, April 1991

Acknowledgements

The IAU Colloquium #126 "ORIGIN AND EVOLUTION OF INTERPLANETARY DUST" was held in August 27~30, 1990 at the Kyoto Auditorium.

The "Interplanetary Dust Community" had held four meetings in the past; Honolulu (1967), Heidelberg (1975), Ottawa (1979) and Marseille (1984). Kyoto Colloquium is the 5th meeting in this series.

This Colloquium was sponsored by IAU Commissions 15, 21 and 22 and also COSPAR B-1. The supporting organizations were Kyoto Univ., Institute for Cosmic Ray Research, Univ. of Tokyo (ICRR), National Astronautical Observatory (NAO), Institute of Space and Astronautical Science (ISAS) and Kanazawa Institute of Technology (KIT). The meeting was supported by grants from the Commemorative Association for Japan World Exposition, Nishina Memorial Foundation, Yoshida Foundation for Science and Technology, the Murata Science Foundation, Yamada Science Foundation, Shimadzu Science Foundation, Konica Corporation, Space Development Group of Fujitsu Ltd., Central Research Laboratory of Sumitomo Metal Mining Co.Ltd., Sanpa Industries Ltd., Arakawa Technical Research Co.Ltd. and Osaka Sangyo University. We would like to express our deep appreciation for their valuable support.

We hope that all participants gathered at the Colloquium were able to carry on fruitful discussion about the origin and evolution of interplanetary dust and perhaps found new scientific subjects to work on. We also hope that the discussion created a basis for future collaborations. Finally, we are grateful for the advice and help by the members of the Scientific Organizing Committee and the staff in the office of the Local Organizing Committee.

Prof. Dr. Hiroichi Hasegawa
The Chairman of Local Organizing Committee
of IAU Colloquium #126

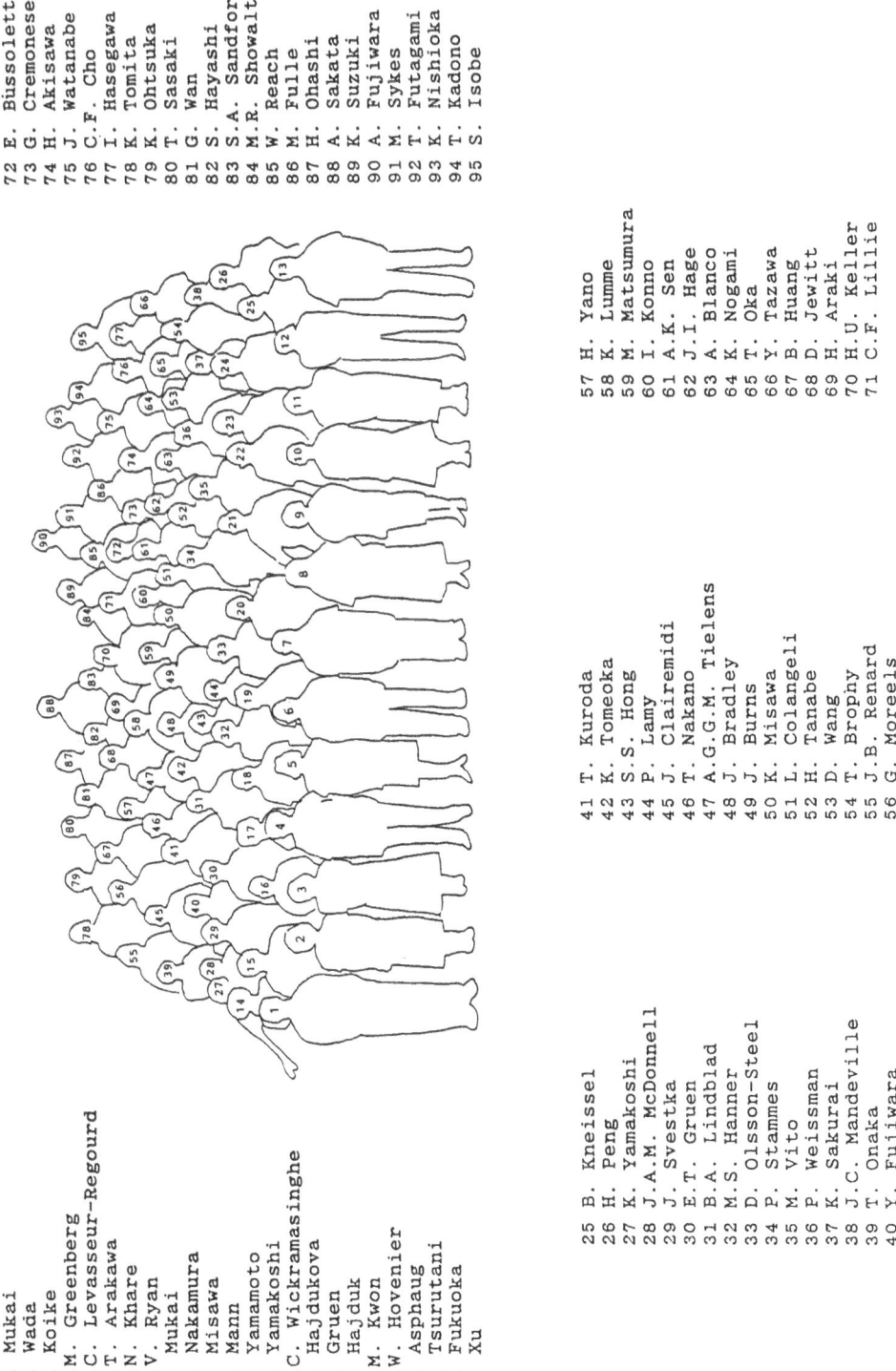

1 T. Mukai
2 S. Wada
3 C. Koike
4 J.M. Greenberg
5 A.C. Levasseur-Regourd
6 E.T. Arakawa
7 B.N. Khare
8 E.V. Ryan
9 S. Mukai
10 A. Nakamura
11 T. Misawa
12 I. Mann
13 T. Yamamoto
14 Y. Yamakoshi
15 N.C. Wickramasinghe
16 M. Hajdukova
17 U. Gruen
18 A. Hajduk
19 S.M. Kwon
20 J.W. Hovenier
21 E. Asphaug
22 B. Tsurutani
23 T. Fukuoka
24 Y. Xu

25 B. Kneissel
26 H. Peng
27 K. Yamakoshi
28 J.A.M. McDonnell
29 J. Svestka
30 E.T. Gruen
31 B.A. Lindblad
32 M.S. Hanner
33 D. Olsson-Steel
34 P. Stammes
35 M. Vito
36 P. Weissman
37 K. Sakurai
38 J.C. Mandeville
39 T. Onaka
40 Y. Fujiwara

41 T. Kuroda
42 K. Tomeoka
43 S.S. Hong
44 P. Lamy
45 J. Clairemidi
46 T. Nakano
47 A.G.G.M. Tielens
48 J. Bradley
49 J. Burns
50 K. Misawa
51 L. Colangeli
52 H. Tanabe
53 D. Wang
54 T. Brophy
55 J.B. Renard
56 G. Moreels

57 H. Yano
58 K. Lumme
59 M. Matsumura
60 I. Konno
61 A.K. Sen
62 J.I. Hage
63 A. Blanco
64 K. Nogami
65 T. Oka
66 Y. Tazawa
67 B. Huang
68 D. Jewitt
69 H. Araki
70 H.U. Keller
71 C.F. Lillie

72 E. Bussoletti
73 G. Cremonese
74 H. Akisawa
75 J. Watanabe
76 C.F. Cho
77 I. Hasegawa
78 K. Tomita
79 K. Ohtsuka
80 T. Sasaki
81 G. Wan
82 S. Hayashi
83 S.A. Sandford
84 M.R. Showalter
85 W. Reach
86 M. Fulle
87 H. Ohashi
88 A. Sakata
89 K. Suzuki
90 A. Fujiwara
91 M. Sykes
92 T. Futagami
93 K. Nishioka
94 T. Kadono
95 S. Isobe

LIST OF PARTICIPANTS

No.	Family Name	Given Name	Nationality
1	OLSSON-STEEL	Duncan	AUSTRALIA
2	VISVANATHAN	N.	AUSTRALIA
3	PENG	Hanchang	CHINA
4	WANG	Daode	CHINA
5	WAN	Gucun	CHINA
6	HUANG	Bojun	CHINA
7	XU	Yiwen	CHINA
8	HAJDUK	Anton	CZECHOSLOVAKIA
9	SVESTKA	Jiri	CZECHOSLOVAKIA
10	HAJDUKOVA	Maria	CZECHOSLOVAKIA
11	El-NAWAMY	M.S.	EGYPT
12	LUMME	Kari	FINLAND
13	CLAIREMIDI	Jacqes	FRANCE
14	LAMY	Philippe	FRANCE
15	LEVASSEUR-REGOURD	A. Chantal	FRANCE
16	MANDEVILLE	Jean-Claude	FRANCE
17	MOREELS	Guy	FRANCE
18	RENARD	Jean-Baptiste	FRANCE
19	PERRIN	Jean-Marie	FRANCE
20	FANG	Yanling	FRANCE
21	CREMONESE	Gabriel	FRANCE
22	GRUEN	Eberhard	FRG
23	KELLER	H.U.	FRG
24	KNEISSEL	Berhnard	FRG
25	RICHTER	K.	FRG
26	HUEDEPOHL	Axel	FRG
27	IGLSEDER	H.	FRG
28	MANN	Ingrid	FRG
29	SEN	Asoke Kumar	INDIA
30	PRASAD	C. Debi	INDIA
31	LEIBOWITZ	Elia M.	ISRAEL
32	BLANCO	Armando	ITALY
33	BUSSOLETTI	Ezio	ITALY
34	COLANGELI	Luigi	ITALY
35	FULLE	Marco	ITALY
36	MELLELLA	Vito	ITALY
37	Di MARTINO	M.	ITALY
38	AKISAWA	Hiroki	JAPAN
39	FUJIWARA	Yasunori	JAPAN
40	HASEGAWA	Ichiro	JAPAN
41	HAYASHI	Shinji	JAPAN
42	ISOBE	Shuzo	JAPAN
43	KAMIJO	Fumio	JAPAN
44	KOIKE	Chiyoe	JAPAN
45	KURODA	Toshio	JAPAN
46	MATSUMURA	Masafumi	JAPAN
47	MISAWA	Keiji	JAPAN
48	MIZUTANI	Hitoshi	JAPAN
49	MUKAI	Sonoyo	JAPAN
50	NAKANO	Takenori	JAPAN
51	OHASHI	Hideo	JAPAN
52	OHTSUKA	Katsuhito	JAPAN
53	OKA	Takuma	JAPAN
54	ONAKA	Takashi	JAPAN
55	SAKATA	Akira	JAPAN
56	SAKURAI	Kunitomo	JAPAN

No.	Family Name	Given Name	Nationality
57	SASAKI	Sho	JAPAN
58	SUGAWARA	Ken	JAPAN
59	SUZUKI	Kazuhiro	JAPAN
60	TANABE	Hiroyoshi	JAPAN
61	TOMEOKA	Kazushige	JAPAN
62	TOMITA	Koichiro	JAPAN
63	TORII	Shoji	JAPAN
64	TSUCHIYAMA	Akira	JAPAN
65	UESUGI	Kuninori	JAPAN
66	WADA	Setsuko	JAPAN
67	WATANABE	Jun-ichi	JAPAN
68	YAMAKOSHI	Kazuo	JAPAN
69	YAMAMOTO	Tetsuo	JAPAN
70	YANO	Hajime	JAPAN
71	FUJIWARA	Akira	JAPAN
72	FUKUOKA	Takaaki	JAPAN
73	MUKAI	Tadashi	JAPAN
74	NAGASAWA	Ko	JAPAN
75	NOGAMI	Ken-ichi	JAPAN
76	TAZAWA	Yuji	JAPAN
77	SHIMAOKA	Taro	JAPAN
78	SASAKI	Takashi	JAPAN
79	CHO	Chang Fee	JAPAN
80	KAITO	Chihiro	JAPAN
81	NAKAMURA	Akiko	JAPAN
82	FUTAGAMI	Tuneji	JAPAN
83	KAMIMURA	Kunio	JAPAN
84	TAGUCHI	Yasuo	JAPAN
85	ARAKI	H.	JAPAN
86	MOURI	H.	JAPAN
87	NAKAMURA	Takuji	JAPAN
88	NISHIOKA	Kimihiko	JAPAN
89	SEKI	Munezo	JAPAN
90	KURIHARA	Hiroshi	JAPAN
91	SUZUKI	Bunji	JAPAN
92	INAGAKI	T.	JAPAN
93	NAKAGAWA	Yoshitugu	JAPAN
94	FUJII	Naoyuki	JAPAN
95	SUGAWARA	Chikako	JAPAN
96	HONG	Seung Soo	KOREA
97	KWON	Suk Minn	KOREA
98	LINDBLAD	Bertil-Anders	SWEDEN
99	HAGE	J.I.	The NETHERLANDS
100	HOVENIER	J.W.	The NETHERLANDS
101	STAMMES	Piet	The NETHERLANDS
102	GREENBERG	J.M.	The NETHERLANDS
103	MARTELLI	Guiseppe	UK
104	McDONNELL	J.A.M.	UK
105	WICKRAMASINGHE	N.C.	UK
106	ARAKAWA	E.T.	USA
107	BURNS	Joseph A.	USA
108	HANNER	Martha S.	USA
109	JEWITT	David	USA
110	KHARE	Bishun N.	USA
111	KONNO	Ichishiro	USA
112	REACH	William T.	USA

```
===================================================================
No.    Family Name        Given   Name      Nationality
===================================================================
113    SANDFORD           Scott A.           USA
114    SHOWALTER          Mark R.            USA
115    SYKES              Mark V.            USA
116    TSURUTANI          Bruce              USA
117    BRADLEY            John P.            USA
118    CHOKSHI            Arati              USA
119    DOYLE              Laurence R.        USA
120    LILLIE             Charles F.         USA
121    TIELENS            A. G. G. M.        USA
122    KELSALL            Thomas             USA
123    BOICE              Danial C.          USA
124    ASPHAUG            Erik               USA
125    LUHMANN            Janet G.           USA
126    RUSSELL            Christpher T.      USA
127    WEISSMAN           Paul R.            USA
128    DAVIS              Donald             USA
129    RYAN               E.                 USA
130    BROPHY             Thomas             USA
```

ORGANIZING COMMITTEES

Scientific Organizing Committee

Chairperson
A.C. Levasseur-Regourd(CNRS, France)

General Secretary
T. Mukai(KIT, Japan)

Members

D.E.Brownlee	(USA)	S.S.Hong	(Korea)
J.A.Burns	(USA)	D.Hughes	(UK)
H.Fechtig	(FRG)	P.L.Lamy	(France)
J.M.Greenberg	(Netherlands)	J.A.M.McDonnell	(UK)
E.Gruen	(FRG)	J.Rahe	(FRG)
M.S.Hanner	(USA)	H.Tanabe	(Japan)
H.Hasegawa	(Japan)		

Local Organizing Committee

Chairperson
H. Hasegawa(Osaka Sangyo Univ., Japan)

Secretary
K. Yamakoshi(Tokyo Univ., Japan)

Members

A.Fujiwara	(Japan)	K.Nakazawa	(Japan)
T.Fukuoka	(Japan)	A.Sakata	(Japan)
I.Hasegawa	(Japan)	H.Tanabe	(Japan)
T.Maihara	(Japan)	T.Yamamoto	(Japan)
T.Mukai	(Japan)		

R.H. Giese (1931-1988)

Richard H. Giese was born in Stuttgart/Germany. While in High School he was already enthusiastic about Astronomy and Space Science. He studied Physics and Astronomy at the Universities of Stuttgart and Tübingen. His thesis was on "Scattering of Electromagnetic Radiation at Absorbing and Dielectric Spheric Particles" and in 1961 he graduated with the well known astronomer Prof. Siedentopf, Tübingen. Three years later he became a University Teacher through Habilitation on "Optical Determination of the Rotation Axis of Artificial Satellites". After a few years of scientific work at the Institute of Extraterrestrial Physics in the Max-Plank-Institut für Physik und Astrophysik, Munich, he was asked to join the Ruhr-University of Bochum. There he founded the Bereich Extraterrestrische Physik as part of the Faculty of Physics and Astronomy of the Ruhr-University.

From this time onwards (1963) Richard H. Giese worked with young scientists and students in the field of Interplanetary Dust. His special interest was the scattering of solar radiation on dust grains, the so-called Zodiacal Light. He has promoted this field in two directions: experimentally by using microwaves and laser- light and theoretically by modelling the dust distribution in the solar system. We had a continuous close collaboration in interpreting our joint results from space observations of the zodiacal light and from direct dust experiments on space probes. In particular he strongly cooperated as a Co-experimenter in the dust experiments of the space mission Helios, GIOTTO, Galileo and Ulysses.

Richard H. Giese's international cooperation both within the International Astronomical Union (President of Commission 21 from 1982 to 1985) and COSPAR (commission B.1) brought him high prestige. During his IAU-presidentship he was Chairman of the IAU-Colloquium No.85 on "Properties and Interactions of Interplanetary Dust" which was held in Marseille/France in 1984, the most recent IAU-Meeting devoted to our field of research.

At the Ruhr-University, Bochum, he was a highly esteemed teacher and lecturer. We all have lost not only an excellent scientist but also a good friend. We will never forget his integrity and friendly personality with a good sense of humor.

Hugo Fechtig, Christoph Leinert

M. Huruhata (1912-1988)

Masaaki Huruhata was born in Nagano-ken/Japan. In his school days he was already very active in observation of the variable stars. Immediately after graduation from the Astronomy Department, University of Tokyo in 1938, he worked at the Harvard College Observatory in the United States for three years. His research at this observatory was mainly concerned with variable stars and meteors. He returned to Japan in 1941 and joined the Astronomy Department and the Tokyo Astronomical Observatory of the University of Tokyo. During this period his interests expanded to encompass, besides variable stars and meteors, the zodiacal light and airglow, making use of the then newly developed technique of photoelectric photometry. He made significant contributions to these research fields with various pioneering works.

Masaaki Huruhata obtained his Doctor of Science in 1955 and he was appointed Professor of the Tokyo Astronomical Observatory in 1957. At the time of the International Geophysical Year (1957-58), he organized a cooperative program among the airglow observatories in Japan. He also took charge of the World Data Center C2 for Airglow. His efforts in the promotion of Japanese airglow observations and of international cooperation were continued until his retirement.

From 1968 to 1973 Masaaki Huruhata assumed the position of Director of the Tokyo Astronomical Observatory. In addition, he held many important and responsible posts, among which are Vice-President of the Astronomical Society of Japan (1961-63), President of Commission 21 of the International Astronomical Union (1967-70), Senator of the University of Tokyo (1968-73) and member of the Science Council of Japan (1969-72).

In 1973, Masaaki Huruhata retired from the Tokyo Astronomical Observatory and was conferred the title of Emeritus Professor of the University of Tokyo. After the retirement, he re-started observation of the variable stars with his own telescope and, sometimes, gave useful advice to young scientists and amateur astronomers.

He passed away on 23 November 1988 at the age of seventy-six. We express our deepest regret to the loss of this pioneer and fine gentleman.

Hiroyoshi Tanabe

Peter Mackenzie Millman 1906 – 1990

Peter M. Millman died at Ottawa, Canada, on December 11, 1990 at age 84. As the son of a missionary, he spent most of his youth in Japan and remained fluent in Japanese. After an undergraduate degree at the University of Toronto, he completed a PhD in 1932 at Harvard with a study of meteor spectroscopy. His working career as an astronomer was spent at the University of Toronto, the Dominion Observatory, Ottawa, and until his retirement in 1971, as head of upper-atmosphere research at the National Research Council, Ottawa. Even after retirement he continued to come into his office at NRC until about a month before his death. His special field was the spectroscopy of meteors but he also carried out visual and radar studies.

He received many awards for his contributions to science including the J. Lawrence Smith Medal of the US National Academy of Sciences, the Gold Medal of the Czechoslovak Academy of Sciences, and most recently, he was honoured by the naming of Minor Planet No. 2904 as "Millman".

In later years he was interested in planetary-system nomenclature and served as chairman and member of the IAU Working Group for some 15 years.

Bruce A. McIntosh

I

INTERPLANETARY DUST :
SPACE AND EARTH ENVIRONMENT STUDIES

PARTICULATE DETECTION IN THE NEAR EARTH SPACE ENVIRONMENT ABOARD THE LONG DURATION EXPOSURE FACILITY LDEF: COSMIC OR TERRESTRIAL?

J.A.M. McDonnell, K. Sullivan, T.J. Stevenson & D.H. Niblett
Unit for Space Sciences, University of Kent at Canterbury,
Canterbury,
Kent,
CT2 7NR,
United Kingdom

ABSTRACT. Examination of surfaces exposed for more than five and a half years, from detectors with unique attitude stabilisation relative to the orbital velocity vector, offers scope for examining definitively the sources of hypervelocity space particulates. Surfaces reveal discrete crater morphologies, crater size distributions and incident flux distributions. Discrete crater studies will later also reveal the chemistry of residues which can, especially via the capture cell principle, lead to elemental analysis of micron dimensioned particles.

First analyses of the flux data from the thin foil perforation experiments (MAP) involve a study of the statistics of the forward (ram) direction, the rear (trailing) direction and the space pointing direction. Modelling of the dynamics of geocentrically bound and unbound orbits yields evidence that the characteristics of the particles, and hence probably their source, change over the particle size range measured by the experiment. Smaller particles (< 1 μm diameter) have lower velocities which could include geocentrically bound particulates, whereas the larger particles (5-10 μm diameter) can be identified with "cosmic" particles of interplanetary or interstellar origin.

1. The LDEF Opportunity

LDEF's launch delay and its subsequent perilous approach towards uncontrolled re-entry and total loss in January 1990 can now be seen to have provided a highly significant exposure record and at a unique epoch for current space developments.

The exposure of 5.778 years (from launch on 7th April 1984 to recovery on 12th January 1990) at a 28.5° inclination offers excellent application to LEO satellites and to Space Station Freedom designs, and at a period when the presence of space debris has caused a high level of interest and, indeed, concern. The attitude stabilisation provides very adequate compensation for what could otherwise be a convoluted and confused record of differing exposure geometries (Figure 1). In section 3 we will see also that three faces on LDEF also promise to offer almost exclusive exposure to extraterrestrial particles whereas other surfaces will show high efficiency for the collection of orbital particulates. They may, or may not, prove to be orbital debris.

1.1 ORBITAL PARAMETERS

The temporal mean altitude (H) of LDEF (Figure 1) over its entire exposure duration was 458km and the total exposure time was 5.778 years (1.822×10^8 seconds). The orbital velocity at this altitude, assuming a circular orbit (LDEF's initial eccentricity was approximately 0.00015), is 7.64 km s^{-1} using 6371km for the mean radius of the Earth

3

A.C. Levasseur-Regourd and H. Hasegawa (eds.), Origin and Evolution of Interplanetary Dust, 3–10.
© *1991 Kluwer Academic Publishers.*

(R$_E$); the escape velocity at this mean altitude is 10.81 km s^{-1}. A value of 185km for the effective atmospheric height (h$_a$) is used, based on atmospheric drag calculations on a particle of 10^{-11}g, corresponding to the capture of a typical interplanetary particle within one Earth revolution.

1.2 EXPOSURE FACTORS

TABLE 1. LDEF altitude and Earth obscuration angle history .

Year	Altitude H [km]	H+R$_E$ [km]	Fraction of year	Horizon angle [deg]	Effective flat plate solid angle [ster]
1984	478	6849	0.745	106.82	2.141
1985	473	6844	1.000	106.68	2.137
1986	470	6841	1.000	106.60	2.136
1987	468	6839	1.000	106.54	2.134
1988	459	6830	1.000	106.28	2.124
1989	410	6781	1.000	104.80	2.077
1990	340	6711	0.033	102.34	1.995

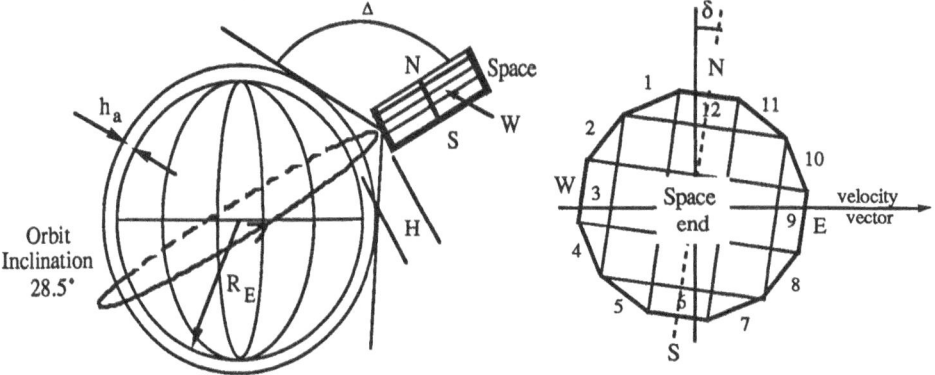

Figure 1. The orbit and the relationship of the actual attitude of LDEF to nominal.The actual attitude departed from the nominal by an angle δ ≈ 8°.

1.3 SOLID ANGLES

The effective solid angle of a flat plate parallel to the Earth's radius vector is given by [Δ-0.5sin(2Δ)] steradians, where Δ (radians) is the angle from the horizon to the zenith (Figure 1). This corresponds to $\pi/2$ steradians effective solid angle for $\Delta=\pi/2$ radians, namely a very low orbit, and π steradians for a space facing plate. The effective solid angle for a cone of θ radians half angle from the normal to the surface is $\pi(1-\cos^2\theta)$ steradians. Δ is given by $\sin^{-1}(A/R)$, where $A=R_E + h_a$ and $R=R_E + H$ (Figure 1). The mean LDEF effective peripheral tray exposure solid angle is 2.125 steradians.

1.4 PENETRATION RELATIONSHIPS

The crater depth to crater diameter ratio (d_c/D_c) for a semi-infinite aluminium target was established for the Solar Maximum craters in aluminium as 0.62 (Laurance and Brownlee 1986). The ratio of marginal thin foil perforation thickness to the crater depth of a semi-infinite target (f/d_c) was given as 1.15 (McDonnell 1970). Therefore, for marginal perforation, the ratio of the foil thickness limit to the semi-infinite crater diameter is given as $f=0.71D_c$. Relevant penetration relationships are published in McDonnell et al (1990). We use these relationships currently to compare thick target data and thin foil perforation data for the same type of material (aluminium). More detailed comparisons will at a later stage be established from the actual data. Such detailed comparisons will also incorporate specific material properties but currently these differences should not preclude general conclusions from data which cover a very wide range of magnitudes.

2. Data Available from LDEF's Exposure

Data on LDEF's impact environment are available from spacecraft recovery and de-integration procedures and also from Principal Investigator experiments.

Principal Investigator impact data exist in published form from the A0023 MicroAbrasion Package (MAP) experiment (McDonnell et al, 1990) and from the A0201 Interplanetary Dust Experiment (IDE) (Singer et al, 1990). The plots shown in Figure 2 for the MAP experiment result from a large number of independent measurements on aluminium surfaces of different thicknesses. This is to be contrasted with the cumulative distribution from *one* surface where the flux at one thickness is partially correlated with an adjacent flux measurement and a negative cumulative slope is excluded. It is to be noted that the Space end face has currently only one data point referring to one foil thickness. The form of the marginal penetration distribution on this most valuable pointing direction is not all lost, however, because 5 micron thick brass foil was also flown on this face. Once a calibration of the brass foil has been undertaken from laboratory hypervelocity impact simulations, the Space face will then have two marginal data points, thus yielding valuable information on the interplanetary flux; in the future, supra-marginal data will also be offered.

A co-ordinated and comprehensive approach to the spacecraft operations and de-integration was made by the formation of the Special Investigator Groups (SIGs). The Meteoroid and Debris SIG operations documented impact sites above a diameter of some 300 µm on most surfaces of the entire spacecraft. Although the publication of impact data from dedicated impact experiments remains the prerogative of the Principal Investigator, summary M-D SIG data which incorporate some of these data from the initial operations are now published (See et al, 1990); some of these are included (courtesy of the Principal Investigators) in Figure 2. For example, the first results from the larger craters on the S0001 Space Debris Impact Experiment (SDIE) (Humes, 1984) scanned by the M-D SIG team at Kennedy Space Center are shown by courtesy of D. Humes, but do not represent his analysis of the experiment, which must await his more comprehensive analysis programme. A major potential source of impact data (some 18 square metres) is also offered by the covers of the A0178 Ultra-heavy Cosmic-Ray Experiment (UHCRE) (O'Sullivan, 1984), comprising silvered Teflon sheet 120 µm thick backed by some 80 µm of paint; these showed the first spectacular evidence of sustained impact erosion damage in space, clearly visible to the STS 43 astronauts. Preliminary data relating to the marginal perforation of these covers are included in Figure 2 (courtesy F. Levadou). Figure 2 also includes data from the A0138 Frecopa experiment (Mandeville, 1984).

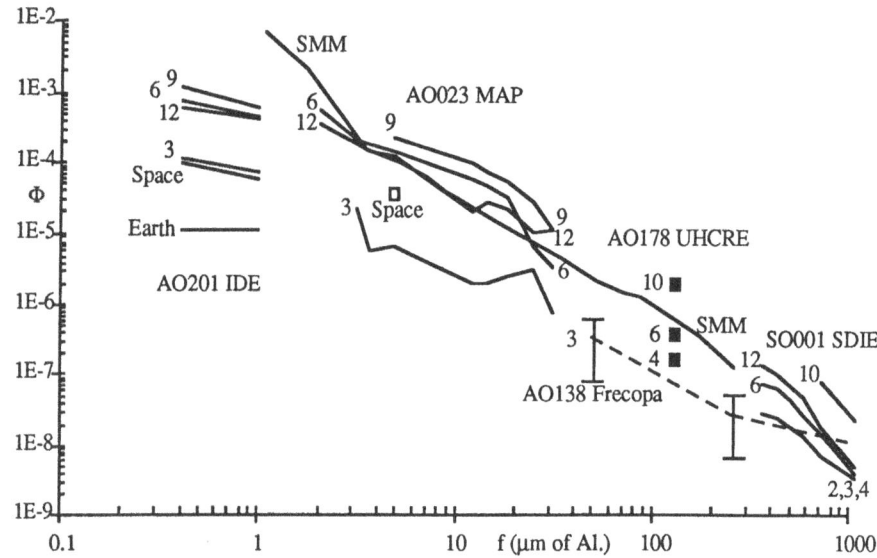

Figure 2. Flux Φ (in $m^{-2}s^{-1}$) versus aluminium foil penetration thickness from the Interplanetary Dust Experiment (IDE), the Micro-Abrasion Package (MAP), the Ultra-Heavy Cosmic Ray Experiment (UHCRE), the Space Debris Impact Experiment (SDIE) and the FRECOPA experiment, compared to the Laurance and Brownlee data from the Solar Maximum Mission (SMM). Numbers represent the nominal LDEF bay angle for each detector.

3. Features of the LDEF Impact Data

Differences between the North and South MAP detectors (Figure 2), representing directions approximately at right angles to LDEF's trajectory, can be examined very closely in the context of the absence of experimental bias, namely in view of the accurately known exposure areas and geometry and calibrated foil thicknesses. Clearly seen is that the South flux exceeds the North flux for small particulates but that the trend reverses for the thicker foils. We must also take note of the *actual* pointing direction of LDEF which does not correspond to the nominal (Figure 1). All LDEF impact experiment data shows a high East flux corresponding to the RAM direction, and a decrease towards the West. We might perhaps expect the North and South fluxes to be somewhat "equal" in LDEF's swathe through various uncorrelated orbits; in this situation the effect of this spacecraft pointing offset would be an increase in the *North* flux relative to the South. This is seen to be true only for larger foil thicknesses, but clearly this trend reverses for smaller particulates. The true bias to the South for small particulates is considerably greater therefore, because of the offset, than that shown in Figure 2 and is clearly of great significance; as yet the cause of this is undetermined!

What should we expect regarding the question of isotropy from existing knowledge of either orbital or hyperbolic (interplanetary) particles? Concerning *orbital particulates,* recent papers by Olsson-Steel and McDonnell (1990), Kessler (1990) and Zook (1990) and supported by simple, but firm, concepts show that intercepting of any two bound circular orbits lead to an exact equality on the (true) North and South faces for any inclination of the two orbits. It extends also to eccentric orbits if the arguments of perigee of the particles are randomly distributed. If the orbital inclination is equal to 63.4° there is no advance of the line of apsides for an elliptical orbit as is the case for the significant number of Molniya-type spacecraft which utilise this property to remain above the horizon of the Soviet Union for as long as possible. If this type of spacecraft produced a debris cloud it could in turn produce a North-South anisotropy in LDEF impact rates. Thus, North-South asymmetry is not explained by inclination distributions unless firstly, orbital eccentricities are significant and, secondly, some quite special non-randomised orbital parameters pertain to the overall distribution.

Concerning *unbound (hyperbolic)* particulates, the distribution of impact rates also reflects the initial (heliocentric) orbital parameters. Although the distribution of larger particles is highly non-isotropic (meteor streams) due to their source in comets, smaller interplanetary grains of the type detected by MAP are unlikely to have retained all the information of their parent bodies' orbits due to non-gravitational forces such as Poynting-Robertson and radiation pressure. Anisotropy for large grains could result from the non-random interception of meteor streams by the Earth; this is less likely for the smaller grains whose orbits are likely to be more evenly distributed in the ecliptic. Arguments on the dynamics of unbound particulates and interception with LDEF (Olsson-Steel, 1990, Zook 1990) show two features of relevance. The first point is that the radius vector of LDEF's 28.5° geocentrically inclined orbit is swept through a wide range of ecliptic latitudes, and can perhaps in the first instance be considered "random". LDEF's orbital plane will have an average ecliptic referenced inclination of +23.5° with a swing of ± 28.5°. The Space end will then point to ecliptic particulates over a very wide range of ecliptic latitudes throughout its orbit, namely over ± 52° which is further convolved with the acceptance angle of a flat plate detector. We should view therefore the extraterrestrial flux on the Space end as an "average" of all ecliptic latitudes and longitudes. The second point is that the Space and West faces have a very low probability of interception with Earth-orbital particulates. Despite LDEF's directional excursions relative to the ecliptic reference plane, the North and South faces *do* have a bias towards ecliptic North and South; always pointing to within 52° of their respective poles. Only if the number of interplanetary particulates in interplanetary space at 1 A.U. is symmetrical with regard to ascending and descending node would there be symmetry (Olsson-Steel, personal communication). The LDEF MAP data would suggest there are more particulates in the ascending node for small masses *if* these North and South fluxes on LDEF were interplanetary in origin. The question of resolving North South (and East) fluxes into a two component model (interplanetary and terrestrial) has not yet been resolved on LDEF, however, and must await the inputs from chemical microanalysis of the residues, in addition to the dynamical arguments presented here.

4. Application of Dynamical Modelling Techniques to Data Interpretation

In all modelling, examination and interpretation of impact/flux data from differing spacecraft attitudes or pointing directions, care must be exercised in distinguishing between flux variations either at *constant mass* or alternatively at *constant crater size*. Crater size is, of course, directly related to the marginal perforation foil thickness (Section 1.4).

Flux enhancement at constant mass is the "sweeping-up" effect of the satellite into the particulate cloud and leads to an enhancement of numbers intercepted compared to the trailing face. A consequential effect of this, but quite separate physically, is that those particles will also have a different relative velocity for the two faces, and hence, will upon impact lead to different crater dimensions; because most impact observations (and observed crater flux distributions) refer to a particular crater dimension, the experiment detector surfaces receiving greater numbers of particles will yield a flux value which is relevant to smaller and invariably more numerous particles. The latter sensitivity enhancement depends on the size distribution of particulates which, fortunately, can be deduced from the data.

Relationships in this sensitivity enhancement have been established (McDonnell et al, 1991). Since the West face flux distribution can be assumed to be the result of non-Earth bound particles, a prediction of the interplanetary flux as encountered by any other LDEF face can be achieved by dynamical modelling. Two transformations must be carried out: one to account for the effective equivalent foil thickness because of the differing relative impact velocities and another to account for both the exposure geometry and also the sweeping effect due to the spacecraft velocity vector.

The West and Space face flux plots can be compared and also can be transposed to give an expected value for any other LDEF face, because of the implicit assumption that these fluxes are due to interplanetary particulates only. By varying the geocentric particle velocity (V_{PE}) in the model input, the West and Space face distributions can be transposed to coincide: this yields a value of $V_{PE} = 16 \pm 2$ km s^{-1}. Incorporating the gravitational attraction, we can then derive a value for the interplanetary approach velocity to the Earth (V_∞) to be $V_\infty = (V_{PE}^2 - 10.81^2)^{1/2} = 12 \pm 3$ km s^{-1}. If then the interplanetary flux derived from the West and Space faces is transformed to calculate the expected East interplanetary component, a measure of the excess East flux due to orbital particulates can be derived. Such an excess is found for the smaller masses only. If the transformation is performed for the thicker foils, the predicted interplanetary flux appears to *exceed* the observed East flux. This argues against orbital particulates in this range ($\approx 10^{-9}$g) and indeed at higher masses. It could call for higher particle velocities at these masses, which would be closer to interplanetary approach velocities deduced from meteor studies. Further aspects are developed in an accompanying paper (McDonnell et al, 1991).

5. Conclusions

Impact and penetration data from the LDEF spacecraft shows already the high value of the concept of the return to Earth of the largest area-time product of space age detector. Even the preliminary data shows a high level of definition regarding the size distribution and anisotropy of the flux. The completion of this morphological analysis and its extension into the dimension of particulate chemistry and isotopic abundances will lead to a "gold standard" in the definition of the near Earth space particulate environment.

At this stage we are able to see that data from several key experiments, and from the spacecraft as a whole, is leading to a coherent size distribution for craters ranging from sub-micron to millimetre dimensions. Detectors on the East (ram) face universally demonstrate "high" fluxes, but the maximum appears shifted towards the South for small particulates, indicating a departure from the North-South symmetry which would be expected *a priori* . Surprisingly, the asymmetry is reversed for particles capable of penetrating some 20 to 30 μm of aluminium or greater, which show an excess on the North, and this feature is shared by the penetration of the UHCRE cover and craters observed from the M-D SIG data from the spacecraft.

It can be shown that the Space and West faces are accessible to extraterrestrial particles rather than to orbital particles and, by invoking modelling, a geocentric velocity can be deduced for this extraterrestrial component of about 16 km s^{-1} corresponding to an approach velocity of 12 km s^{-1} to the Earth for particulates of mass 10^{-10}g. This velocity is slightly lower than that anticipated from extrapolations based on data from the interplanetary meteoroid flux; the small difference could, however, be very significant and point to lower eccentricities and inclinations of the orbits of these smaller particles.

Within the measurement range of LDEF, therefore, dynamical modelling applied to the data shows that Earth orbital particulates are significant only at the small (sub-micron) dimensions and that these are detected dominantly on forward faces. At larger dimensions than the 50 micron range, interplanetary extraterrestrial particulates are found to dominate the impact scenario on all LDEF's surfaces, and are hence the major penetration and erosion risk for millimetre dimension spacecraft surfaces. The space pointing, trailing and very probably the Earth facing directions appear to be accessible only to these extraterrestrial components at all dimensions.

The identification of an Earth orbital component with space debris can thus be demonstrated as a possible hypothesis only at small dimensions (sub-micron). This hypothesis has been strongly indicated by crater residues on the Solar Max mission by Laurance and Brownlee (1986). However, it is shown that the choice of penetration formula referring those data to particulate mass (Pailer and Grün 1980) is at variance with other calibration data. The raw data of Laurance and Brownlee compare excellently with LDEF data; it is argued therefore that their data do not indicate a dominance of space debris below 10^{-9} g. It *could yet* be found to be significant, and correspond to the excess flux seen on the East penetrations below approximately 25 μm.

Arguing against this source being space debris, the temporal behaviour of the near Earth satellite flux is seen to be remarkably stable over the development of a growing space population (McDonnell et al, 1991). If this micron dimensioned particulate flux is supposed to be debris related, as concluded by Laurance and Brownlee's association of residual elemental comparisons with typical spacecraft material, then we could be forced to conclude that there was as much space debris in the early 1960s as in 1984, despite the linear growth of space launches and the orbital satellite population.

Acknowledgements
    We acknowledge all members of the LDEF MAP team in the Unit for Space Sciences at the University of Kent ; NASA Project Management (W.Kinard, J.Jones); Alison and Margaret, the Science and Engineering Research Council (U.K.) and Auburn University Space Power Institute (U.S.A.), Prime Contract N660921-86-C-A226, subcontract 87-212.

References
Humes, D. (1984). 'Space debris impact experiment (S0001)', in: *LDEF Mission 1 Experiments*. eds, Clark, L.G., Kinard, W.M., Carter, D.J.,Jones, J.L. NASA SP473, 136-137.
Kessler, D.J. (1990). 'Orbital debris predictions for LDEF', NASA communication SN3-90-18.
Laurance, M.R., and Brownlee, D.E. (1986). 'The flux of meteoroids and orbital space debris striking satellites in low Earth orbit', Nature, 323, 136-138.
Mandeville, J.C. (1984). 'Study of meteoroid impact craters on various materials' (AO138)', in: *LDEF Mission 1 Experiments*. eds, Clark, L.G., Kinard, W.M., Carter, D.J.,Jones, J.L. NASA SP473, 124-126.
McDonnell, J.A.M. (1970). 'Factors affecting the choice of foils for penetration experiments in space', Space Res., X, 314-325.
McDonnell, J.A.M., Deshpande, S.P., Green, S.F., Newman, P.J., Paley, M.T., Ratcliff, P.R., Stevenson, T.J., and Sullivan, K. (1990). 'First results of particulate impacts and foil perforations on LDEF', Adv. Space Res. (in press).
McDonnell, J.A.M., Sullivan, K., Stevenson, T.J., Niblett, D.H., and Green, S.F. (1991). 'Dynamic Modelling Transformations for the Low Earth Satellite orbit

particulate environment', Proc. IAU Coll. on 'Origin and Evolution of Interplanetary Dust', Kyoto, Japan.

Olsson-Steel, D., and McDonnell, J.A.M., (1990).   'Face-dependent collision probabilities upon the Long Duration Exposure Facility for particles in geocentric orbits', Planet. Space Sci. (in press).

Olsson-Steel, D. (1990). 'Face dependent impact probabilities, velocities and angles by space debris upon LDEF', Adv. Space Res. (in press).

O'Sullivan, D. (1984).   'A high resolution study of ultra-heavy cosmic-ray nuclei (A0178)', in: *LDEF Mission 1 Experiments*. eds, Clark, L.G., Kinard, W.M., Carter, D.J.,Jones, J.L. NASA SP473, 101-104.

Pailer, N., and Grün, E. (1980). 'The penetration limit of thin films', Plan. Space. Sci., 28, 321-331.

See, T., Allbrooks, M., Atkinson, D., Simon, C., and Zolensky, M. (1990). 'Meteoroid and debris impact features documented on the Long Duration Exposure Facility', M-D SIG, Prelim. Report., NASA.

Singer, S.F., Stanley, J.E., Kassel, P.C., Kinard, W.H., Wortman, J.J., Weinberg, J.L., Mulholland, J.D., Eichhorn, G., Cooke, W.J., and Montague, N. (1990). 'First spatio-temporal results from the LDEF Interplanetary Dust Experiment', Adv. Space Res. (in press).

Zook, H.A. (1990). 'Flux vs. direction of impacts on LDEF by meteoroids and orbital debris', Proc. 20th Lun. Plan. Sci. Conf., (in press).

STUDY OF COSMIC DUST PARTICLES ON BOARD LDEF AND MIR SPACE STATION

J.C. MANDEVILLE
ONERA/CERT Space Technology Department
P.O. Box 4025
31055 Toulouse Cedex
France

ABSTRACT. Interplanetary and near-earth space contains solid objects whose size distribution continuously covers the interval from submicron sized particles to km sized asteroids or comets.
Two French experiments partly devoted to the detection of cosmic dust have been flown recently in space. One on the NASA Long Duration Exposure Facility (LDEF), and one on the Soviet MIR Space Station. A variety of sensors and collecting devices will make possible the study of cosmic particles after recovery of exposed material. Flux mass distribution is expected to be derived from craters counts, with a good accuracy. Remnants of particles, suitable for chemical identification are expected to be found within stacked foil detectors. Discrimination between extraterrestrial particles and man-made orbital debris will be possible.

1. Introduction

Our present knowledge of the occurence and of the physical properties of micrometeoroids is based primarily on earth bound observation of meteors, comets, zodiacal light, data from infrared satellites (IRAS) as well as on board measured flux by instrumented spacecrafts (Pegasus, Explorers, Vega, Giotto, Space Shuttle), study of lunar samples and dust collection in upper atmosphere (Leinert and Grün, 1990). Part of meteoroids originate from comets (mainly dust ejected at perihelion), part originate from collisions within asteroid belt. The relative contribution of these two sources is still a matter of debate, a majority of particles are likely coming from comets but recent data from Infrared Astronomy Satellite (IRAS) indicate than asteroids could be a source larger than expected. In addition to natural particles, a significant and growing number of particles has been added by human activity, in near earth space.
Extensive modeling of the distribution of dust particles has been made, and the NASA 1969 Meteoroid model (Cour-Palais, 1969) is still widely accepted. Recent modeling for artificial debris has been proposed by D. Kessler (Kessler et al.,1989). Data from recent space experiments could possibly lead to a reassessment of earlier models (Alexander et al.,1963).

Originally launched in April 1984 for a nine months mission the NASA Long Duration Exposure Facility (LDEF) has been retrieved, in January 1990, after 2105 days in terrestrial orbit, between 450 and 330 km altitude.

The MIR Soviet Space Station has been in a 350 km circular orbit since February 1986. It is composed of a main module (15 meter long, 4 meter in diameter), associated with other specialised modules. The French experiment, named "Echantillons" was deployed outside the station, during the Aragatz Mission in december 1988; it was retrieved 13 months later.

A.C. Levasseur-Regourd and H. Hasegawa (eds.), Origin and Evolution of Interplanetary Dust, 11–14.
© 1991 Kluwer Academic Publishers.

2. Experimental Approach

Part of the LDEF tray allocated to french experiments, known as FRECOPA payload, has been devoted to the study of dust particles. The tray was located on the face of LDEF directly opposed to the velocity vector (west facing direction, location B3, according to LDEF description).

Two passive experiments have been flown : one composed of a set of glass and metallic samples and one composed of multilayer thin foils detectors. Collection area was about 2000 cm^2. In addition of these dedicated experiments a broad variety of materials has been exposed to the bombardment of microparticles and are expected to provide additional data. Thick target experiment comprises selected metallic (Al,Au,Cu, W, Stainless Steel, thickness: 250 μm) and glass surfaces 1.5 mm thick.

Crater size distribution from these thick target experiments will enable, with the aid of laboratory calibration by solid particle accelerators, the evaluation of the incident microparticle flux in the near earth environment. A more critical issue is the determination of the chemical composition of the impacting particles. In general they are physically destroyed and mixed with target material in the process of crater formation, however chemical identification of remnants has proven possible (Warren et al.,1989).

The aim of the multiple foil penetration and collection experiment is primarily to investigate the feasability of multilayer thin film detectors acting as energy sorters in order to collect micrometeoroids, if not in their original shape , at least as "break-up" fragments suitable for chemical analysis.

Foil thickness ranges from 0.75 μm to 5 μm of aluminium; such foils are expected to slow down particles with diameters between 1 and 10 μm diameter, without complete destruction.

The experiment on MIR carried basically the same passive sensors, with an addition, an active dust detector. This detector was based on the monitoring of the discharge upon impact of a thin foil capacitor. Due to the stabilization mode of the station, sensors were facing the velocity vector direction during aproximately half of the orbit. Detailed description of hardware has been given elsewhere (Mandeville, 1984, Mandeville and McDonnell, 1984).

3. Preliminary Results.

Both experiments have been recovered in good conditions after exposure to space. As expected, the number of impact craters vary significantly with the location on spacecraft surface (with respect to velocity vector). During its mission LDEF was stabilized with the long axis continually pointed toward the center of the earth, and surfaces perpendicular to this axix pointed at fixed angles with respect to the direction of orbital motion. Preliminary comparison with data from similar experiments located on different sides of LDEF has been made. Materials not specifically dedicated to dust detection have provided usefull data, mostly because the large area time exposure. The spacecraft has been the first one purposely designed to make possible an extensive study of space environment, in peculiar micrometeoroids, after a recovery of the experiments. The detection range goes from submicron to mm sized particles.

Two large large impact features on the Frecopa tray (LDEF) have been found : one full penetration and one marginal penetration of a 1mm aluminium shield. About 100 craters larger than 30 microns have been found on a area of 1 m^2. Most of the large craters are circular in outline, however some small craters do indicate oblique incidence. SEM investigation for small craters is still in progress but preliminary results do indicate apparently a deficiency in micron sized craters. First results from the largest craters on the S0001 Space Debris Impact Experiment (D.Humes) are shown by courtesy of M-D SIG Team (See et al.,1990), in comparison with our data. The impact site survey yields a crater size distribution, which should be converted to a particle mass distribution by using relevant relationships between crater sizes and particles mass and velocity. The discussion is out of the scope of this paper, however assuming an average impact velocity of 15 km/s, the value of the crater diameter to the particle diameter ratio, could be chosen as D/d = 5, for aluminium targets.

The figure 1 shows the preliminary size distribution of craters found on some detectors for the LDEF and for the MIR experiments. Detailed SEM analysis has been made on a few aluminium targets. A large density of submicron craters has been found on detectors exposed on the MIR station. The data indicate the presence of particles with a diameter of a few tenths of microns (mass of 10^{-15} gm) in earth bound orbits. Data from IDE experiment on LDEF are consistent with these results (Singer et al.,1990), as well as measurements reported by Singer (Singer and Stanley,1980).

The size distribution shows a good agreement with comparable near-earth data obtained by McDonnell (McDonnell et al. 1991); flux on the west face of LDEF is about 20 times lower than on the east face, for large particles. Most of the particles impacting this face should be interplanetary dust particles, not orbital debris. This fact will be substantiated further by the chemical identification of projectile remnants inside craters.

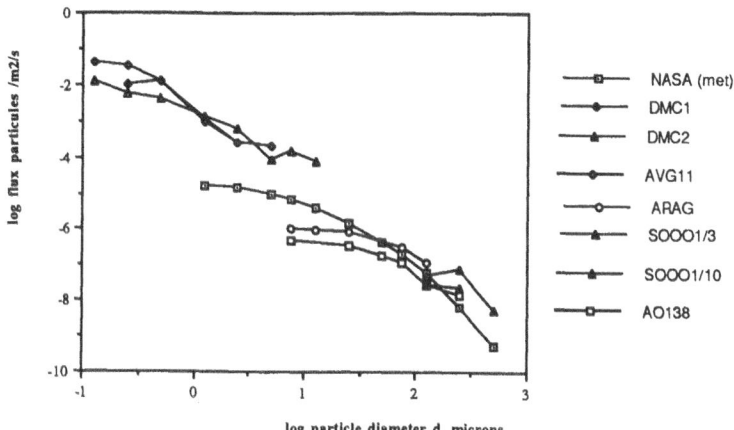

Comparison MIR/LDEF/NASA Model

Figure 1. Flux (particles/m^2/s) of dust particles versus diameter (microns). Comparison between space experiments (near earth) and NASA model is shown. Symbols are as follows :
NASA(met) is the NASA 1969 Meteoroid model.
DMC1, DMC2 and AVG11 are data from dust detectors on MIR (for small particles).
ARAG are data from MIR experiment (for large particles).
AO138 are data from LDEF FRECOPA experiment (for large particles)
S0001/3 and S0001/10 are data from NASA Langley (D. Humes) LDEF dust experiment (Large particles, west and east faces, respectively).

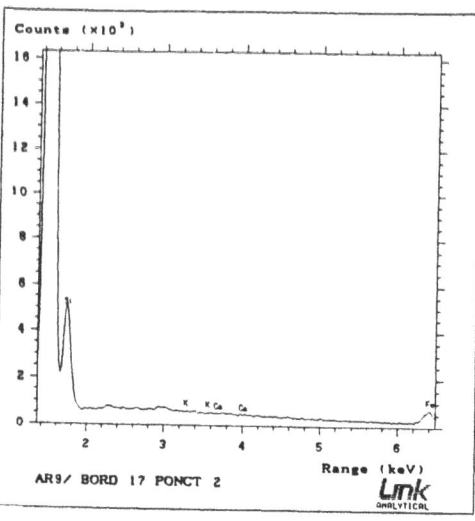

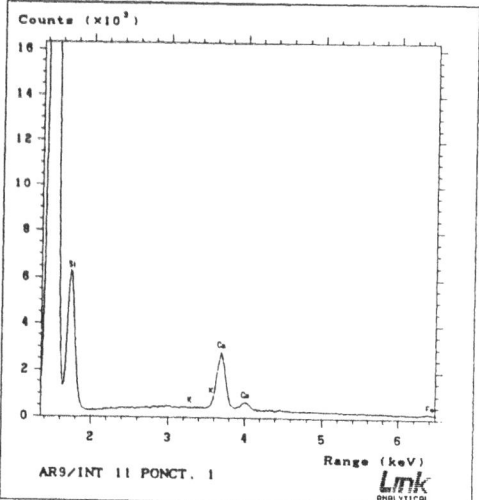

Figure 2. Energy dispersive X-Ray spectra from points located inside impact craters

The slope of the size distribution of particles is consistent with data obtained elsewhere on LDEF, the change of slope at large sizes could be an artifact due to the limited time-area of sampling, or more likely real as shown by McDonnell .

As the inclination of the orbit of the MIR station is 58°, a comparison with LDEF data (inclination 28°) is expected to provide some information on the orbital distribution of artificial debris. Preliminary results indictate a somewhat higher flux of micron-sized particles on MIR orbit.

The first X ray analysis of some small craters has shown evidence of elements Ca, K, Ti, Fe and S. The figure 2 shows the occurence of calcium and of iron inside two small craters found on MIR samples. Detailed analysis will doubtless give insight on the origin of impacting objects and some information on the relative contribution of cosmic dust particles and man-made debris.

4. Conclusion.

As shown by the preliminary investigation of experiments and materials retrieved on LDEF and on MIR station, better knowledge of the micrometeoroid environment in near-earth orbits will be obviously obtained from dedicated in-flight, retrievable experiments. Large time area exposure of detectors will make possible reliable statistics on size distribution of particles. Discrimination between man-made orbital debris and cosmic particles is a difficult but necessary task in order to assess the extent of the pollution of near space.

Collection in space of undisturbed particles will remain generally difficult, perhaps impossible for the highest meteoritic velocities. However the lower velocity window for multiple layer foil deceleration would prove sufficient to expect the retention of material suitable for identification. Investigation of the near-earth region of space is a necessity as well as for scientific or technical purposes.

REFERENCES.

Alexander, W.M. et al. (1963) 'Review of direct measurements of interplanetary dust from satellites and probes', in W. Priester (ed.), Proc. Third Intern. Space Sci.Symp., J.Wiley Publishers, p.891.
Cour-Palais, B.G. (1969) 'Meteoroid environment model', NASA SP 8013.
Kessler, D.J., Reynolds, R.C. and Anz-Meador, P.D. (1989) 'Orbital debris environment', NASA TM 100471
Leinert, Ch. and Grün, E. (1990) 'Interplanetary Dust', in Physics and Chemistry in Space, Springer Publishers, (in press).
Mandeville, J-C. and McDonnell, J.A.M. (1980) 'Micrometeoroid multiple foil penetration and particle recovery experiments on LDEF', in IAU Symposium 90 Proceedings, D. Reidel Publishers, Dordrecht.
Mandeville, J-C. (1984) 'AO138-1 and AO138-2 Experiments', in L.G.Clark et al. (eds.), LDEF Mission 1 Experiments, NASA SP-473.
Mandeville, J-C. (1990) 'Aragatz Mission Dust Collection Experiment', Adv.Space Res. 10, pp.397-401.
McDonnell, J.A.M. et.al. (1991) 'Particulate detection in the near earth space environment aboard the long duration exposure facility LDEF: Cosmic or Terrestrial', Proc. IAU Coll. on 'Origin and Evolution of Interplanetary Dust', Kyoto, Japan.
See T., et al. (1990). 'Meteoroid and debris impact features documented on the Long Duration Exposure Facility', M-D SIG, Prelim.Report.,NASA.
Singer, S.F. et al. (1990) 'First spatio-temporal results from the LDEF Interplanetary dust experiment (IDE)', Preprint XXVIII COSPAR 1990.
Singer, S.F. and Stanley, J.E. (1980) 'Submicron particles in meteor streams', in IAU Symposium 90 proceedings, D.Reidel Publishers, Dordrecht.
Warren, J.L. et al. (1989) 'The detection and observation of meteoroid and space debris impact features on the Solar-Max satellite', in Proceedings XIXth Lun.Plan.Sci.Conf., pp. 641-657.

Acknowledgements.
Support from CNES (F) and Glavcosmos (USSR) for experiments on the MIR Space Station and from NASA (USA) for experiments on LDEF is greatly acknwoledged.

THE PRESENT STATUS OF THE MUNICH DUST COUNTER EXPERIMENT ON BOARD OF THE HITEN SPACECRAFT

E.Igenbergs, A.Hüdepohl [1], K.Uesugi ,T.Hayashi [2], H.Sved-
hem [3], H.Iglseder [4], G.Koller [5], A.Glasmachers [6],
E.Grün [7], G.Schwehm [3], H.Mizutani, T.Yamamoto, A.Fujimura,
N.Ishii, H.Araki [2], K.Yamakoshi [8], K.Nogami [9]

[1] Lehrstuhl für Raumfahrttechnik, Technische Universität
    München, Richard Wagner Str. 18, 8000 München 2, FRG
[2] Institute of Space and Astronautical Science, J
[3] European Space Research and Technology Centre of ESA, NL
[4] Center for Applied Space Technology and Microgravity,
    Universität Bremen, FRG
[5] Lehrstuhl für Prozeßrechner, Technische Universität
    München, FRG
[6] Mikroelektronik-Zentrum, Ruhr-Universität Bochum, FRG
[7] Max-Planck-Institut für Kernphysik, Heidelberg, FRG
[8] Institute for Cosmic Ray Research, University of Tokyo, J
[9] Dep. of Physics, Dokkyo University School of Medicine, J.

ABSTRACT. The Munich Dust Counter (MDC) is a scientific experiment on
board the MUSES-A mission of Japan measuring cosmic dust. The satellite
HITEN of this mission has been launched on January 24th, 1990 from Kagoshima Space Center. Here the present status of the MDC experiment is
summarized. The number of dust particles measured so far is presented
together with first and preliminary results of flux calculations and
spatial as well as directional distributions of cosmic dust particles
measured until July 25, 1990. A clear evidence of particles coming from
the inner solar system (beta-meteoroids) already has been found. These
are compared to particles coming from the apex direction.

1. Introduction

The Munich Dust Counter (MDC) determines mass and velocity of cosmic
dust particles by measuring the charges generated by the high velocity
impacts of these particles on a gold target. It is installed on the
Japanese HITEN spacecraft launched on January 24, 1990 in a high elliptic orbit around the earth. The details of the instrument and of the
mission of the spacecraft are discussed by Igenbergs et al. (1990) and
Uesugi et al. (1990).

A.C. Levasseur-Regourd and H. Hasegawa (eds.), Origin and Evolution of Interplanetary Dust, 15–20.
© 1991 Kluwer Academic Publishers.

Generally the MDC can measure particles with masses between 10^{-15} to 10^{-7} grams and velocities between 1 and 70 km/s. As it is mounted on the perimeter of the spinning spacecraft it scans the whole ecliptic plane during one revolution of the satellite, the spin axis direction of which is maintained perpendicular to the ecliptic plane. The effective sensor area is 100 cm^2, the field of view is 148^0.

After the instrument switch-on on January 31, 1990, and initial tests and optimizations the nominal operation of the MDC started on March 3, 1990. Since then the MDC is measuring continuously. So far no quantitative evaluation of the data, the determination of dust particle mass or velocity, has been carried out.

2. Signal Types

Included in this paper are data gathered up to July 25, 1990. In these first 144 days of nominal operation since March a large number of signals have been measured of which about 67 are supposed to be particle impacts. The signals measured so far can be arranged into eight types which are shown in fig. 1.

Every measurement made by the MDC consists of two curves, one for the negative impact charges, labeled EC (electron channel), and one for the positive impact charges, labeled IC (ion channel). In the case of a particle impact one or both curves should rise, with risetimes between some μs and 100 μs, see Igenbergs et al. (1990).

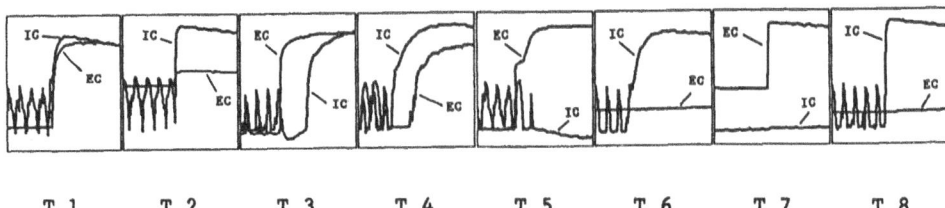

   T.1   T.2   T.3   T.4   T.5   T.6   T.7   T.8

Figure 1.   Signal types encountered by the MDC experiment

Signals of type 1 to 4 show the typical shape expected for dust particle impacts. While type 1 is just perfect, the type 2 signals have some noise on the EC. Type 3 signals show a remarkable delay on the IC, while type 4 show a delay on the EC. Type 5 signals show only a signal on the EC, type 6 only a signal on the IC. All these effects are due to certain impact locations inside the sensor and could be verified during the calibration of the instrument. Type 7 signals have a very fast burst on the EC, type 8 a very fast burst on the IC. These signals are supposed to be noise and not due to particle impacts. Not included in fig. 1 is the noise generated by the sun shining into the detector opening, which has a very flat rise and has to be deleted by the MDC microprocessor system about once every revolution of the satellite.

The following graphs and diagrams have been made by using all signals of types 1 to 6.

3. Particle Flux

The MDC is measuring cosmic dust particles in its normal operational state since March 3, 1990. The diagram in fig. 2 shows the particle flux rate in impacts per day over the time in days from launch on January 24, 1990. During the initial tests dedicated to the adaptation of the experiment software to the noise generated by the sun already four particle impacts have been measured. The 67 dust impacts measured in about 140 days of nominal operation give an average of 0.48 impacts per day, indicated as dotted line in fig. 2, which corresponds to a flux rate of $5.5 \cdot 10^{-4}$ $m^{-2}s^{-1}$ for particle masses between 10^{-15} and 10^{-7} grams.

The flux diagram may suggest events like "groups" or "swarms" found with previous dust experiments, Hoffmann et al. (1975), Fechtig et al. (1979), but there is not yet enough data sampled for a good statistical treatment. Therefore no further evaluation is presented here.

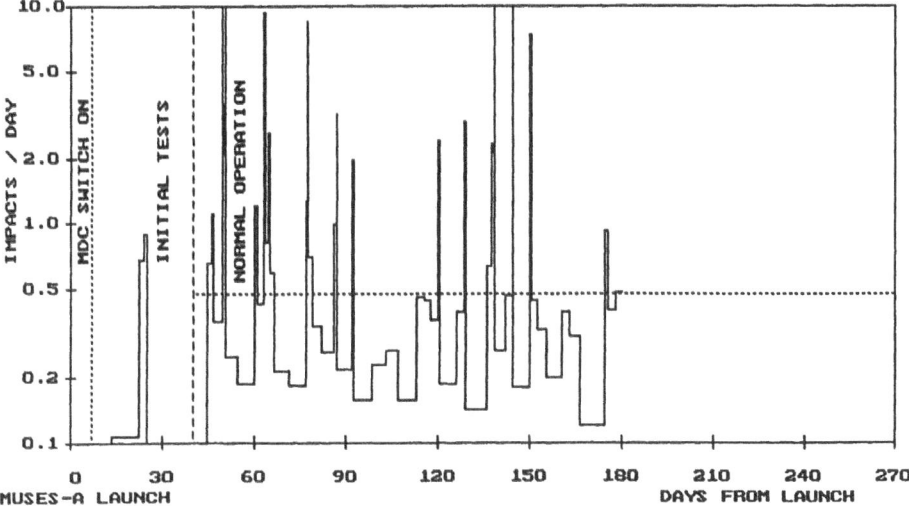

Figure 2.   Cosmic dust flux as measured by the MDC in impacts per day

4. Spatial Distribution of Dust Particles

The orbit of the HITEN spacecraft covers nearly the whole sphere of influence of the earth with perigees ranging from some 1000 km to about 100000 km and apogees between 300000 km and more than 1 million km.

Figure 3 shows a plot of the orbits of HITEN and the moon with dust particle measurements indicated. The orbits are plotted in a rotating coordinate frame with the line connecting the sun and the earth fixed and pointing left. The lines marking every impact indicate the direction in which the MDC was pointed at the time of measurement.

Here the spatial distribution of the dust particles seems to be quite uniform, but it has to be kept in mind that due to the high-eccentric orbit of the satellite the MDC is for a much longer time far away from earth than it is near the earth. A better approach will be to define appropriate regions of distances from earth and calculate the flux rates accordingly. This will be done when enough data are gathered.

This plot shows also that more particles have been measured from the apex direction and sun direction than from the anti-apex direction. This will be discussed in detail in the next section.

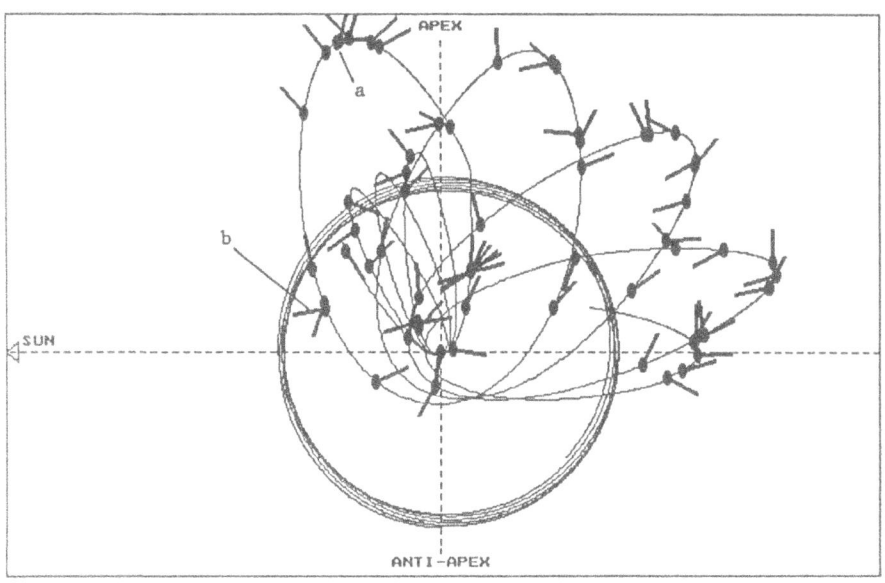

Figure 3. Orbit plot of the HITEN satellite. Positions of dust impact measurements are indicated, the lines show the MDC orientation. A typical apex-particle measurement is marked with "a", a typical beta-meteoroid measurement is marked with "b", see section 5.

5. Directional Distribution of Dust Particles

As the MDC is scanning the ecliptic plane about every three seconds, it has the possibility to gather information about the particle flight directions. Of course, due to the field of view of 148^0 of the MDC, both in ecliptic and in north/south directions, the particle flight direction determination for a single measurement is not accurate at all. But by putting together many measurements a quite clear directional distribution can be obtained. The diagram shown in fig. 4 gives the number of impacts measured for different directions. This graph was obtained by calculating the impact probability according to the projection of the sensor area to the incident angles spread over 148^0 and adding up all measurements gathered so far.

With about 67 particle measurements displayed in the graph of fig. 4 no good statistics are possible, therefore all results have to be treated as preliminary. Up to now it seems that about 50% of the particles come roughly from the apex direction, with the peak offset outwards away from the sun. But, this effect may be due to noise which periodically occurs at about this angular position and which is currently under detailed investigation. Another 27% come from the sun direction, with the peak offset towards the anti-apex direction. The minimum is quite clear in the anti-apex direction with less than 5% of all measurements.

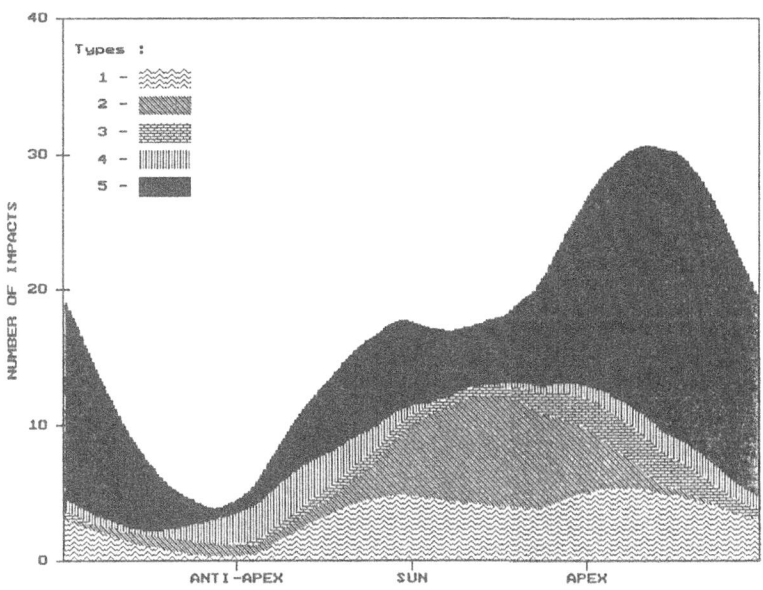

Figure 4.   Spin phase angle distribution of dust particles measured by the MDC.

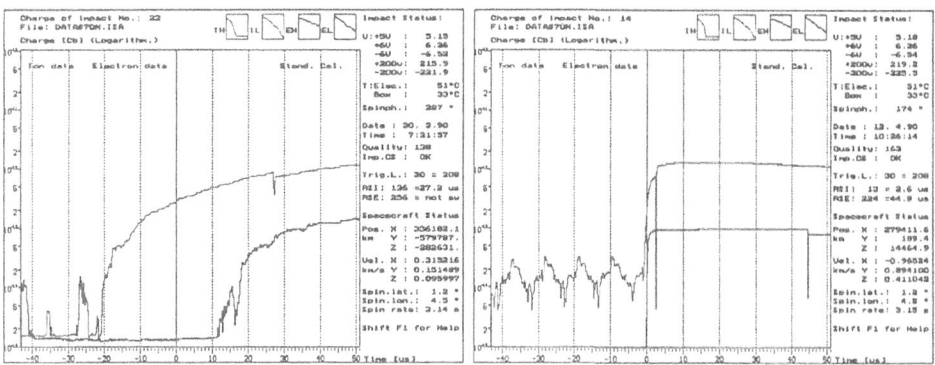

Figure 5a.   Typical impact signal of an apex-particle, position of this measurement is marked in fig. 3 with "a".

Figure 5b.   Typical impact signal of a beta-meteoroid, position of this measurement is marked in fig. 3 with "b".

By comparing the signals of a particle coming from the inner solar system (beta-meteoroid), Zook and Berg (1975), with a signal of an apex-particle, Grün and Zook (1980), the typical features of these two classes can be distinguished. Figure 5a shows the typical impact signal of an apex-particle, marked in fig. 3 with "a". The long risetimes of the signals give a particle velocity in the order of 3 km/s. In fig. 5b the typical signal generated by the impact of a beta-meteoroid is shown, this measurement is marked in fig. 3 with "b". The short risetimes of the curves indicate a particle velocity in the order of 50 km/s. As the satellite velocities were quite low, these values can be taken as an indication for the velocities of the particles relative to the earth system. The masses of the particles are considered to be in the order of 10^{-10} g in the case of the apex- particle and 10^{-14} g in the case of the beta-meteoroid for these specific measurements, which may not be typical for the two populations.

6. Conclusions

The cosmic dust experiment Munich Dust Counter is successfully measuring cosmic dust particles in the mass range between 10^{-15} and 10^{-7} grams and in the velocity range between 1 and 70 km/s on board of the Japanese spacecraft HITEN. From March 3 until July 25, 1990, about 67 dust particles have been measured in total, which do not yet allow a good statistical treatment of the data.

By a preliminary classification of the measurements made so far informations about the flux and spatial distribution of the cosmic dust particles in the earth-moon-system could be gained. The angular distribution of the dust measured by the MDC already seems to confirm the presence of apex-particles and beta-meteoroids.

The primary phase of the mission of the HITEN satellite will end in March 1991, then a thorough data evaluation on the basis of hopefully more than 150 measurements will be possible.

References

Fechtig H., Grün E., Morfill G. (1979) 'Micrometeoroids within Ten Earth Radii', Planet. Space Sci., 27, 511-531.

Grün E., Zook H.A. (1980) 'Dynamics of Micrometeoroids', in Halliday I. and McIntosh B.A. (eds.), Solid Particles in the Solar System, Reidel Pub. Co., London, 293-298.

Hoffmann H.-J., Fechtig H., Grün E., Kissel J.(1975) 'Temporal Fluctuations and Anisotropy of the Micrometeoroid Flux in the Earth-Moon System Measured by HEOS 2', Planet. Space Sci., 23, 985-991.

Igenbergs E., et al. (1990) 'The Munich Dust Counter - A Cosmic Dust Experiment on Board of the MUSES-A Mission of Japan', Proc. of the IAU Colloquium No. 126 "Origin and Evolution of Interplanetary Dust", Kyoto.

Uesugi K., Matsuo H., Kawaguchi J. Hayashi T. (1990) 'Japanese first Double Lunar Swingby Mission HITEN', Proc. of the 41st Congress of the IAF, Dresden.

Zook H.A., Berg O.E. (1975) 'A Source for Hyperbolic Cosmic Dust Particles', Planet. Space Sci., 23, 183-203.

IN-SITU EXPLORATION OF DUST IN THE SOLAR SYSTEM AND INITIAL RESULTS FROM THE GALILEO DUST DETECTOR

E. GRÜN[1], H. FECHTIG[1], M. S. HANNER[2], J. KISSEL[1],
B.-A. LINDBLAD[3], D. LINKERT[1], G. MORFILL[4] and H. A. ZOOK[5]

[1] Max-Plank-Institut für Kernphysik, 6900 Heidelberg, Germany
[2] Jet Propulsion Laboratory, Pasadena, CA 91103, U.S.A.
[3] Lund Observatory, 221 Lund, Sweden
[4] Max-Planck-Institut für Extraterrestrische Physik,
    8046 Garching, Germany
[5] NASA Johnson Space Center, Houston, TX 77058, U.S.A.

ABSTRACT. In-situ measurements of interplanetary dust have been performed in the heliocentric distance range from 0.3 AU out to 18 AU. Due to their small sensitive areas (typically 0.01 m^2 for the highly sensitive impact ionization sensors) or low mass sensitivities ($\geq 10^{-9}$g of the large area penetration detectors) previous instruments recorded only a few 100 impacts during their lifetimes. Nevertheless, important information on the distribution of dust in interplanetary space has been obtained between 0.3 and 18 AU distance from the Sun. The Galileo dust detector combines the high mass sensitivity of impact ionization detectors (10^{-15} g) together with a large sensitive area (0.1 m^2). The Galileo spacecraft was launched on October 18, 1989 and is on its solar system cruise towards Jupiter. Initial measurements of the dust flux from 0.7 to 1.2 AU are presented.

1. Introduction

There are several methods to study various aspects of interplanetary dust from the Earth. Observations of the scattered light (zodiacal light) from interplanetary dust and its thermal emission reveal the large scale spatial distribution of particles in the 10 μm to 1 millimeter size range. Meteor observations refer to mm and larger objects, the orbits of which intersect the Earth. Interplanetary dust particles collected in the stratosphere allow us to obtain compositional and morphological information. From lunar micro-crater studies, the size distribution of sub micron to mm sized particles was determined (for a review see Leinert and Grün, 1990).

Complementary to the above mentioned methods are in-situ studies by dust impact detectors on board interplanetary spacecraft. The purpose of this paper is to give a survey of previous interplanetary dust instruments and their major findings, and to present first results of the dust detector on board the Galileo spacecraft.

A.C. Levasseur-Regourd and H. Hasegawa (eds.), Origin and Evolution of Interplanetary Dust, 21–28.
© 1991 Kluwer Academic Publishers.

2. Characteristics of Recent In-Situ Dust Detectors.

In-situ measurements of interplanetary dust have been performed in the heliocentric distance range from 0.3 AU out to 18 AU (Table 1). We have included in the list of interplanetary dust detectors also the two Earth satellites HEOS 2 (Hoffmann et al., 1975) and Hiten (Igenbergs et al., 1990a) because they performed significant portions of their measurements outside the range of influence of the Earth.

Two types of impact detectors were used for interplanetary dust measurements: impact ionization detectors with detection thresholds of 10^{-16} to 10^{-13} g and penetration detectors with detection thresholds of 10^{-9} and 10^{-8} g. These detection thresholds refer to a typical impact speed of 20 km/s. The penetration detectors on board Pioneers 10 and 11 (Humes, 1980) have large geometric factors, i.e. sensitive areas and effective solid angles (for reference, a flat plate has π sr effective solid angle). Most impact ionization detectors have sensitive areas of only 0.01 m^2, except the Galileo instrument (Grün et al., 1990) which has a ten times larger sensitive area. Because the effective solid angles of the HEOS 2, Helios 1/2 (Dietzel et al., 1973) and Galileo detectors are significantly less than that for a flat plate, they are able to provide better directional information. E.g. in an isotropic flux half of the particles are recorded within a cone of 32 degrees half angle around the axis of the Galileo sensor. The Pioneer 8 and 9 detectors (Berg and Richardson, 1969) could record approximate directions for a few time-of-flight events, however, most of the impacts were recorded by the wide angle front film sensor (Berg and Grün, 1973).

The dynamic range of the instruments describes the range over which particle masses can be determined at a given impact speed. For larger particles the instrument reaches saturation and only lower mass limits can be stated. Dynamic range of 1 for the Pioneer 10 and 11 instruments implies that only lower mass limits for all impacts can be stated. The last two columns of table 1 show theoretical performance parameters of the different

TABLE 1: In situ dust detectors in interplanetary space
  (1) heliocentric distance (AU)   (2) mass threshold at 20 km/s (g)
  (3) sensitive areae (m^2)   (4) effective solid angle (sr)
  (5) dynamic range   (6) number of impacts at 1 AU (yr^{-1})
  (7) largest particle in 1 yr at 1AU (g)

Spacecraft	(1)	(2)	(3)	(4)	(5)	(6)	(7)
Pioneer 8	0.97-1.09	$2 \ 10^{-13}$	0.0094	2.9	200	17	$5 \ 10^{-10}$
Pioneer 9	0.75-0.99	$2 \ 10^{-13}$	0.0074	2.9	200	14	$3 \ 10^{-10}$
HEOS 2	1	$2 \ 10^{-16}$	0.01	1.03	10^4	370	$3 \ 10^{-11}$
Pioneer 10	1 - 18	$2 \ 10^{-9}$	0.26*	3**	1	19	$3 \ 10^{-7}$
Pioneer 11	1 - 10	$1 \ 10^{-8}$	0.26*	3**	1	10	$3 \ 10^{-7}$
Helios 1/2	0.3 - 1	$9 \ 10^{-15}$	0.012	1.23	10^4	40	$9 \ 10^{-11}$
Galileo	0.7 - 5.2	$1 \ 10^{-15}$	0.1	1.4	10^6	1500	$2 \ 10^{-8}$
Hiten	1	$1 \ 10^{-15}$	0.01	2**	$3 \ 10^4$	220	$2 \ 10^{-10}$

* initial area, actual area decreased as cells were punctured
** estimated values

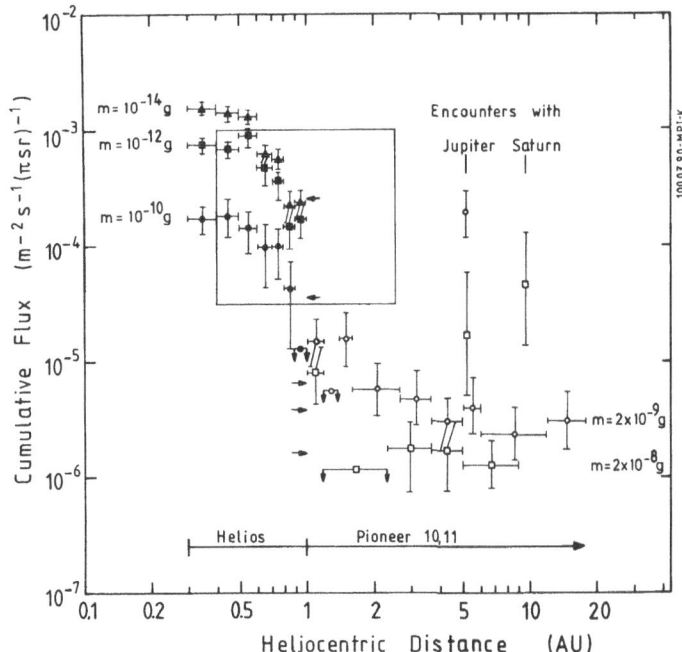

Figure 1. Flux of interplanetary dust particles measured by the Helios and Pioneer 10 and 11 spaceprobes. The different symbols represent flux values for five different mass thresholds. Zero impacts recorded in specific distance intervals are presented by upper flux limits. For comparison corresponding fluxes determined at 1 AU are indicated by horizontal arrows. The rectangular insert shows the ordinate and abscissa ranges used in Figure 4 where the Galileo data are displayed.

instruments. For the interplanetary dust flux at 1 AU given by Grün et al. (1985b) the following values are calculated: column 6, the number of impacts per year above the detection threshold and column 7, the largest particle that would be recorded in one year. In these calculations only the geometric factors have been considered but not the orbits of the spacecraft and the actual orientations of the sensors with respect to the Sun and the ecliptic plane. Therefore, the actually measured quantities may deviate somewhat from the given values.

3. Interplanetary Dust Measurements

In this section we describe major results from the various interplanetary dust measurements. A radial profile of the dust flux in the inner solar system between 1 and 0.3 AU distance from the Sun has been determined by the Helios 1 and 2 spaceprobes (Grün et al., 1985a). Figure 1 shows radial flux profiles for 3 threshold masses adjusted for a flat plate sensor. Helios measurements allowed us to identify two dynamically different interplanetary dust populations in the inner solar system (Grün et al., 1980): (1) particles which orbit the sun on low eccentric orbits (e < 0.4) with rather small semi major axes (a ≤ 0.5 AU) - this population had already been noticed by the previous Pioneer 8/9 (Berg and Grün, 1973) and HEOS 2 (Hoffmann et al., 1975) dust experiments - and (2) particles on highly eccentric orbits (e > 0.4) which have also large semi major axes (a > 0.5 AU).

The Pioneer 8 and 9 experiments detected a dominant flux of small particles from approximately the solar direction. Existence of these particles was recently confirmed by Hiten (Igenbergs et al., 1990b) measurements. A radial profile of the dust flux (Zook, 1975) could not conclusively be determined from the Pioneer 8/9 measurements since McDonnell et al. (1975) could explain the same data as variation with heliocentric longitude. With the Pioneer 8/9 and HEOS 2 measurements it was possible to determine the flux of small (≤ 10^{-12}g) interplanetary particles at 1 AU (cf. Grün et al., 1985).

In the outer solar system the penetration detectors on board Pioneers 10 and 11 determined the interplanetary dust flux (Humes, 1980). Figure 1 shows radial flux profiles for the two threshold masses. No sign of an increase in the asteroid belt was detected and outside about 3 AU they recorded a flat flux profile, except for strong increases of the flux near Jupiter and Saturn. The Pioneer 11 data obtained between 4 and 5 AU are best explained by meteoroids moving on highly eccentric orbits.

For comparison cumulative flux values measured by detectors in Earth-like orbits are shown in Figure 1 for the same five mass thresholds. The obvious discrepancy of the fluxes at 1 AU can not be explained only by the large statistical uncertainties of the spaceprobe measurements. The measurements presented were obtained by detectors which moved on different heliocentric trajectories at 1 AU. Therefore, the different fluxes measured reflect different relative speeds with respect to the interplanetary dust cloud. Other differences are the viewing directions of the sensors which have an influence as well since the flux is not isotropic. These considerations make it difficult to determine the radial profile of the spatial density of dust at 1 AU.

4. The Galileo Dust Experiment

The Galileo spacecraft was launched on 18 October 1989 on its trajectory to Jupiter. However, before the Galileo spacecraft reaches Jupiter it has to perform a six year journey through the solar system before it will become the first man made satellite of the giant planet. Figure 2 shows the first portion of the Galileo trajectory through the inner solar system. A Venus swing by took place on 8. February 1990 and the spacecraft flew by the Earth in December 1990 for the first time.

The dust detector instrument is mounted to the spinning section of the spacecraft. During this initial phase of the mission the spin vector points away from the Sun. The viewing direction of the dust detector forms an angle of 55 degrees with the spin vector. This angle has been chosen for optimum dust detection geometry at Jupiter. The Galileo dust instrument is a multi-coincidence impact ionization detector (Grün et al., 1990), with geometric factors given in Table 1. From the released impact charge and the time relations of different charge signals, the mass, speed and electric charge of impacting dust particles can be determined. From each impact or noise

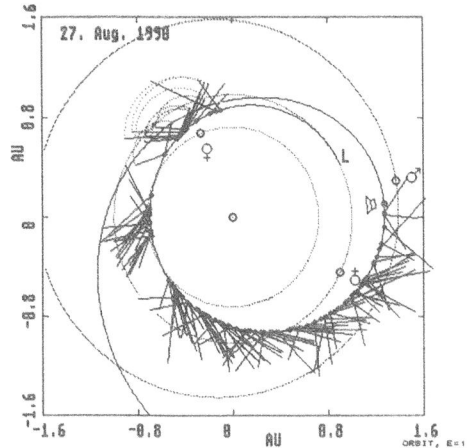

Figure 2. View onto the ecliptic plane with the orbits of Venus, Earth, Mars and the Galileo spacecraft. Launch (L) took place on 18. October 1989. The positions of the planets and Galileo are shown for the given date. Discrete dust impacts are indicated by dots on the Galileo orbit and the corresponding sensor pointing directions at impact (projection onto the ecliptic plane) are give by the attached bars. The length of the bars indicate the magnitude of the impact charge. A fan presents the full range of sensor pointing directions for a time when no pointing information was available.

event all the measured quantities are transmitted to ground and each event is counted in one of several accumulators.

5. First Results

On 28. December 1989 the dust detector was switched on for the first time, when Galileo was at a distance of 0.88 AU from the Sun. During a three day configuration period the instrument was set to a state where noise was sufficiently suppressed and impact events could be reliably distinguished from noise events. Around the time of the Venus fly-by an increased noise rate was encountered, so that the sensitivity threshold of the instrument had to be increased in order to cut the noise down to ≤ 5 events per day.

The recorded discrete impacts are displayed in Figure 2. Dust impacts are overlaid on Galileo's orbit. The impact direction is indicated as well as the magnitude of the impact. During most of the time data were transmitted in small packets only once or twice a week. Apparent gaps in the distribution of impacts are due to two approximately 4-weeks long gaps in data transmission. Therefore, information on discrete impacts is only complete to 55 %. However, from the accumulated data the impact rate can be reconstructed over the whole period without any gaps (Figure 3). The highest short-term rate of 3.5 impacts per day was observed shortly after perihelion passage. The impact rate presented here corresponds to an operational mass threshold which is approximately a factor of 10 above sensitivity threshold

26

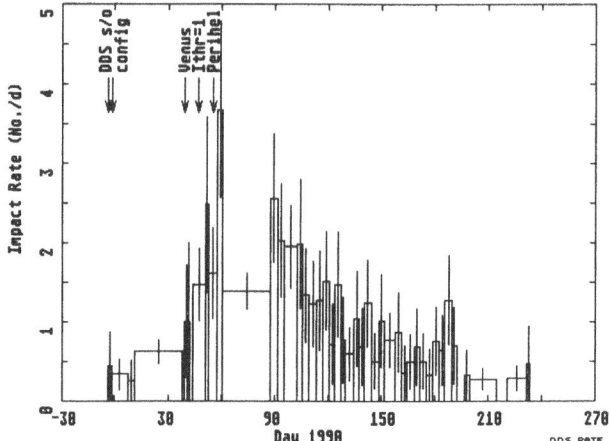

Figure 3. Impact rate observed by the Galileo dust detector until 27. August 1990. Major events are indicated: instrument switch-on (DDS s/o), termination of instrument configuration (config), Venus fly-by (Venus), change of the sensitivity threshold to suppress excessive noise (Ithr=1) and perihelion passage (Perihel).

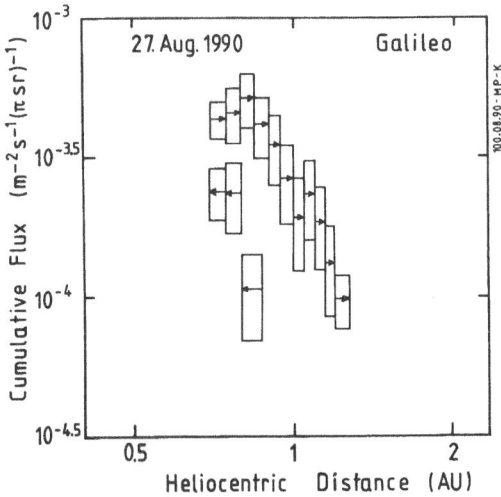

Figure 4. Cumulative flux of dust particles recorded by the Galileo dust detector until the indicated date. The flux corresponds to an operational mass threshold of 10^{-14} g (for an assumed impact speed of 20 km/s) and to a flat plate sensor (cf. Figure 1). The arrows represent the averaged fluxes and indicate whether Galileo was moving inward or outward. Each box represents the distance interval over which the flux was averaged and the 1-σ uncertainty of this flux.

given in Table 1. The full sensitivity will only be reached when higher a data transmission rate can be established in mid 1991.

In order to allow the comparison with other flux data we have rebinned the data in about equal sized distance intervals and calculated the flux onto a flat plate sensor by assuming an isotropic flux. Figure 4 shows the measured flux during the initial Galileo mission phase. The general decrease of the flux with heliocentric distance is due to the decrease of the dust population with increasing distance from the sun (cf. Figure 1). No significant deviation from a smooth slope at 1 AU is observed. The inbound fluxes are factors two to five smaller than the outbound fluxes at the same heliocentric distance. This can be understood by the orientation of the viewing cone of the dust sensor relative to the orbit of the spacecraft. In the outbound section the fluxes are comparatively higher because the spacecraft motion is outward towards the hemisphere the sensor is facing. A first analysis of these data indicates that the dust population observed by Galileo is compatible with dust particles on low eccentric (e $\leq$ 0.3) orbits.

ACKNOWLEDGEMENTS. The authors are especially indebted to G. Linkert, D. Maas, G. Matt, G. McSmith, C. Polanskey and N. Siddique for their personal involvement during the initial phases of the Galileo mission. This work has been supported by the Bundesminister für Forschung and Technologie, under the grant 01 QJ 900400.

References

Berg, O. E. and Grün, E. (1973) Evidence of hyperbolic cosmic dust particles. Space Research XIII, 1047-1055.

Berg, O. E. and Richardson, F. F. (1986) The Pioneer 8 cosmic dust experiment, Rev. Sci. Instrum. 40, 1333-1337.

Dietzel, H., Eichhorn, G., Fechtig, H., Grün, E., Hoffmann, H. J. und Kissel, J. (1973) The HEOS 2 and Helios micrometeoroid experiments, J. Phys. (E) Scientific Instrum., 6, 209-217.

Grün, E., Pailer, N., Fechtig, H. and Kissel, J. (1980) Orbital and physical characteristics of micrometeoroids in the inner solar system as observed by Helios 1, Planet. Space Sci., 29, 333-349.

Grün, E., Fechtig, H. and Kissel, J. (1985a) Orbits of interplanetary dust particles inside 1 AU as observed by Helios, in Properties and Interactions of Interplanetary Dust, R. H. Giese and P. Lamy (eds.), Reidel, Dordrecht, pp. 105-111.

Grün, E., Zook, H. A., Fechtig, H., and Giese, R. H. (1985b) Collisional balance of the meteoritic complex, Icarus, 62, 244-272.

Grün, E., Fechtig, H., Hanner, M. S., Kissel, J., Lindblad, B. A., Linkert, D., Morfill, G. E. and Zook, H. A. (1990) The Galileo dust detector, submitted to Space Sci. Rev.

Hoffmann, H. J., Fechtig, H., Grün, E. und Kissel, J. (1975) Temporal fluctuation and anisotropy of the micrometeoroid flux in the earth-moon system, Plant. Space Sci., 23, 985-991.

Humes, D. H. (1980) Results of Pioneer 10 and 11 meteoroid experiments: Interplanetary and near-Saturn. Journ. Geophys. Res., 85, 5841-5852.

Igenbergs, E., Hüdepohl, A., Uesugi, K. T., Hayashi, T., Svedham, H., Igelseder, H., Koller, G., Glasmachers, A., Grün, E., Schwehm, G., Mizutani, H., Yamamoto, T., Fujimura A., Ishii, N., Yamakoshi, K. and Nogami, K. (1990a) The Munich dust counter - A cosmic dust experiment on board of the Muses-A mission of Japan, in Origin and Evolution of Interplanetary Dust, This issue.

Igenbergs, E., Hüdepohl, A., Uesugi, K. T., Hayashi, T., Svedham, H., Igelseder, H., Koller, G., Glasmachers, A., Grün, E., Schwehm, G., Mizutani, H., Yamamoto, T., Fujimura A., Ishii, N., Yamakoshi, K. and Nogami, K. (1990b) The present status of the Munich dust counter experiment on board of the Hiten spacecraft, in Origin and Evolution of Interplanetary Dust, This issue.

Leinert, C. and Grün, E. (1990) Interplanetary dust, in Physics of the Inner Heliosphere, R. Schwenn and E. Marsch (eds.), Springer-Verlag, Heidelberg, pp. 207-275

McDonnell, J. A. M., Berg, O. E., and Richardson, F. F. (1975) Spatial and time variations of the interplanetary microparticle flux analyzed from deep space probes Pioneers 8 and 9, Planet. Space Sci. 23, 205-214.

Zook, H. A. (1975) Hyperbolic cosmic dust: its origin and its astrophysical significance, Planet. Space Sci. 23, 1391-1397.

THE NASA SOLAR PROBE MISSION: IN SITU DETERMINATION OF INTERPLANETARY OUT-OF-THE ECLIPTIC AND NEAR-SOLAR DUST ENVIRONMENTS

BRUCE T. TSURUTANI
Solar Probe Study Scientist
Jet Propulsion Laboratory   and
California Institute of Technology
4800 Oak Grove Drive
Pasadena, California 91109

JAMES E. RANDOLPH
Solar Probe Study Manager
Jet Propulsion Laboratory
California Institute of Technology
4800 Oak Grove Drive
Pasadena, California 91109

ABSTRACT. The NASA Solar Probe mission will be one of the most exciting dust missions ever flown and will lead to a revolutionary advance in our understanding of dust within our solar system. Solar Probe will map the dust environment from the orbit of Jupiter (5 AU), to within 4 solar radii of the sun's center. The region between 0.3 AU and 4 R_S has never been visited before, so the 10 days that the spacecraft spends during each (of the two) orbit is purely exploratory in nature. Solar Probe will also reach heliographic latitudes as high as ~ 15° to 28° above (below) the ecliptic on its trajectory inbound (outbound) to (from) the sun. This, in addition to the ESA/NASA Ulysses mission, will help determine the out-of-the-ecliptic dust environment. A post-perihelion burn will reduce the satellite orbital period to 2.5 years about the sun. A possible extended mission would allow data reception for 2 more revolutions, mapping out a complete solar cycle. Because the near-solar dust environment is not well understood (or is controversial at best), and it is very important to have better knowledge of the dust environment to protect Solar Probe from high velocity dust hits, we urgently request the scientific community to obtain further measurements of the near-solar dust properties. One prime opportunity is the July 1991 solar eclipse.

1. Introduction

In the late 1970's and early 1980's, NASA studied a mission to go close to the sun, called Starprobe. In this study: 1) fields and charged particles, 2) solar imaging, and 3) drag compensation experiments to study relativistic effects and the solar gravitation quadrupole moment were considered. However, it became apparent that the latter two categories of experiments necessitated complex spacecraft designs and escalated overall costs. Thus, in 1985 a Space Science Board committee of the National Academy of Sciences, recommended that "....a Solar Probe mission whose primary objective is to carry out the first in-situ observations of the solar wind plasma and fields near the source of the wind in the solar atmosphere. Included will be a detailed study of energetic particles which will yield important diagnostic data on particle acceleration processes and coronal structure" [1].

In 1988, NASA selected a Solar Probe science advisory panel, and from 1989 to the present, this group has carried out such studies [2, 3]. NASA chose experts from many areas of science to cover the anticipated goals. However, a person for dust science was not included, because Solar Probe was initially thought to be primarily a coronal plasma science mission. With help from M. Hanner and H. Zook, a dust instrument is now part of the payload. Similar effort is presently underway towards the possible inclusion of a hard (> 20 keV) X-ray experiment and a 3-dimensional coronal imager.

Solar Probe is presently in NASA's strategic plan for the 1990s. The preliminary design studies are scheduled in the first-half of the 1990s, phase A from 1991-1992 and phase B from 1993-1994. A Project Start is currently scheduled for 1995, completed spacecraft design and fabrication is scheduled for 1999, leading to a launch in September 2000. The first perihelion would occur in June 2006 and a second in December 2008. The satellite will remain in a solar orbit with a 2.5 year period, so an extended mission is possible.

2. The Solar Probe Mission

2.1. EARTH-JUPITER GRAVITY ASSIST (ΔV-EJGA) TRAJECTORY

29

A.C. Levasseur-Regourd and H. Hasegawa (eds.), Origin and Evolution of Interplanetary Dust, 29–32.
© 1991 *Kluwer Academic Publishers.*

Solar Probe will receive two gravitational boosts and three major velocity increments in the proximity of the Earth, Jupiter and the Sun (Figure 1). Following launch, the spacecraft will receive a deep space velocity increment (ΔV), orbit the sun once and encounter the Earth to have a gravity-assisted flyby and another ΔV. The spacecraft will then travel to Jupiter (5 AU) experiencing another gravity assist and will drop in towards the sun. At this point Solar Probe will be in a highly elliptical polar orbit with a perihelion of 4 R_S, an aphelion of about 5 AU, and a period of about 4 years. After perihelion flyby, a burn will be commanded to decrease the spacecraft velocity and shorten the orbital period. This burn will be commanded when the spacecraft is 10 to 15 R_S past the sun to minimize perihelion science data loss.

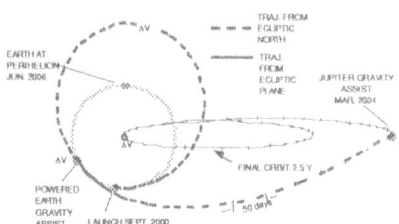

Figure 1. The Solar Probe Trajectory

The near-perihelion passage is shown in Figure 2. Besides monitoring the interplanetary environment out to 5 AU, Solar Probe will enter an exploratory phase as it goes inside 0.3 AU, a region that has never been penetrated before. The spacecraft will spend ~ 10 days in this region. Close to the Sun, Solar Probe will first cross the north pole at ~ 7 R_S, the equator at 4 R_S and then the south pole at ~ 7 R_S all in a span of about 14 hours. The spacecraft orbital plane has been designed to be nearly orthogonal to the Earth-Sun line, so that constant communication with the spacecraft can be maintained. A Ka-band downlink telemetry system will be used to maintain a 70 kb/s data stream through the solar coronal (interference) environment near perihelion. The downlink telemetry has higher rate capability (over 200 kb/s) further from the sun. Some of the near perihelion data will be stored on-board and will be telemetered down after perihelion passage.

At 1 AU on the inbound pass, Solar Probe will be at an out-of-the-ecliptic latitude of 15°. This latitude increases with decreasing distance from the sun, reaching +28° at 0.3 AU and +90° seven hours prior to perihelion. The outbound pass will sample the dust environment at latitudes below the ecliptic plane, symmetric to the inbound pass.

Solar Probe will therefore give dust information from 5 AU to 4 R_S using a single instrument. Solar Probe, together with the Galileo mission, will solve one of the current problems of instrument intercalibration (see E. Grün [4], for discussion). If the extended mission is approved, then possible solar cycle dust fluence dependences could also be measured.

2.2. SOLAR PROBE SCIENCE

The prime Solar Probe mission objective is to determine the mechanisms by which the solar wind plasma is accelerated and heated. For this objective, Solar Probe is targeted to reach or go inside of 4 R_S, the nominal solar wind sonic point. Inside this theoretical point, the solar wind is subsonic, and outside, it is supersonic. This perihelion distance is also close to the theoretical dust free cavity boundary [5] (< 4 to 5 R_S) so that the dust properties in and near this boundary will also be explored.

Figure 3 gives the current spacecraft configuration. The heat shield is made of carbon-carbon and will reach surface temperatures up to 2100 K at perihelion. The heat shield will point toward the sun and be designed so that the umbra (shadow) region where the scientific instruments are located, will remain near room temperature (20 - 30° C).

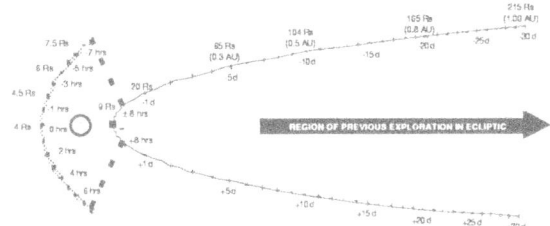

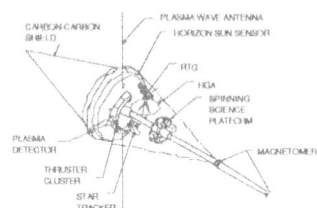

Figure 2. The near perihelion trajectory. The orbit plane is orthogonal to the Earth-Sun line.

Figure 3. The Solar Probe spacecraft configuration.

Most of the science instruments are located on the science platform, which is moved away from radioisotope thermoelectric generators (RTGs) by a mast. The platform will spin at 60 rpm to allow full three-dimensional distribution measurements once per second. The dust detector is currently located on the platform. It is also possible to place it on the body of the spacecraft, if so desired. The strawman payload is given in [3]. The mass, power, data rate, range, type of detector, field of view, location and heritage for each instrument is given in the reference. At perihelion the spacecraft will be travelling at over 300 km/s (and the solar wind only 100-200 km/s), so Solar Probe will be passing through magnetic, plasma and dust structures quite rapidly. It will be necessary to have high time resolution to be able to adequately identify these structures, thus the very high spacecraft data rates.

3. Dust Hazards

There are substantial dust hazards, especially in the region near perihelion. At 4 R_S, circularly orbiting dust particles have a velocity of 220 km/s while the spacecraft velocity will be over 300 km/s. It is particularly important to know the direction of the dust orbits. As an example, if they are all in a circumsolar ring in the elliptic plane, then the spacecraft and dust velocity vectors will be orthogonal to each other and the relative velocity will be 380 km/s. Protection will be needed on only one side of the spacecraft. If, on the other hand, the near solar dust flux is isotropic, then all sides of the spacecraft will have to be protected. The dust impact velocities will range from 90 to 530 km s^{-1}.

The near solar dust environment is presently a controversial issue. From balloon-borne and solar eclipse observations of the solar corona [6-10] ,enhanced near-infrared emissions have been used to argue for the presence of a circumsolar dust ring at 4 R_S (and rings at larger distances). Mizutani et al. [9] has presented a model of a toroidal-shaped cloud 8 R_S in diameter and 1 R_S in radius. The particles have a size of 10^{-2} cm and a number density of 10^{-14} cm^{-3}. Mukai et al. [11] have discussed spatial concentration of dust grains caused by the stabilization of their orbits by sublimation. On the other hand, Mann and Kneissel [12] demonstrate that the near-infrared emission enhancement at 4 R_S could be explained by optical diffraction instead.

To estimate the dust environment at 4 R_S for Solar Probe, we take two different approaches: 1) we make an extrapolation from 1 AU, and 2) we use the dust ring values of Mizutani et al. [9]. We compare the two sets of numbers below.

To extrapolate from 1 AU, we use the interplanetary mass density model of Grün et al. [13]. The radial gradient in densities are assumed to vary as $r^{1.3}$, where r is the distance from the sun [14]. This dependence was based on Helios zodiacal light observations made for distances in as close as 19 R_S from the sun. (However, we note that recent IRAS results indicate that the particle albedo increases with decreasing solar distance [15-17]. A $r^{1.0}$ scaling law has been derived [17]). The spatial number density at 4 R_S extrapolated from the Grün et al. 1 A.U. values is given in Table I. This corresponds to the number densities in the ecliptic plane.

Table I. The dust number density extrapolated to 4 R_S assuming the Grün et al. [13] 1 AU spectrum and a $r^{-1.3}$ radial dependence.

Extrapolation to 4 R_S

Mass Range	Spatial Number Density
3×10^{-7} to 3×10^{-6} gm	1.1×10^{-15} cm^{-3}
3×10^{-6} to 3×10^{-5} gm	1.8×10^{-16} cm^{-3}
3×10^{-5} to 3×10^{-4} gm	1.1×10^{-17} cm^{-3}

Comparison of the above values to those of Mizutani et al. indicate that the ring model is about 15 times higher in density than the extrapolated values. If a $r^{-1.0}$ model is used for the extrapolation, then the difference between the ring model and the extrapolation is even greater.

4. Conclusions

The Solar Probe mission will be a revolutionary advance in our understanding of dust within the solar system. It will be able to answer the question as to whether there are dust rings near the sun, determine the particulate mass and velocity distributions in the near-solar region and determine the particle spectrum at the boundary(s) to the dust free zone(s). It will also be able to measure the gradient in particles from 4 R_S to 5 AU and to measure the out-of-the-ecliptic dust distribution. The measurements should provide much of the necessary information to determine the sources of interplanetary dust within the heliosphere.

Acknowledgements. Portions of this research were carried out at the Jet Propulsion Laboratory, California Institute of Technology under contract with the National Aeronautics and Space Administration. We very much appreciate comments and correspondence from M. Hanner, B. Kneissel, I. Mann, A. Levasseur-Regourd, P. Lamy and T. Mukai, and the meeting convenors for their kind invitation to speak at the IAU Colloquium. We also appreciate a reading of this manuscript by B. E. Goldstein, (the Solar Probe Deputy Study Scientist) and J. A. M. McDonnell, the referee of the paper, for helpful comments and suggestions.

5. References

1. 'An Impl. Plan for Priorities in Sol.-Syst. Space Phys.' (1985), Com. Sol. and Space Phys. of the Space Sci. Bd., Nat. Acad. Press, Wash., D.C.
2. Sol. Probe Miss. Sys. Design Concepts 1989' (1989), ed. J. E. Randolph, JPL Internal Document D-6798.
3. 'Solar Probe Scientific Report' (1989), JPL Internal Document D-6797.
4. Grün, E., et al. (1991), In-situ space expl. of dust in sol. syst. and init. results from Galileo dust det., in Orig. and Ev. Interpl. Dust, Kluwer, Dordrecht.
5. Lamy, Ph.L. (1974), 'Interact. of interpl. dust grains with sol. rad. field', Astron. and Astrophy., 35, 197-207.
6. MacQueen, R. M. (1968) 'Infrared obs. of outer sol. cor.', Astrophys. J., 154, 1059-1076.
7. Peterson, A. W. (1969) 'Exp. det. therm. rad. interpl. dust', Astrophys. J., 148, L37-39.
8. Peterson, A. W. (1971), Bull. Amer. Astron. Soc., 3, 500.
9. Mizutani, K. et al. (1984), 'Near-infrared obs. circumsol. dust emis. during 1983 sol. eclipse', Nature, 312, 134-136.
10. Mukai, T. (1985), 'On the sol. dust rings', in Props. Interacts. Interpl. Dust, R. H. Giese and P. Lamy, (eds.), D. Reidel, 59-62.
11. Mukai, T. (1974), 'On circum. grain mat.', Publ. Astron. Soc. Japan, 26, 445-458.
12. Mann, I. and Kneissel, B. (1991), 'Interpl. dust close to the sun', in Orig. Ev. Interpl. Dust, A. C. Levasseur-Regourd (ed.), Kluwer, Dordrecht.
13. Grün, E., et al. (1985), 'Coll. bal. of meteoric compl.', Icarus, 62, 244-272.
14. Leinert, C., Richter, I., Pitz, E., and Planck, B. (1981) 'The zod. light from 1.0 to 0.3 AU as obs. by Helios space probes', Astron. Astrophys., 103, 177-188.
15. Good, J. C., et al. (1986), 'IRAS obs. of zod. backgrd.' Adv. Space Res., 6, 83-86.
16. Kneissel, B. and Giese, R. H. (1986), 'The impact of IRAS results on 3-D models glob. distr. interpl.. dust', Adv. Space Res., 6, 79-82.
17. Levasseur-Regourd, A. C. and Dumont, R. (1990) 'IRAS obs. zod. backgrd.' Adv. Space Res., 6, 83-86.
18. Levasseur-Regourd, A. C., et al. (1990), 'Dust op. prop.: comp. betw. comet. interpl. grains', Adv. Space Res.'

A BALLOON-BORNE DETECTOR FOR STRATOSPHERIC COSMIC DUST DETECTION

WAN GUCUN, OUYANG ZIYUAN, XU YIWEN, WU XIGUANG
Institute of Geochemistry, Academia Sinica,
Guiyang, Gaizhou Province,
People's Republic of China, 550002

ABSTRACT. This paper introduces a kind of cosmic dust collecting technique anddescribes in detail the structure of the collector, the balloon-basket system and concerning experimental skill.

1. INTRODUCTION

Cosmic dust is one of the material sources of the earth. Studies of cosmic dustis of great importance not only in exploring the origin and evolution of thesolar system and some small celestial bodies such as comets, but also in dis-cerning the environmental catastrophic events and extinction of living thing in the history of the earth, providing the scientific basis for the stratigraphic livision and correlation and calculating the flux of cosmic dust particles. It is still of practical significance in space exploration through the study of the composition, flux and distribution of cosmic dust.[1]

   Although great advances have been made in modern aerospace technology, the high-altitude scientific balloon is still an important approach to collecting cosmic dust [2].

   From 1984 to 1987, in order to establish some practical and effective experi-mental methods in collection of cosmic dust from the stsatosphere, we had done two flights in which our experiments were done on the trial basis and the collectors were attached to the balloon baskets for other experiments. Then , we did three flights in which we self-designed balloon baskets and collectors were used and the whole experiment process was established. The balloon lauching site is in Xianghe County, Hebei Province. Usually the flight altitude for col-lecting work were in 25000~35000M high. The radius of balloon level-flight was controled by telecontrol device in range of 150~200KM.

2. THE TYPE OF COLLECTOR AND ITS LOCATION

According to the conditions of laboratory measurement and analysis, we are engaged mainly in the collection of cosmic dust particles measuring 10 μ or more in grain size. So it is advised to use the settling-plate type collectors as a major approach. [2]

   The location of the collectors in the balloon-basket system has a great bearing on the whole layout of the collectors, so at the initial stage of de-signing the location must be determined.Fig.1.shows the location of our colle-ctors.

   At the time the balloon goes up and the collectors open to work, the col-lectors are put on the two racks fixed on both sides of the basket.   Once the collection is finished, the collectors are closed automatically. Before the basket and balloon are separated, the collectors are pulled into the

A.C. Levasseur-Regourd and H. Hasegawa (eds.), Origin and Evolution of Interplanetary Dust, 33–36.
© 1991 *Kluwer Academic Publishers.*

framework of the basket. Thus, at the time the basket goes down, there will occur no "rotor blade effect" (which is harmful to the work of parachute) and if the basket turns over or rolls after landing, the collectors still can be effectively protected by the framework of the basket. The experiments on high-altitude balloon collection of cosmic dust have shown that this approach is feasible and effective with respect to the function and recovery of the collectors.

3. THE STRUCTURE OF THE SETTLING-PLATE COLLECTOR

According to the flux introduced by D. E. Brownlee [2] and L. Hemenway [3], the surface collection area is determined to be $0.8 \, M^2$.

The collector takes the form of double cassettes with the covers which can open and close. The collection surfaces made of a kind of transparent polyester film are coated on the inner surfaces of the covers and the bottoms made of hard aluminium plate (2 mm thick). The cover and bottom of the collecor are equipped with the rectangular frame made of angle-shaped aluminium material and sealing material. When the cover of the collector cassette revolves and closes, the frame on the cover just hoods the rectantular frame on the bottom, constituting a labyrinthine structure. The sealing material is a kind of clean plastic foam. Before the balloon is launched, the foam should be smeared with clean silicone oil. Thus, when the cassette colses and the basket falls to the ground from the space, the ventilation of plastic foam makes the inner pressure of the cassette keep up with the outside pressure. On the other hand, plastic foam smeared with silicone oil can effictively prevent pollutant particles from finding their way into the collector cassette.

Journal bearings are adopted in the mechanical tranmission system of the collectors. Inthe system there is no need to apply lubricant so as to avoid some possible troubles brought about by lubricating-oil at lower temperature.

As an adhesive agent to the collector, freezeproof clean silicon oil is coated on the collection surfaces.

As the collectors are easy to dismount from the basket on the landing spot, the collectors can be recovered without pollution.

4. THE STRUCTURE OF THE BASKET AND THE BALLOON-BASKET SYSTEM

The basket has a rectangular frame made of angle-shaped steel. According to the technical requirements, the racks are fixed on both sides of the basket frame. The rigid slideways are welded in the basket on the frame and linked up with the racks. There still fix limit switch and rope towed mechanism which can pull the collectors into the basket.

We fix miniature permanent magnets on the ends of the base plates and racks. Thus, even though the basket slants at 30°, the collectors will not slide. The towed force of the motor is enough. The motor which tows the ropes is a torque motor manufactured in Beijing Micromotor Factory. Through the deceleration of the transmission system, the base plate of the collectors can get a speed of about one metre per minute to remove smoothly. The ropes we used are those of 8020-type parachute. So under lower temperature and high altitude conditions, their strength is reliable.

Fig.1 Shows the positions of all the apparatuses in the basket. Some heavy apparatuses such as storage battery and sand-throwing device are located at the bottom of the basket.

The lower chamber of the basket in which electrical facilities are located is sealed with 50mm thick polystyrene foam plastic plate. The batteries are most sensitive to lower temperature so every group of batteries is sealed with 20mm thick polystyrene foam plastic plate again.

According to the weight of the basket, we choose a kind of balloon whose volume is $30000 M^3$. In order to reduce the "shielding effect of balloon" as small as possible in the flight and in regard to the bearing capacity of the balloon, the length of the ropes have to be 70M. When the balloon goes up to the high altitude, its maximum diametre is 42.4M and its half height is 21.5M [4]. So the distance from the collection surface to the half height of the balloon is 105M. The shadow half spread angle in the collector is 11.36° . Obviously this is acceptable.

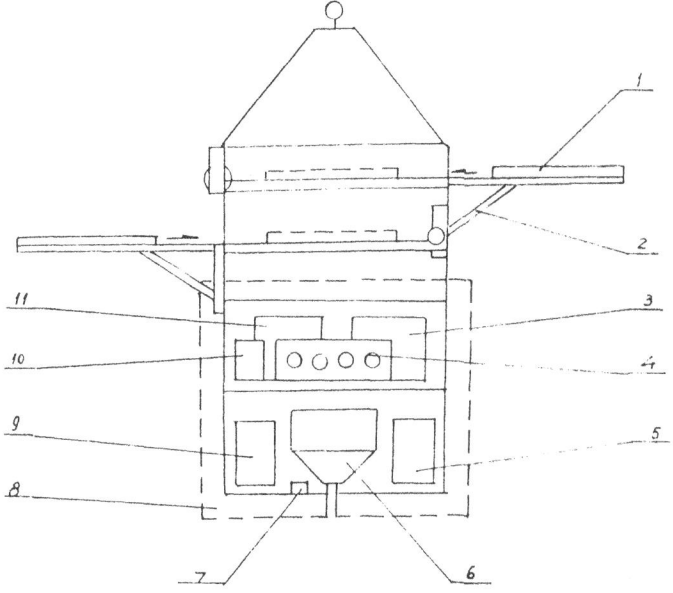

Fig. 1. Shows the positions of all the apparatuses in the basket.

1. settling plate collector   2. cracks   3. sigal transmitting set   4. timer
5. power A   6. sand-thowing device   7. antenna
8. thermal insulating layer of polystyrene foam   9. power B
10. collectors controller   11. signal recriver

5. CONCLUSIONS

The high-altitude settling-plate cosmic dust collector designed by us has some outstanding advantages: It is simple in structure; it has a big enough collecting surface area; the surface of organic thin film is easy to clean; it is convenient to observe and select specimens under the microscope. During the flying of the balloon, the collecting facilities work normally. After the collectors are closed and pulled into the basket and land with it, under the protection of the frame, the collectors could be recoverd without any damage and the specimens will not be polluted. We have obtained some specimens in this way. Further studies will be published in anothe paper [5].

The collectors we designed and the method of locating them on the basket can also be used in the same practice in the future time. A set of feasible and effective experimental methods and techniques concerned are established.

6. ACKNOWLEDGEMENTS

Our experiments on high-altitued cosmic dust collection were carried out under the great help of the balloon engineering groups of the Institute of High Energy Physics and the Institute of Atmospheric Physics, Academia Simica. They provided the relevant devices of balloon engineering. The work on balloon launching, flight controlling and recovery was accomplished by them. Here we express our appreciation to them.

The project is financially supported by the National Natural Science Found-ation of China.

References

[1] 欧阳自远(1988) 天体化学, 科学出版社,北京PP.273.

[2] Brownlee,D.E.(1978) `Studies by sampling techniques ',in J.A.McDonnell
     (edit),Cosmic Dust,John Wiley and Sons,New York. PP.225-335.
[3] Hemenway, C.L. et.al.(1967)`High altitude balloon-top collection of
     cosmic dust',in Space Research,Vol.7.North-Holland, PP.1423-1431.
[4] 中国科学院高空科学气球技术总体组,中国科学院万M³级高空科学气球系统鉴定会
     文件--气球及发放,飞行分系统技术文件,中国科学院高能物理所,北京,PP.12-15,34.
[5] 黄伯均, 欧阳自远, 付平秋, 高振敏(1990)`平流层尘粒的初步研究'中国科学(待刊)

DYNAMIC MODELLING TRANSFORMATIONS FOR THE LOW EARTH ORBIT SATELLITE PARTICULATE ENVIRONMENT

J.A.M. McDonnell, K. Sullivan, S.F. Green, T.J. Stevenson & D.H. Niblett
Unit for Space Sciences, University of Kent at Canterbury,
Canterbury,
Kent,
CT2 7NR,
United Kingdom

ABSTRACT. A simple dynamic model to investigate the relative fluxes and particle velocities on a spacecraft's different faces is presented. The results for LDEF are consistent with a predominantly interplanetary origin for the larger particulates, but a sizable population of orbital particles with sizes capable of penetrating foils of thickness <30μm. Data from experiments over the last 30 years do not show the rise in flux expected if these were space debris. The possibility of a population of natural orbital particulates awaits confirmation from chemical residue analysis.

1. Dynamic modelling

The Long Duration Exposure Facility (LDEF) has provided an unprecedented source of information on extraterrestrial particulates after almost 6 years in low Earth orbit. Some initial results are presented by McDonnell et al (1990; 1991). Any instrument to detect interplanetary or orbital particles will not immediately reveal the true space density or flux in Earth orbit. The *detected* flux is dependent on the spacecraft's orbital velocity and the instrument pointing history. Because of the range of possible particle orbits the problem cannot be uniquely solved by inversion techniques. A model particle distribution must be produced, used to predict the observed flux, and modified until a match is obtained. Most previous penetration experiments yielded no directional information, but LDEF, with its controlled attitude with respect to the orbital radius vector, provides valuable constraints on the contribution to the observed fluxes of interplanetary and orbital populations.

In order to determine the particulate population in Low Earth orbit, a detailed model of the distribution of trajectories of incoming interplanetary particles and the geocentric orbits of bound particles (natural or debris) is required. In practice, valuable results may be obtained from a simple model, employing the same methodology, which does not implicitly distinguish between interplanetary or orbital particles, but utilises the very low probability of impact of an orbital particle on LDEF's space-pointing and trailing (West) faces. In the geocentric reference frame the flux distribution is defined by a particle mass distribution having an isotropic velocity distribution with a single (average) velocity (V_p) at each mass. The resultant velocity vectors (V_{face}) impacting normal to a face of a spacecraft with orbital velocity (V_s) are then calculated taking into account shielding by the Earth, from which a mean V_{face} can be determined. The foil perforation fluxes observed in a capture cell experiment such as LDEF A0023 MAP (Micro-Abrasion Package) are a function of this normal velocity (foil thickness marginally penetrated, $f \propto V_{face}^\beta$ - see McDonnell et al, 1990 for details).

For the case of an isotropic velocity distribution and the known spacecraft orbital velocity, altitude and effective atmospheric altitude, a geometry factor K can be defined for

37

A.C. Levasseur-Regourd and H. Hasegawa (eds.), Origin and Evolution of Interplanetary Dust, 37–40.
© 1991 *Kluwer Academic Publishers.*

each spacecraft face such that $K_{face} = n_{face}(V_s) / n_{face}(0)$ i.e. the ratio of the number of particles impacting that face, travelling through the particle population with velocity V_s, to that seen when it is at rest. Due to Earth shielding, certain trajectories are forbidden, so a further ratio, the Earth shielding correction Ω is required where $\Omega_{face} = n_{face}(0) / n(0)$. $n(0)$ is the number of particles impacting a stationary plate with no Earth shielding.

Capture cell experiments do not in general provide information on particle velocities. It is possible however, for LDEF data, to derive the mean velocity V_p which is consistent with results obtained from different faces (i.e. different pointing directions) of the spacecraft. Figure 1 illustrates the technique on a plot of log particle flux (Φ) vs log foil thickness (f). α is the cumulative particle mass distribution index, Φ_1 and Φ_2 are foil penetration fluxes observed on faces 1 and 2 with (initially unknown) normal impacting velocities V_1 and V_2. The flux ratio at constant foil thickness ($\varepsilon|_f$) can be directly obtained from the figure. However, the flux ratio at constant mass ($\varepsilon|_m$) is of more interest since it relates to the actual population of particles and their mean velocity. The transformation of data from one face (1) to the other (2) consists of two components: i) $\Delta \log f|_m = \log[(V_2/V_1)^\beta]$ accounts for the difference in mass sensitivity due to the difference in impacting velocity, ii) $\Delta \log \Phi|_m = \log[K_2\Omega_2/K_1\Omega_1]$ accounts for the geometrical factors described above. Since both the K factors and the normal velocities V depend on the initial particle velocity V_p its value can be derived by application of the model.

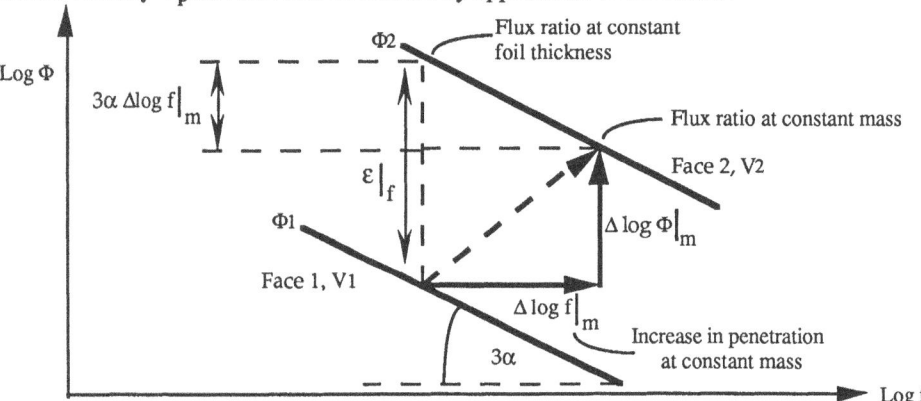

Figure 1. Method of transposition of the cumulative flux plot from face 1 to that of face 2. The flux ratios at constant crater size (or foil thickness, given as the enhancement at constant foil thickness, $\varepsilon|_f$) and at constant mass $\varepsilon|_m$ are indicated. See text for details.

2. Model Results and Application to LDEF Data

Using LDEF's orbital velocity (7.64 km s^{-1}) and mean altitude (458 km) and an effective atmospheric height of 185 km, the following relationships for LDEF's SPace, West (trailing) and East (Leading) faces have been derived from fits to results (for $v_p > 10$ km s^{-1}) from the isotropic single velocity model:

$$V_E = 0.713V_p + 4.19 \qquad K_E = (V_p+7.71)/(V_p-1.38) \qquad \Omega_E = 0.676$$
$$V_{SP} = 0.671V_p - 0.370 \qquad K_{SP} = (V_p-7.57)/(V_p-7.35) \qquad \Omega_{SP} = 1.0$$
$$V_W = 0.710V_p - 5.32 \qquad K_W = (V_p-7.92)/(V_p+1.32) \qquad \Omega_W = 0.676$$

These relationships can be applied to LDEF data (McDonnell et al 1991) for the Space and West faces which are likely to consist only of interplanetary particles. A particle velocity V_p of 16±2 km s^{-1} is required to satisfy both data sets, consistent with unbound particles. If there were no orbital particles impacting LDEF this velocity should also satisfy the data detected on all other faces. Figure 2 shows LDEF foil penetration data (or equivalent - see McDonnell et al, 1991) for East, West and Space. Using $V_p=16$ km s^{-1} the West data have been transformed to produce a predicted East flux. Although this is close to the observed East data at foil thicknesses > 30 μm, an additional component is required for thinner foils. Since the East face is readily accessible to orbital particulates (but not the Space or West faces) this may be considered to be the result of the presence of a population of orbital particulates. If this additional population of bound particles is a result of man-made space debris, it would be expected to have increased over the period in which experiments have been deployed in Low Earth Orbit.

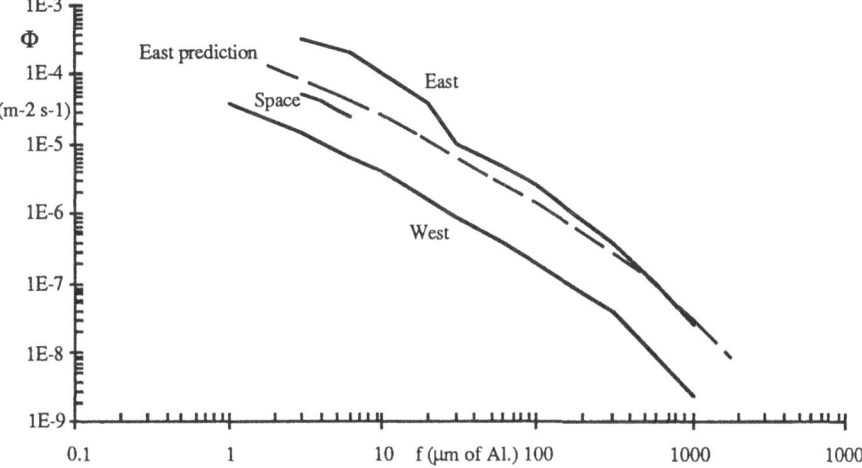

Figure 2. The predicted LDEF East flux is shown (dashed line) using a constant geocentric particle velocity of 16 km s^{-1} obtained from the fit to the West and Space fluxes. This is compared with LDEF data of the East face (full line) (see McDonnell et al 1991).

3. Temporal Stability of the Near Earth Particulates

Spacecraft data from experiments which employ the same detection technique, (thereby avoiding intercalibration problems) are plotted in Figure 3. In the mass range considered, temporal changes in the observed flux over some 30 years are less than a factor of ±2. In

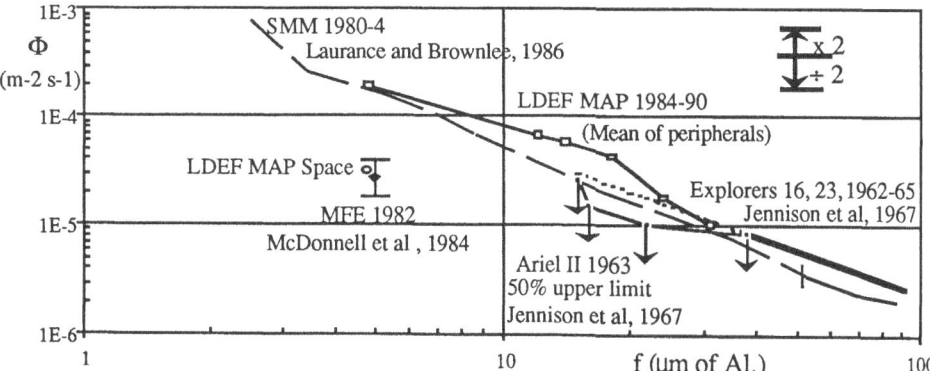

Figure3. Flux measurements from perforation experiments scaled to their equivalent aluminium foil thicknesses. All data are corrected for Earth shielding ($\Omega = 0.631$ for Solar Max Mission, SMM). SMM data is identified as space debris (dashed) and natural particulates (full line).

this same period space launches and the potential debris source have increased fairly steadily at a mean rate of 110 to 120 objects per year from 1961 to 1989, including the removal of objects by re-entry (NASA Ref 2 in Report of the ESA Space Debris Working Group).

4. Conclusions

Application of a simple dynamic model to LDEF data indicates a predominance of interplanetary particulates at large masses but a sizeable population of orbital particles for foils <30μm thick. Contrary to the possibility of this being space debris, foil penetration experiments do not show an appreciable increase in flux over a 27 year period, despite rapidly increasing space traffic.

The possibility of an Earth orbital component of extraterrestrial captured material has been proposed before (Whipple 1961, McCracken et al 1961) and dismissed many more times (Singer 1961, Nilsson 1966)! This "Dust Belt", as proposed, applied to considerably larger dimensions (of nanogram mass) than is now considered. Although arguments based on particulate dynamics demonstrate the low efficiency of capture in orbit by the Earth for interplanetary particulates (due to the spread of inclination and eccentricity) it should be noted that the "excess" of orbitals we now suspect is both much smaller and at much smaller particle masses than previously postulated. A trend to lower values of i and e for interplanetary orbits at these smaller dimensions would increase the capture efficiency. SMM chemical residue analysis by Laurance and Brownlee (1986) for smaller particles was consistent with terrestrial sources. Clearly, resolution of this problem awaits chemical analysis of the impact residues on LDEF.

Acknowledgements
We acknowledge all members of the LDEF MAP team in the Unit for Space Sciences at the University of Kent ; NASA Project Management (W.Kinard, J.Jones); Alison and Margaret; S.E.R.C. (U.K.); and Auburn University Space Power Institute (U.S.A.), Prime Contract N660921-86-C-A226, subcontract 87-212.

References
Jennison, R.C., McDonnell, J.A.M., and Rodger, I. (1967). 'The Ariel II micrometeorite penetration measurements', Proc. Roy. Soc. A., 300, 251-269.
Laurance, M.R., and Brownlee, D.E. (1986). 'The flux of meteoroids and orbital space debris striking satellites in low Earth orbit', Nature, 323, 136-138.
McCracken, C.W., Alexander, W.M., and Dubin, M. (1961). 'Direct measurements of interplanetary dust particles in the vicinity of Earth', Nature, 192, 441-442.
McDonnell, J.A.M., Carey, W.C., and Dixon, D.G. (1984). 'Cosmic dust collection by the capture cell technique on the Space Shuttle', Nature, 309, 237-240.
McDonnell, J.A.M., Deshpande, S.P., Green, S.F., Newman, P.J., Paley, M.T., Ratcliff, P.R., Stevenson, T.J., and Sullivan, K. (1990). 'First results of particulate impacts and foil perforations on LDEF', Adv. Space Res. (in press).
McDonnell, J.A.M., Sullivan, K., Stevenson, T.J., and Niblett, D.H., (1991). 'Particulate detection in the near-Earth space environment aboard the Long Duration Exposure Facility LDEF: cosmic or terrestrial?', Proc. IAU Coll. on 'Origin and Evolution of Interplanetary Dust', Kyoto, Japan.
Nilsson, C.S. (1966). 'Some doubts about the Earth's dust cloud', Science, 153, 1242-1246.
Singer, S.F. (1961). 'Interplanetary dust near the Earth', Nature, 192, 321-323.
Whipple, F.L. (1961). 'The dust cloud above the Earth', Nature, 189, 127-128.

FACE-DEPENDENT IMPACT PROBABILITIES UPON LDEF FOR HELIOCENTRIC PARTICLE ORBITS

DUNCAN STEEL
Department of Physics and Mathematical Physics
The University of Adelaide
G.P.O. Box 498, Adelaide, SA 5001
Australia

ABSTRACT. If the impact record upon LDEF is to be interpreted so as to determine the flux, orbits, sizes and compositions of natural meteoroids and dust, and space debris, then it is necessary to relate the microcraters and perforations recorded to the likely source orbit of the particle in each case. Here a single-particle approach is used to calculate the relative impact probabilities upon six orthogonal faces of LDEF for particles coming from heliocentric orbits confined to the ecliptic; the results are presented as functions of impact velocity and impact angle for each face. The flux from geocentric orbits to the Space-and Earth-pointing faces is much lower than to the other faces; experiments positioned on those faces are thus likely to be less contaminated by space debris. Particles from heliocentric orbits can impact both the Space and Earth faces, but the latter is less likely to be hit due to the shadowing effect of the planet. The cratering ratios for the East (or leading) face compared to the West (or trailing) and the Earth-directed faces are strongly dependent upon the velocities of the particles and can therefore indicate of the velocity distribution of meteoroids and interplanetary dust.

1. Introduction

The Long Duration Exposure Facility (hereafter LDEF) was recovered in January 1990 having spent close to six years in a near-circular orbit of inclination 28°.5 and initial altitude of about 477 km; increased atmospheric drag meant that this had decayed to about 340 km at the time of its recovery. Since LDEF maintained the same orientation in its orbit throughout its flight, the amount of cratering on its different faces can tell us much about the flux and orbital distribution of the various classes of impactor. These comprise, as a broad division, man-made space debris in geocentric orbits and natural meteoroids and interplanetary dust from heliocentric orbits. This paper deals predominantly with the second of these classes. Throughout the faces of LDEF under consideration are limited to the six orthogonal faces known as the East (or leading) face, which is taken to be the plane perpendicular to and facing the direction of motion of LDEF, the West (or trailing) face, the Space face which always points directly away from the Earth, the Earth face which is the opposite to this, and the North and South faces which are constrained to LDEF's orbital plane.

The case of particles (such as man-made space debris) in geocentric orbits has been investigated elsewhere (Olsson-Steel, 1990; Steel and McDonnell, 1991) and it was shown that very few impacts are expected upon either the Space- or Earth-directed faces due to

41

A.C. Levasseur-Regourd and H. Hasegawa (eds.), Origin and Evolution of Interplanetary Dust, 41–44.
© 1991 *Kluwer Academic Publishers.*

the low eccentricity (and hence low radial velocity) of such particles. The range of impacts velocities and angles were also presented therein, and it was found that in terms of the amount of cratering, for geocentric orbits we expect for LDEF:

Space/Earth/West ≪ North/South < East

In this paper we are interested in how the fluxes to these faces from heliocentric orbits might compare with this relationship.

2. Method

Work in this area, using velocity distributions (as opposed to the single particle approach used here), has been carried out by Zook and co-workers (Warren et al, 1989; Zook, 1987, 1990). The method used in this paper will be detailed elsewhere (Steel and Cervera, in preparation). Briefly, the technique we have developed is as follows. At the present stage of our program we are able to calculate impact probabilities, including the velocity and angle of impact probability distributions, for particles of different initial velocities approaching the Earth *along the ecliptic only*. The computations carried out involve considering all possible positions of LDEF in its orbit to be equally likely, and all possible directions of the pole of its orbit also to be equally likely over its orbital lifetime; this is valid in view of its high precession rate. In all cases gravitational focussing is included, and this is to a certain extent responsible for the exposure of the Earth face of LDEF which would otherwise be shielded from meteoroids: if there was no gravitational focussing then the Space:Earth ratio would be the same as the ratios of the solid angles of the sky visible from each of these faces, this being about 20:1.

3. Results

The results for different meteoroid velocities are shown in Table 1. The altitude of LDEF has been taken as 477 km, and any particle with a perigee height below 150 km has been assumed to be absorbed by the atmosphere; particles with perigees between 150 and 477 km can hit LDEF as they recede from the planet, and thus may strike the Earth-directed face of the satellite. In each case the relative probabilities, normalized to the East face, are given for the West, Space, Earth, and North/South faces. For the last-named pair the exposure would be expected to be identical, but for the others the exposures will be different (*e.g.* the East and West faces are expected to contrast strongly).

Table 1. Relative impact probabilities upon the different faces of LDEF for particles coming from heliocentric orbits confined to the ecliptic, as a function of velocity.

Velocity at infinity (km/sec)	Velocity at LDEF (km/sec)	East	West	North/ South	Earth	Space
5.0	11.9	1.0	0.074	0.267	0.056	0.539
10.0	14.7	1.0	0.164	0.230	0.042	0.737
15.0	18.5	1.0	0.251	0.258	0.045	0.854
20.0	22.7	1.0	0.332	0.285	0.050	0.943
30.0	31.9	1.0	0.462	0.326	0.057	1.069
40.0	41.4	1.0	0.555	0.353	0.062	1.152
50.0	51.2	1.0	0.622	0.371	0.065	1.208
60.0	61.0	1.0	0.672	0.384	0.068	1.250
70.0	70.8	1.0	0.710	0.395	0.069	1.283

The number of impacts upon different faces of LDEF by particles coming from heliocentric orbits will depend upon their velocity relative to the Earth (*i.e.* the orbit of the particle). In order to interpret the LDEF data correctly, it will be necessary to know both the collision velocities and impact angles to be expected for a variety of particle orbits, since these affect the sizes of the craters produced (or the thickness of material perforated) for particles of constant mass. In the Figures we present the relative impact probabilities as a function of velocity (Fig. 1) and impact angle (Fig. 2) for an initial meteoroid velocity of 20 km/sec. As can be seen from Fig. 1, not only does the East face have the highest impact probability (*cf.* Table 1) but also these impacts would mostly occur at higher velocities (above 29 km/sec); in addition, they occur at near-normal incidence upon that face (impact angles below 30°: Fig. 2), which means that we may expect much higher cratering/perforation rates on the East face. The Space face would receive impacts across the full range from 14 to 31 km/sec, with increasing likelihood at higher velocities, and with an angular distribution peaking near 30°. This contrasts with the Earth face which according to these calculations receives impacts only at low velocities (< 17 km/sec) or high velocities (> 29 km/sec), all of these being at oblique incidence (impact angles > 65°); however, these results are to a certain extent artefacts of the model used whereby the impactor is assumed to be coming from a trajectory in the ecliptic. Nevertheless, it is clear that the Space:Earth microcratering/perforation ratio will be affected by the rather different impact velocity and angle characteristics, so that the ratio derived from the LDEF surfaces will be rather different to the values given in Table 1.

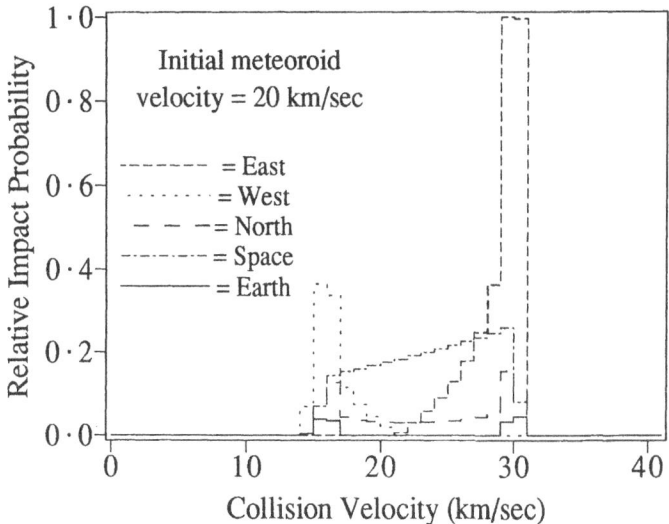

Figure 1. The relative impact probabilities upon the different faces of LDEF as a function of collision velocity for a particle coming from a heliocentric orbit confined to the ecliptic and with an original geocentric velocity of 20 km/sec (equivalent to 22.7 km/sec at the height of LDEF).

4. Conclusion

As noted above, these represent only preliminary results. For particles coming from heliocentric orbits a full consideration of the exposure of LDEF requires an expansion of the present program such that the effect of the actual trajectories of the incoming meteoroids and the obliquity of the ecliptic is included. For both geocentric and heliocentric orbits the velocity- and impact angle-dependence of the craters/perforations produced needs to be included if the LDEF data are to be properly interpreted. These facets of the problem will be dealt with in later papers.

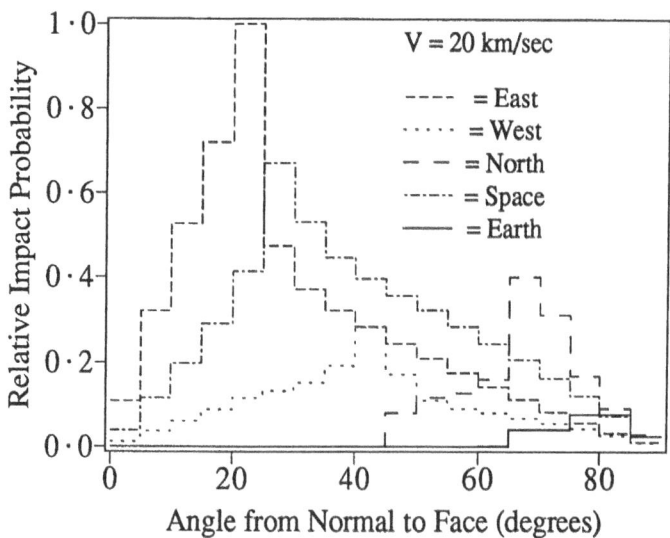

Figure 2. As Figure 1 except showing the relative impact probabilities as a function of impact angle upon each face.

Acknowledgements: This work was supported by the Australian Research Council and the Science and Engineering Research Council (U.K.).

5. References

Olsson-Steel, D.I. (1990). 'Face-dependent impact probabilities, velocities and angles upon LDEF by space debris and natural meteoroids', *Adv. Space Res.* (in press).

Steel, D.I. and McDonnell, J.A.M., (1991). 'Face-dependent collision probabilities upon the Long Duration Exposure Facility for particles in geocentric orbits'. *Planet. Space Sci.* (submitted).

Warren, J.L., Zook, H.A., Allton, J.H., Clanton, U.S., Dardano, C.B., Holder, J.A., Marlow, R.R., Schultz, R.A., Watts, L.A. and Wentworth, S.J. (1989). 'The Detection and Observation of Meteoroid and Space Debris Impact Features on the Solar Max Satellite'. *Proc. 19th Lunar and Planetary Science Conference*, 641-657.

Zook, H.A. (1987). 'The velocity distribution and angular directionality of meteoroids that impact on an Earth-orbiting satellite'. *Proc. 17th Lunar and Planetary Science Conference*, 1138-1139.

Zook, H.A. (1990). 'Flux vs direction of impacts on LDEF by meteoroids and orbital debris'. *Proc. 20th Lunar and Planetary Science Conference* (in press).

THE MUNICH DUST COUNTER — A COSMIC DUST EXPERIMENT ON BOARD OF THE MUSES-A MISSION OF JAPAN

E.Igenbergs, A.Hüdepohl [1], K.Uesugi ,T.Hayashi [2], H.Sved-
hem [3], H.Iglseder [4], G.Koller [5], A.Glasmachers [6],
E.Grün [7], G.Schwehm [3], H.Mizutani, T.Yamamoto, A.Fujimura,
N.Ishii, H.Araki [2], K.Yamakoshi [8], K.Nogami [9]

*[1] Lehrstuhl für Raumfahrttechnik, Technische Universität
München, Richard Wagner Str. 18, 8000 München 2, FRG
[2] Institute of Space and Astronautical Science, J
[3] European Space Research and Technology Centre of ESA, NL
[4] Center for Applied Space Technology and Microgravity,
Universität Bremen, FRG
[5] Lehrstuhl für Prozeßrechner, Technische Universität
München, FRG
[6] Mikroelektronik-Zentrum, Ruhr-Universität Bochum, FRG
[7] Max-Planck-Institut für Kernphysik, Heidelberg, FRG
[8] Institute for Cosmic Ray Research, University of Tokyo, J
[9] Dep. of Physics, Dokkyo University School of Medicine, J.*

ABSTRACT. The Munich Dust Counter (MDC) is a scientific experiment on
board of the MUSES-A mission of Japan. It is the result of a cooperation
between the Institute of Space and Astronautical Science (ISAS) of Japan
and the Chair of Astronautics of the Technische Universität München
(TUM) of Germany. The MDC is an impact ionization detector designed to
determine mass and velocity of cosmic dust. Here a short overview over
the MUSES-A mission is given to show the measurement situation of the
MDC experiment. The measurement principle of the instrument together
with a discussion of the scientific objectives and the design of the
experiment is summarized.

1. Introduction

In 1987 ISAS offered TUM the opportunity to fly a scientific experiment
on the MUSES-A mission, the 13th scientific satellite developed and
launched by ISAS. The MUSES-A mission has been launched on January 24,
1990, from Kagoshima Space Center, Japan. ISAS and TUM agreed that TUM
would develop and built a space experiment to measure cosmic dust particles by impact ionization within the specifications and the time frame
set by ISAS. The weight of the instrument was limited to 800 g, the
power consumption to 2.5 W. This experiment has been named Munich Dust
Counter (MDC) and has been delivered to ISAS in September 1989.

A.C. Levasseur-Regourd and H. Hasegawa (eds.), Origin and Evolution of Interplanetary Dust, 45–48.
© 1991 *Kluwer Academic Publishers.*

The MUSES-A satellite is a spin stabilized type spacecraft composed of a cylindrical main body, 1.4 m in diameter and 0.8 m in height. The MDC is installed on the edge of the instrument platform with an aperture of 12 × 12cm. As the spin axis is maintained perpendicular to the ecliptic plane and the spin rate is 20 rpm, the MDC therefore scans within 3 sec. over 360^0 in the ecliptic plane. The field of view of the experiment is about 148^0. Details of the mission of MUSES-A, which includes multiple lunar swingby maneuvers, are given by Uesugi et al. (1990).

2. Scientific Objectives of the Experiment MDC

The sensitivity and effective sensor area of the MDC are similar to the dust detectors on board of the spacecraft Pioneer 8, Berg and Richardson (1969), and HEOS-2, Hoffmann et al. (1975). Phenomena to be investigated in detail will include the interplanetary dust flux, dust in the earth's magnetosphere and magnetotail, and the lunar dust environment.

3. MDC Measurement Principle

The MDC experiment measures the electrical charges generated by the impact of small masses on a gold surface. The basic setup of the MDC as shown in fig. 1 consists of a target and two charge collectors which are biased by positive and negative high voltage, separating the impact plasma into positively and negatively charged components. Connected to the charge collectors are charge sensitive amplifiers which convert the input charge into an output voltage.

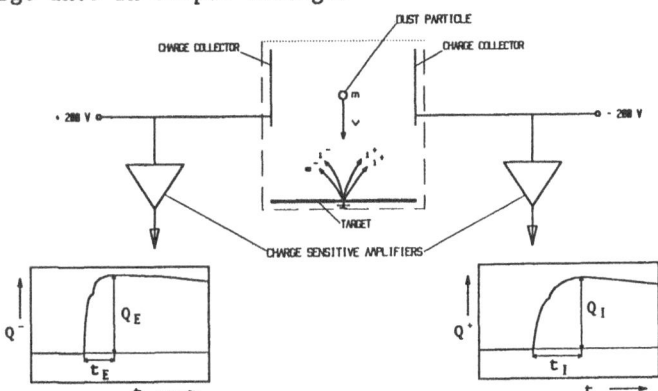

Figure 1. MDC Measurement Principle

For each charge signal amplitude and risetime are evaluated. The mass m and velocity v of the dust particles can be determined by using the following empirical formulae, first found by Friichtenicht and Slattery (1962) and adapted for the particular setup by Iglseder (1976):

$$Q_{maxK} \ / \ m = C_{rK} \cdot v^{\beta_K} \qquad \text{and} \qquad t_K = C_{gK} \cdot v^{\eta_K},$$

valid for the ion channel (K = I) and electron channel (K = E) with:
Q_{max}: maximum charge in ion/electron channel
t   : rise time of signal in ion/electron channel

The constants C_{rI}, C_{rE}, β_I, β_E, η_I, η_E have been calibrated by using the Electrostatic Dust Accelerator of the Max-Planck-Institut für Kernphysik in Heidelberg with particle masses between 10^{-15}g to 10^{-10}g and velocities between 2 km/s and 58 km/s, and by using the Plasma Accelerator of TUM with masses between 10^{-10}g to 10^{-7} g and velocities up to 20 km/s. The sensitivity range of the MDC experiment is shown in fig. 2. The accuracy of the calculation of the mass is at a factor of 0.4 to 2.5, the accuracy of the calculation of the velocity is at a factor of 0.7 to 1.5 .

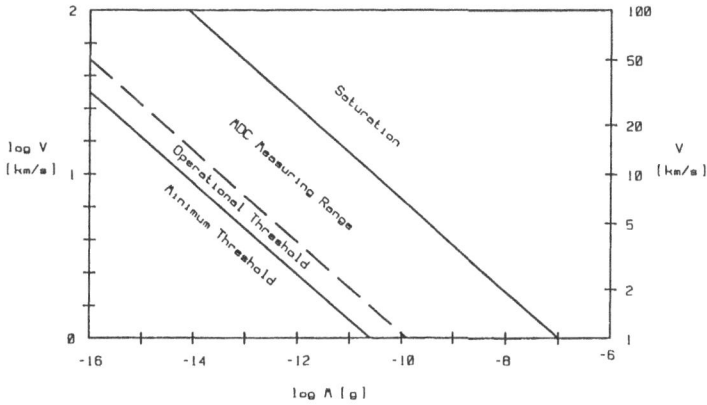

Figure 2. Sensitivity range of the MDC experiment for normal impacts in the target center.

4. Design of the Experiment MDC

The actual dimensions of the MDC are 105 × 110 × 160 mm, the weight of the flight model is 605 grams. The effective sensor area is 100 cm^2. Figure 3 gives a general outline of the MDC design.

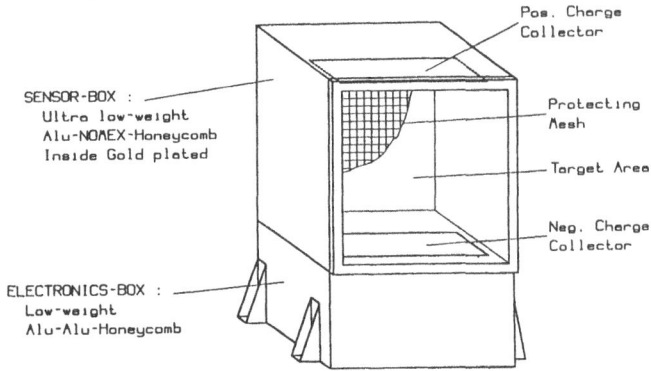

Figure 3. Outline of MDC mechanical design

The impact charges are converted into a voltage signal by charge sensitive amplifiers with a sensitivity from $5 \cdot 10^{-15}$ Cb to $5 \cdot 10^{-11}$ Cb. The output of the charge sensitive amplifiers (electron and ion channel) is digitized and stored in a transient recorder consisting of two 8 bit A/D converters, clocked with 5 MHz each and using a FIFO memory.

A microprocessor system, 80C85 CPU with 2 KByte ROM and 32 KByte RAM, processes measured signals, controls the spacecraft interface and monitors voltages and temperatures. A switching power supply uses +29 V input to generate a ±6 V output for the charge sensitive amplifiers and ±200 V for the bias voltages of the charge collectors. The power consumption of the MDC flight model is 1.8 Watts.

Generally this design allows to record and store the whole particle impact signals. The amount of impact data is 1 KByte per signal or 512 byte per channel, giving a total measurement time of $100 \mu s$. Together with the impact data calibration and housekeeping data valid for the time of impact are stored and transmitted down to ground, giving the opportunity to inspect on the ground the real signals measured in space.

5. Conclusion

Within a cooperation between the Institute of Space and Astronautical Science of Japan and the Technische Universität München of Germany the scientific experiment Munich Dust Counter has been developed and manufactured within two and a half years from start of the project until integration on the satellite MUSES-A. By measuring dust particles in a high-elliptical earth orbit and in the vicinity of the Moon, valuable information about the mass, velocity and the flight direction of cosmic dust will be obtained. Preliminary results of the operation of the MDC from January to July 1990 are described by Igenbergs et al. (1990).

References

Berg O.E., Richardson F.F (1969) 'The Pioneer 8 Cosmic Dust Experiment', Rev. Sci. Instrum. 40, 1333-1337.

Früchtenicht J.F., Slattery J.C. (1962) 'Ionization Associated with Hypervelocity Impact', NASA Technical Note D-2091.

Hoffmann H.-J., Fechtig H., Grün E., Kissel J.(1975) 'First Results of the Micrometeoroid Experiment S 215 on the HEOS 2 Satellite', Planet. Space Sci., 23, 215-224.

Igenbergs E. et al.(1990) 'The Present Status of the Munich Dust Counter Experiment on Board of the HITEN Spacecraft', Proc. of the IAU Coll. No. 126 "Origin and Evolution of Interplanetary Dust", Kyoto.

Iglseder H. (1986) 'Ladungsemission beim Hochgeschwindigkeitseinschlag', Dissertation, Lehrstuhl für Raumfahrttechnik, Technische Universität München.

Uesugi K., Matsuo H., Kawaguchi J. Hayashi T. (1990) 'Japanese first Double Lunar Swingby Mission HITEN', Proc. of the 41st Congress of the IAF, Dresden.

COLLECTION OF STRATOSPHERIC MICROPARTICLES ABOVE THE SULFATE LAYER USING BALLOON-BORNE COLLECTORS

J. R. STEPHENS[1], Y. NAKADA[2], T. ONAKA[2], F. J. M. RIETMEIJER[3]
[1] *Los Alamos National Laboratory, Los Alamos, New Mexico 87545, U.S.A.*
[2] *Department of Astronomy, Faculty of Science, Univ. of Tokyo, Tokyo 113, Japan.*
[3] *Department of Geology, University of New Mexico, Albuquerque, NM 87131, USA.*

ABSTRACT. We report preliminary analytical electron microscope (AEM) analysis of nearly 300 stratospheric particles collected using balloon-borne collectors at 34-36 km altitude. The particles are predominantly silica, plagioclase feldspar, Mg, Fe-silicates and rare barite, metal oxides, and unidentified Fe, Ni, Zn, and Pb particles. The majority of these generally submicron-sized particles are comparable to volcanic particles collected at 20 km altitude from the 1982 eruption of the El Chichon volcano. Because of the uniqueness in altitude and collected particle sizes the collection may also contain interplanetary dust particles of types poorly represented in present collections.

1.   Introduction

Particles may be injected into the stratosphere from large volcanic eruptions [D'Altorio and Visconti, 1983], form by gas to particle transformations, or by introduction of extraterrestrial particles or vapors [Hughes, 1978]. Above the sulfate layer (>33 km) it is expected that extraterrestrial particles dominate the stratospheric particle population. Most of the small extraterrestrial particles that enter the earth's atmosphere vaporize at altitudes between 80 and 100 km and are believed to form refractory grain condensates. Hunten et al., 1980 modeled the formation of condensates from meteoritic vapors. They considered the chemical reactions regulating particle formation from vapors including nucleation, condensation, and coagulation.

Volcanic particles are periodically injected to high altitudes. For example, the El Chichon volcanic event injected particles to between 33 and 35 km during 1982 [Coulson et al., 1982]. Small, micron-sized particles have settling times on the order of years in the stratosphere [Mackinnon et al., 1984].

In May 1985 stratospheric particles were collected at 34-46 km using a high-altitude balloon, the initial results of which have been reported [Testa et al., 1990]. The balloon-borne collector captured solid particles (<0.045 - 10.0 µm diameter) using cascade impactors and filters. Particle concentration, size distribution, and bulk elemental composition were measured on a major portion of the collected particles using scanning electron microscope (SEM) and proton-induced X-ray emission (PIXE) instruments. The particles in the collection complement the existing particle collections now obtained up to 20 kilometers using high altitude aircraft such as the Lockheed U-2 [Brownlee, 1978].

A.C. Levasseur-Regourd and H. Hasegawa (eds.), Origin and Evolution of Interplanetary Dust, 49–52.
© 1991 *Kluwer Academic Publishers.*

We report here a detailed analysis of several hundred of the collected particles including particle morphology, major element data, and electron diffraction data. Comparison of the elemental and morphological data with characteristics of known extraterrestrial and volcanic particles is made in an attempt to deduce the origin of the collected particles.

2. Experimental

The particles were collected using a balloon-borne instrument that employed three parallel collectors each consisting of a combination of a single stage cascade impactor followed by a Nuclepore Membrane Filter (NMF) filter. One collector, used to collect particles for particle counting, used NMF's for both the impact surface and filter. A second collector, used to collect particles for analytical electron microscope (AEM) analysis, consisted of nine holey-carbon filmed beryllium AEM grids glued to a NMF impact surface in a cross pattern followed by a NMF used as a filter. The third "flight blank" collector had the same configuration as the first collector but was not attached to a pump and received no flow during the flight. Upon collection the samples were placed in sealed containers and stored in a laminar-flow clean bench.

Particles measured in this study were present on 3 TEM grids. No sample preparation was used prior to inserting the grids into microscope. The dust grains occur dispersed on holey-carbon thin films that are supported by Be TEM grids. Particles were examined using a JEOL 2000FX AEM which is operated at an accelerating voltage of 200 kV. The TEM grids are housed in a Gatan low-background, double-tilt specimen holder during analysis using secondary and transmitted electrons for TEM and SEM imaging and X-rays for *in situ* chemical micro-analysis using a probe size of ~20 nm in diameter. Semi-quantitative chemical data were obtained for Na and heavier elements using a Tracor-Northern TN5500 energy dispersive spectrometer (EDS). Particles with morphologies characteristic of poorly graphitized carbon, which are formed during production of the carbon film on the grids, were rejected from the particle counts based on their unique morphology and lack of X-ray spectra [Reitmeijer, 1985].

3. Results

Concentrations of particles between 0.045 and 1.0 μm were at least 50 times the handling blank levels and 30 to 10^4 times the concentrations predicted by the model of Hunten et al., 1980. The observed particle concentration is similar to concentrations measured by balloon-borne particle counters and SAGE II satellite extinction observations between November 29, 1984 and October 11, 1986 [Osborn et al., 1989].

The majority of the particles are angular shards of silica, plagioclase ($CaAl_2Si_2O_8$), Mg, Fe-silicates (including pyroxenes and olivines), rare grains of barite ($BaSO_4$), and oxides of Pb, Fe, Bi, and Al. The silica and silicate minerals constitute 83% of all particles collected. Individual particle sizes ranged from 0.5 to 5.0 μm. In addition to the particles listed above, unidentified particles containing Fe, Ni, Zn and Pb, and occasionally sulfur [Rietmeijer, 1990] were seen. Approximately 74% of the particles by number are <~1.0 μm in size with 17% of particles <~0.2 μm. Clusters of silica, silicate, and barite grains, in variable proportions are common. In addition nanometer-size NaCl and KCl crystallites and sulfuric acid droplets are seen adhering to the particle surfaces. An example of a cluster particle is shown in Figure 1.

4. Discussion

Our particles show platey shard morphology, a high concentration of clusters and also silica particles with adhered salts and sulfuric acid droplets that are most similar to stratospheric aircraft collections of volcanic particles from the El Chichon event [Mackinnon et al., 1984]. Volcanic particles show characteristic mineralogy and chemistry consisting of angular silica-rich grains, particle clusters, and particles containing large amounts of metals including Cu and Pb. The presence of a high concentration of micron-sized volcanic particles after several years requires an efficient mechanism for lofting submicron particles in the stratosphere. Extraterrestrial cluster particles collected by aircraft in the stratosphere predominantly consist of pyroxenes and olivines plus carbonaceous materials and typically have a chondritic major element composition. We have not yet found this type of particle in our collection.

Some of our particles could be a unique class of extraterrestrial particles smaller than 1 μm that impact the stratosphere. For example, iron oxide particles in our collection may be of meteoric origin. Such particles would not be collected in aircraft flights due to the aircraft collector size cutoff above this size. A larger flux of submicron micrometeoroids than previously thought is in accord with high particle counts found on an aircraft collection flag down to the lowest size collection limit (~1 μm) [Zolensky and Mackinnon, 1985]. Alternatively, some of our particles may have formed by breakup of larger porous micrometeoroids composed of low strength aggregates of submicron grains. More collections are necessary to resolve the genesis of the small particles in our collection.

REFERENCES

D. E. Brownlee, Microparticle studies by sampling techniques, in: Cosmic Dust, J. A. M. McDonnell, ed., pp. 295-366, John Wiley, New York, 1978.

K. E. Coulson, T. E. Defour, J. Deluis, LIDAR and optical polarization measurements of stratospheric cloud in Hawaii, EOS, Trans. Am. Geophys. Union, 63, 897, 1982.

A. D'Altorio and G. Visconti, LIDAR observations of dust layers transient in the stratosphere following the El Chichon volcanic eruption, Geophys. Res. Letters, 10, 27-30, 1983.

D. W. Hughes, Meteors, in: Cosmic Dust, J. A. M McDonnell, ed., pp. 123-186, John Wiley, New York, 1978.

D. M. Hunten, R. P. Turco and O. B. Toon, Smoke and dust particles of meteoric origin in the mesosphere and stratosphere, J. of Atmos. Sci., 32, 1342-1357, 1980.

I. D. R. Mackinnon, J. L. Gooding, D. S. McKay, and U. S. Clanton, The El Chichon stratospheric cloud: Solid particulates and settling rates, J. Volcanol. and Geotherm. Res., 23, 125-146, 1984.

M. T. Osborn, J. M. Rosen, M. P McCormick, P. Wang, J. Livingston, and T. J. Swissler, SAVE II aerosol correlative observations: profile measurements, J. Geophys. Res. 94, 8353-8366, 1989.

F. J. M. Rietmeijer, A poorly graphitized carbon contaminant in studies of extraterrestrial materials, Meteoritics 20, 43-48, 1985.

F. J. M. Rietmeijer, El Chichon dust a persistent problem, Nature, 344, 114-115, 1990.

J. P. Testa, J. R. Stephens, W. W. Berg, T. A. Cahill, T. Onaka, Y. Nakada, J. R. Arnold, N. Fong, and P. D. Sperry, Collection of Microparticles at High Balloon Altitude in the Stratosphere, Earth and Planet. Sci. Lett, 98, 287-302, 1990.

M. E. Zolensky and I. D. R. Mackinnon, Accurate stratospheric particles size distributions from a flat plate collection surface, J. Geophys. Res. 90, 5801-5808, 1985.

Figure 1:

A composite cluster of plagioclase, tridymite, and Mg, Fe-silicate (probably low-Ca pyroxene) shards. The background is a holey-carbon support film. The filaments are glue used to adhere the grids to the impact surface.

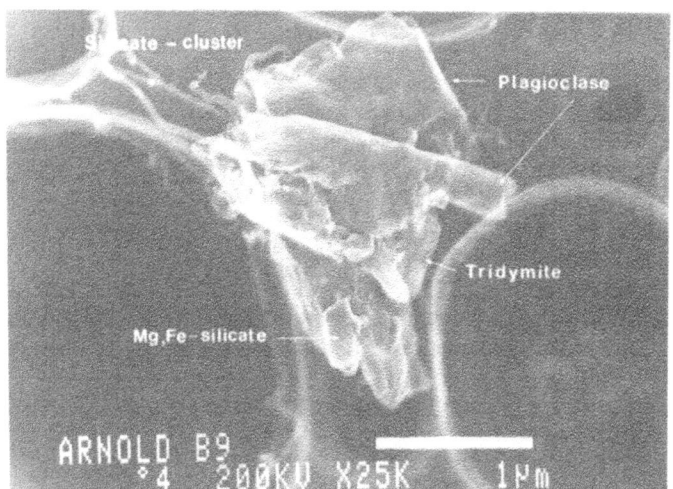

A COSMIC MATTER ACCRETION EVENT AROUND 660,000 YEARS BEFORE PRESENT FOUND IN TWO DATED, CENTRAL PACIFIC CORES

Kazuo Yamakoshi[1], Ken'ichi Nogami[2], Rie Omori[2],

Ma Jianguo[3] & Ma Shulan[3]

[1] Institute for Cosmic Ray Research Univ.Tokyo,
    Tanashi Tokyo 188
    Japan.

[2] Phys.Lab. Dokkyo Univ.School Medicine
    Mibu Shimotsugagun 321-2
    Japan.

[3] Appl.Nucl.Tech.Div. Inst.High Energy Phys.
    Chinese Academy of Science Beijing
    China.

ABSTRACT In order to study large-scaled cosmic matter accretion events in the past, Ir enriched layers at C-T and other geological boundaries and dated sedimental cores have been searched by many scientists.   In this work, Iridium contents and the ratios of (Co/Fe) in two dated, respective layers of the cores are determined. These samples were dated fortunately with the paleo-magnetic and also with the cosmogenic Be-10 methods. Ir enrichments are found at (0.660 ± 0.030) My before present.

1.Introduction

The origin and evolution of the cosmic meteoroids in the solar system are important problems. It is well known, siderophile elements including noble metals, such as Pt, Os, Ir and so on, are contained much in cosmic meteoroids, which are fallen onto the Earth recently. Of course, it is not so clear, chemical compositions of cosmic meteoroids were also the same as those of the recent ones in the past.

A.C. Levasseur-Regourd and H. Hasegawa (eds.), Origin and Evolution of Interplanetary Dust, 53–56.
© 1991 Kluwer Academic Publishers.

Ir anomalies at the C-T and other geological boundaries have been studied by many scientists for the studies of relationships between accidental accretion events of meteoroids and geo-and biological disturbances. In sedimental cores also enrichments of the siderophile elements have been investigated.

In this work, the assumption: enrichments of siderophile elements are a good indication for meteoroids, is applied, so that the siderophile elements are measured with INAA in the whole length of the cores covered for a few million years in the past.

2. Sample Description and Experimental

In this work, two core samples were examined, which were collected by a piston corer on board R/V Hakuho-Maru, Ocean Research Institute, Univ.Tokyo.

The collection sites of the used cores were located on the longitude of 160°W and and the Sample A [KH-68-4,St-15] was obtained at 12°00'N(depth; 5775 m and also the Sample B [KH-68-4,St-18] at 1°59'N (depth; 5360 m).

These two cores are dated fortunately by paleomagnetic (Kobayashi and Kitazawa 1971) and also Be-10 (Tanaka and Inoue 1979) method. Ir contents and the ratios of (Co/Fe) in the respective layers of the cores, which are shown in Fig.1 and Fig.2 (Yamakoshi 1988). Each sample was analyzed with INAA in non-destructive forms. The detection limits for Ir inthis work are $\sim$ 1 ppb for Sample A and $\sim$ 0.5 ppb for Sample B.

The neutron irradiations were carried out by a TRIGA II reactor, whose neutron flux was 0.7x10E11 (n/sec.cm²).The irradiation times were 10 hours for Sample A and 18 hours for Sample B. The processed Canyon Diablo and JB-1 were used as the reference samples of the chemical component determinations. A few layers of the sample A and B were lost, thus their compositions could not determined.

In this work, the detailed profiles around the Ir anomalies are obtained with additional samples, which are divided for shorter intervals. These are shown in Fig.3. The time of the meteoroid accretion was revised as (0.660 ± 0.030) My.

3.Discussion

[³He/⁴He] ratios in deep sea sediments are found to be abnormaly high. [³He/⁴He] ratios were measured in the same core, [KH-68-4, St-18], specially around the time of the Ir enrichment the ratio looks as increased.

The higher value of [^{3}He/^{4}He] is interpreted as sudden increase of accreted meteoroids and/or increase of ^{3}He particle implantation due to the solar activity (Takayanagi and Ojima 1987).

The event found in this work might be local and not so large-scaled ones. No mass extinction, biological disturbance was reported. However, systematic surveys of dated core samples from shallow and deep sea sediments will give us fruitful information on the evolution of the solar system.

Fig.1 Ir and (Co/Fe) ratios in the core sample A.

Fig.2 Ir and (Co/Fe) ratios in the core sample B.

SAMPLE A (KH-68-4, St-15)

Paleomagnetic age(1) mean (my)	Be-10 age(2) mean (my)	Depth in Core (cm)	(Co/Fe)ratio in Core
	0.11	0 ~ 45	
	0.19	45 ~ 95	no Sample
	0.27	95 ~ 145	
	0.42	145 ~ 195	
	0.56	195 ~ 245	
0.69	0.68	245 ~ 297	
	0.81	297 ~ 342	
0.89	0.94	342 ~ 387	
0.95	1.01	387 ~ 420	
	1.11	420 ~ 470	
	1.29	470 ~ 525	
	1.41	525 ~ 575	
	1.55	575 ~ 625	
1.65	1.73	625 ~ 671	
1.85	1.80	671 ~ 714	
	1.92	714 ~ 755	
	2.05	755 ~ 795	
2.43	2.17	795 ~ 835	
	2.26	835 ~ 880	
	2.37	880 ~ 925	
	2.48	925~974(I)	Ir in Core
	2.48	925~974(II)	(→ ⊢⊣) pg mg,10^3 y

Ir in Core (→ ⊢⊣) 0 2 4 6 8 10 12 14 16

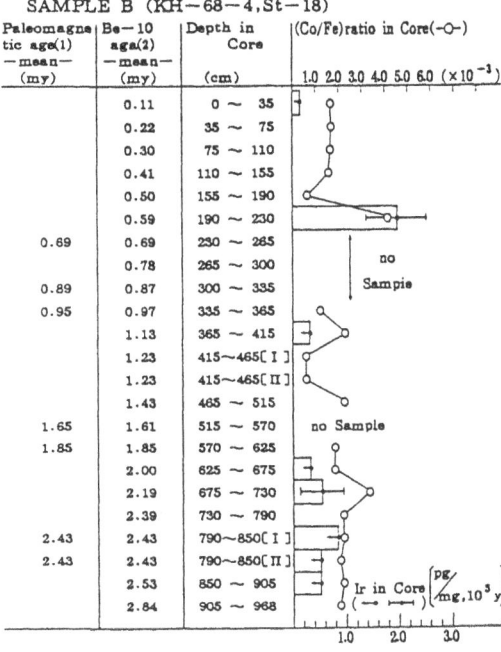

SAMPLE B (KH-68-4, St-18)

Paleomagnetic age(1) mean (my)	Be-10 age(2) mean (my)	Depth in Core (cm)	(Co/Fe)ratio in Core
	0.11	0 ~ 35	
	0.22	35 ~ 75	
	0.30	75 ~ 110	
	0.41	110 ~ 155	
	0.50	155 ~ 190	
	0.59	190 ~ 230	
0.69	0.69	230 ~ 285	
	0.78	265 ~ 300	no Sample
0.89	0.87	300 ~ 335	
0.95	0.97	335 ~ 365	
	1.13	365 ~ 415	
	1.23	415~465[I]	
	1.23	415~465[II]	
	1.43	465 ~ 515	
1.65	1.61	515 ~ 570	no Sample
1.85	1.85	570 ~ 625	
	2.00	625 ~ 675	
	2.19	675 ~ 730	
	2.39	730 ~ 790	
2.43	2.43	790~850[I]	
2.43	2.43	790~850[II]	
	2.53	850 ~ 905	Ir in Core (→ ⊢⊣) (pg/mg,10^3 y)
	2.84	905 ~ 968	

(Co/Fe)ratio in Core(-○-) 1.0 2.0 3.0 4.0 5.0 6.0 (×10^{-3})

Ir in Core (→ ⊢⊣) 1.0 2.0 3.0

Fig.3 Ir and (Co/Fe) ratios in the core B (in detail).

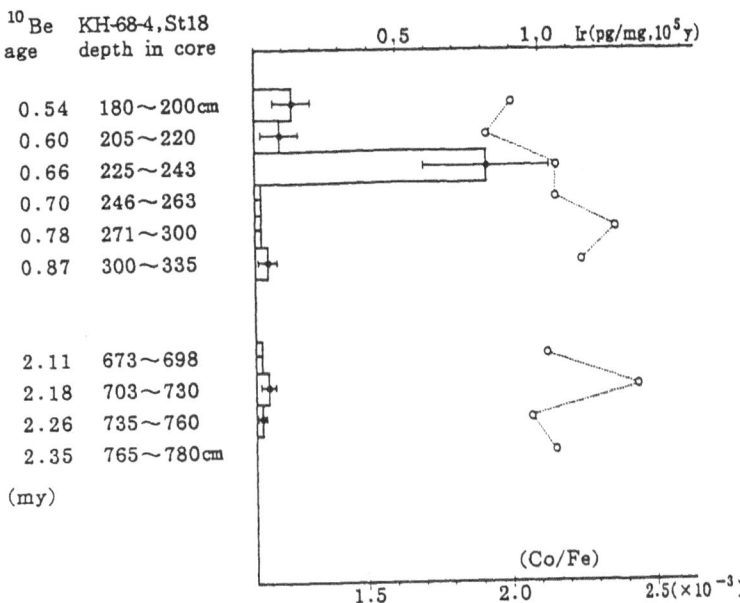

References

Kobayashi K and Kitazawa K 1971 Deep Sea Res. 18 1045

Takayanagi M and Ojima M 1987 J.Geophys.Res. 92 12531

Tanaka S and Inoue T 1979 Earth Planet.Sci.Lett.45 181

Yamakoshi K 1988 Abstracts LPSC 19th(Houston) 130

PRELIMINARY STUDY ON NEOGENE MICROTEKTITES IN THE CORE COLLECTED FROM NORTH PACIFIC*

PENG HANCHANG (1), LIU ZHENKUN(2),ZHUANG SHIJIE(3), MAO XUEYING(4)AND CHAI ZHIFANG(4)
(1) First Institute of Oceanography, State Oceanic Administration, Qingdao P.O.Box 98,China, (2) Institute of HUNAN Coloured Metallurgical, Changsha, China, (3) China National Nonferrous Metals Industry Corporation, Research Institute of Geology for Mineral Resources, Guilin, China, (4) Institute of High Energy Physics, Academia Sinica, Beijing, China

ABSTRACT: A great number of microtektites were found in core collected from North Pacific. Because abundant microtektites are restricted to a 20-30 cm thick zone of core, we called this zone is miicrotektite layer. The age of sediments concentrated microtektites is from the Pliocene to the Pleistocene epoch. Research results indicate that these microtektites are similar to the North American tektites and the Australasian tektites, but they have the unique property on the chemistry.

1. INTRODUCTION

During the North Pacific Manganese Noduls Investigation in 1983, we collected a large-diameter core (8°00.15'N,176°10.65'W; 3991 m depth of water). This gravity core is 420cm long in total. It is composed of calcureous ooze at 0-320 cm and siliceous ooze at 320-420cm. The magnetic iron cosmic spherules were contained in each sediment samples from top to bottom of the core (average content is 2.3 spherules per 100 gram sediment). A great number of microtektites were found from 270 cm to 300 cm (average content is 15 microtektites per 100 gram sediment). Because abundant microtektites are restricted to a 20-30-cm-thick zone of deep-sea core, we called this zone is microtektite layer.

This core have performed the palaeomagnetism analysis. The age of sedimentary stratum contained mictotektites is from pliocene epoch to pleistocene epoch by the palaeomagnetic and the palaeonotogic analysis (SOA,1986).

Using various analysis technicalities conducted the study on microscopic chracteristics, major-oxide compositions and microstructures of the Neogene microtektites, and pursued their origin.

2. MICROSCOPIC CHARACTERISTICS

Under the binocular steroscopic microscope with as much as 50 X - 70 X magnification, most of microtektites appeared as light yellowish-green small glassy spherules, and individual grain is dumb-bell shape (Fig. 2-(A)). Their average size is 137 μm in diameter, a largest dumb-bell grain's size is 652 X 223 X 151 μm. The average refractive indices of microtektites determined by the oil immersion method is 1.52.

* The Project Supported by the National Natural Science Foundation of China.

A.C. Levasseur-Regourd and H. Hasegawa (eds.), Origin and Evolution of Interplanetary Dust, 57–60.
© 1991 Kluwer Academic Publishers.

3. MAJOR-OXIDE COMPOSITON

The major-oxide compositions of five Neogene microtektites were determined by electron probing(Table 1).

Table 1. Major-oxide compositions of five Neogene microtektites in a core collected from North Pacific.

	MN-1	MN-2	MN-3	MN-4	MN-5	Site 149*	DSDP 612**
SiO_2	70.54	69.24	69.34	69.13	69.31	64.52	72.30
TiO_2	1.09	1.25	1.07	1.06	1.34	0.70	0.85
Al_2O_3	16.19	16.12	17.40	17.17	15.89	19.38	15.10
FeO	4.16	4.24	3.78	3.92	3.45	6.39	4.80
MgO	2.78	3.21	2.92	3.00	3.62	3.29	1.20
CaO	2.60	2.33	1.93	2.29	5.06	2.35	0.85
Na_2O	0.17	0.13	0.18	0.15	0.14	0.81	0.30
K_2O	2.78	1.95	2.49	2.52	1.24	2.56	3.80

* From B. P. Glass and M. J. Zwart, 1979.
** From C. Koeberl and B. P. Glass, 1988.

These microtektites have SiO_2 contents of about 70%, and the SiO_2 content of each sample are closer. The SiO_2, Al_2O_3, FeO, MgO, CaO and K_2O contents are all similar to the microtektites from Site 149 in the Caribbean sea belong to the North American tektite strewn field (Glass et al., 1979). The chemistry property of Neogene microtektites can not only be seen from the Table 1, but also clearly reflected by the spectral pattern of the energy-dispersive X-ray analysis (Fig.1).

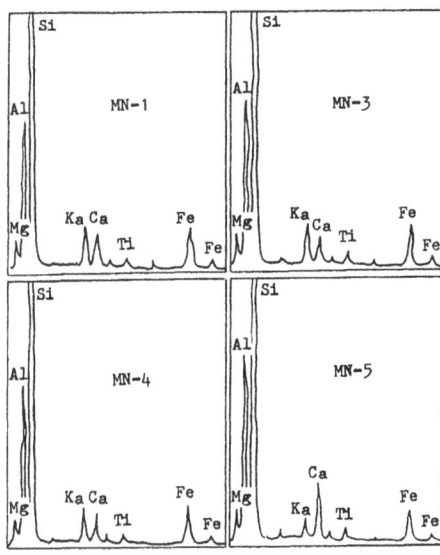

Fig. 1. The Spectrum Patterns of the Energy-dispersive X-ray analysis of the Neogene microtektites in core collected from North Pacific.

Although the contents of most oxide-compositions of Neogene microtektites are similar to the North American microtektites, but the Na_2O and TiO_2 contents are quite different from the late. Five Neogene microtektites have great low Na_2O contents. The peaks of sodium element of five samples almost can't seek from the energy-dispersive spectrum patterns in the Fig.1. But the TiO_2 contents are much higher than North American microtektites and other microtektites(O'Keefe, 1976; Hanchang, 1984).

4. MICROSTRUCTURES

Several ten Neogene microtektites were selected for further studies using scanning electron microscope (SEM). Five photographys were showed in Fig.2 (from B to F). These photographys showed various sculpturing characteristics. The pits, grooves, mounds, etc. were appeared on the surface of each microtektite, which are similar to the Australasian tektites (Glass, 1974). Some of them are similar to the crater on the Earth(B), which suggest that the microtektites were impacted by one or many small particles with superhigh-velocity impact and strong kinetic energy in outer space, and proved an important evidence for extraterrestrial origin. Some Neogene microtektites have circular, flat-topped, elevated area at top (C and D), which suggest that a sharp mound was pared by other particles when flying. A deep groove was appeared on the microtektites (E and F), perhaps which was as the result of attact by particles.

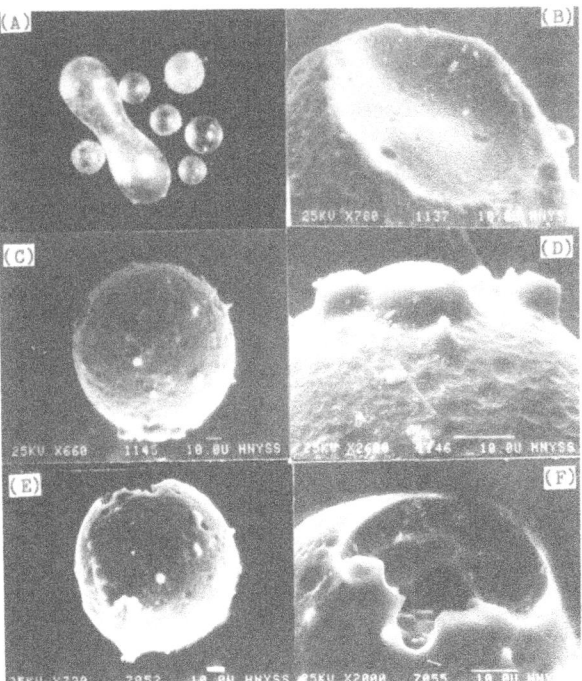

Fig. 2. Microscope and SEM photographs of Neogene microtektites. (A) Microscope photograph, size of a largest dumb-bell grain is 652 X 223 X 151 μm; (B) Like-meteorite crater rebound elasticity microimpact crater on microtektite; (C and D) A Neogene microtektite with circular, flat-topped, elevated area at top; (E and F) A Neogene microtektite with a great number of grooves and a largest groove on microtektite.

5. DISCUSSION

Based on present these data we obtained following two preliminary views.

(1) These Neogene microtektites found in core collected from North Pacific are similar to the North American microtektites on major-oxide composition and similar to the Australasian tektites on microstructures, but they have great low Na_2O content and much high TiO_2 content, which indicate that perhaps these microtektites not belong to any known tektite strewn field.

(2) Based on some special microstructures on the surface of Neogene microtektite and other analysis data (for example, presence of traces elements Os, Ir, Ni, Co, Cr, etc. determined by INAA) we suggest that these microtektites come from outer space.

REFERENCES:

State Oceanic Administration, 1986, *Comprehensive Investigation Report on Manganese Nodules in the Northern Central Pacific*, CHINA OCEAN PRESS, 65-69.

B.P. Glass and M.J. Zwart, 1979, North American microtektites in Deep Sea Drilling Project cores from the Caribbean Sea and Gulf of Mexico, *Geological Society of America Bulletin*, Part I, V.90, 595-602.

C. Koeberl and B.P. Glass, 1988, Chemical composition of North American microtektites and tektite fragments from Barbados and DSDP Site 612 on the continental slope off New Jersey, *Earth and Planetary Science Letters*, V.87, 286-292.

J.A. O'Keefe, 1976, *Tektites and their Origin*, Elsevier Scientific Publishing Company, 115-145.

P. Hanchang, 1984, A Study of Microtektites in the Pacific Ocean, Acta *Oceanologica Sinica*, V.3 No.1, 106-113.

B.P. Glass, 1974, Microtektite Surface Sculpturing, *Geological Society of America Bulletin*, V.85, 1305-1314.

II

INTERPLANETARY DUST :
PHYSICAL AND CHEMICAL ANALYSIS

PHYSICAL AND MINERALOGICAL PROPERTIES OF ANHYDROUS INTERPLANETARY DUST PARTICLES IN THE ANALYTICAL ELECTRON MICROSCOPE

J. P. Bradley, McCrone Associates, Westmont,
Illinois 60559, USA

Abstract
    The fine grained mineralogy and petrography of anhydrous
"pyroxene" and "olivine" classes of chondritic interplanetary dust
have been investigated by numerous electron microscopic studies.
The "pyroxene" interplanetary dust particles (IDPs) are porous,
unequilibrated assemblages of mineral grains, metal, glass, and
carbonaceous material. They contain enstatite whiskers, FeNi
carbides, and high-Mn olivines and pyroxenes, all of which are
likely to be well preserved products of nebular gas reactions.
Solar flare tracks are prominent in most "pyroxene" IDPs,
indicating that they were not strongly heated during atmospheric
entry. The "olivine" IDPs are coarse grained, equilibrated mineral
assemblages that have probably experienced strong heating. Since
most "olivine" IDPs do not contain tracks, it is possible that this
heating occurred during atmospheric entry.

Introduction
    Chondritic interplanetary dust particles collected from the
stratosphere are believed to be derived from comets and asteroids
[1,2]. They arrive at Earth-crossing orbits under the influence of
Poynting-Robertson drag, and many of them survive atmospheric entry
without significant modification [2,3]. They include compact
strong objects (similar to CI and CM chondrites), as well as
fragile, porous objects that could not survive atmospheric entry as
larger (millimeter-sized) bodies [1,4]. Identification of the
asteroidal and cometary subsets of interplanetary dust are primary
goals of IDP research, because asteroidal IDPs probably sample a
broader range of parent bodies than conventional meteorites [5],
and cometary IDPs are presently the only available samples of
comets. During the past decade new information about the chemical,
isotopic, and mineralogical nature of IDPs has been obtained [6-8].
At the same time, data about interplanetary dust has been obtained
from ground based and airborne telescopic observations and

A.C. Levasseur-Regourd and H. Hasegawa (eds.), Origin and Evolution of Interplanetary Dust, 63–70.
© 1991 Kluwer Academic Publishers.

spacecraft measurements [9,10]. All of these data have focused attention on the relationship between IDPs collected from the stratosphere and possible parent bodies within the solar system.

This paper describes recent results obtained from studies of IDPs using the analytical electron microscope (AEM). The AEM has proven to be useful for analysis of IDPs because it is designed specifically for chemical and structural microanalysis of materials at the highest possible spatial resolution. Optimum performance of the AEM requires thin specimens, and most of the results described in this paper were obtained from ultramicrotomed thin (< 100 nm) sections [11]. The following text will first introduce the three classes of chondritic IDPs, and then describe the two **anhydrous** "pyroxene" and "olivine" classes. (The third **hydrated** "layer silicate" class is described by Tomeoka (this volume)). Finally, the properties of the anhydrous classes will be evaluated in terms of likely sources.

Chondritic Interplanetary Dust

The most common IDPs collected from the stratosphere belong to a group whose compositions generally agree with those of CI and CM carbonaceous chondrites [1]. The extraterrestrial origins of these chondritic IDPs have been confirmed by measurement of solar noble gases, D/H isotopic ratios, and solar flare tracks [2]. Sandford and Walker [12] first showed using infrared (IR) transmission absorption spectra that chondritic IDPs fall into three major groups referred to as "pyroxene", "olivine", and "layer silicate", after the minerals that provide the best match for the observed IR spectral features. The first two groups contain only **anhydrous** phases while the third contains (**hydrated**) layer silicates in addition to anhydrous phases. Electron microscopic studies have confirmed that pyroxene, olivine, and layer silicates are indeed major constituents of IDPs in their respective IR classes [13,14]. Because the anhydrous "pyroxene" and "olivine" classes contain only anhydrous phases, they have no known (mineralogical) counterparts among carbonaceous chondrites. The (hydrated) "layer silicate" IDPs, on the other hand, are mineralogically similar to CI and CM chondrites (see Tomeoka, this volume).

Anhydrous IDPs

[i] "Pyroxene" class

The "pyroxene" class of IDPs are unique in that they are highly porous aggregates (Fig. 1). Sometimes large ($\approx$1 μm) euhedral mineral grains can be observed embedded within the IDPs. The most distinctive are enstatite whiskers (Fig. 1), which have been the objects of a detailed mineralogical study [15]. Solar flare tracks (Fig. 2) and solar wind radiation damaged rims are usually prominent in "pyroxene" IDPs, indicating that most of them survive atmospheric entry without severe (>600°C) heating. Typical track densities are on the order of 10^{10}-10^{11} cm^{-2} (Fig. 2), which corresponds to an exposure age of $\approx 10^4$ years within the inner solar system [16].

Figure 1   Scanning electron micrograph of the "pyroxene" IDP U220A19. An enstatite whisker is arrowed.

In thin section, each particle is composed of loosely agglomerated, predominantly submicrometer components (Fig. 3). These components can be single mineral grains, glass, carbonaceous material, and microcrystalline aggregates (Fig.4). The most common single mineral grains are enstatite and iron-rich sulfides (pyrrhotite). Less common are olivine (forsterite), troilite, and FeNi alloy or carbides. Glass is ubiquitous throughout pyroxene IDPs. It forms the groundmass of some microcrystalline aggregates, but it also occurs as discrete inclusions. The composition of the glass is variable and, in one IDP alone, three compositionally distinct glasses were identified. Typical glass compositions are (Na), Ca, Fe alumino-silicate, pure alumino-silicate glass, and pure Mg silicate. The carbonaceous material is a non-crystalline phase which can also occur as a groundmass containing embedded mineral grains or as discrete inclusions. Although it is possible that organic material is present, little is known about the molecular constitution of the carbonaceous material. All IDPs are collected in silicone oil, and washed with organic solvents to remove the oil. Bulk carbon contents as high as 16 wt % have been reported [17], but contamination and analytical difficulties could complicate interpretation of these abundances.

Microcrystalline aggregates are discrete 0.1-0.5 μm objects composed of nanometer-sized grains embedded in a glassy groundmass (Fig. 4). (They have also been referred to as "tar balls" [11] and

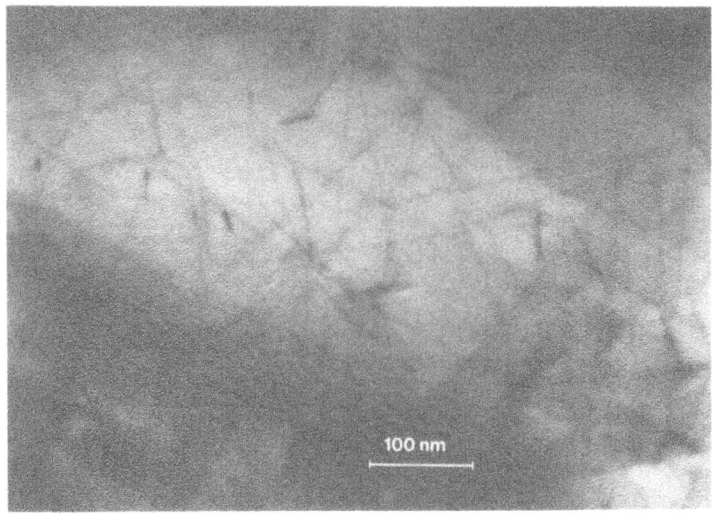

Figure 2 Darkfield electron micrograph of solar flare tracks in an enstatite crystal in IDP W7027C4.

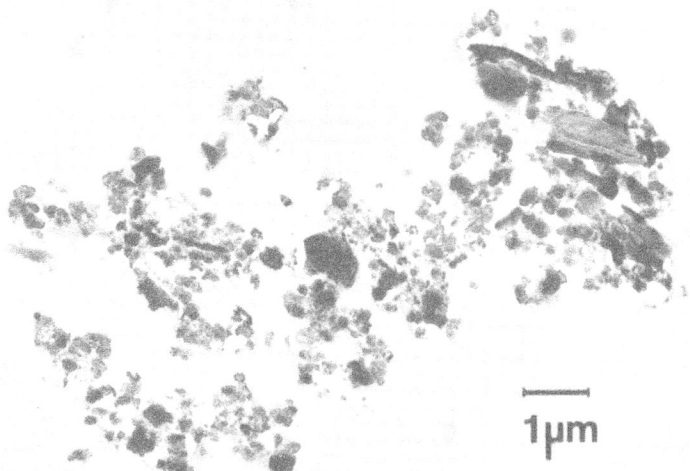

Figure 3 Brightfield electron micrograph of an ultramicrotomed thin section (≈80 nm thick) of IDP U222B42.

"granular units" [18]). They are perhaps the most fascinating components of chondritic IDPs, because they are discrete chondritic objects that have been incorporated into IDPs. Clearly, their formation predates that of the IDPs. The major crystalline component is body centered cubic kamacite (FeNi alloy), and most of the metal in "pyroxene" IDPs is contained as kamacite in

microcrystalline aggregates. These (alloy) grains range from less
than 5 to over 20 nm in diameter. Mg-rich pyroxene, Fe-sulfide,
and possibly olivine are also present as nanometer-sized
inclusions. Magnetite is sometimes present on the outer surfaces
of the aggregates.

Crystallographic studies have indicated that the enstatite
whiskers (Fig. 1) were formed by gas phase condensation [15], and
FeNi carbides in "pyroxene" IDPs are biproducts of Fischer-Tropsch
type catalytic reactions [19]. Klock et al [6,20] have studied the
crystal chemistry of olivines and pyroxenes in "pyroxene" IDPs.
All of the IDPs contained LIME (low iron manganese enriched)
olivines and/or pyroxenes, whose high Mn abundances have been
interpreted in terms of direct vapor phase condensation of olivine
and pyroxene [6]. Most "pyroxene" IDPs exhibit a wide range of
Fe/Mg ratios in olivine and pyroxene, indicating that these IDPs
are truly unequilibrated [20].

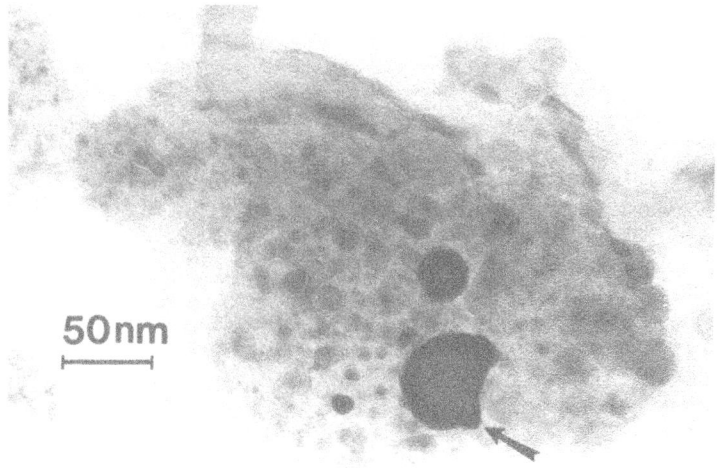

50nm

Figure 4   Brightfield electron micrograph of a microcrystalline
aggregate in a thin section of U222B42. An unusually large
kamacite (FeNi alloy) grain is arrowed.

[ii] "Olivine" class

Although the bulk compositions of "olivine" IDPs are similar to
those of "pyroxene" IDPs, SEM images suggest that "olivine" IDPs
are less porous and more coarse grained than the "pyroxene" class.
In thin section most grains are between 0.1 and 1.0 μm in diameter
(Fig. 5), although in some particles finer grained material is also
present. The most common minerals are olivine and iron-rich
sulfides. Less common minerals include pyroxenes, chromite,
magnetite, Fe-Ni alloy (kamacite), carbonaceous material, and
silica-rich glass. Glass is very abundant in some "olivine" IDPs

where it forms a matrix between embedded mineral grains.

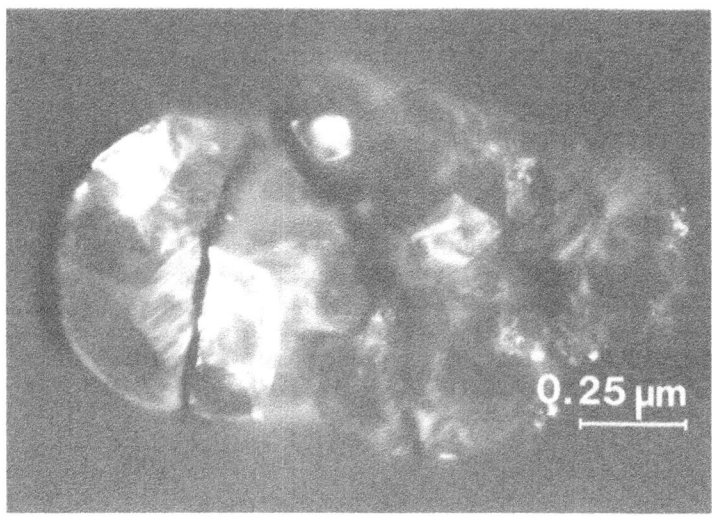

Figure 5   Darkfield electron micrograph of a thin section of part of the "olivine" IDP U220A14.

The mineral chemistry of "olivine" particles suggests that most of them contain equilibrated mineral assemblages. Several authors have noted that olivines and pyroxenes show a relatively narrow range of Mg/Fe ratios [13,20,21]. Moreover, "olivine" IDPs apparently do not contain LIME silicates or pyroxene whiskers and platelets, whose occurrences would otherwise indicate the preservation of pristine components [6,15].   Solar flare tracks have been found in only two "olivine" IDPs [13], but it is not yet clear whether the two IDPs that do contain tracks are the same types of materials as the IDPs without tracks.   The fact that only two "olivine" IDPs contain tracks while most do not, together with the mineralogical diversity exhibited by this group [13,21], suggest that the "olivine" class of IDPs probably incorporates more than one genetic group.   In any case, it seems that most "olivine" IDPs have experienced strong heating, which has produced both equilibrated silicate mineralogy and coarse grain size.

Sources of anhydrous IDPs
   Since there are now morphological [4], isotopic [7], chemical [19], and crystallographic [15] data suggesting that "pyroxene" IDPs are among the most primitive meteoritic materials yet encountered, comets are prime candidates for their source(s). Mass spectrometry data from the PIA and PUMA instruments on the comet Halley probes yielded compositional data that are compatible with "pyroxene" IDPs but incompatible with "olivine" (and "layer silicate)" IDPs [11].   However, infrared spectra recorded from

Halley and other comets seem to require the presence of significant pyroxene and olivine to account for the observed spectral features [9]. Since solar flare tracks are conspicuously absent in almost all of the "olivine" IDPs so far examined, their status as a legitimate class of unmodified IDPs is uncertain. However, if the tracks are erased by heating during atmospheric entry, then comets are also the more likely source(s) for "olivine" IDPs [3]. The paper by Sandford (this volume) deals exclusively with the sources of chondritic IDPs.

Acknowledgment

This research was supported by NASA grant NAS-9-17749. Helpful comments were provided by S. Sandford and K. Tomeoka.

References

[1]   Schramm, L. S., Brownlee, D. E. and Wheelock, M. M. (1989) Major element composition of statospheric micrometeorites. Meteoritics, 24, 99-112.

[2] Sandford, S. A. (1987) The collection and analysis of extraterrestrial particles. Fund. Cosmic Phys., 12, 1-73.

[3] Sandford, S. A. and Bradley, J. P. (1989) Interplanetary dust particles collected in the stratosphere: observations of atmospheric heating and constraints on their interrelationships and sources, Icarus, 82, 146-166.

[4] Bradley, J. P. and Brownlee, D. E. (1986) Cometary particles: thin-sectioning and electron beam analysis, Science, 231, 1542-1544.

[5] Germani, M. S.., Bradley, J. P. and Brownlee, D. E. (1990) Automated thin-film analyses of hydrated interplanetary dust particles in the analytical electron microscope, Earth Planet. Sci. Lett., in press.

[6] Klock, W., Thomas, K. L., McKay, D. S. and Palme, H.(1989) Unusual olivine and pyroxene composition in interplanetary dust and unequlibrated ordinary chondrites, Nature, 339,126-128.

[7] McKeegan, K. D., Walker, R. M. and Zinner, E. (1985) Ion microprobe isotopic measurements of individual interplanetary dust particles, Geochim. Cosmochim. Acta, 49, 1971-1987.

[8] Mackinnon, I. D. R. and Rietmeijer, F. J. M (1987) Mineralogy of chondritic interplanetary dust particles, Rev. Geophys., 25(7), 1527-1553.

[9] Bregman, J. D., Campins, H., Witteborn, S. C., Wooden, D. H., Frank, D. M., Allamandolla, L. J., Cohen, M. and Tielens, A. G. G. M. (1987) Airborne and ground based spectrophotometry of comet

P/Halley from 5-13 micrometers, Astron. Astrophys., 187, 616-620.

[10] Jessberger, E. K., Christoforidis, A. and Kissel, J. (1988) Aspects of the major element composition of Halley's dust, Nature, 332, 691-695.

[11] J. P. Bradley (1988) Analysis of chondritic interplanetary dust thin-sections, Geochim. Cosmochim. Acta, 52, 889-900.

[12] Sandford, S. A. and Walker, R. M. (1985) Laboratory infrared transmission spectra of individual interplanetary dust particles from 2.5 to 25 microns. Astrophys. J., 291, 838-951.

[13] Christoffersen, R. and Buseck, P. R. (1986) Mineralogy of interplanetary dust particles from the "olivine" infrared class, Earth Planet. Sci. Lett., 78, 53-66.

[14] Tomeoka, K. and Buseck, P. R. (1985) A carbonate-rich, hydrated, interplanetary dust particle: possible residue from protostellar clouds. Science 231, 1544-1546.

[15] Bradley, J. P., Brownlee, D. E. and Veblen, D. R. (1983) Pyroxene whiskers and platelets in interplanetary dust: evidence of vapor phase growth, Nature 301, 473-477.

[16] Bradley, J. P., Brownlee, D. E. and Fraundorf, P. (1984) Discovery of nuclear tracks in interplanetary dust, Science 226, 1432-1434.

[17] Blanford, G. E., Thomas, K. L. and McKay, D. S. (1988) Microbeam analysis of four chondritic interplanetary dust particles for major elements, carbon, and oxygen, Meteoritics, 23, 113-122.

[18] Rietmeijer, F. J. M. (1989) Ultrfine-grained mineralogy and matrix chemistry of olivine-rich chondritic interplanetary dust particles, Proc. 19th Lunar Planet. Sci. Conf., 513-521.

[19] Christoffersen, R. and Buseck, P. R. (1983) Epsilon carbide: a low temperature component of interplanetary dust particles, Science, 222, 1327-1329.

[20] Klock, W., Thomas, K. L., McKay, D. S. and Zolensky, M. E. (1989) Olivine compositions in anhydrous and hydrated IDPs compared to olivines in matrices of primitive meteorites (abstract), Lunar Planet. Sci. XXI, 637-638.

[21] Bradley, J. P., Germani. M. S. and Brownlee, D. E. (1989) Automated thin-film analyses of anhydrous interplanetary dust particles in the analytical electron microscope, Earth Planet. Sci. Lett., 93, 1-13.

AQUEOUS ALTERATION IN HYDRATED INTERPLANETARY DUST PARTICLES

Kazushige TOMEOKA
Mineralogical Institute
Faculty of Science
University of Tokyo
Hongo, Tokyo 113, Japan

ABSTRACT. Interplanetary dust particles (IDPs) characterized by chondritic composition can be divided into two principal groups, anhydrous and hydrated. This paper summarizes recent results of mineralogical and petrological studies dealing with the IDPs of hydrated type. Studies on mineralogical characteristics, infrared absorption spectra, and isotopic properties of the hydrated particles have suggested that they are primitive and may contain surviving interstellar material. The hydrated IDPs consist in major part of layer silicates and resemble CI and CM carbonaceous chondrites. Mineralogical and chemical data of both IDPs and carbonaceous chondrites have accumulated, and it is now possible to compare the mineralogies of the IDPs and the meteorites in considerable detail. Evidence was found that a significant proportion of the hydrated IDPs have been processed by aqueous alteration, and the nature of the alteration resembles that of similarly affected meteorites. The mineralogical and chemical data provide important clues to the possible origins of IDPs.

1. INTRODUCTION

Interplanetary dust particles (IDPs) collected in the stratosphere have been a subject of intense laboratory studies for the last ten years. A significant proportion of collected IDPs have a composition close to solar (or chondritic) elemental abundance. Two major groups have been recognized in the chondritic IDPs by using infrared spectroscopy [1], electron microscopy (e.g., ref. 2), and micro-beam analysis [3]. One group consists exclusively of anhydrous minerals, while the other group contains major amounts of hydrated phyllosilicates (layer silicates).

Of particular interest are the mineralogical relationships between the hydrated IDPs and carbonaceous chondrites of CI and CM types. The CI and CM chondrites are the most chemically primitive of known meteorites. These meteorites contain major amounts of layer silicates, and it is now widely accepted that they were produced by aqueous alteration that probably occurred on the meteorite parent bodies [4-6]. There is growing evidence that the hydrated IDPs also have been affected by aqueous alteration. Therefore, it is important to understand the mineralogy of hydrated IDPs and the nature of their aqueous alteration, and thus to

71

A.C. Levasseur-Regourd and H. Hasegawa (eds.), Origin and Evolution of Interplanetary Dust, 71–78.
© 1991 Kluwer Academic Publishers.

clarify the details of their relationship to the carbonaceous chondrites.

2. INFRARED SPECTRAL AND ISOTOPIC PROPERTIES

Infrared spectroscopy has shown that most chondritic IDPs can be divided, based on major silicate types, into three groups, olivine-rich, pyroxene-rich, and layer-silicate-rich [1]. Among them, the layer-silicate-rich (hydrated) particles exhibit infrared spectra closely similar to those of CM chondrites. Spectra of some hydrated particles are also similar to a telescopic spectrum of a protostar W33A [1]. Sandford and Walker [1] tried to fit spectra of IDPs to that of comet Kohoutek, and found that the superposition of roughly equal amounts of spectra of layer-silicate-rich and pyroxene-rich IDPs give rise to an excellent match, suggesting that some comets may contain layer silicates and pyroxene.

McKeegan et al. [7] showed, using ion microprobe, that five out of twelve hydrated particles that they studied have large deuterium enrichments relative to terrestrial D/H values. One particle exhibited particularly large deuterium enrichments, up to +2191‰ relative to standard mean ocean water. Even larger deuterium excesses have been observed in some interstellar molecular clouds, and the deuterium enrichments in IDPs appear to be best explained by the surviving molecules in interstellar molecular clouds [7,8]. These results of both infrared spectroscopy and ion microprobe analysis suggest that hydrated IDPs are primitive and some may contain material that predates the formation of the solar system.

3. MINERALOGY AND COMPOSITION OF HYDRATED IDPs

Hydrated particles have relatively smooth surfaces and compact appearance, while anhydrous particles have a more porous texture. Therefore, hydrated particles are often called "chondritic smooth (or CS)", and anhydrous particles are called "chondritic porous (CP)" [9]. (As an exceptional case, there is a report that minor amounts of layer silicates were found from a CP IDP [10].) Schramm et al. [3] reported that CS and CP particles comprise 37% and 45% of 200 IDPs they studied, respectively. Although both types of particles exhibit almost identical bulk compositions, they are distinct from one another in internal mineralogy and texture. To examine specific, detailed physical and mineralogical properties of IDPs, transmission electron microscopy (TEM) is the most powerful method. So far twenty three particles that belong to the hydrated class have been studied by TEM in various details [2,10-16], and it has been found that the hydrated particles can be further divided into two subgroups, smectite-rich and serpentine-rich. Most particles contain one or the other type of these layer silicates. Constituent minerals of hydrated IDPs are listed in Table 1.

3.1 Smectite-rich IDPs

This group constitutes ~ 70 % (sixteen in total) of the hydrated IDPs studied by TEM [2,10-15]. Smectite occurs in extremely small crystals (<500 Å in diameter) that are commonly poorly crystallized. It shows a wide range of compositions, but predominant is the Mg-Fe-rich smectite, whose composition is represented by $[Mg_{1.9}Fe_{1.0}Al_{0.1}](Si_{3.8}Al_{0.2})O_{10}(OH)_2$. Another prominent phase is glass, which occurs in Fe-rich and Si-Mg-Al-

TABLE 1. Constituent minerals of hydrated IDPs

Layer silicates (hydrated)	Smectite*	$(MgFe)_3Si_4O_{10}(OH)_2$
	Serpentine*	$(MgFe)_6Si_4O_{10}(OH)_8$
	Cronstedtite*	$(FeMg)_6(Fe^{3+}Si)_4O_{10}(OH)_8$
Anhydrous silicates	Pyroxene	$(MgFe)SiO_3$
	Olivine	$(MgFe)_2SiO_4$
	Fassaite	$Ca(Al,Mg)(SiAl)_2O_6$
	Diopside	$CaMgSi_2O_6$
Glass	(Si-rich)	SiO_2
	(Fe-rich)	$Fe_1\text{-}Si_m\text{-}Mg_n\text{-}O_p$
Sulfides	Pyrrhotite	$Fe_{1-x}S$
	Pentlandite	$(FeNi)_9S_8$
Carbonates	Breunnerite*/ Siderite*	$(Mg,Fe,Ca)CO_3$
	Calcite*	$CaCO_3$
Others	Magnetite	Fe_3O_4
	Chromite	$FeCr_2O_4$
	Metal	(FeNi)
	Tochilinite*	$Fe_{1.3}Ni_{0.1}SO_{1.4}(OH)?$
	Carbon (Amorphous/ Graphitized)	C
	Organic polymers	$C_1\text{-}H_m\text{-}O_n$

* Minerals exclusively found in the hydrated class of IDPs

TABLE 2. Comparison of mineralogies

	IDPs		Carbonaceous chondrite matrices		
	Smectite -rich	Serpentine -rich	CI	CM	CO/CV
Smectite	●	•	●		
Serpentine	•	●	●	●	
Cronstedtite		•		●	
Pyroxene	•	•			•
Olivine	•	•			●
Glass	•	•			
Pyrrhotite/Troilite	•	•	•	•	•
Pentlandite	•	•	•	•	•
Breunnerite/Siderite	•		•		
Dolomite			•		
Calcite		•		•	
Magnetite	•	•	•	•	•
Chromite	•	•	•	•	•
Metal	•	•		•	•
Tochilinite		•		●	
Ferrihydrite			●		
Sulfates			•		
Carbon/Organic polymers	•	•	•	•	•

The black filled circles indicate the presence of minerals. The large circles indicate predominant minerals.

rich compositions [2,14]. The Mg-rich and Fe-rich smectites and the glass are commonly intimately intergrown on a scale <1000 Å.

Ca-Mn-bearing Mg-Fe-rich carbonates, primarily breunnerite and siderite, are abundant constituents and characteristic of this class of IDPs, [13,14]. They occur as relatively large (500 to 3000 Å in diameter) rhombohedral grains and also as smaller (100 to 1000 Å) rounded grains. Carbonaceous material including amorphous and graphitized carbon is present [14] and is probably a major carrier of deuterium [7]. There are unusual aggregates of small rounded grains (<100 to 500 Å) of a variety of minerals embedded in a carbonaceous matrix [2].

Most smectite-rich IDPs contain minor amounts of anhydrous silicates such as Mg-rich pyroxene and olivine. Some of the pyroxene crystals have platelet or whisker morphologies like those found in anhydrous particles [2,12]. Minor amounts of Ca-Al-rich clinopyroxenes (fassaite and diopside) are also present [12,14]. Other minerals contained in this class of particles include Fe-Ni sulfides (pyrrhotite and pentlandite), magnetite, chromite, and Fe-Ni metal.

3.2 Serpentine-rich IDPs

Only seven particles of this group ($\sim$30 % of the hydrated population) have been reported to date [2,14-16], and detailed data on this class are limited relative to the smectite-rich class. (An eighth particle was studied by an X-ray powder diffraction method [17], but TEM data are not available.) Serpentine in the IDPs shows a wide range of Mg/Fe ratios. Cronstedtite (very Fe-rich serpentine) was identified from one particle [16]. The serpentines in this class of particles commonly show larger crystal sizes and higher crystallinity than the smectites in the smectite-rich class. Fe-rich and Si-rich glasses occur in some of the particles, but olivine and pyroxene appear to be less common than in the smectite-rich particles. Other minerals include pyrrhotite, pentlandite, magnetite, kamacite (Fe-Ni alloy), and fassaite. Bradley [16] reported tochilinite from one particle, which occurs as intimate intergrowths with cronstedtite. In contrast to the smectite-rich particles, carbonates appear to be rare; only a grain of Ca carbonate, possibly calcite, was identified [14]. Serpentine-rich IDPs, in general, appear to be more homogeneous in internal elemental distributions than smectite-rich IDPs [14].

3.3 Comparison to meteorites

Table 2 compares major minerals in IDPs, both smectite-rich and serpentine-rich, and carbonaceous chondrite matrices. For the comparison, it should be noted that there is a large difference in scale between IDPs and meteorite matrices; most IDPs are <20 μm in size, while the matrices of meteorites are larger than 1 mm and commonly exhibit mineralogical variations on a scale of <1 to 10 μm. Therefore, it is not always possible to compare directly an individual IDP to a meteorite matrix. We need a relatively large group of IDPs to make an appropriate comparison.

Layer silicates in CI chondrites are mostly Mg-rich smectite (saponite) and serpentine [18], and those in CM chondrites are serpentines with various Mg-to-Fe ratios as well as cronstedtite [5,19]. Layer silicates in CI chondrites are commonly very poorly crystallized, while those in CM chondrites are much larger in size and their crystal structures are more ordered. The comparison of the layer silicates and other major minerals appears to indicate that smectite-rich IDPs have a general similarity to CIs and serpentine-rich IDPs to CMs.

4. AQUEOUS ALTERATION: COMPARISON TO METEORITES

CI and CM chondrites show abundant evidence of secondary alteration that probably occurred by the activity of aqueous solutions on their parent body regolith, and that the layer silicates in these meteorites were produced by alteration of previously condensed olivine and pyroxene. It is beyond our scope to discuss here details of alteration experienced by the carbonaceous chondrites. The readers are recommended to refer to some of the following articles for that information: refs. 4-6,18.

Several lines of evidence for aqueous alteration were reported from hydrated IDPs. Tomeoka and Buseck [12] found evidence that smectite was produced by alteration of pyroxene (see Fig.2 in ref.12). Bradley [2] found that glass and layer silicates are intimately intergrown and interpreted that the layer silicates were formed by alteration of the glass. It is known from the studies of meteorites that pyroxene and glass are particularly susceptible to aqueous alteration and are preferentially altered to layer silicates.

A recent discovery of tochilinite and cronstedtite in a serpentine-rich IDP [16] provided strong mineralogical evidence for aqueous alteration. Tochilinite is one of the most mineralogically distinctive minerals that only occurs in CM chondrites [5,20]. Tomeoka and Buseck [5] showed that tochilinite was produced by alteration of metal and sulfides and formed complex intergrowths with cronstedtite during successive alteration.

The effect of aqueous alteration is also reflected in elemental abundances in the IDPs. Schramm et al. [3] performed bulk analysis of 200 IDPs, by using an energy dispersive X-ray (EDX) method, and found that hydrated IDPs show systematic depletion of Ca relative to solar abundance, while anhydrous IDPs do not show such depletion. Ca is water-soluble and tends to be leached at early stages of aqueous alteration in carbonaceous chondrite matrices [21], and deposited mostly as carbonates and sulfates.

One of the most important mineralogical differences between hydrated IDPs and meteorite matrix is that the former contain considerable amounts of submicron-to-micron size grains of glass and anhydrous silicates such as pyroxene and olivine, while the latter lacks those phases. The presence of glass and anhydrous silicates may indicate that alteration in the IDPs occurred over a very limited range. The microbeam analysis of hydrated particles at a size scale of <1000 Å [14] indicates that the distributions of Si, Mg, and Fe are much broader than those in the carbonaceous chondrite matrices, suggesting that the IDPs are more inhomogeneous in elemental composition than the meteorite matrices. These observations and analyses support the view that IDPs were less processed by aqueous alteration than the meteorites.

Ca-sulfate, Mg-sulfate, and Na-bearing sulfate are other typical minerals that were deposited during aqueous alteration in CI chondrites [21,22]; thus, Na and S (like Ca) tend to be depleted in these meteorite matrices [21]. Ferrihydrite, poorly crystallized Fe^{3+} hydroxide, is another major phase in the CI matrices, and was interpreted to have been formed together with the sulfates [18]. However, none of these phases and no depletions of Na and S are observed in the hydrated IDPs. The discrepancies, however, appear to be explained by the recent studies of a new Antarctic CI chondrite Yamato-82162. It has been revealed that Y82162 lacks sulfates, and its matrix and layer silicates are more enriched in Na than those in non-Antarctic CI chondrites [23]. (Sulfur is depleted in its matrix, but that is regarded as due to thermal metamorphism.) Tomeoka et al. [23] attributed these features to a much lesser degree of aqueous alteration than that experienced by non-Antarctic CI chondrites. Therefore, the lack of sulfates and ferrihydrite, and the non-depletions of Na and S in the hydrated IDPs may be other indications that the IDPs were less affected by aqueous alteration than CI chondrites.

5. THE SOURCES OF IDPs

Although there are a number of possible sources for the interplanetary dust, the major ones are considered to be comets and asteroids [e.g., 24]. However, until recently, there has been no way to limit the possibilities for the sources of IDPs. The results of mineralogical and chemical studies on both IDPs and meteorites provided much insight into the question.

As discussed in the previous sections, mineralogy of hydrated IDPs is similar to CI and CM chondrites, and hydrated IDPs have been processed by aqueous alteration analogous to these meteorites. The recent finding of tochilinite and cronstedtite in a serpentine-rich IDP [16] provided especially strong evidence for a link between the IDP and the CM chondrites. It is widely accepted that the most probable sources for meteorites are asteroids [25]. Recent analyses of telescopic reflectance spectra of the main-belt asteroids [26] suggested that aqueous alteration affected surfaces of some primitive-type asteroids.

Various authors have related the highly porous texture of anhydrous IDPs to the prior presence of volatile ices, and thus suggested that the anhydrous IDPs are cometary [e.g., 17,27]. On the other hand, the low porosity of the hydrated IDPs may have resulted from compaction process that occurred in relatively large parent bodies, i.e., asteroids. Fe/(Fe+Mg) distributions in pyroxene-rich anhydrous particles at submicron scale resemble Comet Halley dust, while those distributions in hydrated particles are similar to the Orgueil CI meteorite (see Fig. 9 in ref. 2).

The mineralogical and chemical information suggest that the hydrated IDPs are from asteroids and the pyroxene-rich anhydrous IDPs are from comets. Observational and experimental studies on atmospheric heating of IDPs [28] appear to indicate that main-belt asteroids, and comets having low inclinations and perihelia outside 1.2 AU, are the best candidates for the sources of these two types of dust. Sandford and Bradley [28] suggested that many of the olivine-rich IDPs may have been strongly heated during atmospheric entry, and thus they came from comets having more eccentric orbits.

Although it has not been established that extensive alteration can occur on comets, McSween and Weissman [29] recently proposed a hypothesis that significant aqueous alteration may occur in comet nuclei. Rietmeijer [30] has pointed out the possibility of cryogenic (<0 °C) alteration in comet nuclei. If the alteration indeed occurs in comets, the link between hydrated IDPs and meteorites would be somewhat blurred. The possibility of alteration in comets needs to be further evaluated and clarified in the future.

6. SUMMARY AND CONCLUSION

1. Hydrated IDPs can be divided into two subgroups, one dominated by smectite and the other by serpentine. The smectite-rich IDPs are mineralogically similar to CI chondrites, and the serpentine-rich IDPs are similar to CM chondrites, although they are not perfect matches.
2. There is evidence that the layer silicates in the IDPs were formed by aqueous alteration of glass and pyroxene. However, the presence of anhydrous silicates and glass and the compositional heterogeneity on a scale <1000 Å in the IDPs suggest that the alteration occurred over a very limited range. The degree of alteration in the IDPs was much lesser than those in meteorites.
3. It appears to be a current consensus of the majority of IDP researchers that most hydrated IDPs were derived from asteroids and most anhydrous IDPs were derived from comets.

Aknowledgements - I thank Dr. J.P. Bradley for helpful discussions and providing some of his unpublished data. This work was supported by a Grant-in-Aid of the Japan Ministry of Education, Science and Culture, No. 02640622.

REFERENCES

1 Sandford, S.A. and Walker, R.M. (1985) Laboratory infrared transmission spectra of individual interplanetary dust particles from 2.5 to 25 microns, Astrophys. J. 291, 838-851.
2 Bradley, J.P. (1988) Analysis of chondritic interplanetary dust thin-sections, Geochim. Cosmochim. Acta 52, 889-900.
3 Schramm, L.S., Brownlee, D.E. and Wheelock, M.M. (1989) Major element composition of stratospheric micrometeorites, Meteoritics 24, 99-112.
4 Bunch, T. E. and Chang, S. (1980) Carbonaceous chondrites--II. Carbonaceous chondrite phyllosilicates and light element geochemistry as indicators of parent body processes and surface conditions, Geochim. Cosmochim. Acta 44, 1543-1577.
5 Tomeoka, K. and Buseck, P.R. (1985) Indicators of aqueous alteration in CM carbonaceous chondrites: Microtextures of a layered mineral containing Fe, S, O and Ni, Geochim. Cosmochim. Acta 49, 2149-2163.
6 Zolensky M.E. and McSween, H.Y. (1988) Aqueous alteration, in J.F. Kerridge and M.S. Matthews (eds.), Meteorites and the Early Solar System, University of Arizona Press, pp. 114-143.
7 McKeegan, K.D., Walker, R.M. and Zinner, E. (1985) Ion microprobe isotopic measurements of individual interplanetary dust particles, Geochim. Cosmochim. Acta., 49, 1971-1987.
8 Geiss J. and Reeves H. (1981) Deuterium in the solar system, Astron. Astrophys. 93, 189-200.
9 Brownlee D.E., Olszewski E. and Wheelock M. (1982) A working taxonomy for micrometeorites, Lunar Planet. Sci. XIII, 71-72.
10 Rietmeijer, F.J.M. and Mackinnon, I.D.R. (1985) Layer silicates in a chondritic porous interplanetary dust partilce, Proc. 16th Lunar Planet.

78

Sci. Conf., J. Geophys. Res. 90:D149-155.

11 Tomeoka, K. and Buseck, P.R. (1984) Transmission electron microscopy of th "LOW-CA" hydrated interplanetary dust particle, Earth Planet. Sci. Lett. 69, 243-254.

12 Tomeoka, K. and Buseck, P.R. (1985) Hydrated interplanetary dust particle linked with carbonaceous chondrites? Nature 314, 338-340.

13 Tomeoka, K. and Buseck, P.R. (1986) A carbonate-rich, hydrated, interplanetary dust particle: Possible residue from protostellar clouds, Science 231, 1544-1546.

14 Germani, M.S., Bradley, J.P. and Brownlee, D.E. (1990) Automated thin-film analyses of hydrated interplanetary dust particles in the analytical electron microscope, Earth Planet. Sci. Lett. (in press).

15 Thomas, K.L., Zolensky, M.E., Klock, W. and McKay, D.S. (1990) Mineralogical descriptions of eight hydrated interplanetary dust particles and their relationship to chondrite matrix, Lunar Planet. Sci. XXI, 1250-1251.

16 Bradley, J.P. (1991) An interplanetary dust particle linked directly to type CM meteorites and an asteroidal origin, Science (submitted).

17 Brownlee, D.E. (1978) Interplanetary dust: possible implications for comets and presolar interstellar grains, in T. Gehrels (ed.), Protostars and Planets, University of Arizona Press, pp. 134-150.

18 Tomeoka, K. and Buseck, P.R. (1988) Matrix mineralogy of the Orgueil CI carbonaceous chondrite, Geochim. Cosmochim. Acta 52, 1627-1640.

19 Barber, D.J. (1981) Matrix phyllosilicates and associated minerals in C2M carbonaceous chondrites, Geochim. Cosmochim. Acta 45, 945-970.

20 Mackinnon, I.D.R. and Zolensky, M. (1984) Proposed structures for poorly characterized phases in C2M carbonaceous chondrite meteorites, Nature 309, 240-242.

21 Richardson, S.M. (1978) Vein formation in the CI carbonaceous chondrites, Meteoritics 13, 141-159.

22 Fredriksson, K. and Kerridge, J.F. (1988) Carbonates and sulfates in CI chondrites; Formation by aqueous activity on the parent body, Meteoritics 23, 35-44.

23 Tomeoka, K., Kojima, H. and Yanai, K. (1989) Yamato-82162: A new kind of CI carbonaceous chondrite found in Antarctica, Proc. NIPR Symp. Antarct. Meteorites 2, 36-54.

24 Sandford, S.A. (1987) The collection and analysis of extraterrestrial dust particles, Fundamentals of Cosmic Physics 12, 1-73.

25 Wetherill, G.W. (1985) Asteroidal sources of ordinary chondritic meteorites, Meteoritics 20, 1-22.

26 Vilas, F. and Gaffey, M.J. (1989) Phyllosilicate absorption features in main-belt and outer-belt asteroid reflectance spectra, Science 246, 790-792.

27 Bradley, J.P. and Brownlee, D. E. (1986) Cometary particles: Thin sectioning and electron beam analysis, Science 231, 1542-1544.

28 Sandford, S.A. and Bradley, J.P. (1989) Interplanetary dust particles collected in the stratosphere: Observations of atmospheric heating and constraints on their interrelationships and sources, Icarus 82, 146-166.

29 McSween, H.Y. and Weissman P.R. (1989) Cosmochemical implications of the physical processing of cometary nuclei, Geochim. Cosmochim. Acta 53, 3263-3271.

30 Rietmeijer, F.J.M. (1985) A model for diagenesis in proto-planetary bodies, Nature 313, 293-294.

THE EFFECT OF TOTAL PRESSURE ON VAPORIZATION OF ALKALIS FROM PARTIALLY MOLTEN CHONDRITIC MATERIAL

Taro SHIMAOKA[1] and Noboru NAKAMURA[1,2]

[1]*Department of Science of Material Differentiation, Graduate School of Science and Technology, Kobe University,* [2]*Department of Earth Sciences, Faculty of Science, Kobe University, Nada, Kobe 657, Japan*

ABSTRACT. In order to examine the effect of total pressure on vaporization of alkalis (Na, K, Rb) from a partially molten chondritic material, heating experiments were carried out under various He gas pressures ($\sim 10^{-5}$-$\sim 10^{-1}$torr) at 1300°C. The rate of vaporization decreased in the order of Na > K > Rb with the increasing of the pressure, and reached a minimum at $\sim 10^{-1}$torr.

1. Introduction

Alkali metals are moderately volatile at high temperatures and their vaporization behavior can constrain melting conditions of lunar and meteoritic materials (Gibson and Hubbard, 1972; Kreutzberger et al., 1985; Tsuchiyama et al., 1981; Matsuda et al., 1990).

We have conducted a series of vaporization experiments to study vaporization behavior of alkalis from a partially molten chondritic material (Shimaoka and Nakamura, 1989). Our results suggest that the vaporization is controlled mainly by the loss of elements from the partial melt, which is similar to the case of total melt (Tsuchiyama et al., 1981). However, there exists a systematic difference in vaporization rates between the two results. The vaporization rate can be influenced by total pressure, oxygen fugacity, and the chemical composition of charge as well as temperature. In this work, a heating experiment of a chondritic material was carried out under various total pressures ($\sim 10^{-5}$-$\sim 10^{-1}$torr) at a constant temperature (1300°C) to investigate the effect of total pressure on vaporization of alkalis.

2. Experimental

The apparatus and the general procedure used for the vaporization experiment were described previously (Shimaoka and Nakamura, 1989). In this work, the starting material with fine grain-size ($\phi < 10$um), similar to sample A in our previous work, was newly prepared from the Etter (L5) chondrite. Total pressure ($\sim 10^{-5}$-$\sim 10^{-1}$ torr) was adjusted using helium

79

A.C. Levasseur-Regourd and H. Hasegawa (eds.), Origin and Evolution of Interplanetary Dust, 79–82.
© 1991 *Kluwer Academic Publishers.*

gas with high purity (99.999%). The sample was heated at a constant rate (~30°C/min) up to 1300°C. After maintaining this temperature for 40min, the power supply of the furnace was cut off and then the sample was quenched with a cooling rate of ~600°C/min. A run product was broken into fragments and divided into two parts. One part was used for preparation of thin section. The other part was crushed to fine powder and subjected to analyses of Na by atomic absorption spectroscopy, and of K and Rb by isotope dilution mass spectrometry. The proportions of holes and melt in the run products were estimated from the SEM back-scattered electron images (Figure 1) of run products (Shimaoka and Nakamura, 1989).

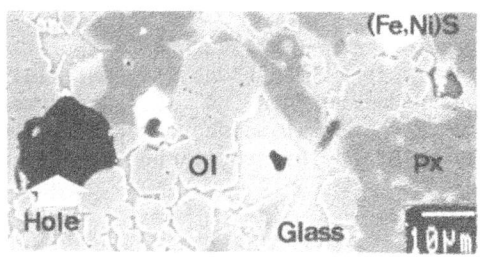

Figure 1. Back-scattered electron image of one of run products (1300°C, 40min). Ol: olivine, Px: pyroxene, (Fe,Ni)S: pentrandite

3. Result and Discussion

3.1. PREVIOUS STUDY

Previously, we obtained the following equation applicable to the vaporization of Na from a partial melt (Shimaoka and Nakamura, 1989):

$$\ln(C/C_0) = -k(f \cdot S_{tot}/V_p)t \qquad (1)$$

where C and C_0 are the concentrations in run product and starting material respectively, k, S_{tot} and V_p are the vaporization rate, the overall surface area and the volume of partial melt of charge respectively, t is the heating duration and f is the overall proportion of effective surface area of melt to total surface area of charge. The f value is tentatively estimated to be 0.2. Assuming that f, S_{tot} and V_p are constant during the heating, a linear correlation between t and $\ln(C/C_0)$ is expected. In Figure 2, results of an additional heating experiment for Na of this work are compared with those of previous work. In this diagram, all data points distribute along a linear curve, indicating a good reproducibility of vaporization rate in our heating experiments. As previously mentioned, the vaporization rate (k) calculated from the slope of regression lines are systematically high compared to those of the same temperature range obtained by extrapolating Tsuchiyama et al.(1981)'s values.

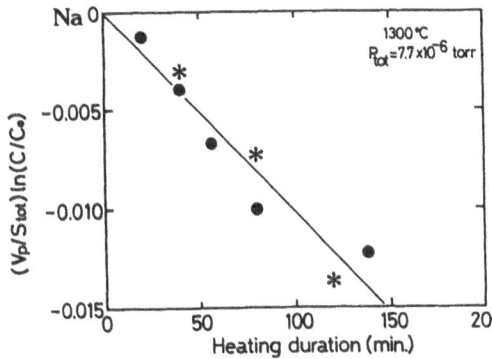

Figure 2. Plot of relative concentration of Na in heated samples corrected for effects of holes and partial melts as a function of heating duration. Symbols of stars and circles represent the present and previous results, respectively (Shimaoka and Nakamura, 1989).

3.2. EFFECT OF TOTAL PRESSURE

Applying equation (1), the vaporization rates of alkalis at various total pressures are obtained (Table 1 and Figure 3): in general, the vaporization rates of alkalis decrease with increasing total pressures. It is noted that the rate of vaporization decreases gradually for Na but abruptly for K and Rb at $\sim 10^{-1}$ torr. Actually, almost no vaporization loss of Rb was detected for the run product at $\sim 10^{-1}$ torr. It is interesting to note, that the pressure effect is greatest for Rb and smallest for Na.

TABLE 1. Results for run products at 1300°C

Run No.	89A	88A	86A	87A	90A
Duration (min)	40	40	40	40	40
Total pressure (torr)	7.7×10^{-2}	7.7×10^{-3}	7.7×10^{-4}	7.7×10^{-5}	7.7×10^{-6}
Holes (%)*1	10.4	9.2	11.0	8.3	9.5
Deg. of melt (%)*1	41.9	48.9	43.6	55.9	39.2
Olivine Fa (mol%)	27.5	26.9	25.1	25.0	26.3
Na (ppm)	2074	790	239	384	548
K (ppm)	454	315	235	238	286
Rb (ppm)	1.32	0.962	0.836	0.822	0.979
k_{Na} (cm^3/min·cm^2)*2	1.07×10^{-4}	1.89×10^{-4}	2.43×10^{-4}	2.93×10^{-4}	2.81×10^{-4}
k_{K} (cm^3/min·cm^2)*2	9.83×10^{-6}	4.18×10^{-5}	5.59×10^{-5}	7.69×10^{-5}	6.39×10^{-5}
k_{Rb} (cm^3/min·cm^2)*2	0	2.79×10^{-5}	3.33×10^{-5}	4.84×10^{-5}	3.34×10^{-5}

*1 See Shimaoka and Nakamura (1989).
*2 Vaporization rates of alkali metals (Na, K, Rb) calculated using eq.(1) with f=1 (see text).

Concentration of alkalis in the starting material;
    Na=6550ppm, K=506ppm, Rb=1.32ppm

82

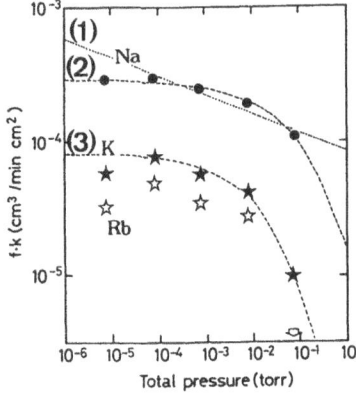

Figure 3. Plot of the vaporization rate of alkali metals as a function of total pressure. The regression lines are:
(1) $\log(f \cdot k) = -0.142\log P_{tot} - 4.08$
(2) $\log(f \cdot k) = -1.34\exp(\log P_{tot}) - 3.54$
(3) $\log(f \cdot k) = -2.79\exp(\log P_{tot}) - 4.08$

Inspection of the least square fits indicates that an exponential curve ((2) and (3) in Figure 3) is best fitted to the data points. However, the physical meaning of exponential function is not clear. For simplification, we tested a linear curve ((1) in Figure 3) to compare with the data obtained by Tsuchiyama et al.(1981). The vaporization rate of Na was calculated to be 1.6×10^{-4} $cm^3/min \cdot cm^2$ at 1300°C (f=0.2) by extrapolating this line to 1 atm. This value seems to be consistent with the value ($0.5-1.7 \times 10^{-4} cm^3/min \cdot cm^2$) calculated from Tsuchiyama's experiment at the same temperature and oxygen partial pressure ($10^{-11.6}$ atm). Since data points (particularly at high pressure) are still limited, more work may be needed to confirm the present result.

We thank Mrs T. Baba and N. Miyano for the setting up the programmable controller, and Dr. K. Yamamoto for the support of this work.

REFERENCES

Gibson, E. K., Jr. and Hubbard, N. J. (1972) 'Thermal volatilization studies on lunar samples', Proc. Lunar Conf. 3rd, 2003-2014.
Kreutzberger, M. E., Drake, J. V. and Lewis, V. A. (1985) 'Origin of the Earth's Moon: Constrains from alkali volatile trace elements', Geochim. Cosmochim. Acta 49, 91-98
Matsuda, H., Nakamura, N. and Noda, S. (1990) 'Alkali(Rb/K) abundances in Allende chondrules: Implication for the melting conditions of chondrules', Meteoritics 25, 137-143.
Shimaoka, T. and Nakamura, N. (1989) 'Vaporization of sodium from a partially molten chondritic material', Proc. NIPR Symp. Antarct. Meteorites 2, 252-267.
Tsuchiyama, A., Nagahara, H. and Kushiro, I. (1981) 'Volatilization of sodium from silicate melt spheres and its application to the formation of chondrules', Geochim. Cosmochim. Acta 45, 1357-1367.

CONDENSATION EXPERIMENTS OF MG–SILICATE MINERALS

A. TSUCHIYAMA
Institute of Earth and Planetary Sciences
College of General Education
Osaka University
Toyonaka 560
Japan

ABSTRACT. Condensation experiments were performed in the simple but most fundamental system Mg-Si-O-H with forsterite vaporization source. At temperatures above about $1000^{\circ}C$, euhedral crystals of forsterite (Mg_2SiO_4) of a few μm were formed. These crystals are similar to olivines in Allende matrix. At temperatures below about $1000^{\circ}C$, whiskers of forsterite and enstatite $(MgSiO_3)$ were formed by vapor-liquid-solid growth mechanism. These whiskers are different from enstatite whiskers in interplanetary dust, which were probably formed at small super coolings.

1. INTRODUCTION

Condensation is the most essential process for producing solid grains in the solar nebula and circumstellar regions. Many condensation experiments have been done for studying condensation processes in circumstellar regions (*e.g.*, Nuth and Donn, 1982; Koike and Tsuchiyama, 1991), but important silicate minerals, such as olivine and pyroxene, have not been formed. On the other hand, only a few experiments appropriate to condensation in the nebula were done (Nagahara *et al.*, 1988; Tsuchiyama *et al.*, 1988). In these experiments, some important silicate minerals were formed, but the total pressure was too low compared to conditions in the nebula.

In the present study, condensation experiments were carried out in the simple but most fundamental system of Mg-Si-O-H to obtain Mg-silicate minerals under conditions near those expected in the primordial solar nebula. Condensation products are compared with natural condensates.

2. EXPERIMENTS

Experiments were done with the condensation furnace described by Tsuchiyama (1989) (Fig.1). Powders of forsterite (Mg_2SiO_4) placed at the bottom of the crucible were vaporized at $1480-1545^{\circ}C$ for 15-90 hrs,

A.C. Levasseur-Regourd and H. Hasegawa (eds.), Origin and Evolution of Interplanetary Dust, 83–86.

and condensation took place from the gas on a cold finger with a temperature gradient from 1455°C to room temperature. A constant H_2 pressure of 1.4 Pa, which is similar to the nebula pressures, was obtained by bleeding of H_2 gas into the chamber during evacuation. Redox conditions were controlled by the material of the crucible; a reduced condition near the $Ta-Ta_2O_5$ buffer and neutral condition were obtained by the Ta and Mo crucibles, respectively. The ratio of Si/H was about 10^{-2}, which is larger than that in the nebula (4×10^{-5}). About 500 mg or more of condensates were recovered as a function of the redox condition and temperature.

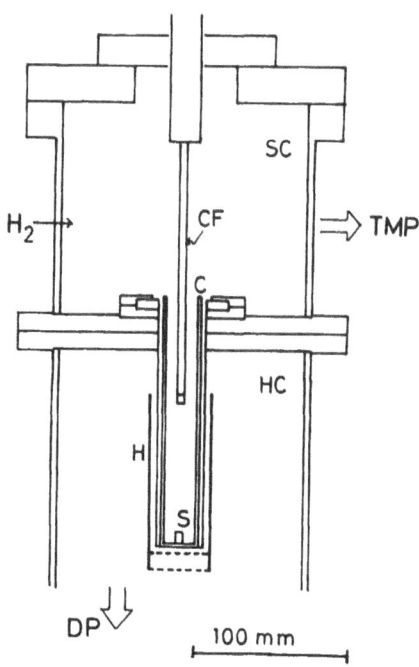

Fig. 1. A schematic drawing of the condensation furnace used in the present experiments. SC = sample chamber, HC = heater chamber, CF = cold finger (Mo), C = crucible (Ta or Mo), H = heater, S = vaporization source (forsterite).

3. RESULTS

Forsterite starts to condense at about 1200°C joined by silicon crystals at about 950°C in the Ta crucible, while forsterite at about 1200-1250°C jointed by enstatite ($MgSiO_3$) at about 900°C in the Mo crucible. In both experiments, condensates become amorphous at about 500°C. The condensation sequences are qualitatively explained by thermochemical calculations of solid-gas equilibria (Tsuchiyama, 1989).

At high temperatures (1250-1000°C) forsterite condenses as euhedral

crystals usually elongated in the c-axis direction (Fig.2). At lower
temperatures (<1000°C) forsterite and enstatite condense as whiskers
which always have amorphous droplets on their tops (Fig.3). The
enstatite whiskers are clinoenstatite elongated to the c-axis direction.

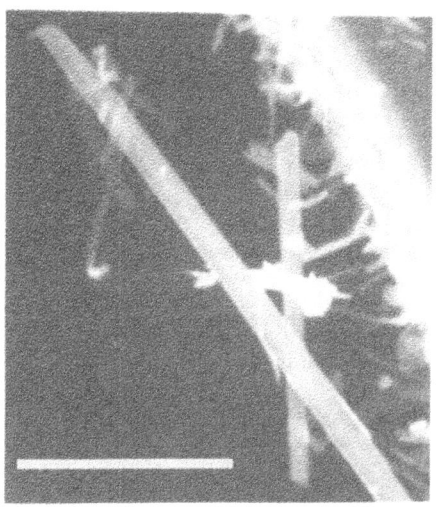

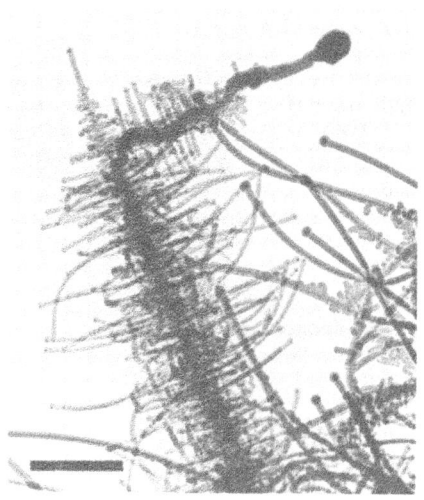

Fig. 2. A scanning electron
micrograph of euhedral forsterite
condensed at about 1120°C in the
crucible. Scale bar is 10 μm.

Fig. 3. A transmission electron
micrograph of forsterite whiskers
condensed at about 850°C. Each Ta
whisker always has an amorphous
droplet on its top. Scale bar is
1 μm.

4. DISCUSSION

4-1. VLS growth mechanism of whiskers

The amorphous droplets of the whiskers must be melted during the
experiments. Thus, the formation of the whiskers can be explained as
follows; (1) the melt droplets condensed first as a metastable phase,
and (2) gas molecules are incorporated into the melt, and then
forsterite, a stable phase, grows from the melts as whiskers. This is a
kind of vapor-liquid-solid (VLS) growth mechanism (Wagner and Ellis,
1965). The metastable condensation of the melt can be explained by
condensation at low temperatures with large super coolings.
    It was pointed out that chondrules were formed by direct
condensation of metastable melts in the cooling nebula (Blander and
Katz, 1967). However, the present experiments indicate that direct
condensation of melts would produce VLS-grown whiskers instead of
chondrules. Such whiskers could be formed under large super coolings in
dense parts of the primordial solar nebula although the VLS-grown

whiskers have not been found in natural samples at present.

4-2. Comparison with natural condensates

Clinoenstatite whiskers in interplanetary dust are considered to be condensates formed in the solar nebula or presolar environments because of their unique morphology and microstructure, such as elongation along the a-axis with screw dislocations (Bradley et al., 1983). These whiskers were not formed in the present experiments. Probably, they were formed directly from a gas at small super coolings.

Fine olivine crystals elongated to the c-axis direction found in the matrix of Allende meteorite are also considered as condensates because of their euhedral morphology (Green et al., 1971). Forsterite crystals with the similar morphology and size were formed in the present experiments although the olivines in Allende contain some iron.

Acknowledgements. This work is partly supported by a Grant-in-Aid for Scientific Research from the Japan Ministry of Education.

REFERENCES

Blander, M. and Katz, J. L. (1967) 'Condensation of primordial dust' *Geochim. Cosmochim. Acta* 31, 1025-1034.
Bradley, J. P., Brownlee, D. E. and Veblen, D.R. (1983) 'Pyroxene whiskers and platelets in interplanetary dust: evidence of vapor phase growth', *Nature* 301, 473-477.
Green, H. W., Radcliffe, S. V. and Heuer, A. H. (1971) 'Allende meteorites: A high voltage electron petrographic study' *Nature* 172, 936-939.
Koike, C. and Tsuchiyama, A. (1991) 'The infrared spectra of synthesized amorphous silicates of olivine and pyroxene', this volume.
Nagahara H., Kushiro, I., Mysen, B. O. and Mori, H. (1988) 'Experimental vaporization and condensation of olivine solid solution', *Nature*, 331, 516-518.
Nuth, J. A. and Donn, B. (1982) 'Experimental studies of the vapor phase nucleation of refractory compounds. I. The condensation of SiO', *Jour. Chem. Phys.* 77, 2639-2646.
Tsuchiyama, A. (1989) 'Condensation experiments in the system Mg-Si-O-H' *Lunar Planet. Sci.* XX, 1136-1137.
Tsuchiyama A., Kushiro, I., Mysen, B. O. and Morimoto, N. (1988) 'An electron microscopy of gas condensates in the system Mg-Si-O-H' *Proc. NIPR Sym. Antarct. Meteorites* 1, 185-196.
Wagner, R. S. and Ellis, W. C. (1965) 'The vapor-liquid-solid mechanism of crystal growth and its application to silicon', *Trans. Met. Soc. AIME* 223, 1053-1064.

ULTRAVIOLET-INDUCED AMORPHIZATION OF CUBIC ICE AND ITS IMPLICATION FOR THE EVOLUTION OF ICE GRAINS

A. Kouchi* and T. Kuroda
Institute of Low Temperature Science, Hokkaido University,
Sapporo 060, Japan
* present address
Laboratory Astrophysics, University of Leiden,
Niels Bohrweg 2, 2333 CA, Leiden, The Netherlands

ABSTRACT   We found that cubic ice is transformed below 70 K to amorphous ice by ultraviolet irradiation, whereas no change in structure is observed at temperatures above 70 K, regardless of the irradiation time.   Experimental results can be interpreted by theoretical consideration of nucleation and growth of cubic ice in amorphous ice. We also discuss the evolution of ice grains in space on the basis of the experimental results.

1. INTRODUCTION

Despite current interest in effects of ultraviolet radiation and high energy particles on ice and on solid mixtures of water and other molecules, most attention has so far been paid to chemical reactions alone.   On the other hand, even pure substances are affected by irradiation as was shown by Lepault et al. (1983) who found that 100 kV electron-beam bombardment will convert ice crystals to amorphous ice below 70 K.   Accordingly, we have studied the effect of UV radiation on ice, using electron diffraction method.
    Phase transformation of H_2O ice at low temperature and at low pressure is especially important for understanding the evolution of ice grains in space.   At temperatures lower than 140 K, amorphous ice transforms continuously and irreversibly from one metastable state to the other with increasing temperature (Hagen et al., 1981; Kouchi, 1990).   At temperatures around 140 K, amorphous ice crystallizes to cubic ice, then cubic ice transforms to hexagonal ice by further heating.   Since all these transformations are irreversible, it has so far been considered that cubic and hexagonal ice formed at high temperatures will persist, even when they have been subsequently cooled to lower temperature.   However, this ignores the effect of UV radiation which is normally present in space.   We found that UV irradiation transforms cubic ice into amorphous ice below 70 K (Kouchi and Kuroda, 1990).

A.C. Levasseur-Regourd and H. Hasegawa (eds.), Origin and Evolution of Interplanetary Dust, 87–90.

2. EXPERIMENTAL METHOD

A thin film of cubic ice was prepared at 135 K by vapor deposition in a vacuum chamber at a pressure of 5 x 10^{-7} Pa. After deposition, the cubic ice was cooled to the desired temperature and was irradiated with UV radiation produced by a D_2 lamp ($110\langle\lambda\langle400$ nm; Φuv$\sim$$10^{12}$ photon cm^{-2} s^{-1}). Structural changes in the ice were examined in-situ by reflection electron diffraction (20 kV).

3. RESULTS

It is found that cubic ice at temperatures lower than 70 K transforms to amorphous ice by UV irradiation, whereas no change in structure occurs at temperatures above 70 K regardless of the irradiation time. At temperatures between 50 and 70 K, irradiation for 30 to 60 min was required for amorphization, whereas only several tens of seconds were required at 10 K. Figure 1 shows reflection electron diffraction patterns of cubic ice and UV-induced amorphous ice. The halo diffraction pattern in Fig. 1b indicates that the irradiated ice is amorphous.

Lepault et al. (1983) found that crystalline ice was transformed below 70 K into amorphous ice after irradiation with 100 kV electrons. However, we did not observe structural change after irradiation with 30 kV electrons (maximum energy of our electron gun) even at 10 K. We were not able to exclude amorphization with 30 kV electrons even at 10 K. The only explanation for this was that the electron flux was too low, but the need for further investigation is indicated.

Our finding of amorphization by UV-irradiation is supported by the latest measurements of infrared absorbance spectra (Kouchi and Greenberg, in prep.). Spectral features and the peak frequency of the OH-stretching mode of cubic ice around 3 μm are substantially changed by the UV irradiation. The structure of the UV-induced amorphous ice absorption is similar to that of the amorphous ice which was directly obtained by vapor deposition at 10 K.

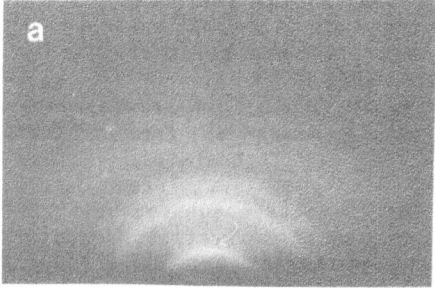 

Figure 1. Reflection electron diffraction patterns of a) cubic ice produced at 135 K, and b) UV-induced amorphous ice after 10 min of irradiation of cubic ice at 10 K.

4. THEORETICAL CONSIDERATION

Here we briefly mention some theoretical aspects of amorphization. There are two competitive processes relevant to amorphization by irradiation. The first one is the process of destroying the cubic lattice caused by displacement or dissociation of molecules by irradiation, and the second one is the restoration process from unstable amorphous structure to stable crystal structure leading to lowering of the free energy of the system. The latter process is controlled by the rate J of nucleation of stable crystalline clusters and their growth rate v. The self-diffusion coefficient Da of amorphous ice is involved with J and v through the rearrangement of molecules from amorphous to crystalline structure. Accordingly, the restoration process is exponentially activated with increasing temperature, so that the resistance of ice to amorphization by irradiation is lost at a critical temperature Tc. In order to quantitatively understand Tc, measurement of Da and the interfacial free energy γ between amorphous and cubic ice are required. It is to be noted that there are several forms of metastable amorphous structures (Kouchi, 1990), and the values of Da and γ may depend on the structure. Thus, strictly speaking, Tc may depend on experimental conditions.

5. ASTROPHYSICAL IMPLICATION

Our results have several implications for ice grains in space. The interstellar UV radiation fluxes in diffuse and dense molecular clouds are 10^8 photon cm^{-2} s^{-1} (Hagen et al., 1979) and 10^{3-4} (Prasad and Tarafdar, 1983), respectively. A UV flux of 10^4 photon cm^{-2} s^{-1} in dense clouds is probably more relevant when including cosmic ray, solar wind and penetrating sources together. The time scale equivalent to one hour of irradiation in the laboratory is about 1 year in the diffuse medium and 10^4 years in a dense cloud. We may conclude that if grains of cubic ice or of amorphous ice are formed at around 100 K in a stellar atmosphere (Seki and Hasegawa, 1983), they will be transformed by UV irradiation to amorphous ice of low temperature form (Kouchi and Greenberg, in prep.) in an astronomically short time after cooling to 10 K after injection into space.

6. ACKNOWLEDGMENTS

We thank Professors H. Hasegawa for discussions and J.M. Greenberg for discussions and critical reading of the manuscript. This work was partly supported by a Grant-in-Aid for scientific research on Priority Area (Origin of the Solar System) from the Japanese Ministry of Education, Science and Culture, and by the Yamada Science Foundation, Japan.

7. REFERENCES

Hagen, W., Allamandola, L.J. and Greenberg, J.M. (1979) 'Interstellar molecule formation in grain mantles: the laboratory analog experiments, results and implications', Astrophys. Space Sci.,65, 215-240.

Hagen, W., Tielens, A.G.G.M. and Greenberg, J.M. (1981) 'The infrared spectra of amorphous solid water and Ic between 10 and 140 K', Chem. Phys., 56, 367-379.

Kouchi, A. (1990) 'Evaporation of H_2O-CO ice and its astrophysical implications', J. Cryst. Growth, 99, 1220-1226.

Kouchi, A. and Kuroda, T. (1990) 'Amorphization of cubic ice by ultraviolet irradiation', Nature, 344, 134-135.

Lepault, J., Freeman, R. and Dubochet, J. (1983) 'Electronbeam induced vitrified ice', J. Microsc., 132, RP3-4.

Prasad, S.S. and Tarafdar, S.P. (1983) 'UV radiation field inside dense clouds: its possible existence and chemical implications', Astrophys. J., 267, 603-609.

Seki, J. and Hasegawa, H. (1983) 'The heterogeneous condensation of interstellar ice grains', Astrophys. Space Sci., 94, 177-189.

SIMULATION IN LABORATORY OF SOLID GRAINS PRESENT IN SPACE

L. COLANGELI[1], E. BUSSOLETTI[2,3] & V. MENNELLA[3]

(1) *Dipartimento Ingegneria Industriale, Università di Cassino, via Zamosch 43, 03043 Cassino, Italy*
(2) *Istituto Universitario Navale, via Acton 38, 80133 Napoli, Italy*
(3) *Osservatorio Astronomico di Capodimonte, via Moiariello 16, 80131 Napoli, Italy.*

ABSTRACT. Laboratory data on cosmic dust analogue materials are compared with recent results obtained by means of spectroscopy and mass spectrometry on cometary dust, meteorites and interplanetary dust. Their actual chemical and physical properties can be further clarified, as well as possible links with interstellar dust.

1. Introduction

Recent observations of cometary dust, meteorites and interplanetary dust particles (IDPs) have shown that the identification of their nature can allow to obtain important hints about the composition and the evolution of cosmic dust. In particular, the correct interpretation of spectroscopic and mass spectrometry results is essential to clarify some open questions. We recall here some relevant points:

1) PUMA 1/2 and PIA experiments have detected: a) silicate grains with elemental abundance close to that of carbonaceous chondrites; b) "CHON" particles containing mainly light elements; c) mixed particles characterized by the presence of both light and heavy elements (Jessberger et al. 1988);

2) spectrophotometry of several comets have evidenced a broad 9.7 μm band associated with amorphous and/or hydrated silicates. Comets Halley and Bradfield show also a sharp peak at 11.3 μm, which is presently attributed to the additional presence of crystalline olivine (Hanner et al. 1990);

3) comets Halley, Wilson and Bradfield show a structured emission feature around 3.4 μm, similar to that observed towards the galactic center source IRS7, which seems to indicate some link between interstellar dust and cometary material. This band is representative of a class of CH_n (n = 1,2,3) resonances which should show additional signatures between 3 and 13 μm. However, as shown by Chyba and Sagan (1987) and by Colangeli et al. (1990), these other *fingerprints* could not be observed in Halley's spectra, recorded when the comet was at a solar distance of about 1 AU, due to the blanketing effect produced by the continuum emission;

4) several laboratory analyses on meteorites have shown the presence of diamond structures, SiC and amorphous carbon. Very recently Cronin and Pizzarello (1990) have analyzed the aliphatic content of Murchison meteorite (CM2) eliminating carefully most of the possible terrestrial contaminants. In contrast with previous

91

A.C. Levasseur-Regourd and H. Hasegawa (eds.), Origin and Evolution of Interplanetary Dust, 91–94.
© 1991 *Kluwer Academic Publishers.*

results, the absorption spectrum does not show evidence of the presence of long chains of n-alkyl groups and olefinic structures. Intense features appear at 3.4 - 3.6 μm and at 7.2 μm (methyl and methylene signatures) and around 6 μm ($C=O$ resonances). Though rare, SiC is observed as small grains in meteorites in two allotropic forms: type 1 (d = 0.06 - 0.2 μm) and type 2 (d = 0.1 - 1 μm), with occasional particles with diameter up to 12 μm (Anders et al. 1989);

5) IR spectra of IDPs present dominant silicate features at 10 and 20 μm. Although their extraterrestrial origin is not confirmed, some carbonaceous features have also been identified around 3.38 μm (methyl stretch), 3.4 μm (methylene feature) and 3.5 μm (CH_2 and CH_3 symmetric stretching). Raman spectroscopy of single IDPs shows the presence of two bands at 1350 and 1600 cm^{-1} typical of disordered carbonaceous materials. The bands relative intensity indicates that graphitic platelets not larger than 25 Å should be present in the grains (Allamandola et al. 1987).

2. Laboratory Analogue Materials

Since 1980 our group has produced, characterized morphologically and measured the optical properties from VUV to FIR of several materials able to simulate cosmic solid particles such as amorphous carbon, SiC, PAHs, and silicates. Details on the subject are reported in Bussoletti et al. (1987).

2.1. SPECTRAL PROPERTIES OF CARBONACEOUS GRAINS

Hydrogenated amorphous carbon (HAC) grains with average radius of $\simeq$ 40 Å show in IR a class of absorption bands which match, in wavelength position, some of the so-called *unidentified infrared bands* observed in space. The IR spectrum of HAC grains closely resembles that reported for *uncontamined* Murchison meteorite by Cronin and Pizzarello (1990); in addition, most bands detected in absorption fall at the same wavelength. Furthermore, HAC grains show a 3.4 μm band very similar in profile to that observed, in absorption, towards the source GC-IRS7 and, in emission, from comet Halley. Colangeli et al. (1990) have shown that the fit of the cometary feature by HAC grains optical properties is rather satisfactory and it is consistent with carbon abundance constraints imposed by PUMA 1/2 and PIA observations. Therefore, it exists the possibility that cometary solid material may be similar to HAC grains. It is interesting to note that (see section 1) both diamond particles in meteorites and some IDPs show features in the 3.4 - 3.5 μm range, but not around 3.3 μm. This evidence can be diagnostic of the materials short-scale structure. In fact, features at 3.39, 3.42, and 3.51 μm are typical of -CH_2 and -CH_3 radicals bound to a diamond-like (sp^3) carbon structure. On the contrary, sp^2-CH_n (n=1,2,3) resonances in graphitic-like structures fall mainly at shorter λ. The same behavior is observed for HAC grains produced in laboratory. Here the presence of diamond-like bonds in an overall amorphous structure is confirmed by the *optical gap* evidenced by extinction data in the far-UV (Colangeli et al. 1990).

Other interesting similarities between IDPs and amorphous carbon grains arise from the comparison of Raman spectra. As for many IDPs, Raman spectroscopy on amorphous carbon grains evidences typical features at 1350 and 1600 cm^{-1} (Fonti et al. 1990). Both these features are Raman active modes in randomly oriented structural units which constitute the grains. According to Allamandola et al. (1987), these features show a striking similarity to interstellar IR emission spectrum towards the Orion nebula. In conclusion, HAC grains contain, on a short-scale, both graphitic and diamond bonds. The similarities with observations of cometary material, meteorites and IDPs indicate that these particles resemble

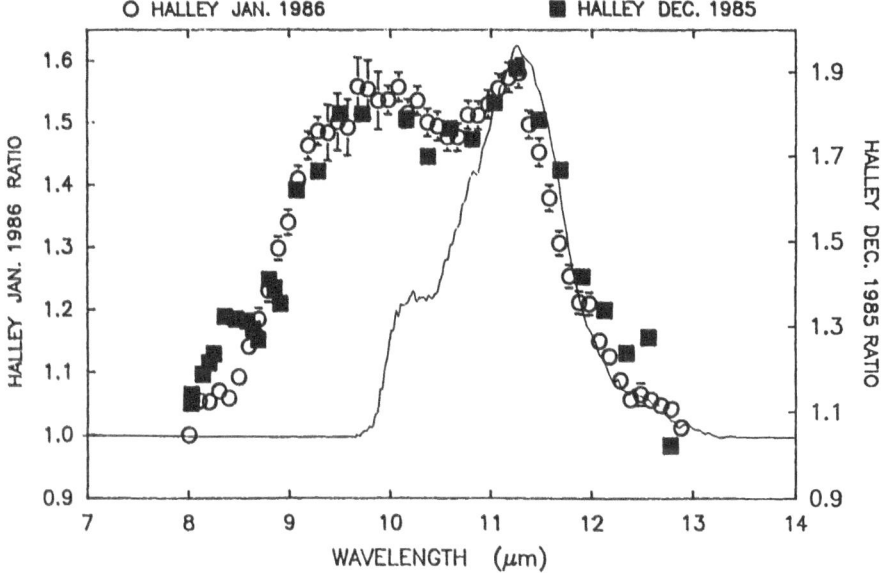

Figure 1: Fit of the cometary band at 11.3μm by means of the optical properties of SiC grains.

HAC grains to some extent.

2.2. SIMULATION OF THE 11.3 μm EMISSION BAND IN COMETS.

As already mentioned in section 1, the 11.3 μm emission band from comets Halley and Bradfield suggests the presence of crystalline silicatic grains in addition to the amorphous and/or hydrated component. To check this possibility, Blanco et al. (this book) have produced in the laboratory and analyzed spectroscopically various forms of crystalline and amorphous synthetic and natural silicates. The fit of the cometary features at 9.7 and 11.3 μm appears satisfactory. Nevertheless, IR spectral laboratory mesurements on various kinds of SiC grains (Borghesi et al. 1985) show a strong 11.3 μm band. We have to recall that silicon carbide solid grains are expected to form mainly in the atmosphere of carbon rich cool stars (C/O > 1). The presence of SiC particles has been confirmed by several astronomical observations, as a definite 11.3 μm band has been seen towards a wide variety of sources. Frenklach et al. (1989) have shown that carbon and silicon carbide could occur in core-mantle structures since their formation in cool stars atmospheres. Therefore, if cometary material is, to some extent, representative of the *primordial* composition of the pre-solar nebula we cannot exclude *a priori* the presence of SiC in comets. In support to this idea comes the detection of small amounts of *exotic* SiC in meteorites (section 1). On this ground we have simulated the emission spectrum of P/Halley at around 11.3 μm with the optical properties of α-SiC grains produced in the laboratory. As shown in Fig. 1, the fit of the peak and the long wavelength-side of the band profile appears rather good. Of course, at shorter wavelengths the contribution from silicates is needed to get a complete simulation of the observations. However, we note that only a few percent (< 10%), by mass, of SiC is required to account for the observed intensity. Therefore, on a qualitative basis, we have to include SiC grains as a possible candidate to cometary grains. This suggestion needs further investigation in order to define abundance constraints on this class of particles. If the detection of SiC grains in comets is confirmed, we will have a further evidence of the relation existing between comets, interplanetary and interstellar dust.

3. Acknowledgements

This work has been supported under contracts by CNR, ASI and MPI.

4. References

Allamandola, L. J., Sandford, S. A. and Wopenka, B. (1987) 'Interstellar polycyclic aromatic hydrocarbons and carbon in interplanetary dust particles and meteorites', Science 237, 56-59.

Anders, E., Lewis, R. S., Ming, T., Zinner, E. (1989) ' Interstellar grains in meteorites: diamond and silicon carbide', in L. J. Allamandola and A. G. G. M. Tielens (eds.), Interstellar Dust, Kluwer Academic Publishers, Dordrecht, 389-402.

Borghesi, A., Bussoletti, E., Colangeli, L. and De Blasi, C. (1985) 'Laboratory study of SiC submicron particles at IR wavelengths: a comparative analysis', Astron. Astrophys. 153, 1-8.

Bussoletti, E., Colangeli, L., Borghesi, A. and Orofino, V. (1987) 'Tabulated extinction efficiencies for various types of submicron amorphous carbon grains in the wavelength range 1000 Å - 300 μm', Astron. Astrophys. Suppl. Ser. 70, 257-268.

Chyba, C. and Sagan, C. (1987) 'Infrared emission by organic grains in the coma of comet Halley', Nature 330, 350-353.

Colangeli, L., Schwehm, G., Bussoletti, E., Fonti, S., Blanco, A. and Orofino, V. (1990) 'Hydrogenated amorphous carbon grains in comet Halley ?', Astrophys. J. 348, 718-724.

Cronin, J. R. and Pizzarello, S. (1990) 'Aliphatic hydrocarbons of the Murchison meteorite', Geochim. Cosmochim. Acta, in press.

Fonti, S., Blanco, A., Bussoletti, E., Colangeli, L., Lugara', M., Mennella, V., Orofino, V. and Scamarcio, G. (1990) ' Raman spectra of different carbonaceous materials of astrophysical interest', Infrared Phys. 30, 19-25.

Frenklach, M., Carmer, C. S. and Feigelson E. D. (1989) 'Silicon carbide and the origin of interstellar carbon grains', Nature 339, 196-198.

Hanner, M. S., Newburn, R. L., Gehrz, R. D., Harrison, T., Ney E. P. and Hayward, T. L. (1990) ' The infrared spectrum of comet Bradfield (1987s) and the silicate emission feature', Astrophys. J. 348, 312-321.

Jessberger, E. K., Christoforidis, A. and Kissel, J. (1988) 'Aspects of the major element composition of Halley's dust', Nature 332, 691-695.

THE INFRARED SPECTRA OF SYNTHESIZED AMORPHOUS SILICATES WITH COMPOSITIONS OF OLIVINE AND PYROXENE

Chiyoe Koike[1] and Akira Tsuchiyama[2]
[1] Kyoto Pharmaceutical University,
    Yamashina, Kyoto 607,Japan.
[2] Osaka University, Toyonaka,
    Osaka 560, Japan.

ABSTRACT. Amorphous olivines synthesized by evaporation method show two very broad bands at 10-11 μm and 17.5-19 μm, which resemble the spectra of symbiotic stars. On the other hand, amorphous pyroxenes produced by the same method show two broad bands at 9.5-10.3 μm and 20-22 μm, which are narrower than that of amorphous olivine. The features of amorphous olivine were easily altered by heating or hydration, and the peak wavelength of 18 μm band was easily shifted to longer wavelengths.

1.INTRODUCTION.

Recently, many unusual spectra of stars were reported from the IRAS data in the low-dispersion spectra(LRS). For instance, very broad features of 10 μm and 18 μm bands were found on symbiotic systems. These features are not explained based on the data measured previously in amorphous silicates, because the width of these bands is significantly narrower than those observed in the stars. Here, we report about the simulation of the infrared spectra of symbiotic stars and very unstable characteristics of the 18 μm band by alteration.

2. SYNTHESIS OF AMORPHOUS SILICATES.

2.1. Experiments

Our experiments are processed in two stages, that is, synthetic experiments and alteration. We synthesized amorphous silicates by an evaporation method using W and Mo boats. Samples were heated at about 1800-2000 °C for several minutes in vacuum. The condensates were obtained on Al substrates several cm apart from the evaporation source. Starting materials of silicates are olivine with different compositions (Fo_{100},Fo_{90},Fo_{80}), Ca-poor and Ca-rich pyroxenes.

2.2. Results and discussion

The condensates from olivine have almost the olivine stoichiometry, while those from pyroxene are enriched in SiO_2. All the condensates contain W or Mo due to contaminations from the boats. These W and Mo are

95

A.C. Levasseur-Regourd and H. Hasegawa (eds.), Origin and Evolution of Interplanetary Dust, 95–98.
© 1991 Kluwer Academic Publishers.

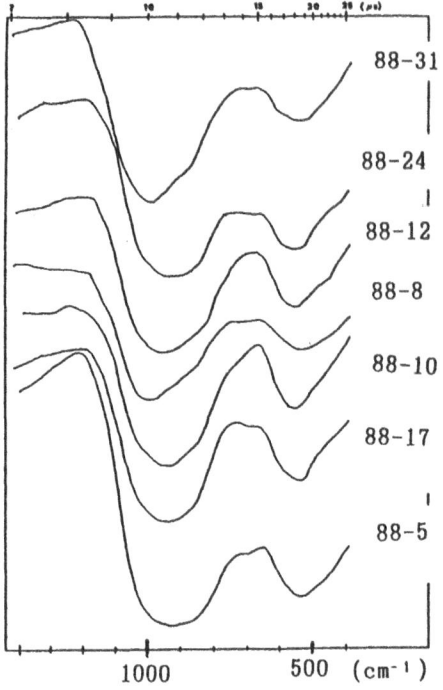

Fig.1. The spectra of amorphous olivines.

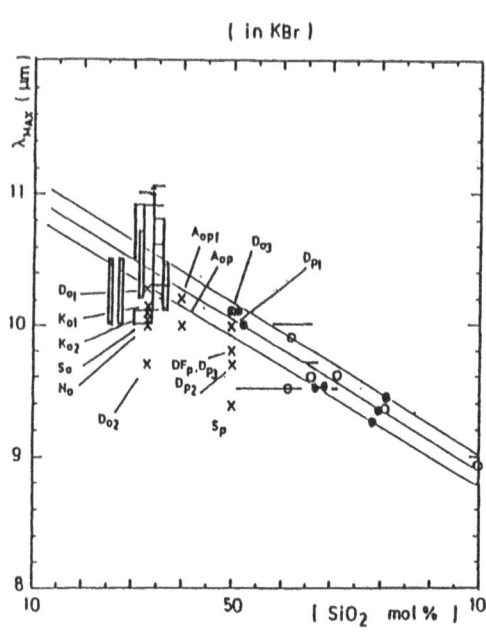

Fig.2. Correlation between the 10 μm peak wavelength λₘ and SiO₂ content. Solid lines and squares are this work,
● natural amorphous silicates,
○ synthetic and fused silictates,
× other laboratory data of synthetic amorphous silicates.

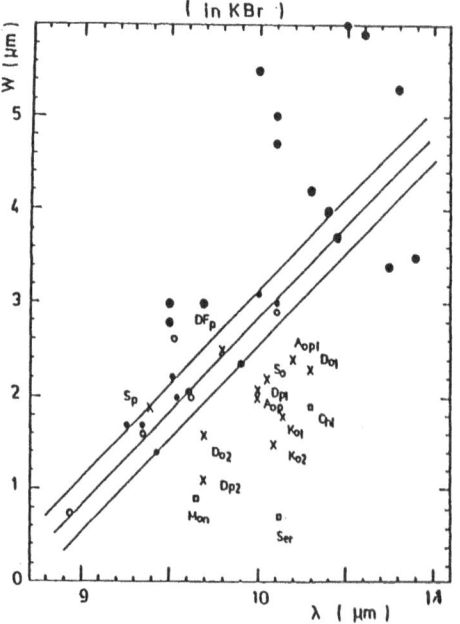

Fig.3. Correlation between the 10 μm band width W and SiO₂ content. ● this work, other notations are the same as in Fig.2.

considered to be metallic based on X-ray and electron diffraction patterns. Therefore, the effect of W or Mo to the IR spectra can be ignored. In fact, oxides of W or Mo were not observed in the IR spectra.

The infrared spectra of amorphous olivine show two broad bands at wavelengths of about 10-11 μm and 18-19 μm (Fig.1), while those of amorphous pyroxene show two broad bands at about 9.5 and 21.5 μm. The correlation of the SiO_2 content with the peak wavelength of 10 μm bands is consistent with the previous relation (Koike et al.1987)(Fig.2). However, the band widths are much broader than the other amorphous silicates (Fig.3). For the 18 μm bands, amorphous olivine shows an absorption peak at about 18 μm, but amorphous pyroxene shows a peak at about 22 μm. The ratio of 10 and 18 μm bands of the present samples are lower than in previous data.

These infrared spectra in KBr are reduced to the spectra in vacuum using dispersion equations to compare with observations. For the first approximations, the corrected features of the amorphous olivine are similar to those of symbiotic stars.

3. ALTERATION OF THE AMORPHOUS SILICATES

3.1. Heating

We examined changes in the spectra of the above amorphous silicates by annealing and hydration. The amorphous silicates were heated to 150 – 600 ℃ for 2 - 105.25 hr. As shown in Fig.4, the spectra of amorphous olivine were not changed up to about 500 ℃, but begun to change distinctly at about 600 ℃. The width of the 10 μm band became wider and the peak wavelengths of 18 μm band shifted to longer wavelengths. The spectra changed to those of crystal olivine by longer heating at 600 ℃.

3.2. Hydration

The hydration effects were examined by keeping the samples in saturated vapor, in water and in boiling water. The spectrum was sligthtly changed in saturated vapor: a subtle dip appeared at 10 μm, and the peak wavelength of 18 μm band shifted from 18 μm to nearly 20 μm. By keeping the samples in water at room temperature for 42 hr, the width of the 10 μm band became slightly narrower and a small shoulder appeared at about 11.8 μm (Fig.5). The peak wavelength of the 18 μm band was significantly shifted from 18 μm to 22 μm. By keeping the sample in boiling water for 10 hr, the spectrum was substantially changed (Fig.5). For this spectrum, the 10 μm band showed sharper structures, especially a clear shoulder at 11.5 μm. The peak wavelength of the 18 μm band was also shifted to 22 μm. From these alteration experiments, we conclude that the 18 μm band is very unstable by annealing and hydration.

REFERENCES

B.G.Anandarao,A.R.Taylor and S.R.Pottasch,(1988)'Dust emission from symbiotic stars: an interpretation of IRAS observations.',Astron.

98

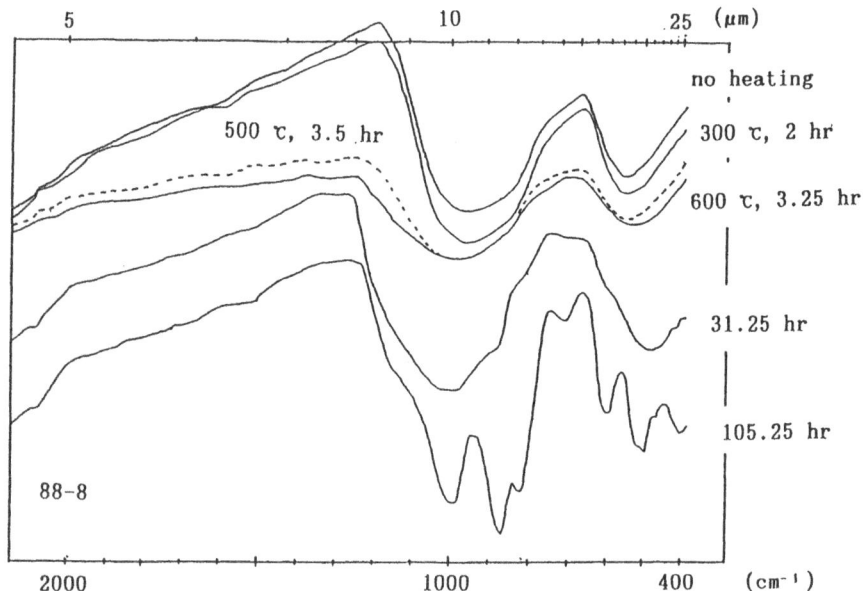

Fig.4. Infrared spectra of an annealed amorphous olivines.

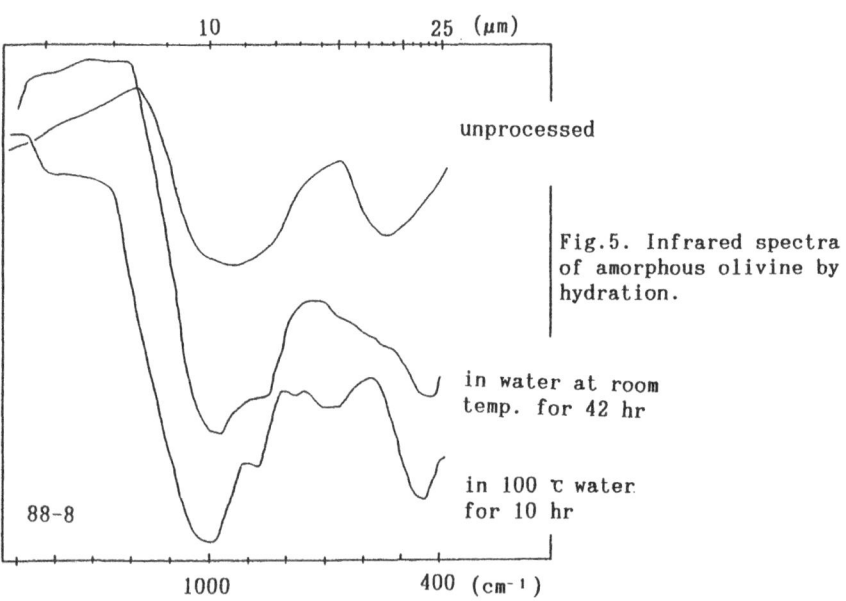

Fig.5. Infrared spectra
of amorphous olivine by
hydration.

Astrophys.203,361-366.
C.Koike and H.Hasegawa,(1987)'Mid-infrared extinction coefficients of
  amorphous silicates', Astrophys. Space Sience 134,361-379.

OPTICAL CONSTANTS OF KEROGEN FROM 0.15 TO 40 μm: COMPARISON WITH METEORITIC ORGANICS

B.N. Khare, W.R. Thompson
and C. Sagan
Cornell University
Ithaca, NY 14853 USA

E.T. Arakawa and C. Meisse
Oak Ridge National Lab.
Oak Ridge, TN 37831

I. Gilmour
The Open University
Milton Keynes MK7 6AA
U.K.

ABSTRACT. A vacuum evaporation technique has been used to produce thin, optical quality films of samples of Type II kerogen and of insoluble organic residue from the Murchison meteorite. Using these films, optical constants have been measured from 0.15 to 40 μm for kerogen, and from 2.5 to 40 μm for the Murchison residue. The infrared absorption properties of these materials show many similarities, although Murchison residue is more opaque throughout the infrared than is kerogen, and shows no distinct aliphatic absorptions.

1. INTRODUCTION

Interplanetary dust is thought to derive from cometary, asteroidal and meteoritic sources along with a possible contribution from planets and presolar dust. Both meteoritic organic residues and kerogens (dark, complex organic materials produced on the Earth primarily by geological processing of biologically derived material) have been used as laboratory models of dark extraterrestrial organic materials. Kerogen-like solids have been proposed as constituents of the very dark reddish surfaces of some asteroids (Gradie and Veverka, 1980), and kerogens and meteoritic organics are also spectrally similar to the Iapetus dark material (Bell et al., 1985). The optical constants of these materials are useful for modeling the scattering properties of interplanetary and cometary dust, and also of asteroid and satellite surfaces.

2. EXPERIMENTAL METHODS

We have measured the optical constants of both Type II kerogen and of a macro-molecular organic residue from the Murchison carbonaceous chondrite via transmission and reflection measurements on thin films. These films, of thickness 0.2-1.3 μm, are produced by vacuum deposition of powdered samples heated to 550-750°C onto sapphire, CaF_2, and CsI substrates. IR spectra of the thin films show that the spectral features of the powder are retained and thus no substantial change in the optical constants occurs upon vacuum deposition.

A.C. Levasseur-Regourd and H. Hasegawa (eds.), Origin and Evolution of Interplanetary Dust, 99–101.
© 1991 Kluwer Academic Publishers.

3. RESULTS

The real part of the refractive index, n, of Type II kerogen is determined by variable incidence-angle reflectance to be 1.60 ± 0.05 from 0.4 to 2.0 μm wavelength. Work extending the measurement of n to longer wavelengths is in progress. The imaginary part of the refractive index, k, of kerogen shows substantial structure from 0.15 to 40 μm (Fig. 1). The values are accurate to $\pm$ 20% in the UV and IR regions and to $\pm$ 30% in the visible. Our measurements of k for Murchison organic residue in the IR are shown and compared with the kerogen results in Figure 2. The Murchison sample, like the kerogen, shows considerable spectral structure, but as expected, does not show any feature associated with aliphatic functional groups. Comparison of the kerogen and Murchison data reveals that between 0.15 and 40 μm, Murchison has a similar structure but no bands as sharp as in kerogen, and that the k values for Murchison are consistently larger than those of kerogen. Further measurements of n and k for Murchison organic residues are in progress.

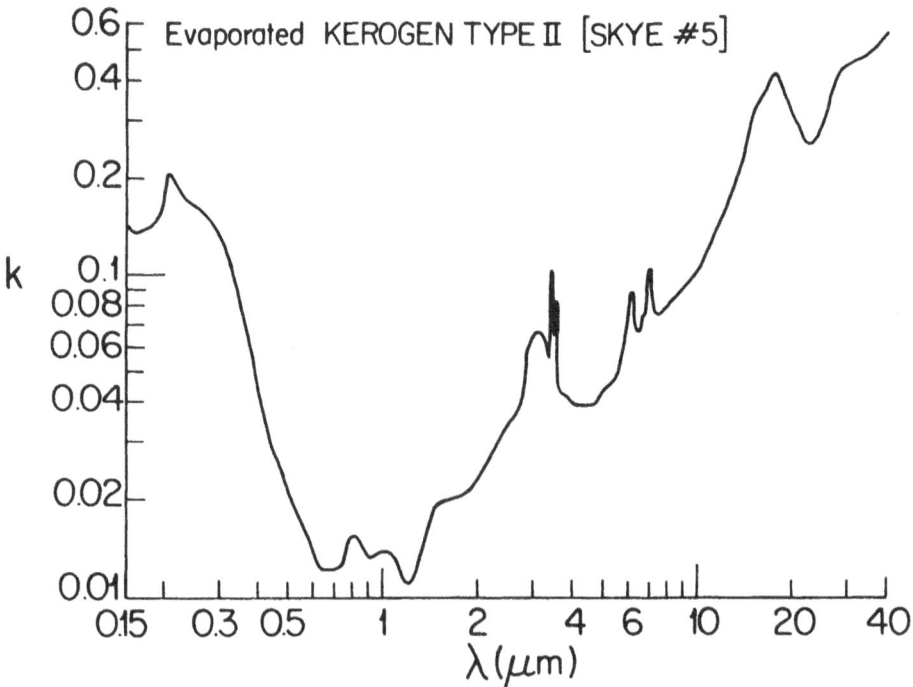

Figure 1. Imaginary part of the refractive index, k, for vacuum-evaporated Type II kerogen. Spectral absorption features (corresponding to maxima in k) include the aliphatic CH_x bands near 3.4 μm.

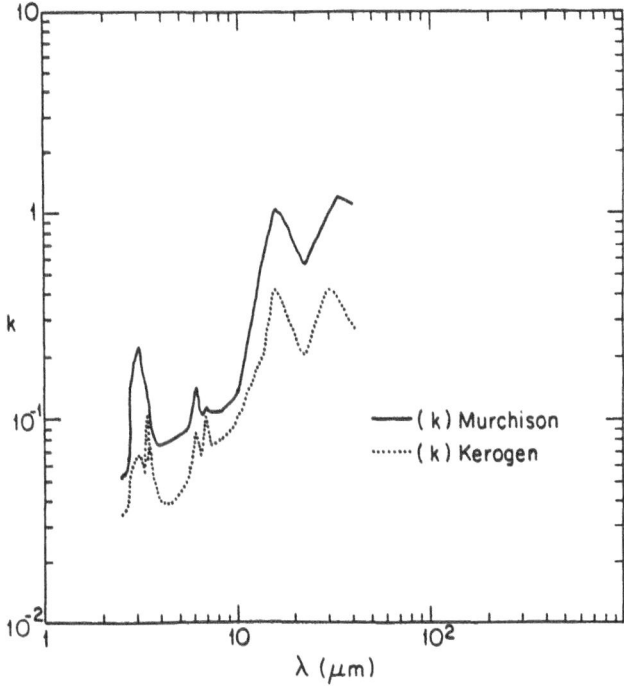

Figure 2. The infrared spectral dependence of k for Type II kerogen is compared with that of Murchison organic residue. The spectra show some similarities, but the Murchison residue is more opaque overall, while the kerogen shows aliphatic absorptions at 3.4 and 7 μm which are weak or absent in the Murchison spectrum.

REFERENCES

Bell, J. F., Cruikshank, D. P., and Gaffey, M. J. (1985) 'The composition and origin of the Iapetus dark material', Icarus 61, 192.
Gradie, J. and Veverka, J. (1980) 'The Composition of the Trojan asteroids', Nature 283, 840.

OPTICAL CONSTANTS OF BASALTIC GLASS FROM 0.0173 TO 50 μm

E. T. Arakawa   J. M. Zhang   B. N. Khare, W. R. Thompson,
D. W. Young   P. C. Eklund   and C. Sagan
Oak Ridge National Lab.   Univ. of Kentucky   Cornell University
Oak Ridge, TN 37831 USA   Lexington, KY 40506 USA   Ithaca, NY 14853 USA

ABSTRACT. Pollack et al. [Icarus 19, 372 (1973)] have reported the optical constants for obsidian, basalt, andesite and basaltic glass over the wavelength range 0.2 to 50 μm, and Lamy [Icarus 34, 68 (1978)] reported the optical constants from 0.10 to 0.44 μm for obsidian, basalt, and basaltic glass. We have revised the former measurements for basaltic glass and extended them into the extreme UV to 0.0173 μm.

1. INTRODUCTION

The major constituents of interplanetary and circumstellar dusts are silicates and carbonaceous materials. Analysis of stratospheric interplanetary dust particles shows chondritic elemental abundances and confirms the silicate mineral identification. The spectra of the majority of collected interplanetary dust particles are dominated by olivines, pyroxenes and layer-lattice silicates (Sandford and Walker, 1985). Comets also contain mixtures of the different crystalline silicates which may vary from comet to comet and even within a comet (Sandford, 1988). Thus the measurement of the optical constants of naturally occurring rocks and glasses will be valuable for studies of the absorption, emission, and scattering properties of rock surfaces, atmospheric dust and interplanetary and interstellar dust grains.

2. METHOD

The real (n) and imaginary (k) parts of the complex refractive index of basaltic glass have been determined from a combination of measurements.

A Seya-Namoika monochromator is employed to measure near normal reflectivity R_n from 0.1033 to 0.6 μm. An infrared spectrometer is used to measure R_n from 2 to 20 μm. Pollack et al. (1973) measured the reflectance of basaltic glass from 0.2 to 50 μm. To 20 μm, the Pollack et al. data agree with our results. For the short wavelength region in the vacuum and extreme UV from 0.0173 to 0.1689 μm, a McPherson Model 247 grazing incidence monochromator was used. The value R_n is low in this region and therefore, we measured the reflectivity as a function of angle of incidence to obtain n and k. The n and k values were then used to calculate R_n. From all these measurements combined with the R_n values of Pollack et al. from 20 to 50 μm, a composite R_n curve was obtained from 0.0173 to 50 μm. This composite curve was then used to obtain n and k over the entire wavelength range by Kramers-Kronig analysis.

102

A.C. Levasseur-Regourd and H. Hasegawa (eds.), Origin and Evolution of Interplanetary Dust, 102–104.
© 1991 Kluwer Academic Publishers.

Ellipsometry is less affected by non-specularity of the reflecting surface. Using ellipsometry to determine n independently from 0.4 to 2.0 μm allowed us to determine a correction factor for R_n, which was slightly low because of microscopic roughness. Thus all R_n values were multiplied by 1.06 allowing a more accurate computation of the optical constants.

3. RESULTS

In Figure 1 the n and k values obtained by the above analysis are compared with results of Pollack et al. (1973).

Values of n and k presented here for 0.0173 μm $\leq \lambda \leq$ 50 μm show peak values of k = 1.0 at 23 μm, 1.| at 9.9 μm, and 0.76 at 0.080 μm. Normal incidence reflectance at 0.5 μm is 0.066, comparable to values observed for the Moon.

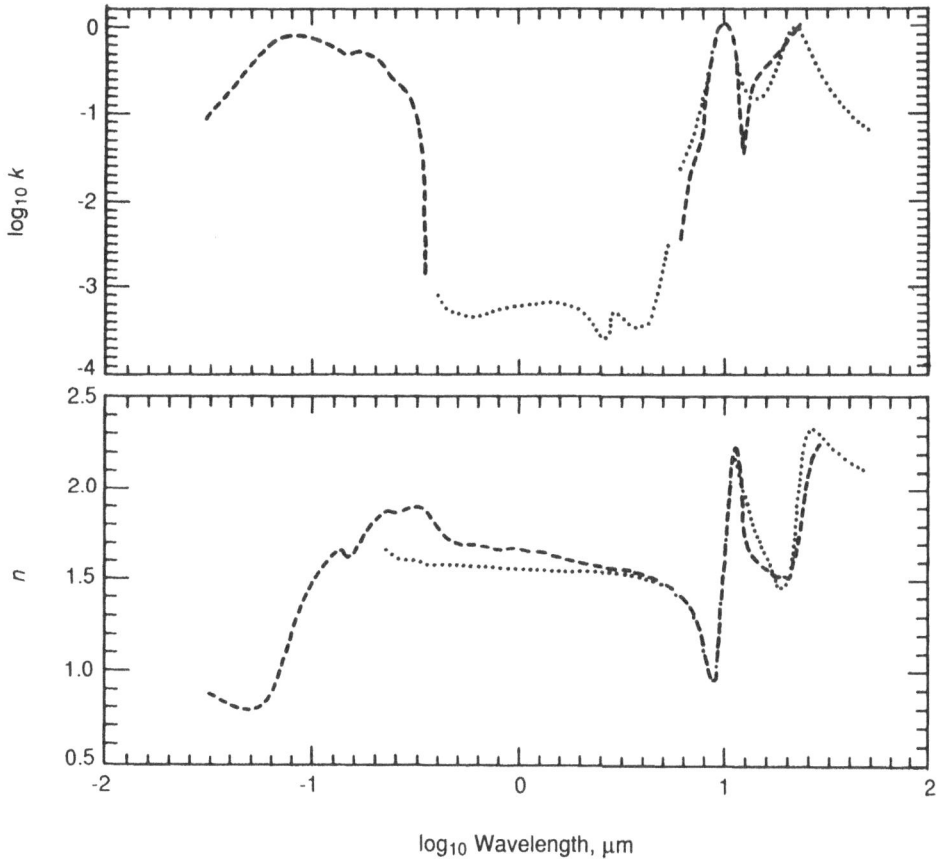

Figure 1. Optical constants of basaltic glass obtained in this work (dashed lines) and those determined by Pollack et al. (1973) (dotted lines) are shown. The present measurements extend the constants to the EUV and thus allow a more reliable determination of n and k via the Kramers-Kronig method.

REFERENCES

Lamy, P. L. (1978) 'Optical Properties of silicates in the far ultraviolet', Icarus 34, 68-75.
Pollack, J. B., Toon, O. B., and Khare, B. N. (1973) 'Optical constants of some terrestrial rocks and glasses', Icarus 19, 372-389.
Sandford, S. A., and Walker, R. M. (1985) 'Laboratory infrared transmission spectra of individual interplanetary dust particles from 2.5 to 25 μm', Astrophys. J. 291, 838-851.
Sandford, S. A. (1988) 'The spectral properties of collected interplanetary dust particles', in M. E. Bailey and D. A. Williams (eds.), Dust in the Universe, Cambridge Univ. Press, Cambridge, pp. 193-197.

NOBLE METAL ENRICHMENTS IN COSMIC SPHERULES

K. NOGAMI[1], K. MISAWA[2], R. OMORI[1], M. JIANGUO[3] & K. YAMAKOSHI[2]
[1]General Education, Dokkyo Univ. School of Medicine
Mibu, Shimotsuga, Tochigi, 321-02, Japan
[2]Institute of Cosmic Ray Research, TOKYO Univ. Tanashi, Tokyo, 188 Japan
[3]Appl, Nucl. Tech. Div., Inst. High Energy Phys. Academia Sinica, Beijing, China

ABSTRACT. In this work, studies on relationships of chemical compositions between fusion crust and nucleus in iron spherules are reported. More than 10% of
the iron spherules which were picked out from deep sea sediment, have cores and
crusts. We were able to divide three of them into cores and crusts. Each cores
and crusts were analyzed individually by INAA. The core mainly consists of iron
and nickel. Other trace elements, especially noble metal Au and Ir were concentrated in the core. The mechanism of core formation in the iron spherules shows
us the origin of them.

1. Introduction

After Murray and Renard's first discovery of cosmic spherules a century ago,
many spherules found out from sea sediments were investigated in recent two decades by means of various new methods. The achievement of this area has been
drawing scientist's attention more and more to it.

Spherules from deep sea sediment were classified into three main groups by
Yamakoshi(1984): iron spherules, silicate spherules and glassy spherules. The
iron spherules are of quite importance for its Ni, Ir and high noble metal contents, which have been determined as extraterrestrial matter. Some of the iron
spherules have a metalic nucleus (Brownlee(1984), Robin(1987)), thereafter we
call it only "core", and in this paper we mainly examine those spherules. How
did the iron spherules with the cores in them come to be? Many hypotheses have
been discussed. A widely believed one is that the spherules are ablation products of larger bodies which have frictionized with the atmosphere. Another one
is that the spherules exist already in interplanetary space as cosmic dusts or
micrometeorites.

2. Experiment

In this work, we examined iron spherules from deep-sea sediments and their
3 cores and its crusts individually with the high sensitive instrumental neutron
activation analysis (INAA). All of our spherules were picked from pelagic clay
sediments. Most of them are of shiny metallic gray or shiny black colors, and

A.C. Levasseur-Regourd and H. Hasegawa (eds.), Origin and Evolution of Interplanetary Dust, 105–108.

small vesicular cavities were found in a few of them. In the iron spherules, small core were found occasionally. Some cracks were in the crusts of 3 iron sperules so we were able to separate the cores from their crusts easily.

Samples of the iron spherules were irradiated with a artificial standard and some powder of Allende chondrite as a reference material in TRIGA II reactor under flux of $0.7-4.0 \times 10^{12}$ n/cm^2/s for 15-24 hours. After the irradiation, each sample was measured for 60000 sec with high sensitive Ge(Li) detector. The obtained result of the Allende chondrite showed a good consistency with that given by the literature materials within ±5% deviation.

3. Results

The element concentrations and sample weights of the cores, crusts of 3 iron spherules are listed in Table 1. Fig.1 shows the ratio of the elemental concentration between core and crust. The data of average five typical ordinary iron spherules and an iron meteorite are listed here together for comparison. The analytical errors for the elements, Fe, Ni, Co, Au, Ir and Os are within ±10% under the confidence level of 90%. But the errors for the other elements for which only core/crust ratio are listed in the table are more than 50%.

The obtained results in Table.1 show that the compositions of 3 cores are simply Fe-Ni and trace of some refractory siderophile element. High contents of the Ni and Ir make them sure as extraterrestrial origin which have often been used as a reliable touchstone to identify the extraterres trial origin for their very low abundance in the earth crust and higher in extraterrestrial materials.

Gold is one of the elements which easily evaporates by ablation. So, the abundance of gold in each iron spherules can be used to know their thermal history, whether they were once melted or not. In addition, the differences between the element contents in cores and in their crusts are so large that it is doubtful whether the cores could formed completely during their passage through the atmosphere without gold depletion.

4. Conclusion

Some spherules from pelagic clay sediments were studied individually with INAA. From 3 of them we can separate their core and crust. Core formation mechanism will show us the history of those spherules. The bulk compositions of them are similar to that of the iron meteorite. Ni contents in the 3 cores are more than 33.6%. This is about the same content (Ni: 13 - 46%) in the taenite in iron meteorite. Almost all measured elements are more abundant in core than in crust. Ni has the highest ratio of elemental abundance between core and crust, (Ni:core/crust)=182 the average of three of them. If the parent body were to be the materials which have the same bulk elemental compositions as spherules with the core, these cores and the crust would have been made by melting somewhere.

The high gold contents, which is easily depleted by high temperature melting, suggests that they were heated by just a little over the melting point. And the dulation of melting shold enough for Ni to move toward the center of the spherules to create cores. And when the cores are created, Ni scavenge the trace shiderophile elements, noble metals and some rare earth elements to the core.

The condensation of noble metals in the core together with volatile gold, imply some special condition for growing the core in the iron spherules. The heating would occure more slowly than the heating by the friction with earth atmosphere when they are rushed into the earth. The artificial ablation experiments will reveals the formation mechanisms of core in the iron spherules or the nuggets in the silicate spherules.

References

Brownlee D. E., Bates B. A. and Wheelock M. M., (1984) 'Extraterrestrial platinum group nuggets in deep-sea sediments', Nature, 309, 693-695

Robin E., Jehanno C. and Maurette M. (1987) 'Characteristics and origin of Greenland Fe/Ni cosmic grains', Proc. 18th LPSC, 593-598

Yamakoshi K. (1984) 'Chemical compositions of magnetic spherules from deep-sea sediments determined by instrumental neutron activation analysis', Geochemical J., 18, 147-152

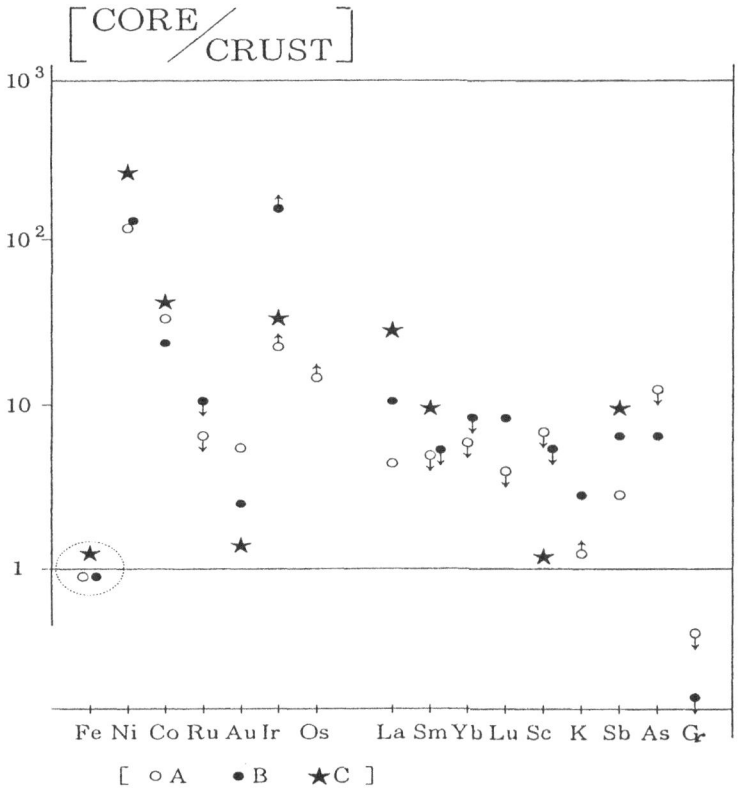

Fig.1 Ratio of the elemental concentration of core/crust.

TABLE 1. Contens of the elements in the core and crust of three iron spherules.

	A CORE	A CRUST	A CORE/CRUST	B CORE	B CRUST	B CORE/CRUST	C CORE	C CRUST	C CORE/CRUST	Iron Sphe. (1)	C.D. (2)
WEIGHT(mg)	0.0128	0.0958		0.0407	0.1269		0.047	0.177			
Fe (%)	64	78.2	0.818	57	70.3	0.81	53.3	43.3	1.22	75.92	92.3
Ni (%)	33.6	0.27	124	44.2	0.33	134	46.0	0.16	288	4.24	7.25
Co (%)	1.6	0.05	32	2.86	0.12	24	1.70	0.044	39	0.206	0.49
Au (ppm)	20.5	3.6	5.7	2.2	1	2.2	7.33	5.5	1.33	0.045	2.1
Ir (ppm)	29.2	<1.3	-	27.2	<0.17	-	42.4	1.3	33	5.2	2.1
Os (ppm)	<20.3	<2.3	-	30.2	<2.0	-	28.3	-	-	17	3.6
Ru	<10.35			<6.8							
Cr	<0.057			<0.45							
K	>1.33			2.63							
La	11.3			4.23					26.0		
Sm	<4.96			<4.67					9.59		
Yb	<8.2			<5.54							
Lu	8.97			<4.05							
Sc	<5.73			<6.71					1.12		
Sb	7.81			2.76					9.93		
As	>12.5			2.76							

(1) Average five typical ordinary iron spherules.

(2) Iron meteorite (Canyon Diabro)

STUDIES ON ISOTOPIC RATIOS OF OSMIUM AND IRIDIUM IN COSMIC SPHERULES USING INSTRUMENTAL NEUTRON ACTIVATION ANALYSIS

Kazuo Yamakoshi[1] & Ken'ichi Nogami[2]

[1] Institute for Cosmic Ray Research   Univ. Tokyo,
    Tanashi   Tokyo 188
    Japan,

[2] Phys.Lab. Dokkyo Univ.School Medicine,
    Mibu   Shimotsugagun   Tochigi 321-2
    Japan.

ABSTRACT   Studies on isotopic anomalies in cosmic meteoroids are expected to reveal the features of nuclear synthesis in various phases of star-evolution. The respective isotopes of noble metals had been produced through various reaction processes as well as in various regions of star-eruptions.   However, isotopic anomalies of Os in extremely refractory inclusions of primordial carbonaceous chondrites have not been found up to now.   In this work, isotopic ratios of the elements having high condensation temperatures, such as Os and Ir, in the cosmic spherules are examined using instrumental neutron activation analysis(INAA).

1. Introduction

The major parts of the magnetic, iron and chondritic spherules collected from deep sea sediments are confirmed to be of extraterrestrial origin (Yamakoshi et al 1981). In previous papers (Nogami et al 1978,Yamakoshi 1982), the isotopic ratios of Os and Ir in the iron spherules from deep sea sediments were measured in non-destructive form. However, no remarkable isotopic anomalies were found. During the entry into Earth's atmosphere, the cosmic meteoroids are suffered the thermal degeneration, lost some volume and changed into rounded bodies. Volatile elemental fractions were evaporated , however, refractory components,

109

A.C. Levasseur-Regourd and H. Hasegawa (eds.), Origin and Evolution of Interplanetary Dust, 109–112.
© 1991 *Kluwer Academic Publishers.*

such as Os and Ir, are enriched in iron spherules. In some iron spherules Os and
Ir are enriched so remarkably, that isotopes of Os and Ir can be well measured
using INAA.

2.Sample Description

Large sized, iron spherules were picked out from the magnetic fractions gathered
in deep sea sediments obtained off Hawaii Islands by a researching vessel
"Hakurei-Maru", which belongs to the Metal Mining Agency of Japan.
The chemical compositions of the used spherules are shown in Table 1. The sample
preparing process for INAA is described in detail elsewhere (Yamakoshi et al
1978).

Table 1. Chemical compositions of the used, iron spherules(Yamakoshi et al
1978). [±σ= error]※

SAMPLE CODE	SIZE [μm]	WEIGHT [μg]	Fe [%]	Ni [%]	Co [%]	Au [ppm]	Ir※※ [ppm]	Os※※※ [ppm]
23	530	293	70.1	10.8±0.6	0.062	0.022±0.011	5.7	23.9±0.5
24	500	303	67.1	5.6±0.2	0.26	0.083±0.018	3.0	9.4±0.2
28	500	255	82.9	1.6±0.1	0.13	< 0.003,	7.0	20.4±0.2
29	540	401	83.7	0.5±0.1	0.11	< 0.01	7.0	20.7±0.1
30	440	250	75.8	2.7±0.2	0.47	0.03±0.01	3.1	9.7±0.4

※In the cases of Fe,Co and Ir, the statistical errors could be neglected.
※※compared with Ir-192 of the reference, ※※※ with Os-191 of the reference.

3.Nuclear Data

The stable isotopes of Os are 184(0.018%), 186(1.59%), 187(1.64%), 188(13.3%),
189(16.1%), 190(26.4%) and 192(41.0%). The neutron induced radioisotopes
used for INAA are 185(93.6 days), 191(15.4 d) and 193(30.0 hours).
Ir isotopes are 191(37.4%) and 193(62.6%). Thus, the radioactive ones are 192
(74.02 days) and 194(19.15 hours). Unfortunately since the thermal neutron cross
section values for such isotopes are not so precisely determined, it is
difficult to calculate the atom numbers of the stable (target) isotopes from the
radioactivities of the induced radionuclides.
In Os isotopes, Os-184 is produced through p-process only and Os-192 is induced
through rapid-process only at super-novae explosions. The other nuclides are

produced through both rapid- and slow-processes.   The ratio of the fractions
of Os-190 produced through r- and s-processes is given as 7(Seeger et al 1965).
In the case of Os, Os-184, -190 and 192 can be determined using INAA, thus we
can take a three-isotope plot for Os. Ir-190 and -192 are also produced through
both r- and s-processes. The ratio of the contributions of r- and s-processes
are 14 and 22.8, respectively. Ir has only two isotopes, so we can not make the
three-isotope plot,thus we can not cancell out the chemical fractionation
effects due to the thermal degenerations.

4.Experimental Results and Discussions
In this work short-lived nuclides were determined not so precisely, because we
had received the irradiated samples delayed for two days from the reactor.
The following corrections were performed here; saturation factors during the
neutron induced reactions, decays, counting efficiencies of detectors,
branching ratios of the measured gamma rays and also side reactions, such as
Ir-193(n,p) and Pt-196(n,alpha).
In this preliminary work,the isotopes of Os in a few iron spherules are measured
by INAA, which are shown in Fig.1.

Fig.1. The three
isotope plot for Os
in iron spherules
(Canyon Diablo is
used for the
reference.)

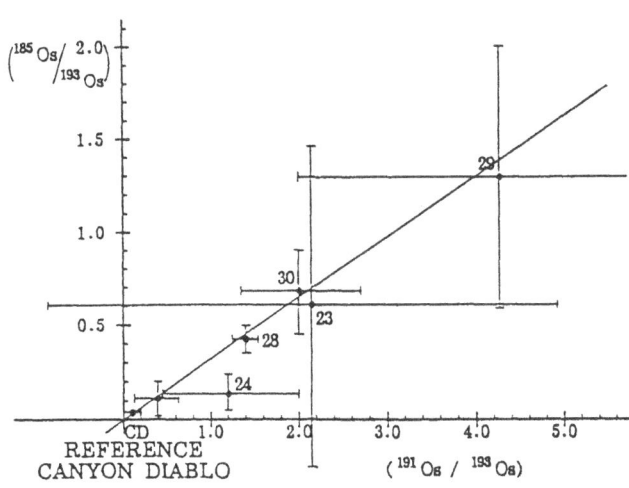

The life-time of Ir-194 is so short, that (Ir-192/Ir-194) ratios in iron
meteorites, metal phases in chondrites and iron spherules were not so well
determined.

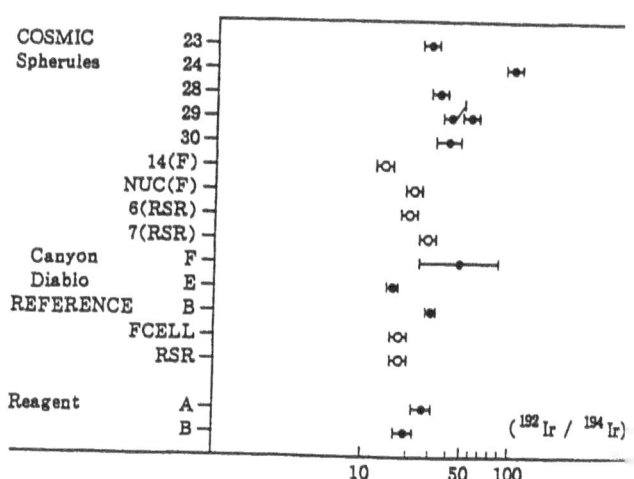

Fig.2. The ratios of
(Ir-191/Ir-193) in various
cosmic meteoroids obtained
with INAA. Iron spherules
are considered as melted
droplets of iron meteoroids
by friction heating in the
upper atmosphere.

If so, during the melting, evaporating, and solidifying of the meteoroids,
the original compositions of chemical and isotopic components will be changed ,
so to say, systematically.

In Fig.1 and Fig.2, preliminary results are shown. No remarkable anomalies
were found. However, if enormous anomalies are found in both elements, further
investigation will be fruitful! In future an ICP mass spectrometry is the most
effective tool for extremely refractory metal studies, however, INAA mass
spectrometry is also useful in frontier researches.

References

Nogami K, Shimamura T and Yamakoshi K 1973 Bull.General Educ.School Medicine
    34 1
Seeger P A, Fowler W A and Clayton D D 1965 Astrophys.J.Suppl.Series 11 121
Yamakoshi K, Nogami K and Shimamura T 1981 J.Geophys.Res.86 B4 3129
Yamakoshi K 1982 Proc.15th ISAS Lunar and Planet.Symposium 326

STRUCTURES OF AMORPHOUS SILICATE DUSTS SIMULATED BY MOLECULAR DYNAMICS METHOD

A. TSUCHIYAMA and K. KAWAMURA*
Department of Earth and Planetary Sciences
College of General Education
Osaka University
Toyonaka 560
Japan
* Dept. Chem., Hokkaido Univ., Sapporo 060, Japan

ABSTRACT. Atomic structures of amorphous silicate dusts with the $MgSiO_3$ composition were simulated by molecular dynamics method as a function of the dust density based on the assumption that the density corresponds to cooling rate of dust formation. The SiO_4 tetrahedra are more polymerized with decreasing density, suggesting phase separation between SiO_2-rich and MgO-rich components in less dense dusts formed by rapid condensation. A mode of atomic vibration probably due to the Si-O bending is different in the amorphous silicates with different densities. This may cause changes of the 20 μm bands of IR spectra of silicate dusts with different cooling rates.

1. INTRODUCTION

It has been proposed that amorphous silicates are present as interstellar and circumstellar dusts based on IR observations of broad and structureless bands at 10 and 20 μm wavelengths (*e.g.*, [1]). There were extensive laboratory works on amorphous silicates with different compositions to reproduce these IR features (*e.g.*, [2]). However, different IR features are expected for amorphous silicates with the same composition if their structures, which are related to conditions of dust formation, are different.

In the present study, structures of amorphous silicates were simulated by molecular dynamics (MD) method (*e.g.*, [3]) to discuss relation between the structures and IR spectra of the silicate dusts. Special attention was paid on the density of the silicate dust because the density is related to the cooling rate of dust condensation, *i.e.*, the density is expected to be smaller with increasing cooling rate.

2. MOLECULAR DYNAMICS (MD) CALCULATION

The composition of the amorphous silicates was chosen as the enstatite composition ($MgSiO_3$) based on the cosmic abundance of major

113

A.C. Levasseur-Regourd and H. Hasegawa (eds.), Origin and Evolution of Interplanetary Dust, 113–116.
© 1991 *Kluwer Academic Publishers.*

refractory elements (Mg/Si=1.06) and the inferred SiO_2 content for astronomical silicates (about 50 wt.%; [4]). Calculations were done in the system with a basic cell containing 800 particles with the ionic pair potentials of Kawamura [5]. In the calculations, initial random structures were equilibrated at 4000 K and 3000 K for 2000-3000 steps

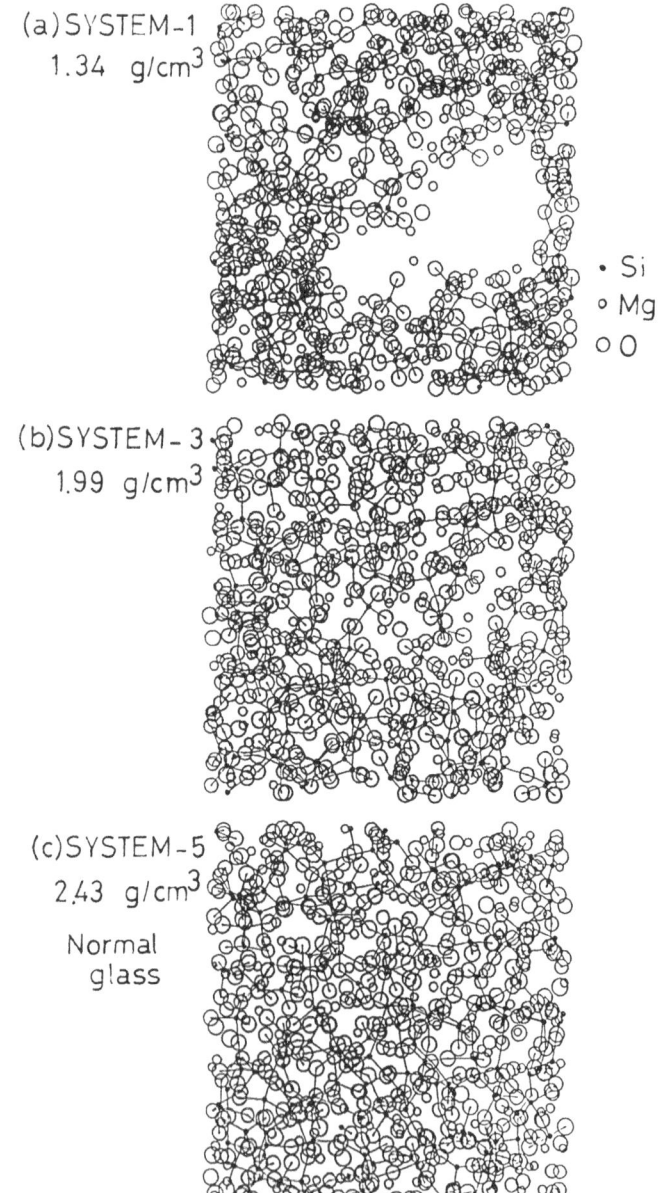

(a)SYSTEM-1
1.34 g/cm³

• Si
∘ Mg
∘ O

(b)SYSTEM-3
1.99 g/cm³

(c)SYSTEM-5
2.43 g/cm³
Normal glass

Figure 1. Instantaneous structures of the amorphous $MgSiO_3$ at 3000 K.

(2×10^{-15} sec/step), and structural properties were obtained at 300 K and 1 bar. Five different systems with different densities (initial density = 1.0, 1.5, 1.8, 2.1 and 2.5 g/cm^3) were examined. The densities of the systems were changed during the calculation by relaxation at the constant pressure (final density = 1.57, 2.02, 2.04, 2.25 and 2.52 g/cm^3, respectively, at 300 K and 1 bar). The MD calculations were also done for crystals of enstatite and forsterite (Mg_2SiO_4) as references.

3. RESULTS

The structures are almost homogeneous except for the least dense system, where vacant parts are clearly present (Fig.1). The amorphous silicates are composed of SiO_4 tetrahedra and Mg ions. The former are polymerized by sharing their corners. Populations of SiO_4-dimer and non-bridging oxygens decreases while those of SiO_4-chains or rings, SiO_4-sheets and bridging oxygens increase with decreasing density. These results indicate that the SiO_4 tetrahedra are more polymerized in the

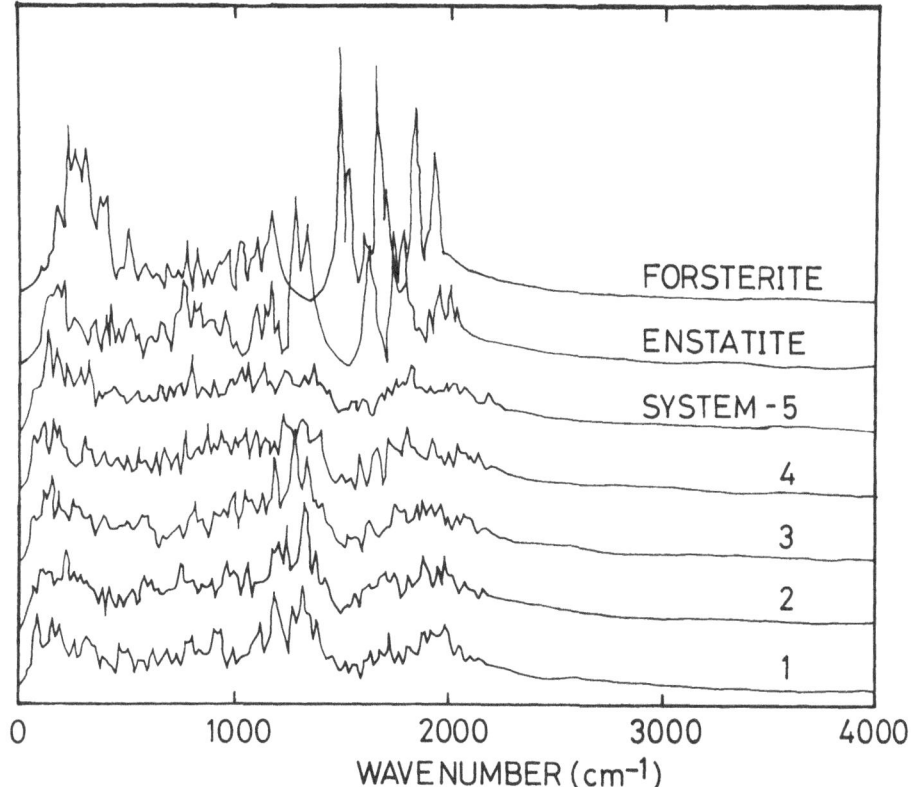

Figure 2. Total vibration spectra of Si atoms in forsterite, ensatite and amorphous $MgSiO_3$ with different densities (1.57, 2.02, 2.04, 2.25 and 2.52 g/cm^3 for systems 1, 2, 3, 4 and 5, respectively).

less dense silicates than the normal glass (2.5 g/cm^3). The polymerization suggests separation between SiO_2-rich (and thus MgO-poor) and SiO_2-poor (MgO-rich) portions in the amorphous silicates.

Total vibration spectra of the Mg, Si and O atoms were obtained to discuss IR spectra (Fig.2). Peaks for O and Si spectra at 1600-2000 cm^{-1} are considered to correspond to the Si-O stretching modes because the correlation between the O and Si spectra is good. In fact, shift of the locations of the 1600-2000 cm^{-1} peaks of the enstatite and forsterite crystals is consistent with that of the IR observations. The deviation of the wave numbers from the real Si-O stretching modes (800-1200 cm^{-1}) is due to the steep potentials used in the present calculation.

Peaks at 800-1400 cm^{-1} are considered as the Si-O bending modes. A peak at about 1300 cm^{-1} is distinctive only in the systems with the density of about 2.0 g/cm^3 or less and enstatite crystal (Fig.2).

4. DISCUSSION

The present MD simulation shows that the structures of the amorphous MgSiO$_3$ and their time variations (vibration spectra) are different with different densities. If amorphous silicate dusts with low densities were formed by rapid condensation, their structures and thus their IR spectra must be different from those of normal glass. If the 800-1400 cm^{-1} peaks (Fig.2) are due to the Si-O stretching and every vibrational peaks are active for the IR adsorption, the peak location of the 20 μm band shifts towards higher wave number with decreasing density or increasing cooling rate. This is consistent with implication from the experiments of Koike and Tsuchiyama (1991).

References

[1] Tielens, A. G. G. M. and Allamandola, L. J. (1987) 'Evolution of interstellar dust', in G. E. Morfill and M. Scholer (eds.), *Physical Processes in Interstellar Clouds*, D. Reidel Pub. Co., pp.333-376.

[2] Dorschner, J., Friedemann, C., Gurtler, J. and Henning, T. (1988) 'Optical properties of glassy bronzite and the interstellar silicate bands', *Astron. Astrophys.* 198, 223-232.

[3] Matsui, Y. and kawamura, K. (1980) 'Instantaneous structure of an MgSiO$_3$ melt simulated by molecular dynamics' *Nature* 285, 648-649.

[4] Koike, C. and Hasegawa, H. (1987) 'Mid-infrared extinction coefficients of amorphous silicates', *Astrophys. Space Sci.* 134, 361-379.

[5] Kawamura, K. (1991) 'Interatomic potential models for molecular dynamics simulations of multi-component oxides', in F. Yonezawa (ed.), *Molecular Dynamics Simulations*, Springer Series in Solid State Sciences (in press).

[6] Koike, C. and Tsuchiyama, A. (1991) 'The infrared spectra of synthesized amorphous silicates of olivine and pyroxene', this volume.

ASTROPHYSICAL INTERESTING COMPOUND GRAINS PRODUCED BY A GAS EVAPORATION METHOD

C.KAITO and Y.SAITO
Department of Electronics and Information Science
Kyoto Institute of Technology
Matsugasaki Sakyo-ku Kyoto 606 Japan

ABSTRACT. Production methods of Fe_3O_4 grain and MgS grain have been introduced. Fe_3O_4 grain was produced in an Ar gas pressure range of 25 to 100 Torr by evaporating FeO powder. The growth of Fe_3O_4 have been discussed as the result of oxidation of Fe grains. MgS grain produced by the reaction of Mg and S vapors grew in the coagulation of tiny cubic sulfide.

1. INTRODUCTION

The most advanced method for producing grains of metal or oxide is the so-called "gas evaporation method", in which a material is heated in inert gas atmosphere(Kimoto et al.(1963)). The heated vapor is subsequently cooled and condensed in the gas atmosphere, resulting in a smoke which looks like that of a candle. The direct evaporation of the oxide or sulfide in the gas was not always given the same components of the evaporant, i.e. decomposition took place (Kaito(1983)). We proposed various methods for producing compounds(Kaito (1981), Kaito et al.(1989)), by using the convection flow of inert gas. New attempts for production of Fe_3O_4 grain and the production of MgS grain are described in this paper.

2. EXPERIMENTAL PROCEDURE

The sample preparation chamber is a glass cylinder of 17 cm in inner diameter and 33 cm in height. A tantalum v-boat (length 50 mm, width 2mm and depth 1 mm) charged with powder of FeO (99.9%) was placed in the chamber. Ar gas at 10-100 Torr was introduced into the chamber and the boat heated up at about 1800°C. Grains in the produced smoke were collected on thin carbon film supported by electron microscopic grids at various position in smoke and observed with a Hitachi H-800 electron microscope. The three-heater method which has been shown in a previous paper (Kaito et al (1990)) was applied for the production of MgS grains.

A.C. Levasseur-Regourd and H. Hasegawa (eds.), Origin and Evolution of Interplanetary Dust, 117–120.
© 1991 Kluwer Academic Publishers.

118

3. RESULTS AND DISCUSSION

Typical smoke which formed by evaporating FeO in Ar gas at 100 Torr is shown in Fig.1(a). The evaporation source was almost perpendicular to the photographic plane. Figs. 1(b) and 1(c) show electron microscopic image and electron diffraction(ED) pattern from the grain collected in smoke shown in Fig.1(a). ED pattern shows the formation of magnetite. External shape of the well-grown grains was cubic octahedron.

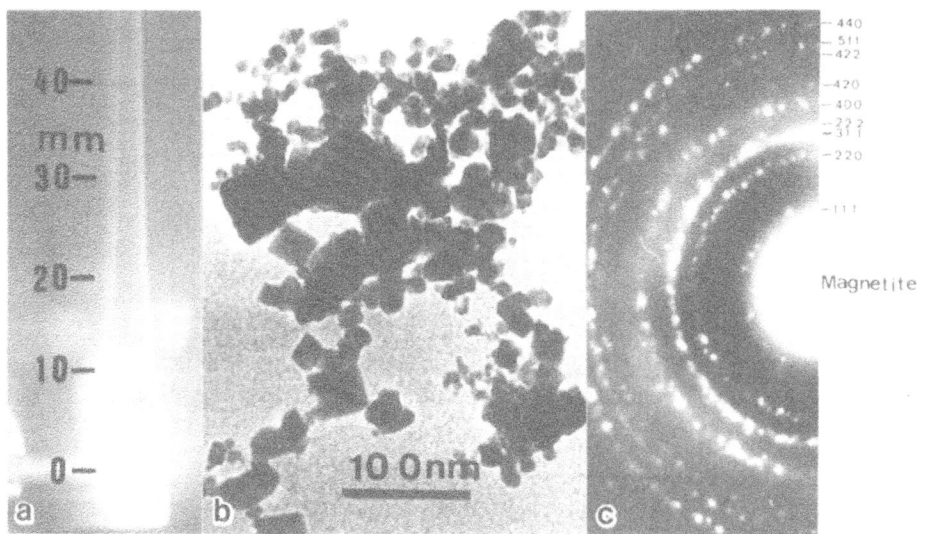

Fig.1. (a) Smoke of Fe_3O_4 grains formed by evaporating FeO powder in Ar gas at 100 Torr, (b) and (c) show electron microscopic image and electron diffraction pattern.

TABLE 1. Results of analyses of iron oxide grains

Ar Gas Pressure (Torr)	10	15	17-20	25-100
Produced Grain	α -Fe	α -Fe (FeO)	FeO	Fe_3O_4

() means the oxide produced on the surface of α-Fe

A summary of the produced grains due to gas pressure is shown in table 1. The magnetite grains were predominantly produced above 25 Torr. At 10 Torr gas pressure, iron grains were produced. At 15 Torr gas pressure, FeO oxide (Wustite) was produced on the surface of iron. The grain of wustite predominantly appeared in gas pressures of 17-20 Torr. Produced grains were changed by the gas pressure of inert gas.

These results show that the evaporated FeO powder was decomposed and oxidation took place in the atmosphere. The temperature of the grains became a few hundred degrees at a point about 10 mm above the evaporation source . The oxidation of iron grain took place near the heat source. Therefore the formation of various oxides by the gas pressure is due to variation of the oxygen gas density per unit volume in smoke. The shape of the smoke changes drastically below the gas pressure of 25 Torr. The width of the smoke at 100 Torr at 10 mm from the evaporation source is about 7 mm as seen in Fig1(a), but the values become about 1.7 and 3 times greater at gas pressure of 20 Torr and 10 Torr, therefore the density of the decomposed oxygen vapors per unit volume becomes smaller. Sine the melting point of iron is 1536°C, the decomposed iron gas was condensed as the iron grain near the evaporation source. The decomposed oxygen can be in the state of gas in the flow of Ar gas. Therefore the oxidation of iron took place and oxide grains can be produced in a high gas pressure.

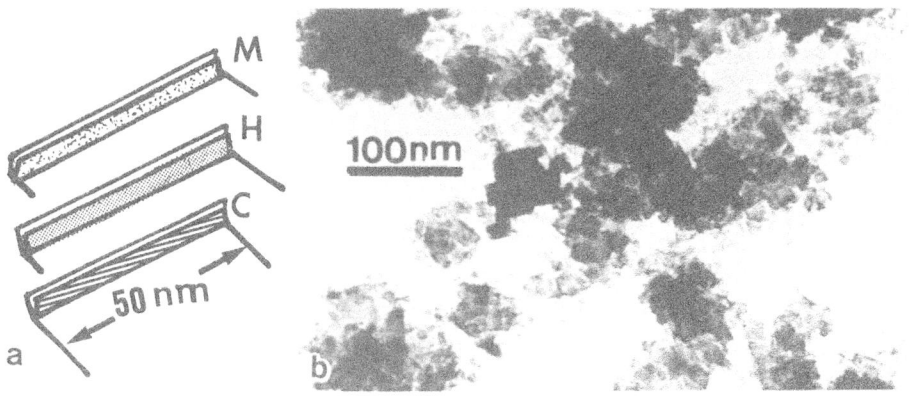

Fig.2. (a) Schematic representation of the production method for MgS.(b) Electron microscopic image.

Fig.2(b) shows MgS grains produced by using the three-heater method which indicated schematically in Fig.2(a). The temperature at locations 5mm above and 5 mm below the heater H which was heated at 1200°C in Ar gas pressure at 100 Torr became 500°C and 300°C (Kaito and Saito (1990)). Heaters M and C which were used as the evaporation source of Mg and S were set at the above locations. In order to control the vapor pressure of both the Mg and the S to bring them closer to the same level, the boat M was heated. When the temperature of the boat M was maintained at about 650°C, the vapor pressure of Mg was 2.04 Torr which was nearly same of the vapor pressure of the S at the boat C (350°C). The evaporated sulfur vapor rose due to the convective flow from the heater H and reacted with the Mg vapor around heater M.

The shape of the smoke was similar to that shown in Fig.1(a). Since MgS has the crystal structure of NaCl type, the external shape of the sulfide became a cubic form (Kaito et al.(1988)). The melting point of MgS was higher, therefore the surface melting coalescence predominantly took place(Kaito, (1985)). Then the small crystallites were coalesced as seen in Fig.2.

4. References

Kaito,C.,(1981) 'Formation of double oxides by coalescence of smoke particles of different oxides', J.Cryst. Growth, 55, 273-280.

Kaito,C.,(1983) 'High resolution electron microscopic study of oxide particles produced by the coalescence of smoke particles of iron and iron oxide', Jpn.J.Appl.Phys., 22, L432-L434.

Kaito,C.,(1985) 'Coalescence growth mechanism of smoke particles', Jpn. J. Appl. Phys., 24, 261-264.

Kaito,C.,Saito,Y.and Fujita.K.,(1989) 'Studies on the structure and morphology of ultrafine particles of metallic sulfide',J. Cryst. Growth,94, 967-977.

Kaito,C.,Saito,Y. and Fujita,F.,(1988) 'Growth of PbSe and PbTe ultra-fine particles by gas evaporation method', Jpn.J.Appl. Phys., 27, 1997-1998.

Kaito C.and Saito, Y., (1990) 'Growth of ultrafine particles of II-VI compounds by a new gas evaporation method, J.Cryst. Growth, 99, 743-746.

Kimoto,K., Kamiya,Y., Nonoyama,M. and Uyeda,R.,(1963) 'An electron microscope study of fine metal particles prepared by evaporation in argon gas at low pressure' Jpn.J.Appl. Phys.,2 ,702-713.

MEASUREMENT OF FAR-INFRARED ABSORPTION FOR AMORPHOUS SILICATES BETWEEN 27 AND 400 μm

Chiyoe Koike[1] and Hiroshi Shibai[2]
[1] Kyoto Pharmaceutical University, Yamashina, Kyoto 607.
[2] Insitute of Space and Astronautical Science,
    Sagamihara 229, Japan.

ABSTRACT. The far-infrared extinction of various silicates was measured in the 27 - 400 μm range of wavelength. The SiO_2 content of our samples distributes between 45 and 100 weight (wt.)%. There is no distinct absorption band in the far-infrared region and the extinction decreases in proportion to $\lambda^{-1.4}$.

1. INTRODUCTION

Amorphous silicates have been considered as one of the candidates minerals for circumstellar, interstellar and cometary grains. Recently, some celestial objects have been observed in detail. For instance, the dust emission around HII region shows a $\lambda^{-1.8\pm0.2}$ dependence (Ward-Thompson and Robson,1990) and for comet P/Halley sharp features were discovered (Herter et al.,1987). However, there are very few laboratory data of far infrared spectra, except synthesized amorphous silicates and glassy bronzite, for candidates minerals. The synthesized amorphous silicates show a $\lambda^{-1.25}$ or $\lambda^{-1.5}$ dependence (Day, 1976), but glassy bronzite shows a λ^{-2} dependence (Dorschner et al.,1988). The study of far infrared spectra of amorphous silicates is required because it is not known whether there are features or not in the far infrared region and how is the spectral index n (assuming a λ^{-n} far infrared absorption law).

In this report we will show the far infrared spectra of various amorphous silicates. The purpose of our measurements of the far infrared spectra is as follows:
(1) are there features or not in the far infrared region ?
(2) the examination of the spectral index n.
(3) the effects of SiO_2 content on the extinction.

2. EXPERIMENTS

Samples are various amorphous silicates with different SiO_2 contents : two synthetic glasses (fused quartz - 100 % <SiO_2 wt.>, basalt glass - 45.2 %) and four natural glasses (obsidian - 75.5 %, tektite - 72.6 %, sanukite - 63.9 %, kilauea volcanic glass - 49.1 %). The SiO_2 content distributes from 100 to 45 wt.%.

121

A.C. Levasseur-Regourd and H. Hasegawa (eds.), Origin and Evolution of Interplanetary Dust, 121–124.
© 1991 Kluwer Academic Publishers.

Fused quartz and basalt glass are obtained from Toshiba Ceramic Co.,Ltd. and Nihon Itaglass Co.,Ltd. respectively. The localities of natural glasses are as follows; Tokachi, Hokkaido, Japan for obsidian, Thailand for tektite, Goshikidai, Kagawa-prefecture, Japan for sanukite and Mt. Kilauea, Hawaii Island for kilauea volcanic glass.

The method of the measurements is the same as that discribed by Koike and Shibai(1990). We measured transmissions of each polyethylene sheets with the BOMEM fourier infrared spectrophotometer at ISAS between 28 and 400 μm in wavelength. Mass extinction coefficients κ are deduced from transmission T;

$$\kappa = \frac{S}{M} \ln \frac{1}{T}$$

where S is the surface area and M is sample mass.

4. RESULT

The results are drawn in Figs. 1 - 6, with together previous data for mid-infrared region (Koike et al.,1987), and they are condensed into Fig.7. The extinction spectra of amorphous silicates are compared with those of chemical synthesized amorphous silicates obtained by Day (1976) drawn in Figure 7.

As for the absorption band, only sanukite shows remarkable absorption band at about 47 μm. On the other hand, obsidian and sanukite show very broad band at about 50 $\sim$ 150 μm and about 110 μm respectively. Tektite, kilauea volcanic glass and basalt glass show no feature and their extinction curves decrease slowly as $\lambda^{-1.2} \sim \lambda^{-1.4}$. As for a λ-dependence, the extinction of fused quartz shows a λ^{-2}-dependence. Five other amorphous silicates show don't have constant spectral index n, but show the extinction of a $\lambda^{-1.4}$ dependence at about 100 μm region. The synthetic amorphous silicates show a $\lambda^{-1.25}$ or $\lambda^{-1.5}$-dependence for wavelength 100$\sim$300 μm (Day,1976). As for the effect with SiO_2 content, the extinction increases as decreasing the SiO_2 content of samples. The extinctions of all samples except fused quartz decrease to nearly same values at about wavelength of 400$\sim$500 μm and are about one order higher than those of fused quartz.

REFERENCES

Day,K.L.,(1976),'Further measurements of amorphous silicates', Ap.J. 210,614.

Dorschner,J.,Friedmann,C.,Gürtler,J. and T. Henning,(1988),'Optical properties of glassy bronzite and the interstellar silicate bands', Astron.,Astrophys., 198,223.

Herter,T.,Campins,H. and Gull,G.E.,(1987),'Airborne spectrophotometry of P/Halleyfrom 16 to 30 microns.', Astron. Astrophys. 187,629.

Koike,C. and Hasegawa H.,(1987),'Mid-infrared extinction coefficients of amorphous silicates.',Astrophys. Space Science,134,361.

Koike,C. and Shibai,H.,(1990),'Optical constants of hydrous silicates from 7 to 400 μm', Mon. Not. R. astr. Soc.,246,332.

Ward-Thompson,D. and Robson,E.I.,(1990),'Dust around HII regions-II. W49A.',Mon. Not. R. astr. Soc.,244,458.

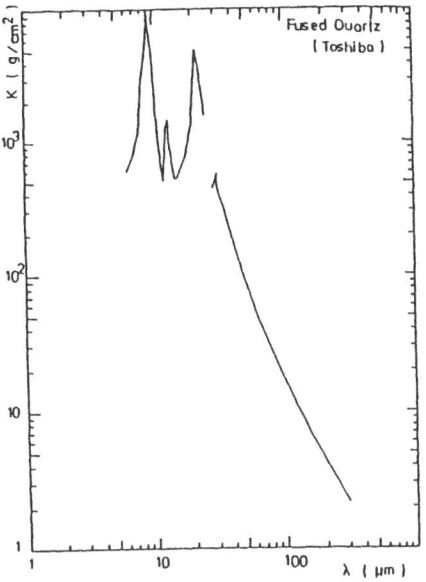

Fig.1. Mass extinction coefficient of fused quartz.

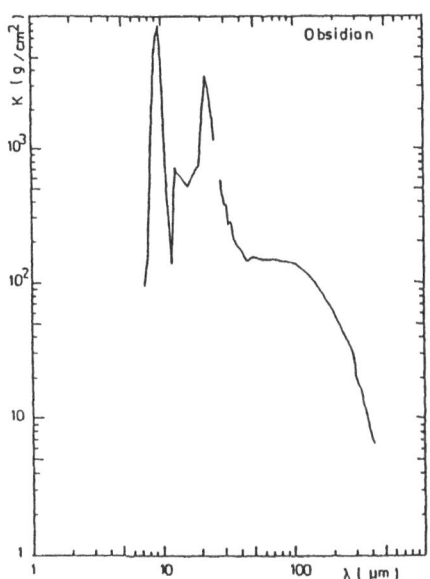

Fig.2. Mass extinction coefficient of obsidian.

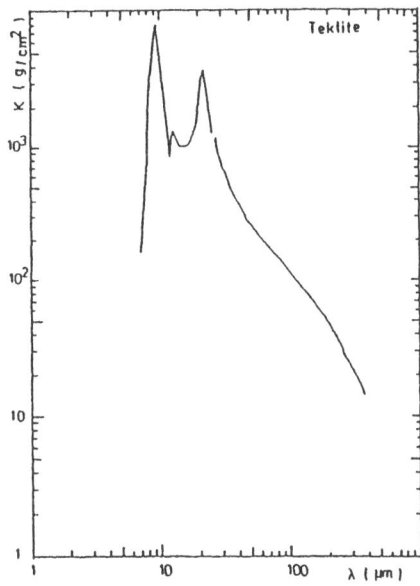

Fig.3. Mass extinction coefficient of tektite.

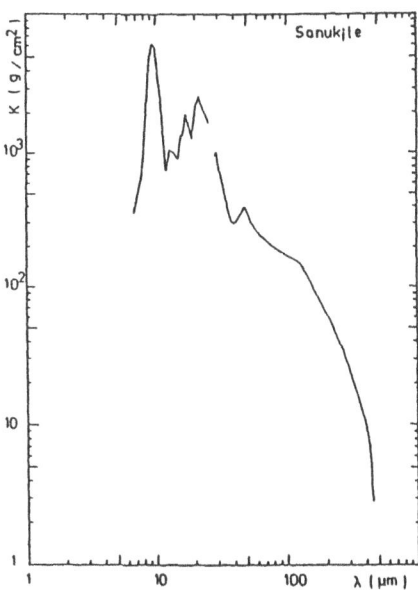

Fig.4. Mass extinction coefficient of sanukite.

124

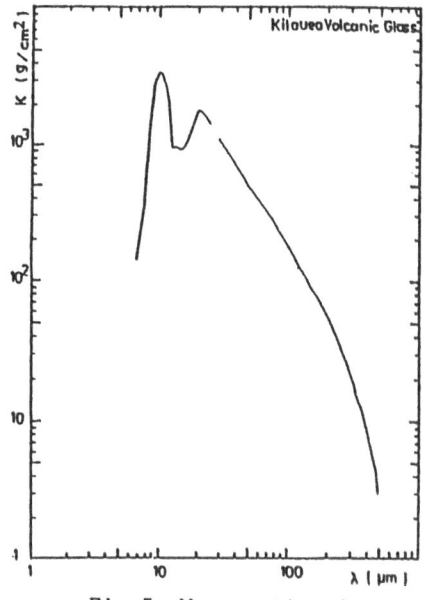

Fig.5. Mass extinction coefficient of kilauea volcanic glass.

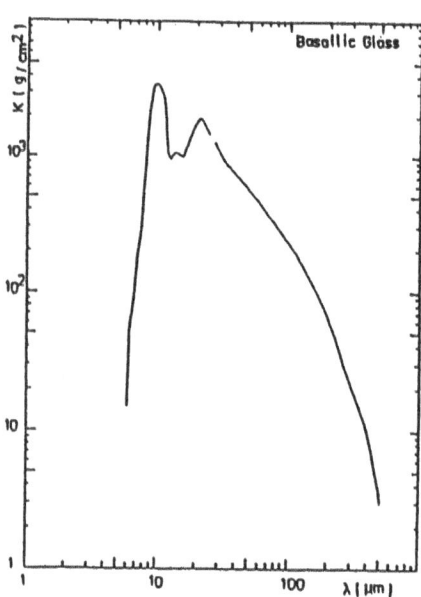

Fig.6. Mass extinction coefficient of basalt glass.

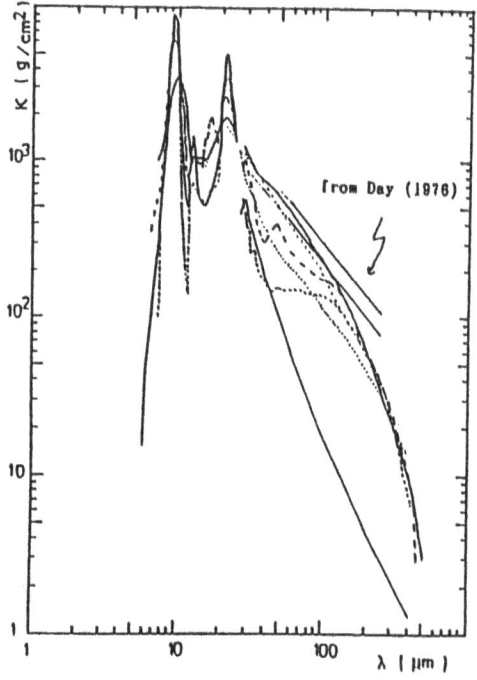

Fig.7. Mass extinction coefficients of various silicates. The thick solid line indicates the extinction spectrum of fused quartz, the broken line is that of obsidian, the dotted line is that of tektite, the dashed line is that of sanukite, the thin dotted line is that of kilauea volcanic glass, and the thin solid line is that of basalt glass.

LABORATORY SPECTRA OF AMORPHOUS AND CRYSTALLINE OLIVINE: AN APPLICATION TO COMET HALLEY IR SPECTRUM

A. BLANCO, V. OROFINO
Department of Physics, University of Lecce,
C.P. 193, 73100 Lecce, Italy

E. BUSSOLETTI, S. FONTI
Institute of Experimental Physics, IUN,
Via Acton 38, 80133 Napoli, Italy

L. COLANGELI
Engineering Faculty, University of Cassino,
Via Zamosch 43, 03043 Cassino, Italy

J. R. STEPHENS
Los Alamos National Laboratory,
Los Alamos, NM 87545, USA

ABSTRACT. Among the various silicates proposed as components of cosmic dust grains, olivine is considered one of the most likely materials. In this work we present the infrared spectra of three different types of olivine grains: crystalline, amorphous and synthetic (also amorphous). While the first and second sample derive from the same natural mineral, the third one has been prepared in the laboratory according to the relative cosmic abundances of the elements. The experimental data are used to fit the emission feature observed in the comet Halley spectrum between 8 and 13 μm. Satisfactory results are obtained by using synthetic olivine mixed with a small amount (5%) of crystalline grains.

1. Introduction

Silicates have been identified as components of the interstellar and interplanetary dust on the basis of agreement in shape and position of observed bands near 10 and 20 μm (for a recent review see Knacke, 1989).

The IR spectra of glassy silicates seem to provide a closer match to astronomical observations than crystalline minerals (Knacke, 1989; Kratschmer and Huffman, 1979). We have therefore undertaken a research program to measure the optical properties of glassy silicates having composition similar to minerals predicted in theoretical condensation sequences (Stephens et al., 1989). Among the samples examined one of the most "popular" is olivine which has also been recently proposed as an important component of cometary dust (Bregman et al., 1987; Campins and Ryan, 1989). In this work we fit the broad emission feature present in the comet Halley spectrum between 8 and 13 μm by using laboratory spectral data of natural (both glassy and crystalline) and synthetic amorphous olivine with

125

A.C. Levasseur-Regourd and H. Hasegawa (eds.), Origin and Evolution of Interplanetary Dust, 125–128.
© 1991 *Kluwer Academic Publishers.*

chemical composition reflecting the cosmic abundances of the various components.

2. Sample preparation and IR spectra

The crystalline samples for IR spectroscopy have been prepared by grinding the natural minerals and embedding the resulting submicron particles in KBr pellets. Amorphous samples have been obtained by vaporizing the same materials by means of a LAMBDA PHYSIK LPX 315i excimer laser operating at a wavelength of 308 nm. The repetition rate was 9 Hz with an energy of 400 mJ/pulse . The condensed smokes were directly collected on KRS-5 substrates inside a sample chamber which was filled with O_2 at a pressure of 10 mbar (this work) or 1 atm (Stephens, 1980). No significant changes were, however, detected by varying the gas in the chamber.

We prepared synthetic samples because available olivine crystals , although they have a metal to silicon ratio very close to that found in typical O-rich stars, exhibit a ratio $Mg/Fe = 9$. This value is about one order of magnitude larger than the cosmic one, and the composition (including the metal atoms) and the metal to silicon ratio are among the most important factors affecting the IR spectra. We produced and analyzed two synthetic olivine-type silicates with the appropriate cosmic elemental content:

$$(Fe_{0.6}Mg_{0.4})_2SiO_4 \qquad \text{and} \qquad (Fe_{0.6}Mg_{0.4}Al_{0.035}Ca_{0.03}Na_{0.025})_2SiO_4$$

In spite of the presence of Al, Ca and Na in the second sample, the IR spectra of the two materials are essentially coincident. The targets for the laser were prepared starting from the following reagent grade materials:

Element	Starting Compound	Final Compound
Mg	$MgCO_3$	MgO
Fe	Fe_2O_3	Fe_2O_3
Si	$Silicic Acid$	SiO_2
Ca	$CaCO_3$	CaO
Na	Na_2CO_3	Na_2CO_3
Al	$Gamma - Al_2O_3$	$Gamma - Al_2O_3$

Each material was heated in successive steps up to 1000 ^{0}C to produce the needed oxides. Na_2CO_3 was not heated, and Fe_2O_3 and Al_2O_3 were heated just to dry the materials. Mixtures of powders for the synthetic samples were pressed in air (10 ton) . Each pellet was then heated in a radio frequency furnace in an Ar atmosphere to temperatures up to 1400 ^{0}C until the sample just melted (approximately 1 minute). The IR spectra of the three different olivine particles are shown in fig. 1.

3. An application to comet Halley

The above mentioned laboratory data have been used to fit the $8 - 13$ μm spectrum of comet Halley.

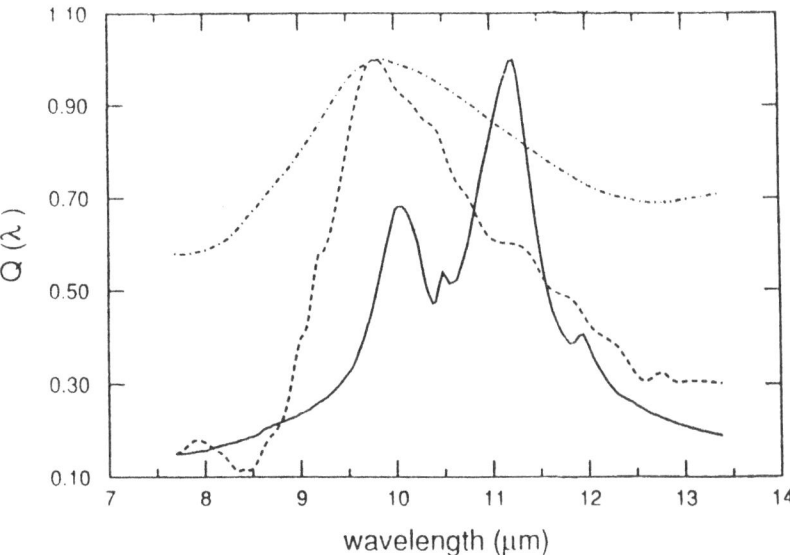

Fig. 1. Extinction efficiencies of the three different types of olivine analyzed in the present work: crystalline (solid line), amorphous (dashed line) and synthetic (dash-dotted line). The curves are normalized at their maximum.

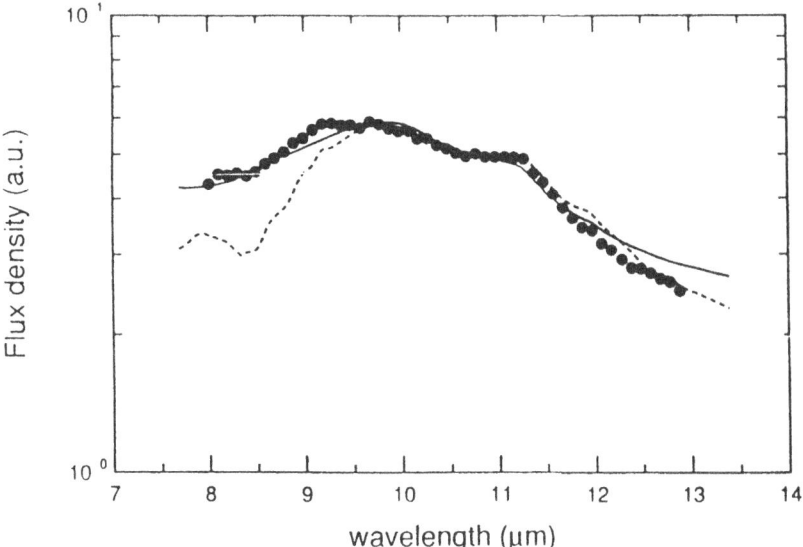

Fig. 2. Best fit of the observed (Campins and Ryan, 1989) Halley spectrum (dots) with laboratory data. The dashed curve is obtained for a 50% mixture of amorphous and crystalline olivine (best fit parameters are: $T_n = 320\ K$, $T_d = 380\ K$, $R_d/R_n = 1000$, $A_c = 1.086\tau = 0.5\ mag$). The solid curve is obtained by using synthetic olivine with only 5% of crystalline olivine. Best fit parameters are the same except for $A_c = 0.25\ mag$.

The model we used to fit the cometary flux is an adaptation of that originally applied to circumstellar envelopes (Orofino et al., 1987; Blanco et al., 1989). The free parameters are: radius and temperature of the nucleus (R_n, T_n), radius and temperature of the dust envelope (R_d, T_d) and the total optical thickness (τ) of the dust at 8 μm. The values of some of these parameters are already available in literature. From our preliminary results, shown in fig. 2, it seems that the Halley 10 μm feature can be satisfactorily fitted by a mixture of amorphous and crystalline olivine. We note however that a better fit is obtained by using the synthetic olivine with cosmic elemental composition. Moreover, in this case, a far smaller amount of crystalline component is needed. This result was also obtained by Greenberg (1990) and can be very important in the study of cometary evolution and its connections with interplanetary dust particles (Campins and Ryan, 1989; Sandford, 1989). It is worthwhile to note that laboratory IR spectra of crystalline olivine are usually taken by embedding dust samples in a matrix. This experimental procedure may change both the shape and the position of spectral features (Bohren and Huffman, 1983). At present extended laboratory work is in progress to account for this effect.

ACKNOWLEDGEMENTS. This work was partially supported by ASI, CNR-GIFCO and MPI. J. R. Stephens gratefully acknowledges the financial support of a NATO Collaborative Fellowship.

REFERENCES

Blanco, A., Borghesi, A., Bussoletti, E., Colangeli, L., Fonti, S., and Orofino, V. (1988) 'Amorphous carbon and silicon carbide grains around carbon stars', in E. Bussoletti, C. Fusco, and G. Longo (eds.), Experiments on Cosmic Dust Analogues, Kluwer Academic Publishers, Dordrecht, pp. 167-173.

Bohren, C. F., and Huffmann, D. R. (1983) Absorption and Scattering of Light by Small Particles, Wiley, New York.

Bregman, J. D., Campins, H., Witteborn, F. C., Wooden, D. H., Rank, D. M., Allamandola, L. J., Cohen, M., and Tielens A. G. G. M. (1987) 'Airborne and groundbased spectrophotometry of comet P/Halley from 5-13 μm' Astron. Astrophys. 187, 616-620.

Campins, H., and Ryan, E. V. (1989) 'The identification of crystalline olivine in cometary silicates' Astrophys. J. 341, 1059-1066.

Greenberg, J. M. (1990) private communication.

Knacke, R. F. (1989) 'Comet dust: connections with interstellar dust' in L. J. Allamandola and A. G. M. Tielens (eds.), Interstellar Dust, Kluwer Academic Publishers, Dordrecht pp. 415-428.

Kratschmer, W., and Huffman, D. R. (1979) 'Infrared extinction of heavy irradiated and amorphous olivine, with application to interstellar dust' Astrophys. Space Sci. 61, 195-203.

Orofino, V., Colangeli, L., Bussoletti, E., and Strafella, F.(1987) 'Amorphous carbon around carbon stars' Astrophys. Space Sci. 138, 127-140.

Sandford, S. A. (1989) 'Interstellar dust in collected interplanetary dust particles' in L. J. Allamandola and A. G. G. M. Tielens (eds.), Interstellar Dust, Kluwer Academic Publishers, Dordrecht pp. 403-413.

Stephens, J. R. (1980) 'Visible and ultraviolet (800-130 nm) extinction of vapor condensed silicate, carbon and silicon carbide smokes and the interstellar extinction curve ' Astrophys. J. 237, 450-461.

Stephens, J. R., Blanco, A., Borghesi, A., Fonti, S., and Bussoletti, E. (1989) 'Infrared spectra of crystalline and glassy silicates and application to interstellar dust' in A. G. G. M. Tielens and L. J. Allamandola (eds.) Interstellar Dust Contributed Papers, NASA CP-3036, pp. 375-380.

III

INTERPLANETARY DUST :
ZODIACAL LIGHT AND OPTICAL STUDIES

THE ZODIACAL CLOUD COMPLEX

A.C. LEVASSEUR-REGOURD and J.B. RENARD   R. DUMONT
Université Paris-6/Aéronomie - BP 3   Observatoire de Bordeaux
91371 Verrières le Buisson   France   33270 Floirac France

ABSTRACT. The physical properties of the interplanetary dust grains are, out of the ecliptic plane, mainly derived from observations of zodiacal light in the visual or infrared domains. The bulk optical properties (polarization, albedo) of the grains are demonstrated to depend upon their distance to the Sun (at least in a 0.1 AU to 1.7 AU range in the symmetry plane) and upon the inclination of their orbits (at least up to 22°). Classical models assuming the homogeneity of the zodiacal cloud are no longer acceptable. A hybrid model, with a mixture of two populations, is proposed. It suggests that various sources (periodic comets, asteroids, non periodic comets...) play an important role in the replenishment of the zodiacal cloud complex.

Introduction

The tridimensional distribution of interplanetary dust grains number densities and physical properties can be, as perfectly summarized a few years ago by the late R.H. Giese (1986), obtained by two methods. "The first method implies direct measurements and analysis of particles impacts on spaceborne detectors... The disadvantages are selection effects and the limited number of impact events... The second method is based on sunlight scattered by interplanetary dust grains... A huge number of particles is involved, which allows to derive the 3D-distribution without statistical problems. However... local number densities can only be derived indirectly i.e. by model computations using some simplifying assumptions, except in a few special cases...".

Interplanetary dust grains have indeed been collected in the Earth's environment (deep sea sediments, polar ices, stratosphere, Earth's orbit with LDEF or HITEN, see for instance McDonnell, 1991 or Igenbergs et al., 1991). Also their impacts on the Moon or on various spacecraft (e.g. Galileo, Grün et al., 1991) have been studied. The chemical composition of the dust grains deduced by analysis of the previous samples supports both the comets or the carbonaceous chondrites origin hypothesis.

However, due to the limited number of impact events (which, anyhow, are up to now restricted to the ecliptic plane), the properties (e.g. density, size, polarization, albedo) of the dust grains have been mainly derived from observations of light scattered or emitted by interplanetary dust. At 1 AU from the Sun, and up to solar distances smaller than 2.5 AU in the ecliptic plane, the so-called zodiacal light is indeed a significant contribution to the light of the night sky, both in the visual domain (scattered solar light) and in the infrared domain (thermal emission).

Intensity and polarization of scattered light have been measured by various groups. A smoothed table of intensities (Levasseur-Regourd and Dumont, "LRD", 1980) and an extensive compilation of intensities and polarizations (Fechtig-Leinert-Grün, "FLG", 1981) are available. Thermal emission from interplanetary dust has been measured from

A.C. Levasseur-Regourd and H. Hasegawa (eds.), Origin and Evolution of Interplanetary Dust, 131–138.

balloons, rockets and IRAS spacecraft, mainly between 60° and 120° solar elongation. Local bulk physical properties, together with some informations about the shape of the cloud, can therefore be derived by model fittings or inversion techniques.

1. Models of the zodiacal dust cloud

1.1. CLASSICAL ASSUMPTIONS

As shown by Dumont (1973), the difference between two consecutive observations made tangentially to the direction of motion (of the Earth or of a moving probe) provides the elemental contribution of the section of the line of sight where the observer is located. For any other direction of observation and section of the line of sight, various assumptions are required.

The zodiacal cloud has been shown to have a warped symmetry surface near the ecliptic plane (Misconi, 1980). From both visual and thermal observations, its parameters have been computed to be of the order of i = 1.5°, Ω = 90° near the Earth's orbit (Dumont and Levasseur-Regourd, 1978, 1987; Hauser et al., 1984; Murdock and Price, 1985). The rotational symmetry assumption with respect to this so-called symmetry plane is fairly acceptable.

It is also generally assumed that the zodiacal cloud is relatively smooth. Local heterogeneities, due to dust from cometary or asteroidal origin, have indeed been suspected in the visual data (Levasseur and Blamont, 1976), and found in the thermal data (Sykes et al., 1986 ; Dermott et al., 1986). They are, however, mainly restricted to cometary orbits and to the asteroid belt and can therefore (at least to the first order) be neglected.

A third assumption is the homogeneity assumption, i.e. the assumption that the physical properties of the grains are the same all over the solar system. With such an hypothesis, the radial dependence of the dust distribution is immediately derived, in the symmetry plane, from that of the intensity. It would increase approximately as $R^{-1.25}$ with decreasing solar distance R (Dumont and Sanchez, 1975 ; Leinert et al., 1977).

1.2. TRIDIMENSIONAL MODELS

With the smoothness assumption, analytical representations of the 3D distribution of interplanetary dust can be derived. As carefully reviewed by Giese et al. (1986, 1989) and Kneissel (1991), several models are available. With the rotational symmetry assumption, isodensity lines (leading to figurative models names) can be drawn in the helioecliptic meridian plane. It has to be emphasized that an homogeneity assumption is almost always made with a density power law R^{-n} (n in a 1 to 1.3 range) in the symmetry plane.

The most simple models developed to fit the observations have been the ellipsoid or blown up ellipsoid models (Giese and v. Dziembovski, 1969 ; Dumont, 1976a). Also widely used have been the fan or flattened fan models which show a depression towards the Sun (Leinert et al., 1976, 1981).

A puzzling result is that the visual observations seem to be better fitted by a bulge than by a depression towards the Sun, with the opposite result for the infrared observations. Various bulge models have been developed, e.g. the sombrero (Dumont, 1976b), cosine (Giese et al., 1985), revised cosine (Rittich, 1986) or modified fan (Lumme and Bowell, 1985) models. Also depression models have been developed, by Murdock and Price (1985), Good et al. (1986) and Lamy and Perrin (1986).

It is likely that such fundamental discrepancies are due to the fact that most of the models assume the physical properties of the grains to be the same everywhere in the solar system. Results obtained both from Helios 1, 2 and Pioneer 10, 11 space probes suggest that this is not the case (Leinert et al., 1981; Toller and Weinberg, 1985). Attempts to some inversion of the observations do confirm that the dust cloud is far from being homogeneous.

1.3. NODES OF LESSER UNCERTAINTY METHOD

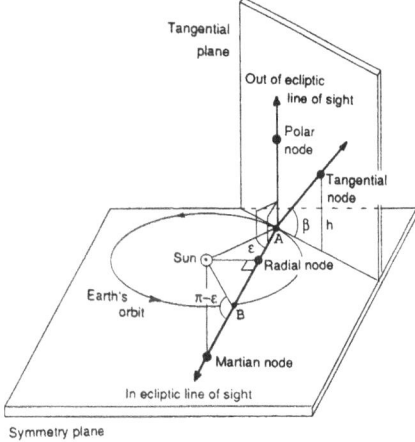

Only assuming the existence of a symmetry plane and the relative smoothness of the cloud (together with the absence of dust at infinity), some inversions are feasible both in the symmetry plane, and in the plane perpendicular to the Sun-Earth line at 1 AU from the Sun, so-called the tangential plane (Fig.1).

It is found that the mathematical smooth functions (polynomial, exponential, lorentzian, Fourier series) which could represent the local brightnesses (integrals equal to the measured brightnesses, asymptotical decrease to zero with increasing solar distance) have to focus in some nodal regions where the local contributions are derived with less uncertainty than elsewhere.

Figure 1 - Geometry of zodiacal observations

In the symmetry plane (Dumont, 1983; Dumont and Levasseur-Regourd, 1985a), two nodes can be found once integrated brightnesses have been measured from A at elongation ε and from B at elongation $180° - \varepsilon$. At the radial node, the phase angle is equal to $90°$ and the solar distance (in AU) is equal to $\sin \varepsilon$. At the martian node, the solar distance R is about 1.54 AU and the phase angle is equal to $\sin^{-1}[(\sin \varepsilon)/R]$.

In the tangential plane, one node is obtained once local brightnesses at 1 AU have been derived from inversion in the symmetry plane, and once integrated brightnesses at ecliptic latitude β are available. The elevation upon the symmetry plane of the tangential (possibly polar) node, equal to $\sin \beta / \text{tg} \alpha$, where α is the phase angle is found to remain smaller than 0.4 AU. A large fraction of the zodiacal cloud can therefore be locally probed and the evolution of the local properties of the grains can be studied all over this domain.

2. New results on interplanetary dust properties

2.1. LOCAL INTENSITY

2.1.1. *Radial dependence in the symmetry plane.* The local intensities are immediately obtained at a phase angle $\alpha = 90°$ from results derived at the radial nodes between $R = 0.09$ AU and 1 AU. There is an excellent agreement between the two sets of data, from Leinert et al. (1976) observations in the inner solar system, and from LRD (1980) smoothing for solar elongations greater than 30°. The error bars (due to the method of inversion) are significantly below 10 %. Also, the local intensities at $\alpha = 90°$ are extrapolated between 1 AU and 1.74 AU from results derived in the symmetry plane at the tangential nodes.

134

There is a fair agreement between the local values of the intensity and the sum of their two polarized components (as derived from Dumont and Sanchez, 1975 or from FLG, 1981). From the values presented in Fig. 2, the local intensity is found to vary as $R^{-1.24}$ (0.999 correlation). We therefore suggest that the local intensity (in $S_{10}(V)$/rad) can be described by a $180 \times R^{-(1.25 \pm 0.05)}$ power law. It is of interest to notice that such a result would perfectly agree with the density power law previously mentioned if the local cross section and albedo were assumed to be constant.

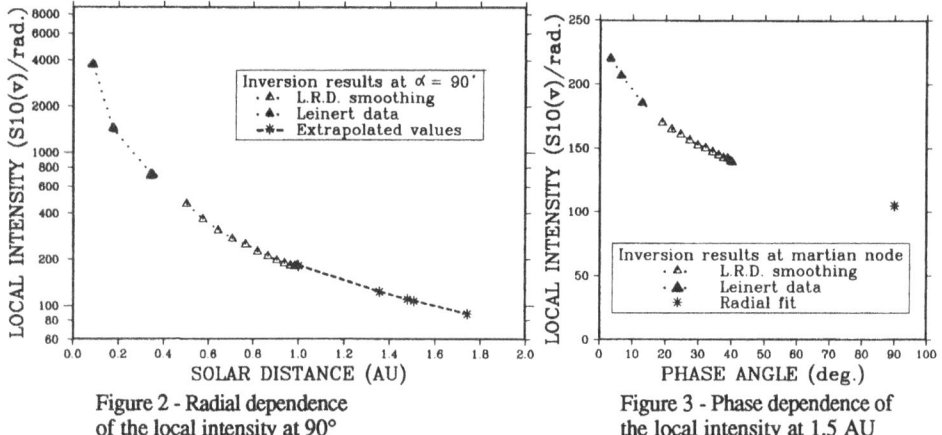

Figure 2 - Radial dependence
of the local intensity at 90°

Figure 3 - Phase dependence of
the local intensity at 1.5 AU

2.1.2. Phase dependence in the symmetry plane. The local intensities are immediately obtained at a solar distance of about 1.5 AU from results derived at martian nodes between $\alpha = 3°$ and 40° (whence the extrapolation mentioned in 2.1.1). The intensity extrapolated at $\alpha = 90°$ agrees well with the intensity extrapolated at 1.5 AU. The relative error bars remain smaller than 1 %.

The decrease in intensity with increasing phase angle (backscattering) is of the order of 0.015 ± 0.005 magnitude per degree, with a slight opposition effect below 7° (Fig. 3). Such a phase function is somewhat reminiscent of the phase function obtained for cometary dust (Meech and Jewitt, 1987), but it is likely that the backscattering function is less steep than in the cometary case.

2.1.3. Elevation dependence above the symmetry plane. From results obtained in the tangential plane, the local intensities at constant solar distance (≈ 1.5 AU) are derived up to an elevation h of about 0.3 AU above the symmetry plane. The relative error bars remain smaller than 2°.

The significant decrease of intensity with increasing elevation (Fig. 4) illustrates the flattening of the dust cloud. It should be noticed that the curvature of the graph is definitely in better agreement with a solar bulge model than with a solar depression model. The best fit is obtained for the cosine models.

2.2. POLARIZATION

2.2.1. Radial dependence in the symmetry plane. The local polarizations at constant phase angle (90°) are also obtained between $R = 0.09$ AU and 1 AU from results at the radial nodes. The discrepancy between the two sets of values is due to the fact that Dumont and demonstrated in this domain. Such a result illustrates the heterogeneity of the dust cloud in the tangential plane, and suggests that the properties of the grains depend upon the inclination of their orbits.

2.3. ALBEDO

2.3.1. *Radial dependence in the symmetry plane.* Inversion of the integrated thermal intensities, as measured by IRAS spacecraft (Hauser et al., 1984) or ZIP rocket (Murdock and Price, 1985) at two wavelengths, allows to estimate local temperatures in the symmetry plane (Levasseur-Regourd and Dumont, 1987; Dumont and Levasseur-Regourd, 1988). From results obtained at the radial and tangential nodes, values can be derived between approximately 0.5 AU and 1.7 AU. The absolute value of the local temperature at 1 AU is of about 255 K from IRAS observations, and of about 295 K from ZIP observations. However, the two sets of data provide the same $R^{-0.35}$ power law (0.990 correlation) for the decrease of temperature with increasing solar distance. The deviation of the temperature gradient from - 0.50 is likely to come from distinct heliocentric changes in the visible and infrared cross sections.

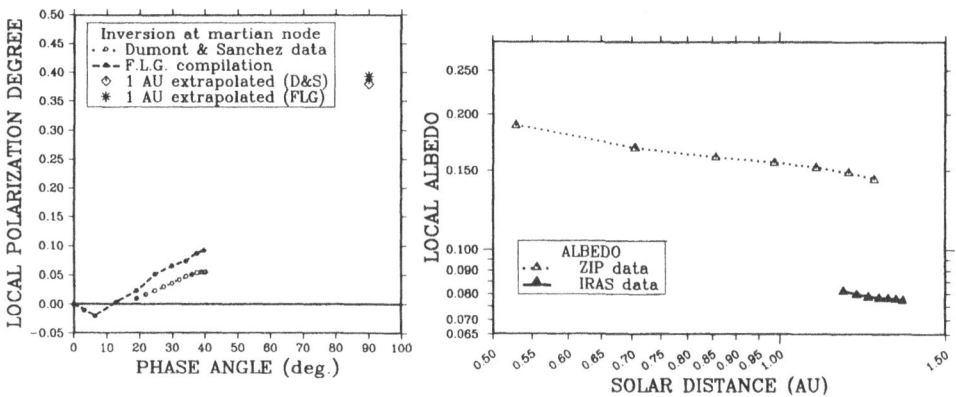

Figure 6 - Phase dependence of the polarization degree

Figure 7 - Radial dependence of the local albedo at 90°

The local bulk albedo at $\alpha = 90°$ (by extension of the definition of the geometric albedo, Hanner et al, 1981) is estimated from results obtained for local intensity (energy scattered in the visual domain) and for local temperature (thermal energy reemitted in the infrared). As can be seen on Fig. 7, the absolute values of the albedo depend upon the calibration of the thermal observations. They are, however, quite low (typically of the order of 0.08 at 1 AU from IRAS data). The trend to an increase with decreasing solar distance is nevertheless fairly well established from 0.5 AU to 1.5 AU, in agreement with previous results derived from other techniques (Fechtig, 1984; Lumme and Bowell, 1985).

The gradient of a power law fitting these local bulk albedos is computed to be of the order of - 0.30 ± 0.10. Since, as shown by Giese and Kinateder (1986), a scattering function gradient - n_1 and an albedo gradient - n_2 lead to a density gradient - n = - (n_1 - n_2), it can be concluded that the density decreases as $R^{-0.90 \pm 0.15}$. Such a value is in perfect agreement with the expected density law (1/R) for interplanetary grains under Poynting-Robertson effect (Hanner, 1980).

136

Sanchez (1975) integrated polarization values are a few percent smaller than those given in FLG (1981) compilation. The relative error bars are below 10 %.

The trend to a decrease with decreasing solar distance, suspected from the decrease in integrated polarizations in the inner solar system (Leinert et al., 1981), is confirmed (Dumont and Levasseur-Regourd, 1985b). In the 0.7 to 1 AU range, the local polarization varies as $R^{0.80}$. A very strong decrease seems to appear below 0.3 AU (Fig. 5). Such a result, based on observations at low solar elongations (Blackwell et al., 1967; Leinert et al., 1976) seems to be confirmed by inversion results obtained in the near coronal region (Dumont and Pelletanne, 1981 ; Mann, 1990). It could suggest a drastic change in the physical properties of the grains.

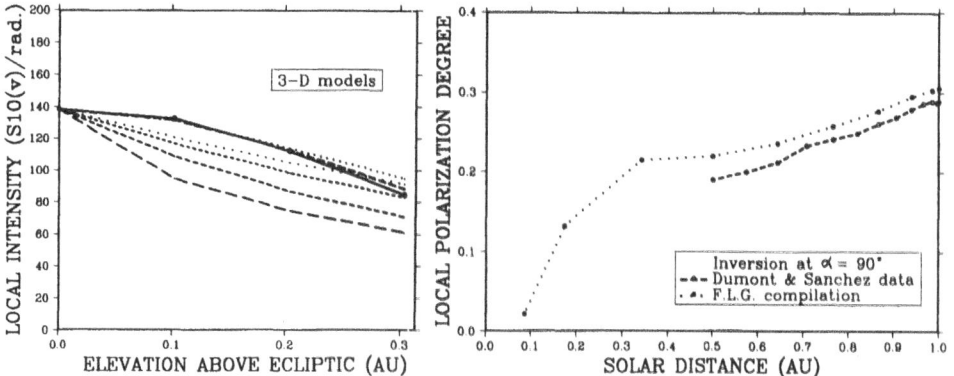

Figure 4 - Comparison of local intensity with computed intensities (dashed lines). From top to bottom, on the left side : ellipsoid, fan, cosine, (Good) infrared, modified fan and (Lamy) depression models

Figure 5 - Radial dependence of polarization at 90°

2.2.2. *Phase dependence in the symmetry plane.* The local polarizations at constant solar distance (of about 1.5 AU) are obtained between 3° and 40° from results at the martian nodes. The error bars are smaller than 2 %, except for the points at 90° which are extrapolated from results at the radial nodes between 0.6 and 1 AU.

Since the integrated polarizations mesured from A at small elongations are uncertain, the local values at small phase angles may be disputable. However, the local polarization is likely to be negative (electric vector in the scattering plane, instead of being perpendicular to it) for α smaller than 10°, to change sign below 20°, and to increase almost linearly up to 40° (Fig. 6). This result confirms the trends obtained from a different approach, using the few available observations in the Gegenschein region (Levasseur-Regourd et al., 1990). Again, the polarization function of interplanetary dust is reminiscent of the function obtained for comets. However, the smaller value of the phase angle at the neutral point (in a 10° to 20° range, instead of 21°± 1° for comet Halley's dust) could suggest that the interplanetary dust grains are not identical to the cometary ones.

2.2.3. *Elevation dependence in the tangential plane.* A map of the polarization degree at α = 90° is derived in the tangential plane from results obtained at the tangential and martian nodes in that plane, after normalization at constant phase angle (Renard et al. 1991, fig. 4). It extends up to 0.4 AU above the Earth towards the pole and as far as 1.5 AU in the symmetry plane.

The polarization decrease (at constant phase angle and solar distance), previously noticed towards the pole of the symmetry plane (Levasseur-Regourd and Dumont, 1990), is

2.3.2. *Elevation dependence above the symmetry plane.* The local bulk albedo at 1 AU and 90° is derived from local intensities and temperatures at the polar nodes of the Earth-pole line of sight, and found to increase with increasing elevation upon the symmetry plane (Renard et al., 1991, fig. 3).

Again, an increase in albedo is found to be correlated with a decrease in polarization. The evolution of albedo with increasing elevation confirms the result obtained for polarization, and suggests that the grains found on inclined orbits are quite different from those which are predominant in the symmetry plane.

Conclusions

To summarize, the heterogeneity of the zodiacal cloud is evident, both in the symmetry plane and in the plane perpendicular to the Sun-Earth line at 1 AU. In the symmetry plane, the local polarization decreases and the local albedo increases with decreasing solar distance. In the tangential plane, the local polarization decreases and the local albedo increases with increasing elevation, i.e. with increasing inclination of the grains orbits upon the symmetry plane. Besides, the slight discrepancies between the phase and polarization functions for comet Halley or interplanetary dust in the symmetry plane need to be emphasized, together with the discrepancies between the various models fitting visible or infrared observations towards the Sun.

The previous results suggest that the cloud is a mixture of two populations of grains, one (1) with low albedo and high polarization, the other (2) the reverse. The former (1) would be more flattened and slower in its radial decrease of density than the latter (2), which could be rather isotropic in its distribution. It is tempting to associate population 1 (to which infrared observations are quite sensitive) to dark grains which could originate in the processed crust of active periodic comets or extinct comets, and in asteroids. Also, population 2 (to which infrared observations are less sensitive) could be associated to new comets on randomly inclined orbits or to β meteoroids. However, without discussing the question of the various origins of the interplanetary dust complex, it is now clear that the 3-D models which do not take into account any spatial change in the properties of the grains are not acceptable and that at least a bimodal population is required to describe the local properties of the zodiacal cloud.

References

Blackwell D.E., Dewhirst D.W., Ingham M.F. (1967), *Adv. Astron. Astrophys.* 5, 1-69.
Dermott S.F., Nicholson P.D., Wolven B. (1986), in C.I. Lagerkvist et al. (eds.), *Asteroids, comets, meteors II*, HSC, Uppsala, 583-594.
Dumont R. (1973), *Planet. Space Sci.*, 21, 2149-2155.
Dumont R. (1976), in H. Elsässer and H. Fechtig (eds.), *Interplanetary dust and zodiacal light*, Springer-Verlag, Berlin, 85-100.
Dumont R. (1976), *IAU General assembly*.
Dumont R. (1983), *Planet. Space Sci.*, 31, 1381-1387.
Dumont R. and Levasseur-Regourd A.C. (1978), *Astron. Astrophys.*, 64, 9-16.
Dumont R. and Levasseur-Regourd A.C. (1985a), *Planet. Space Sci.*, 33, 1-9.
Dumont R. and Levasseur-Regourd A.C. (1985b), in R.H. Giese and P. Lamy (eds.), *Properties and interactions of interplanetary dust*, D. Reidel Publishing co., Dordrecht, 207-213.

Dumont R. and Levasseur-Regourd A.C. (1987), in Z. Ceplecha and P. Pecina (eds.), *Interplanetary matter*, Astron. Inst. of the Czechoslovak Acad. of Sciences, 67, 281-284.

Dumont R. and Levasseur-Regourd A.C. (1988), *Astron. Astrophys.*, 191, 154-160.

Dumont R. and Pelletanne B. (1981), *C.R. Acad. Sci. Paris*, 293, II, 377-380.

Dumont R. and Sanchez F. (1975), *Astron. Astrophys.*, 38, 405-412.

Fechtig H. (1984), *Adv. Space Res.*, 4, 5-12.

Fechtig H., Leinert C., Grün E. (1981) "FLG", in K.Schaifers and H. Voigt (eds.), *Landolt-Börnstein N.S.*, 2a, Springer-Verlag, Berlin, 228-243.

Giese R.H. and v. Dziembowski C. (1969), *Planet. Space Sci.*, 17, 949-956.

Giese R.H., Kinateder G., Kneissel B., Rittich U. (1985), in R.H. Giese and P. Lamy (eds.) *Properties and interactions of interplanetary dust*, D. Reidel Publishing co. Dordrecht, 255-259.

Giese R.H. and Kinateder G. (1986), in R.G. Marsden (ed.), *The Sun and the heliosphere in 3 dimensions*, D. Reidel Publishing co., Dordrecht, 441-454.

Giese R.H., Kneissel B., Rittich U. (1986), *Icarus*, 68, 395-411.

Giese R.H. and Kneissel B. (1989), *Icarus*, 81, 369-378.

Good J.C., Hauser G.M., Gautier T.N. (1986), *Adv. Space Res.*, 6, 83-86.

Grün E. et al. (1991), *(this issue)*.

Hanner M.S. (1980), *Icarus*, 43, 373-380.

Hanner M.S. (1981), *Astron. Astrophys.*, 104, 42-46.

Hauser M.G., Gillett F.C., Low F.L., Gautier T.N., Beichman C.A., Neugebauer G., Aumann H.H., Baud B., Boggess N. Emerson J.P., Houck J.R., Soifer B.T., Walker R.G. (1984), *Astrophys. J.*, 278, L15-L18.

Igenbergs E. et al. (1991), *(this issue)*.

Kneissel B. (1991), *(this issue)*.

Lamy P. and Perrin J.M. (1986), *Astron. Astrophys.*, 163, 269-286.

Leinert C., Link H., Pitz E., Giese R.H. (1976), *Astron. Astrophys.*, 47, 221-230.

Leinert C., Pitz E., Hanner M., Link H. (1977), *J. Geophys.*, 42, 669-704.

Leinert C., Richter I, Pitz E., Planck B. (1981), *Astron. Astrophys.*, 103, 177-188.

Levasseur A.C. and Blamont J.E. (1976), in H. Elsässer and H. Fechtig (eds.) *Interplanetary dust and zodiacal light*, Springer-Verlag, Berlin, 58-62.

Levasseur-Regourd A.C. and Dumont R. (1980), "LRD", *Astron. Astrophys.*, 84, 277-279.

Levasseur-Regourd A.C. and Dumont R. (1987), *Adv. Space Res.*, 6, 7, 87-90.

Levasseur-Regourd A.C. and Dumont R. (1990), *Adv. Space Res.*, 10, 3, 163-170.

Levasseur-Regourd A.C., Dumont R., Renard J.B. (1990), *Icarus*, 86, 264-272.

Lumme K. and Bowell E. (1985), *Icarus*, 62, 54-71.

McDonnell J.A.M. (1991), *(this issue)*.

Mann I. (1990), *Dissertation*, Ruhr Universität, Bochum.

Meech K.J. and Jewitt D.C. (1987), *Astron. Astrophys.*, 187, 585-593.

Misconi N.Y. (1980), in I. Halliday and B.A. McIntosh (eds.), *Solid particles in the solar system*, R. Reidel Publishing co., Dordrecht, 49-53.

Murdock T.L. and Price D. (1985), *Astron. J.* 90, 375-386.

Renard J.B., Levasseur-Regourd A.C., Dumont R. (1991), *(this issue)*.

Rittich U. (1986), *Diplomarbeit*, Ruhr-Universität, Bochum.

Sykes M.V., Lebofsky L.A., Hunten D.M., Low F.J. (1986), *Science*, 232, 1115-1117.

Toller G.N. and Weinberg J.L. (1985), in R.H. Giese and P. Lamy (eds.) *Properties and interactions of interplanetary dust*, D. Reidel Publishing co., Dordrecht, 21-25.

SPATIAL DISTRIBUTION AND ORBITAL PROPERTIES OF ZODIACAL DUST

Bernhard Kneißel and Ingrid Mann[*]
Ruhr-Universität Bochum,
D 4630 Bochum,
RFA

ABSTRACT. Within the recent years the spatial distribution of Zodiacal dust has been subject to a variety of modelling approaches. Whereas models derived from observations in the visual range tend to demand for an increase of interplanetary matter above the solar poles (bulges), models based on infrared measurements and extended to small r seem to favor a decrease there (holes). The models are reviewed, and the dynamical structure implicated in the models is outlined.

1. Spatial Distribution of Zodiacal Dust

The observed brightness of zodiacal dust particles in the optical and infrared wavelength range results from all volume elements along the line of sight (LOS) together with the scattering or thermal properties of grains. Facing the ambiguities in the knowledge of particle properties a variety of different models describing the distribution of interplanetary dust in terms of global properties have been proposed by researchers. They are shown as lines of equal number density n in terms of n_o ($r = 1$ AU, $\beta_o = 0$) in figure 1. The discussed models are based on the assumption that the number density can be expressed by two factors: one factor is a function that increases with solar distance r as a power law of r with exponent ν and another one that varies only with helioecliptic latitude β_o. The additional assumption of a change of the albedo with exponent μ results in an exponent $\nu^* = \nu + \mu$. A Comparison of the different models (references for the compared models are listed in table 1, for further descriptions see Giese and Kneißel 1989) shows that predictions of different authors derived from optical respectively infrared observations have a converging run of the relative distributions close to the Earth. But most of the models, devised on the basis of visible observations, suggest an increase of number density towards the Sun, whereas the models derived from infrared observations demand for a decrease above the solar poles. Even if one questions the validity of models derived from measurements at large elongations for the sunward regions the difference in the absolute scale of particle number densities (see Giese and Kneißel 1989) has to be explained.

2. Orbital Properties of Zodiacal Dust

The number density of particles of each volume element in the interplanetary space is given by the position probability of orbiting grains. Then one can obtain distributions of orbits from the number densities. These distributions are given in terms of the distribution density of orbital

---

[*]New address: Max-Planck-Institut für Aeronomie, D-3411 Katlenburg-Lindau.

A.C. Levasseur-Regourd and H. Hasegawa (eds.), Origin and Evolution of Interplanetary Dust, 139–146.
© 1991 *Kluwer Academic Publishers.*

elements such as major axis: d(a), eccentricity: d(e), and inclination: d(i) (Kneißel and Giese 1987). Due to the rotational symmetry of the cloud, lines of nodes and perihelion distances should be distributed isotropically in space. As there is assumed to be a separation between the in-ecliptic n(r) and the out-of-ecliptic fraction f(β_0) of number density, inversion of dynamical structure leads to a separation between the orbital distribution densities d(a,e) and d(i).

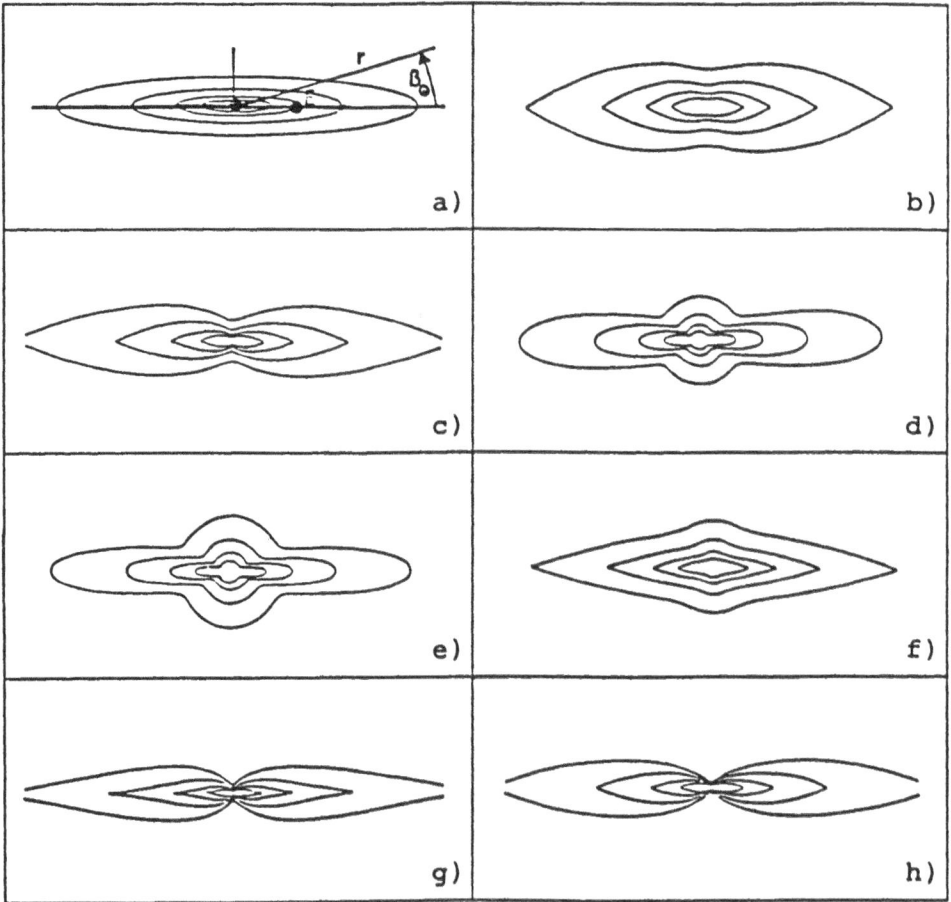

Figure 1: Spatial distribution of the Zodiacal Dust Cloud suggested by a) Giese et al. 1969 b) Leinert et al. 1976, c) Deul & Wolstencroft 1987, d) Dumont 1976, e) Rittich 1986, f) Lumme and Bowell 1985, g) Lamy and Perrin 1986, h) Good et al. 1986

The distribution of particles with solar distance r is related to d(a,e). Dealing with the three-dimensional structure of the cloud the distribution of inclinations d(i) is of most interest. In general, inversion of spatial distributions leads to a bimodal distribution of inclinations. The main component has its maximum close to the ecliptic plane and another one isotropically distributed gives a background to the first component. This background covers to the same amount both the orbits with prograde and the ones with retrograde orientation. The average inclination of orbits derived for the different models and the contribution of retrograde orbits, representing the isotropic component, are listed in table 1. The aligned inclination density distribution for 3 typical models of the spatial distribution is shown in figure 3.b). Since the main component represents the near-ecliptic dust and the second one dust forming the bulge, the isotropic background component is negligible for pole hole models whereas in the case of bulge models the background component contributes considerably. The exponent v^* (including also the change of albedo) amounts 1 to 1.3 for the models listed in table 1.

model	$<i>/°$	$N/N_{ret}/\%$
Giese et al. 1969	30	4
Leinert et al. 1976	36	6
Deul & Wolstencroft 1987	31	3
Dumont 1976	32	8
Rittich 1986	38	10
Lumme & Bowell 1985	37	8
Lamy & Perrin 1986	28	2
Good et al. 1986	23	1

Table 1: average orbital inclination $<i>$ and contribution of retrograde orbits $N/N_{ret}/\%$ (N: whole number of orbits, N_{ret}: number of orbits in retrograde motion) for different models.

3. Comparison with Sources of Interplanetary Dust

The interplanetary dust originates from comets emitting dust during perihelion passage or collision of asteroids, as illustrated in figure 2. The particles make their way in the inner parts of solar system due to the decelerating Poynting-Robertson drag. Especially fragments of colliding meteoroids related to comets and asteroids fill up the zodiacal dust cloud. A Comparison with orbital elements of the meteoroids (with mass $m > 10^{-4}$ g) shows that they are more inclined, with an average inclination $<i> = 39°$ and have a relative strong isotropic component. Thus the zodiacal dust cloud cannot result directly out of this population.

One has to consider the dynamical conditions for the fragments produced by colliding meteoroids. Near perihelion, where most of the collisions should take place (Dohnanyi 1978), the condition for an unbound state is directly dependant on the eccentricity of orbits and the ratio Q_{PR}/d (Q_{PR}: efficiency for radiation pressure, d: bulk density in g/cm^3, see Kneißel and Giese 1987). According to Ceplecha (1977) the isotropic component called C2 meteorids ($m \geq 10^{-3}$ g) among the meteoroid population has randomly distributed orbits with very long semimajor axis and eccentricity ≈ 0.99, amounting to 30% of the whole distribution. Fragments of these particles of the zodiacal dust size will be in unbound states (cf. Kneißel and Giese 1987) and blown off from the solar system. Other meteoroids, with eccentricities $e < 0.9$, will stay in bound orbits after fragmentation (Kneißel 1988) and may contribute to the zodiacal dust cloud as for example the class of the C3 meteoroids ($e = 0.6$-0.7, Ceplecha 1987).

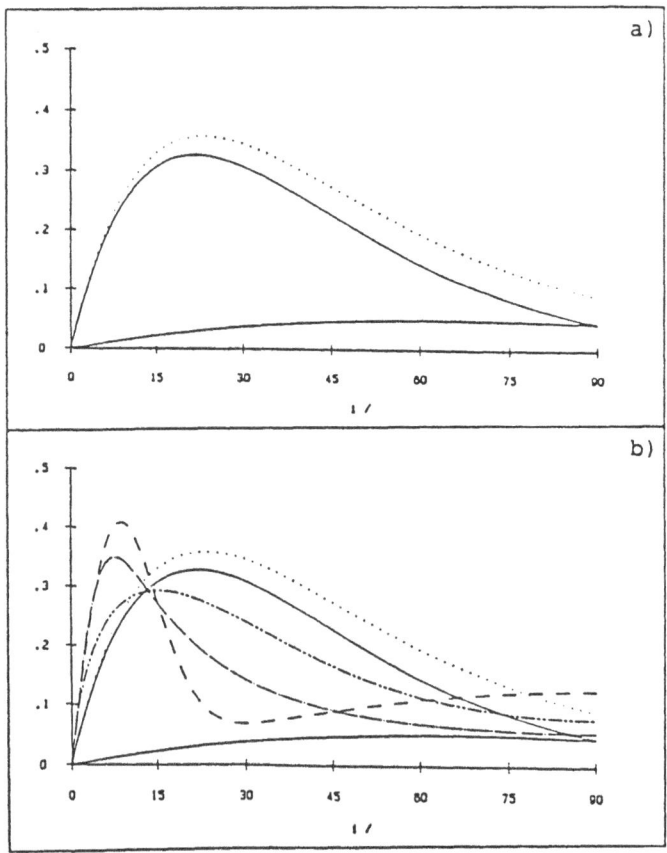

Figure 2 a): Relative distribution of inclinations of meteoroid particles (Andreev and Belkovich 1985), dotted line: whole distribution, upper solid line: distribution for inclinations $0° < i < 90°$, lower solid line: distribution for $90° < i < 180°$ folded into that interval, b): comparison with distributions related to the models suggested by Rittich 1986 (--), by Leinert et al. 1976 (····) and by Giese et al. 1969 (--).

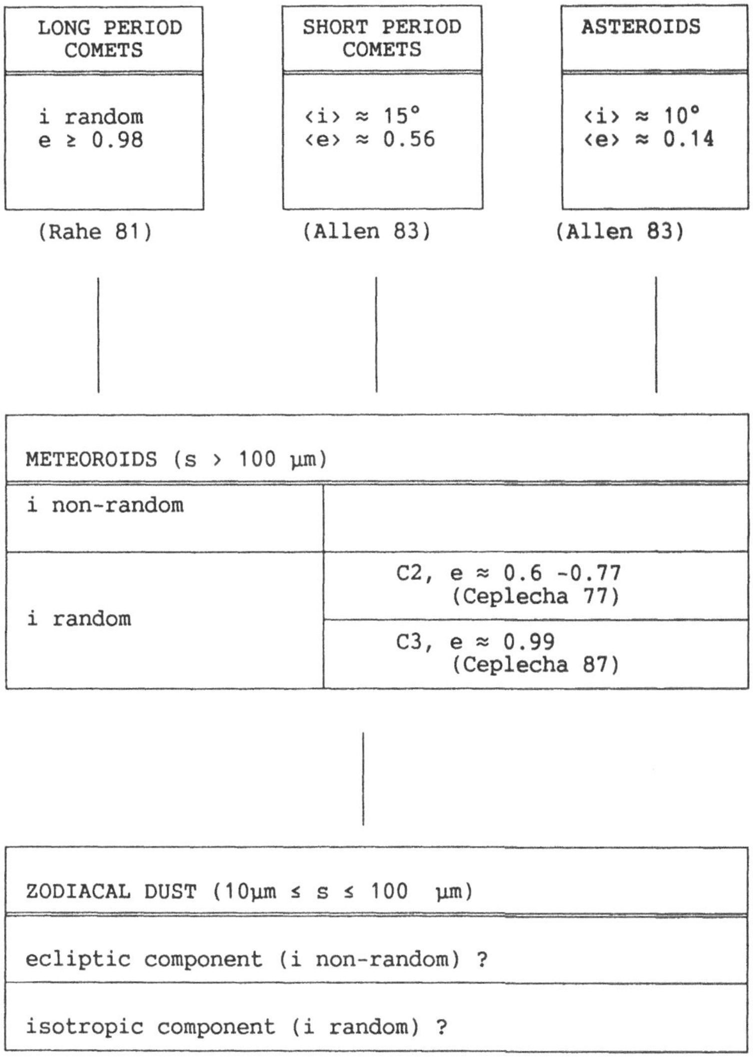

LONG PERIOD COMETS
i random e ≥ 0.98

(Rahe 81)

SHORT PERIOD COMETS
⟨i⟩ ≈ 15° ⟨e⟩ ≈ 0.56

(Allen 83)

ASTEROIDS
⟨i⟩ ≈ 10° ⟨e⟩ ≈ 0.14

(Allen 83)

METEOROIDS (s > 100 μm)	
i non-random	
i random	C2, e ≈ 0.6 -0.77 (Ceplecha 77)
	C3, e ≈ 0.99 (Ceplecha 87)

ZODIACAL DUST (10μm ≤ s ≤ 100 μm)
ecliptic component (i non-random) ?
isotropic component (i random) ?

Figure 3: the zodiacal dust in the meteoritic complex

Deceleration of particles due to Poynting-Robertson effect goes along with a change of eccentricity and a change of the major axis. This orbital elements are related to the variation n(r) of the number density with solar distance. Not only additional sources, but also the reduction of the eccentricity results in an exponent $v > 1$ for the spatial distribution of particles. That is valid for particles that are replenished from particles on high eccentricity orbits. On the other hand particles of asteroid origin have small eccentricities $e < 0.15$ and are mainly in circular orbits when they are decelerated to 1 AU around the Sun.

As according to Fechtig (1989) at least 2/3 of all ecliptic concentrated grains are of asteroid origin, the ecliptic component of the zodiacal cloud might have not only different orbital, but also different optical properties from those of the isotropic component. As cometary dust can be extremely dark (cf. Hanner and Newburn 1989) and may show strong variations of the optical properties due to the particle structure (cf. Greenberg and Grim 1986), especially particles that where not heated before are candidates to explain a change of albedo. This change of albedo was derived from an extensive analysis of different optical and infrared data by Dumont and Levasseur-Regourd (1988, see also Levasseur-Regourd 1991). Based on our discussion of different orbital elements and different sources of the dust particles, one may regard the distribution of particles to be bimodal (Kneißel and Mann 1989). The component concentrated to the ecliptic has a high amount of asteroid particles whereas the second component has randomly distributed orbits (presumably) filled up by long period comets. Whereas the number density of the isotropic component is constant with the heliocliptic latitude, the radial dependant part is increasing with decreasing solar distance. In addition the observed brightness is influenced by the change of albedo. The first approach gives the distribution of the number density weighted with the optical efficiency to be:

$$\frac{<s> n}{<s_o> n_o} = \frac{3}{4} r^{-v_1^*} \cos^{40}\beta_o + \frac{1}{4} r^{-v_2^*} \tag{1}$$

The expression gives the variation of the volume scattering function with respect to the volume scattering function at 1 AU $<s_o> n_o$. For further explanations see Giese and Kneißel (1989). There is no separation included between the change of scattering properties and the change of number density. Modelling the brightness integral with the volume scattering function given with equation (1) including $v_1^* = 1$ and $v_2^* = 2$ the average deviation to observational data is smaller than 10%. That means that the ecliptic component may show a spatial variation with $1/r$ for the particle number density and no significant change of albedo. Nevertheless, particles of the isotropic component, described by the second part of equation 1, contribute to the ecliptic brightness due to their orbits crossing the ecliptic plane. In agreement with results regarding only one population, the combination of the two components gives a run with $v^* \approx 1.3$ within the ecliptic.

4. Summary

The interplanetary dust cloud may be regarded as a superposition of one component mainly concentrated to the ecliptic plane with more or less the regular properties known from the zodiacal dust and a second one that is isotropically distributed and may result from long period comets. Although some interesting work remains to be done on this point, this scenario seems to be compatible with both visual and infrared observations. Beside this separation of components, the different properties of the cometary material itself mentioned by Greenberg (1990), have to be considered for further investigations. One example for this may be the difference between original cometary particles and processed cometary material.

The authors want to thank the referees for the discussion of the paper. This work was supported by the Bundesminister für Forschung und Technologie BMFT (Contract 01 ON 89012).

5. References

Allen, C.W. 1983: Astrophysical Quantities, The Athlone Press, 3. edition

Andreev V.V. and Belkovich O.I., 1985: Distribution of the Meteor Matter in Distance 1 AU from the Sun, IAU Gen. Ass. (New Dehli 1985)

Bandermann, L.W., 1968: Physical Properties and Dynamics of Interplanetary Dust, Ph.D. Thesis, University of Maryland

Ceplecha, Z., 1977: Meteoroid Populations and Orbits, in Comets, Asteroids, Meteorites (Delsemme, A.H. ed.), Univ. of Toledo Press, 143-152

Ceplecha, Z., 1987: Numbers and Masses of Different Populations of Sporadic Meteoroids from Photographic and Television Records, Publ. Astron. Inst. of the Czechosl. Academy of Sciences 67, Proc. Vol. 2, 241

Deul, E.R. and Wolstencroft, R.D., 1987: A Physical Model for Thermal Emission from the Zodiacal Dust Cloud, Astron. Astrophys. 196, 277-286

Dohnanyi, J.S., 1978: Particle Dynamics, in: Cosmic Dust (McDonnell, J.A.M., ed.), John Wiley Sons, New York, 527

Dumont, R., Oral Presentation at the General Assembly IAU (Grenoble, 1976)

Dumont, R. and Levasseur-Regourd, A.-C., 1988: Properties of Interplanetary Dust from Infrared and Optical Observations. I. Temperature, Global Volume Intensity, Albedo and their Heliocentric Gradients, Astron. Astrophys. 191, 154

Fechtig,H., 1989: Dust in the Solar System, Z. Naturforsch. 44a, 877

Giese, R.H. and v. Dziembowski, C. 1969: Suggested Zodiacal Light Measurements from Space Probes, Planet. Space Sci., 17, 949

Giese, R.H. and Kneißel, B., 1989: Threedimensional Models of the Zodiacal Dust Cloud: Compatibility of Proposed Infrared Modells, ICARUS 81, 369

Good, J.C., Hauser, M.G., and Gautier, T.N., 1986: IRAS Observations of the Zodiacal Background, Adv. Space Res., 6, 83

Greenberg, J.M. and Grim, R., 1986: The Origin and Evolution of Comet Nuclei and Comet Halley Results. In: Exploration of Halley's Comet (B. Battrick, E.J. Rolfe, and R. Reinhard,eds.), ESA SP-250, 255

Greenberg, J.M., 1990: Personal Comunication

Grün, E., Zook, H.A., Fechtig, H., and Giese, R.H., 1985: Collisional Balance of the Meteoritic Complex, ICARUS, 62, 244

Hanner, M.S. and Newburn, R.L., 1989: Infrared Photometry of Comet Wilson at two Epochs. Astron.J 97, 254

Humes, D.H., 1980: Results of Pioneer 10 and 11 Meteoroid Experiments: Interplanetary and Near-Saturn. J.Geophys. Res. 79, 5841

Kneißel, B. and Giese, R.H., 1987: The Dynamics of the Zodiacal Dust Cloud on account of Optical and Infrared Observations. Publ. of the Astronomical Institute of the Czechosl. Acad. of Sc. 67, Vol. 2, 241

Kneißel, B., 1988: PH.D. Thesis, University of Bochum

Kneißel, B. and Mann, I., 1990: Three-Dimensional Models of the Zodiacal Dust Cloud, in; Proc. of the 1. COSPAR Colloqium: Physics of the Outer Heliosphere (S. Grziedzielski, ed.),

Pergamon Press PLC

Lamy P.L. & Perrin, J.M., 1986: Volume Scattering Function and Space Distribution of the Interplanetary Dust Cloud, Astron. Astrophys. 163, 269

Leinert, C., Link, H., Pitz, E. & Giese, R.H., 1976: Interpretation of a Rocket Photometry of the Inner Zodiacal Light, Astron. Astrophys.,47

Leinert, C. and Grün, E., 1990: Interplanetary Dust, in: Physics of the Inner Helioshere, (Schwenn, R. and Marsch, E. eds.), Springer

Levasseur-Regourd, A.-C. and Dumont, R., 1980: Absolute Photometry of Zodiacal Light, Astron. Astrophys. 84, 277

Levasseur-Regourd, A.-C., 1991: this issue

Lumme, K. and Bowell, E., 1985: Photometric Poperties of Zodiacal Light Particles, ICARUS, 62, 54

Rahe, J. 1981: in: Landolt-Börnstein (Schaifers, K. and Vogt, H.H., eds.), Gruppe VI, Band 2, Teilband a, Springer, Berlin, 183-198

Rittich, U., 1986: Diploma Thesis, Bochum

ON THE GEGENSCHEIN AND THE SYMMETRY PLANE

S. S. HONG AND S. M. KWON
Department of Astronomy, Seoul National University, 151-742, KOREA

ABSTRACT. Using 3-dim density models of the zodiacal cloud, we have calculated brightness of the zodiacal light over an extended region around the anti-solar point. The isophotal contours of the model Gegenscheins differ from each other, morphologically, to the degree that they can differentiate the competing density models. The recently reduced Gegenschein observations of 2° resolution clearly favour the ellipsoid-type models to the fan-types, and also suggest that the surface of the densest dust concentration in the outer part of the cloud has its ascending node at longitude $100 \pm 20°$ and is inclined $2 \pm 0°.5$ with respect to the ecliptic plane.

1. Introduction

Many models of different nature have been proposed for the 3-dim distribution of the dust particles in the zodiacal cloud (see the thorough discussions by Giese *et al.* [1986] and references therein). According to the late Professor Giese, most models, except for the multi-lobe, surprisingly agree in that the dust density decreases by a factor of 2 within 0.2 to 0.3 AU above the earth orbit. Yet, the morphology of the isodensity contours in the helioecliptic meridian plane looks quite different from model to model. For example, the isodensity contours of the ellipsoid-type models (ellipsoid, sombrero, cosine, ...) are of rounded shape at the ecliptic plane; while the contours of the fan-types (fan, modified-fan, extreme-fan, ...) become very peaked there. The density distribution near the ecliptic ought to be related to the particle dynamics; if so, it is of importance to differentiate the ellipsoid-types from the fan-ones.

In the previous studies, observed profiles of the zodiacal light brightness only along the helioecliptic meridian, the great circle at 90° elongation, and the circles around the sun are compared with the corresponding model profiles. To distinguish the morphological characteristics of the 3-dim models, observed isophotes of the zodiacal light in the $(\lambda - \lambda_\odot, \beta)$ plane should have been compared with their corresponding model isophotes. In order to make meaningful comparisons, however, the zodiacal light distribution has to be known over an extended region of sky with fine angular resolutions. Recently reduced (Kwon, Hong, Weinberg and Misconi 1990) observations of the zodiacal light do provide the basis for such comparisons.

2. Isophotal Contours of the Gegenschein

The usual brightness integral has been numerically calculated to obtain the zodiacal light brightness

A.C. Levasseur-Regourd and H. Hasegawa (eds.), Origin and Evolution of Interplanetary Dust, 147–150.

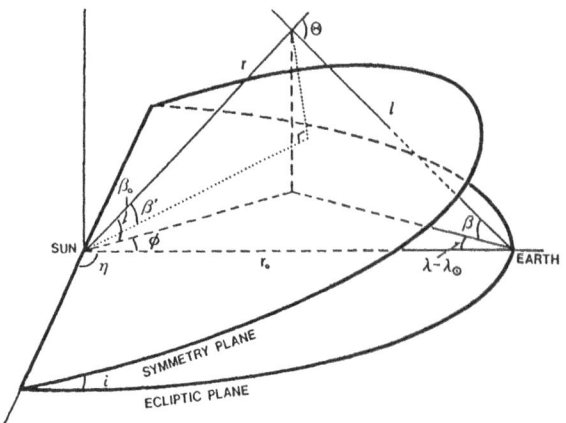

Fig 1 - Geometry involved in the brightness integral.

$Z(\lambda - \lambda_\odot,\ \beta)$ in the $(\lambda - \lambda_\odot,\ \beta)$ plane:

$$Z(\lambda - \lambda_\odot,\ \beta) = \int_0^\infty F_0 \left(\frac{r_0}{r}\right)^2 n_0 \left(\frac{r_0}{r}\right)^{1.3} f(\beta_0)\, \Phi(\Theta)\, dl.$$

The geometry is illustrated in Fig 1; F_0 and n_0 are the solar flux and particle number density at $r_0 = 1$ AU from the sun; and for the scattering phase function $\Phi(\Theta)$ a linear sum of three Henyey-Greenstein functions (Hong 1985) is substituted. As a representative of the ellipsoid-type models, we simply used $[1 + (6.5\sin\beta_0)^2]^{-0.65}$ (Dumont 1976) for $f(\beta_0)$ in the brightness integral, and that of the fan-types, $exp[-2.1\,|\sin\beta_0|]$ (Leinert et al. 1978).

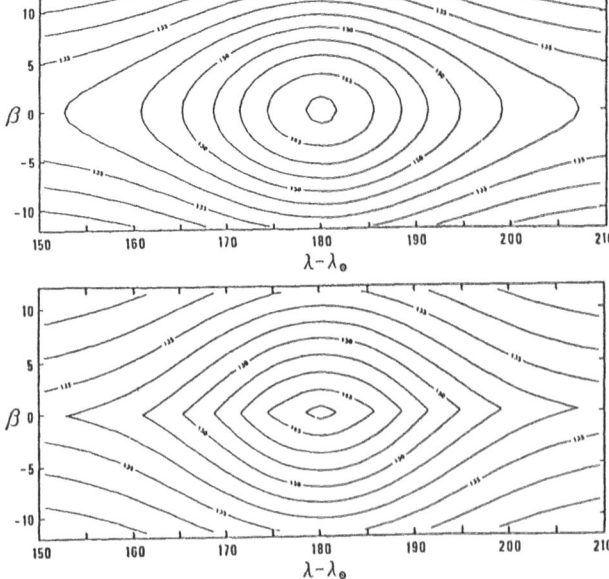

Fig 2 - Isophotes of the model Gegenscheins. The symmetry plane is assumed to be in the ecliptic plane. The ellipsoid-model is used for the upper frame, and the fan-model for the lower one.

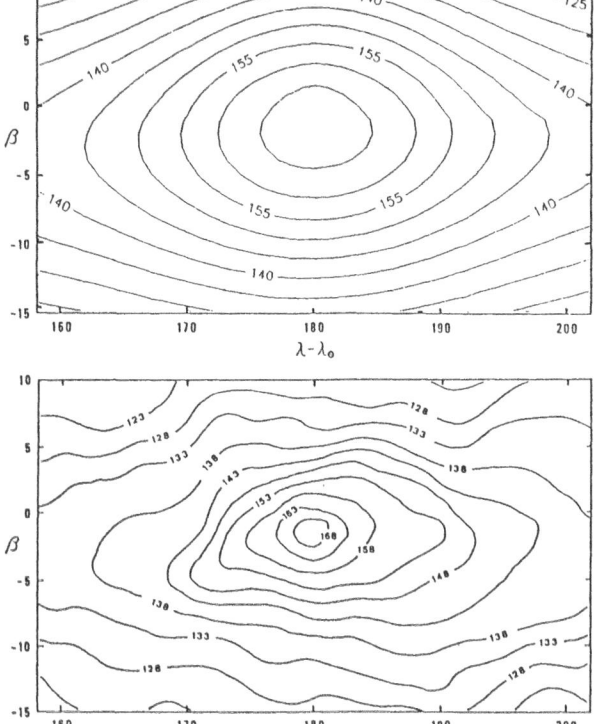

Fig 3 - Isophotes of the model Gegenschein. With the ellipsoid-model for the 3-dim density distribution, we have substituted $i = -2°$ and $\eta = 45°$ for the inclination and line of node of the symmetry plane.

Fig 4 - Isophotal contours of the observed Gegenschein.

The resulting isophotes are shown in the upper frame of Fig 2 for the ellipsoid-type, and in the lower frame for the fan-type. Distinct difference in the contour morphology can be noticed between the two types of cloud model. This figure have been prepared with 1° resolution. If the observed resolution of Gegenschein is better than $\sim 3°$, one could easily differentiate the cloud models on the basis of the comparison made in the figure. Previous observations of the Gegenschein (Tanabe 1965; Dumont 1965; Maucherat et $al.$ 1986) could not quite reach this good a resolution.

To check whether the disalignment of the symmetry plane with respect to the ecliptic might smear the morphological distinctions seen in Fig 2, we have calculated the same brightness integral with $f(\beta_0)$ being simply replaced by $f(\beta')$, where β' is measured from the symmetry plane. As illustrated in Fig 1, the inclination of the symmetry plane is denoted by i, and the angle η orients the line of nodes of the plane with respect to the sun-earth direction. The two angles, β_0 and β', are related through the relation $\sin \beta' = \cos \beta_0 - \sin i \cos \beta_0 \sin(\eta + \phi)$. The distinction survives with an inclined symmetry plane.

To make a comparison with the observation, we have presented in Fig 3 the Gegenschein of the ellipsoid model with $i = -2°$ and $\eta = 45°$. The major axis of the contour "ellipses" is slightly tilted with respect to the ecliptic, and the maximum brightness point is about 2° below the anti-solar point.

The tilt angle becomes the inclination angle when $\eta = 0$. When $i < 0$, the isophotes above the ecliptic plane tend to bunch together; while when $i > 0$, the southern lines do the same. This can also be noticed from the model calculation by Misconi (1981). The characteristics of the contour morphology outlined here hold true for the fan-type models, except that the contour shape becomes very peaked along the major axis.

All sky monitoring observations of night sky brightness was analyzed, with a newly developed reduction method by Kwon, Hong, Weinberg and Misconi (1990). The resulting brightness map has a resolution of $2°$, and the Gegenschein part is reproduced in Fig 4 for a comparison. The maximum brightness point is observed at $\lambda - \lambda_\odot \simeq 180°$ and $\beta \simeq -2°$; the major axis of the elliptical isophotes is somewhat tilted with respect to the ecliptic plane; the brightness gradient is steeper in the northern side than it is in the southern side; and the closed contours are of rounded shape along the direction of their major axes.

3. Conclusion

The observed Gegenschein isophotes in Fig 4 agree, morphologically, with the model shown in Fig 3. Following conclusions may be drawn from the comparison: (1) The ellipsoid-type models describe the 3-dim distribution of dust in the outer part of the zodiacal cloud better than the fan-types do. (2) The Gegenschein observations reported here are not inconsistent with the notion that the surface of the densest dust concentration has an inclination of $2 \pm 0°.5$, and its ascending node at longitude $100 \pm 20°$. (The data were taken, when $\lambda_\odot = 150°$.)

Our result on the location of symmetry plane is in agreement with those by Dumont and Sanchez (1968) ad Dumont and Levasseur-Regourd (1978), and also with the suggestion by Misconi (1980). This study has demonstrated that mappings of the zodiacal light with fine resolution can differentiate the competing models of the 3-dim density distribution. Furthermore, if one monitors the Gegenschein over, say, four seasons, he could trace the symmetry plane(s), quite accurately, at least in the outer part of the zodiacal cloud.

SSH and SMK were supported by the Basic Science Research Institute Program, Korean Ministry of Education. We are grateful to the referees for giving clarifying comments.

REFERENCES

Dumont, R. 1965, *Ann. d'Astrophys.*, **28**, 265
Dumont, R. 1976, in *Lecture Notes in Physics*, vol. 48, eds. H. Elsässer and H. Fechtig
      (Springer-Verlag:Heidelberg), p.85
Dumont, R., and Sanchez, F. 1968, *Ann. d'Astrophys.*, **31**, 293
Dumont, R., and Levasseur-Regourd, A. C. 1978, *Astr. Ap.*, **64**, 9
Giese, R. H., Kneissel, B., and Rittich, U. 1986, *Icarus*, **68**, 395
Hong, S. S. 1985, *Astr. Ap.*, **146**, 67
Kwon, S. M., Hong, S. S., Weinberg, J. L., and Misconi, N. Y. 1990, in this volume
Leinert, C., Hanner, M., and Pitz, E. 1978, *Astr. Ap.*, **63**, 183
Maucherat, A., Llebaria, A., and Gonin, J. C. 1986, *Astr. Ap.*, **167**, 173
Misconi, N. Y. 1980, in *Solid Particles in the Solar System*, eds. I. Holliday, and B. A. McIntoch
      (Reidel:Dordrecht), p.49
Misconi, N. Y. 1981, *Icarus*, **47**, 265
Tanabe, H. 1965, *Publ. Astron. Soc. Japan*, **17**, 339

ULTRAVIOLET OBSERVATIONS OF THE ZODIACAL LIGHT AND THE ORIGIN OF INTERPLANETARY DUST GRAINS

C. F. LILLIE
TRW Space and Technology Group
One Space Park
Redondo Beach, California 90278 USA

ABSTRACT. Surface brightness photometry of the night sky from rocket and satellite experiments shows an increase in the scattering efficiency of interplanetary dust grains in the 1500 to 3000 Å region of the spectrum. This increase is best explained by the presence of small dielectric particles with a mean radius of 0.04 microns. The most likely source of these grains is the dissolution of agglomerates of these particles which are released by comets during their perihelion passage. Many of these agglomerates have been collected in the Earth's atmosphere by high flying aircraft. Submicron particles swept up from interplanetary space may be responsible for the high altitude haze observed in planetary atmospheres.

1. Observations

Ultraviolet observations of the zodiacal light provide a sensitive test for the albedo, phase function and size distribution of interplanetary dust particles. The number of observations in this region are quite limited, however, and show considerable disagreement [1]. These observations must be obtained with instruments on rockets or spacecraft to avoid absorption by the Earth's atmosphere, and are extremely difficult due to the ten-thousand-fold decrease in solar flux between 3000 and 1500 Å. They are also complicated by large contributions from residual airglow, integrated starlight and diffuse galactic light which must be subtracted from the measurements.

Table 1 summarizes the available data for the zodiacal light intensity as a function of wavelength relative to the sun, normalized to 1 at 5500 Å [2,3]. There is good agreement among investigators that the zodiacal light is redder than the sun from 2400 to 2900 Å, with colors ranging from 0.40 to 0.90. And there is general agreement that the zodiacal light is bluer than the sun at wave-

Table 1. Zodical Light Colors [3]

Reference	Wavelength (Å)					
	1680	1800	1920	2200	2600	2900
Sudbury and Ingham (1970)					0.68	
Lillie (1968)				4.1	0.74	0.42
Lillie (1972)	16	15	6.5	0.75	0.54	0.54
Orrall and Speer (1973)	<130	<52	<18.4	0.75		
Morgan et al (1976)					0.76	
Frey et al (1977)				<1		1
Feldman (1977)	<40			0.45		0.90
Pitz, et al (1978)			0.54	0.30	0.40	
Maucherat-Joubert, et al (1979)*	~40			1		
Cebula and Feldman (1979)			<4	0.75		0.68
Tennyson, et al (1988)					0.80	0.80

* Reanalysis of Pitz, et al, (1978)

A.C. Levasseur-Regourd and H. Hasegawa (eds.), Origin and Evolution of Interplanetary Dust, 151–154.

152

lengths below 2200 Å, although there is a large scatter in the measurements. These differences may be due to temporal variations in the interplanetary medium as well as to different viewing geometries, observing equipment and techniques and calibration standards.

Figure 1 shows rocket and satellite data which suggests enhanced ultraviolet scattering. This scattering appears to be quite isotropic [6]. The Aerobee 4.55 data [5] were obtained on 2 September 1964 with photoelectric photometers which scanned the night sky at solar elongations from 40 to 180 degrees, while the data of [7] were obtained with a slitless spectrograph which observed the F and K corona at R=1.07 solar radii during the total solar eclipse of 7 March 1970 (three points are upper limits). The OAO-2 data [6] were obtained with the 8-inch filter photometers of the Wisconsin Experiment Package from February 1969 to March 1970 and cover elongations from 45 to 110

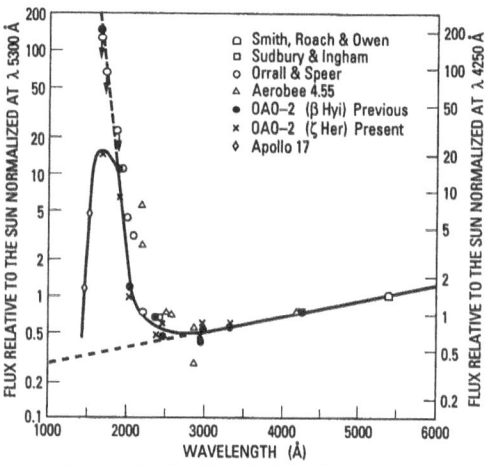

Figure 1. Zodiacal Light Spectrum

degrees. The Apollo 17 data [14,15] were obtained in December 1972 with an ultraviolet spectrometer which scanned the sky from approximately 6 to 180 degrees elongation while in Lunar orbit. Two versions of the OAO-2 data are shown. β Hyi has the spectrum of the sun but ζ Her is closest to the sun in color and ultraviolet flux and is the preferred solar analog. The data in Figure 1 show remarkable agreement, considering the wide variation in observing techniques and instruments which were used to obtain them. All of the observations show an increase below 3000 Å, with a steep rise below 2200 Å and an apparent peak at 1700 Å. The Apollo 17 data points have a large uncertainty, however, due to large contributions from integrated starlight and diffuse galactic light.

3. Analysis

Sudden increases in the scattering efficiency of particles can occur when the particle is much smaller than the wavelength of the incident radiation and the index of refraction is small (Rayleigh scattering), or when the particle is small and the index of refraction of the material is large and real or nearly so (optical resonance)[16]. The strongest resonance peak (magnetic dipole radiation) occurs near $nx = \pi$, where n is the real part of the index of refraction $m = n + ki$ and the particle size parameter $x = 2\pi a/\lambda$. Some examples of optical resonances are the $10\,\mu m$ emission feature in silicates and the ultraviolet extinction feature at 2175 Å.

The optical properties of the small (0.01 and 0.10 μm radius) graphite and silicate spheres which are candidates for interstellar grains [17] were used to compute the ultraviolet scattering efficiency of these grains for comparison with the zodiacal light spectrum. Three spectral features were selected for this comparison: the onset of enhanced scattering at 3000 Å, the mid-point of the steep rise at 2000 Å, and the scattering peak at 1700 Å. The particle radii required to produce each feature are 0.0237, 0.0280 and 0.0487 μm, respectively, for graphite particles or 0.413, 0.400 and 0.407 for "astronomical silicate" particles. The rms variance of 0.0005 for silicate (versus 0.0109 for graphite) suggests silicate particles are responsible for this ultraviolet enhancement. Also, the graphite particles have a secondary scattering feature at 2200 Å, which does not appear in the zodiacal light spectrum.

Figure 2 shows the scattering efficiency required to produce the steep increase in the zodiacal light spectrum for grains of radius 0.036 and 0.057 μm. The scattering curves for non-absorbing spheres for four values of the refractive index [16] are shown for comparison. The inset in Figure 2 shows the variation of the index of refraction with wavelength required to produce the observed scattering for the two radii.

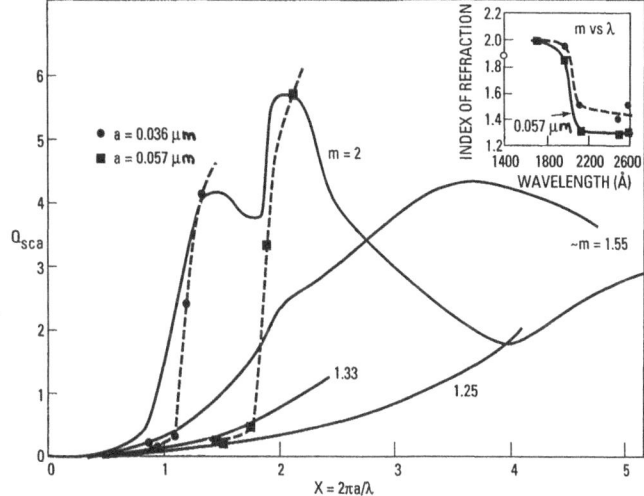

Figure 2. Scattering Efficiency of Small Grains

The peak in the m = 2 curve at x = 1.6 is a magnetic dipole resonance, and a 0.04 μm particle curve would just intersect the peak of this feature. This analysis is only valid for silicate particles [17] whose refractive index is almost entirely real (m ~ 1.73, k < 0.03 at wavelengths longer than 2200 Å, increasing to m = 2.08, k = 0.16 at 1700 Å). ("Astronomical silicate" might be classified a strongly absorbing black glass). The variation of m with wavelength in Figure 2 most nearly resembles the dispersion curve for vitreous quartz or borosilicate glass [18]. Given accurate dispersion curves it may be possible to discriminate between candidate materials for these grains.

4. Discussion

This analysis indicates the steep rise in the zodiacal light spectrum is due to a combination of Rayleigh scattering and magnetic dipole resonance in silicate particles with a mean radius of 0.04 μm. An absorption band at ~ 1400 Å must also contribute to the formation of the scattering peak at 1700 Å.

Particles of this size have a mass of ~ 10^{-15} grams, and will be pushed out of the inner solar system by solar radiation pressure in about 10^7 seconds [19]. The cosmic dust experiments on Pioneer 8 and 9 have observed such particles in hyperbolic trajectories whose apparent radiant is the sun [21]. These particles may be created from cometary fragments (from new and periodic comets) which spiral in toward the sun and partially evaporate at a few solar radii, until radiation pressure exceeds gravity and the particle is ejected radially [20].

Interplanetary particles of probable cometary origin have been collected in the stratosphere at an altitude of 20 km with U-2 aircraft [22]. These "Brownlee Particles" are fragile, highly porous aggregates of mostly submicron (<0.10 μm) polychrystalline grains. They could easily be fragmented by sputtering, differential heating or electrostatic forces in the vicinity of the sun.

Submicron particles believed to be of cosmic origin have also been collected at high altitudes (> 80 km) in the Earth's atmosphere [23]. These particles have a mean diameter of ~ 0.06 μm, with the highest concentrations found in noctilucent clouds. Other evidence for planetary sweeping of submicron particles may be found in the high altitude hazes (with optical depths of .02 to .03 in the violet region of the spectrum) observed on Mars by the Mariner and Viking spacecraft, and in the outer planets by Voyagers 1 and 2. Titan also has high altitude aerosols with a mean radius of 0.05 μm [24]. Large fluxes of dust particles were also detected in-situ by the Voyager plasma wave experiment during equatorial plane crossing at Jupiter, Saturn, Uranus and Neptune.

These observations are consistent with the current core-mantle model for interstellar and cometary grains [25] if we assume the icy coating and refractory organic mantle have been volatilized by solar radiation, leaving relatively bare silicate cores with high scattering efficiency in the 1500 to 3000 Å region of the spectrum.

REFERENCES

[1] Tennyson, P. D., Henry, R.C., Feldman, P.D., Hartig, G.F. 1988, *Ap. J.* **330**, 435.

[2] Smith, L. L., Roach, F. E., and Owen, R. W. 1965, *Planet. Space Sci.*, **154**, 783.

[3] Leinert, C. 1975, *Space Sci. Rev.*,**18**,281.

[4] Sudbury, G. D. and Ingham, M. F. 1970, *Nature*, **226**, 526.

[5] Lillie, C. F. 1968, Ph. D. dissertation, University of Wisconsin.

[6] Lillie, C. F. 1972 in *The Scientific Results from the Orbiting Astronomical Observatory (OAO-2)*, NASA SP-130, p. 95.

[7] Orrall, F. Q., and Speer, R. J. 1973, *Solar Phys.* **29**, 41.

[8] Morgan, D. H., Nandy, K., and Thompson, G. I. 1976, *M.N.R.A.S.*, **177**, 531.

[9] Feldman, P.D. 1977, *Astr. Ap.*, **61**, 635.

[10] Frey, A., Hofmann, W., and Lemke, D. 1977, ibid, **54**, 853.

[11] Pitz E., Leinert, C., Schultz, A. and Link, H. 1978, ibid, **69**, 297.

[12] Maucherat-Joubert, M., Cruvellier, P., Deharveng, J. M. 1979, ibid, **74**, 218.

[13] Cebula, R. P., and Feldman, P. D. 1982, *Ap. J.*, **225**, 987.

[14] Fastie, W. G., Feldman, P. D., Henry, R. C., Moos, H. W., Barth, C. A., Lillie, C. F., Thomas, G. E. and Donahue, T. M. 1974, in *The Apollo 17 Preliminary Science Report*, NASA SP-330, p. 23-1.

[15] Lillie, C. F. 1975, (unpublished).

[16] van de Hulst, H. C. 1957, in *Light Scattering by Small Particles*, Dover, NY

[17] Draine, B. T. 1985, *Ap. J. Suppl.*, **57**, 41.

[18] Jenkins. F. A.and White, H. E. 1957, in *Fundamentals of Optics*, McGraw-Hill, New York, p. 466.

[19] Bierman, L. 1967, in *The Zodiacal Light and the Interplanetary Medium*, NASA SP-150, p. 301.

[20] Belton, M. S. 1967, ibid, p. 279.

[21] Berg, O. E. and Grun, E. 1973, *Space Research XIII*, 1047.

[22] Bradley, J. P, Brownlee, D. E., and Veblen, D. R. 1983, *Nature* **301**, 473.

[23] Soberman, R. K. and Hemenway, C. L. 1965, *J. Geophys. Res.*, **70**, 4943.

[24] Lane, A. L., et. al. 1982, *Science*, **215**, 537.

[25] Greenberg, J. M. and Hage, J. I. 1990, *Ap. J.*, **361**, 260.

LIGHT SCATTERING BY DUST PARTICLES IN THE OUTER SOLAR SYSTEM

J.W. HOVENIER and P.B. BOSMA
Free University, Department of Physics and Astronomy
De Boelelaan 1081
1081 HV Amsterdam
The Netherlands

ABSTRACT. Photometric observations of the zodiacal light performed by Pioneer 10 indicated that there may be very little scattering by dust in the outer solar system. To shed more light on this problem we formulate explicit expressions for interpreting the brightness observed by a spacecraft travelling inside or outside a finite homogeneous cloud of scattering particles. An application is made to the ecliptic zodiacal light brightness as observed by Pioneer 10 and tabulated by Toller and Weinberg (1985). A satisfactory interpretation of these data as well as earthbound observations can be given by means of a model having a particle density distribution or mean scattering cross section which vanishes beyond 2.8 - 3.7 AU. Some implications for the nature and spatial distribution of the interplanetary dust are discussed.

1. Introduction

The main question we wish to address is "How large is the cloud of particles causing the visible zodiacal light?". Indications for very little light scattering by dust in the outer solar system were provided by photometric observations from the interplanetary spacecraft Pioneer 10 [Hanner et al., 1976; Weinberg et al., 1978; Schuerman, 1980]. After making corrections for background sky light Toller and Weinberg (1985) published numbers for the zodiacal light brightness as a function of elongation and Pioneer 10 distance to the Sun for three ecliptic latitudes. This material reveals, within observational inaccuracy, how the zodiacal light tapers off in the solar system during a voyage away from the Sun. A simple theoretical interpretation of this phenomenon is presented in this paper, where we restrict ourselves to dust and space-craft located in the plane of the ecliptic.

2. Method

Let us assume that the cloud of dust particles causing the visible zodiacal light has
 (i) an outer boundary radius, r_m,
 (ii) the same mixture of dust particles everywhere,

155

A.C. Levasseur-Regourd and H. Hasegawa (eds.), Origin and Evolution of Interplanetary Dust, 155–158.
© 1991 *Kluwer Academic Publishers.*

(iii) a particle number density $n(r) = n_0(r_0/r)^\nu$, where r is the distance of the particle to the Sun. Consequently, a spacecraft at a distance R from the Sun detecting the zodiacal light at elongation ϵ observes the brightness [cf. Buitrago and Mediavilla, 1986; Van Dijk et al., 1988]

$$I(\epsilon,R) = F_0 n_0 r_0 \bar{s} \, [(R/r_0)\sin\epsilon]^{-\nu-1} \int_{\theta_1}^{\theta_2} \varphi(\theta)\sin^\nu\theta \, d\theta \qquad (1)$$

where F_0 is the solar flux at $r = r_0$, $\bar{s}$ is the mean scattering cross section, $\varphi(\theta)$ is the mean volume scattering function normalized so that 4π times the average over all directions equals unity and θ_1 and θ_2 are the minimum and maximum scattering angle, respectively. The latter follow from the simple geometric relationships $\sin\theta_2 = (R/r_m)\sin\epsilon$, $\theta_1 = \epsilon$ (if $R < r_m$) and $\theta_1 = \pi - \theta_2$ (if $R \geq r_m$).

We performed model computations of $I(\epsilon,R)$ for various ν and r_m using Eq. (1) and Gaussian quadrature for each value of ν. The necessary input, i.e. $F_0 n_0 r_0 \bar{s} \varphi(\theta)$, was first taken from Hong (1985) for $r_m = \infty$ and then modified to be used for finite values of r_m by requiring that in each case $I(\epsilon, R=1 \text{ AU})$ equals that of Hong (1985). This modification has been reported by Van Dijk et al. (1988) and was employed here to let all our models for an arbitrary ν reproduce the earthbound observations of the zodiacal light equally well as the model with the same value of ν of Hong (1985) does.

Our model computations were compared with the 137 values of $I(\epsilon,R)$ tabulated by Toller and Weinberg (1985) for vanishing ecliptic latitude and in the range $70° \leq \epsilon \leq 180°$ with R (in AU) between 1.011 and 2.939. We used the average over elongation of the values tabulated for R = 1.011 to normalize our computed values of $I(\epsilon,R)$. For the brightness values the usual S10 (V) units were used. The estimated accuracy of the observed data is 2-3 units (Weinberg, 1988).

3. Results and discussion

A specimen of our results is shown in Fig. 1 where I is plotted versus R in AU, both on a logarithmic scale. Here $\nu = 1$ and $r_m = 2.85$ AU for the curve while $\nu = 1$ and $r_m = \infty$ for the straight line. For $r_m = \infty$ always straight lines appear in a plot like this since then $I \propto R^{-\nu-1}$ [see Eq. (1)]. The data points in Fig. 1 clearly do not lie on a straight line with the relevant slope. A similar behaviour is also found for other values of ϵ. Indeed, a statistical analysis of all 137 data points yields an extremely small likelihood for $r_m = \infty$. For finite values of r_m downward bending curves are found in plots like Fig. 1. These are generally in better agreement with the data and show that the effects of a finite r_m value become noticeable for a spaceship long before the outer boundary is reached. Since we are primarily interested in the decline of $I(\epsilon,R)$ for large R, and since the observed values of $I(\epsilon,R=1.011 \text{ AU})$ were used for normalization, we first used the least squares method to the data for $R \geq 1.861$ AU and found the root mean square of the differences between observed and calculated brightness to be at a minimum for $\nu = 1$ and $r_m = 2.85$ AU. Applying the same procedure

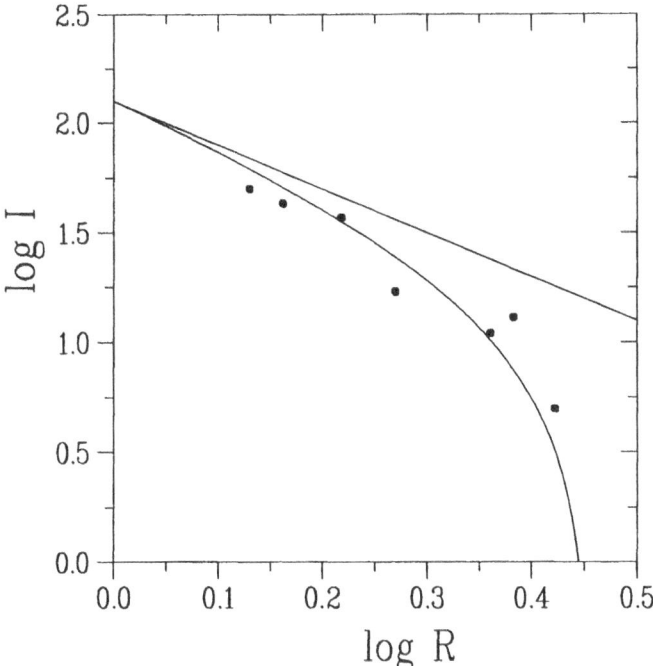

Fig. 1. Brightness of the ecliptic zodiacal light measured at $\epsilon = 125°$ by Pioneer 10 (filled circles) compared with model calculations for $\nu = 1$ and a boundary at 2.85 AU (curve) and at infinity (straight line). At $R = 2.939$ AU ($\log R = 0.4682$) the measured brightness was zero.

to all data yielded $\nu = 1.7$ and $r_m = 3.7$ AU. On the other hand, minimizing the root mean square of the relative differences between all non-vanishing observed and calculated brightnesses yields $\nu = 1.5$ and $r_m = 2.84$ AU. The values of ν mentioned in this section are not unreasonable in the light of analyses of earthbound and Helios data [cf. Hong, 1985; Lamy and Perrin, 1986]. The best we can say about r_m is probably that within the framework of our assumptions its value is presumably 2.8 to 3.7 AU. When more accurate data become available our analysis can be used to set stricter limits on the values of ν and r_m. Such data may be gathered during the CASSINI-mission and plans for such an undertaking have been made.

It should be noted that a model with $r_m = \infty$ and a power law dependence of $n(r)\bar{s}$ would also [cf. Eq. (1)] result in straight lines in plots of log I versus log R and are therefore also not in agreement with the observed data. On the other hand, it may very well be that there are still many particles beyond r_m but that the upper limit in the integral on the right-hand side of Eq. (1) is caused by lack of scattering by those distant particles. Several lines of evidence indicate that this is more likely than the absence of particles beyond r_m [see e.g. Cook (1978), Stanley et al. (1979), Fechtig (1984) and

Levasseur-Regourd et al. (1990)]. Consequently, we need more spacecraft in the outer solar system measuring particle densities as well as properties of the zodiacal light.

Acknowledgments

We are grateful to Drs. Hanner, Leinert, Van der Mee and Van Stokkum for useful comments and criticisms.

References

Buitrago, J. Mediavilla, E. (1986) 'Astron. Astrophys.' 162, pp. 95-98.

Cook, A.F. (1978) 'Icarus' 33, pp. 349-360.

Dijk, M.H.H. van, Bosma, P.B., Hovenier, J.W. (1988) 'Astron. Astrophys.' 201, pp. 373-378.

Fechtig, H. (1984) 'Adv. Space Res.' 4 (9), pp. 5-11.

Hanner, M.S., Sparrow, J.G., Weinberg, J.L., Beeson, D.E. (1976) in Interplanetary dust and zodiacal light, eds. H. Elsässer, H. Fechtig, Lect. Notes Phys. 48, Springer Verlag, Berlin, pp. 29-35.

Hong, S.S. (1985) 'Astron. Astrophys.' (146) pp. 67-75.

Lamy, P.L., Perrin, J.-M. (1986) 'Astron. Astrophys.' 163, pp. 269-286.

Levasseur-Regourd, A.C., Dumont, R., Renard, J.B. (1990) 'Icarus' 86, pp. 264-272.

Schuerman, D.W. (1980) in Solid Particles in the Solar System, Proc. I.A.U. Symp. 90, eds. I. Halliday, B.A. McIntosh, Reidel, Dordrecht, pp. 71-74.

Stanley, J.E., Singer, S.F., Alvarez, J.M. (1979) ' Icarus' 37, pp. 457-466.

Toller, G.N., Weinberg, J.L. (1985) in Properties and interactions of interplanetary dust, Reidel, Dordrecht, pp. 21-25.

Weinberg, J.L., Sparrow, J.G. (1978) in Cosmic dust, ed. J.A.M. McDonnell, Wiley and Sons, New York, pp. 75-122.

Weinberg, J.L. (1988) private communication.

LIGHT SCATTERING BY SOLAR SYSTEM DUST:
THE OPPOSITION EFFECT AND THE REVERSAL OF POLARIZATION

K. MUINONEN* and K. LUMME**

* Lowell Observatory, 1400 West Mars Hill Road,
Flagstaff, Arizona 86001, U.S.A.
** University of Helsinki, Observatory and Astrophysics Laboratory,
Tähtitorninmäki, 00130 Helsinki, Finland

1. Introduction

The opposition effect and the reversal of linear polarization, or negative polarization, at small phase angles have been almost universally observed in light scattered from atmosphereless solar system bodies (e.g., Seeliger 1887, Lyot 1929). Recent investigations have indicated that both phenomena can be qualitatively understood as resulting from a common physical mechanism: coherent multiple backscattering (Shkuratov 1989, Muinonen 1989). These findings have cast doubt on the hitherto accepted explanation that mutual shadowing alone is responsible for the opposition effect, and for the first time offer an acceptable interpretation of the polarization reversal near opposition. As for interplanetary dust, the coherent backscattering mechanism contributes both to the Gegenschein and to the almost certainly existing negative polarization branch (Roosen 1970, Lumme and Bowell 1985).

In the following, theoretical results supporting the coherent backscattering explanation are briefly presented. As future work, we suggest modeling light scattering by a particulate medium to include the first, second and, if necessary, higher orders of scattering in the range below the typical particle size.

2. Coherent Backscattering Mechanism

The mechanism of second-order coherent backscattering is illustrated in Figure 1, in which an electromagnetic plane wave with wave number k is scattered at two scattering centers separated by a distance d. The scattering centers can be individual particles, subparticles in an aggregate particle, or cracks or other optical inhomogeneities.

The phase difference between the wave components that propagate in opposite directions (cyclic passage) determines the interference. In the backward direction (at phase angle $\alpha = 0°$), the phase difference is always zero and the two paths of propagation coincide. This leads to constructive interference and coherent second–order backscattering. For non–zero phase angles, the interference varies from constructive to destructive depending on kd and the orientation of the system. Coherence also occurs in scattering orders higher than the second. Phase reddening and the color opposition effect can also be explained by this mechanism, since it predicts a narrower opposition effect for shorter wavelengths.

Negative polarization near opposition can be understood by calculating the phase difference in the yz–plane in the two scattering geometries shown in Figure 1. Since first–order scattering is predominantly positively polarized (e.g., Rayleigh scattering, Fresnel reflection), the scattering centers sufficiently far from each other interact mainly with the electric field vector perpendicular to the plane defined by the Sun and the scattering centers. The

159

A.C. Levasseur-Regourd and H. Hasegawa (eds.), Origin and Evolution of Interplanetary Dust, 159–162.
© 1991 Kluwer Academic Publishers.

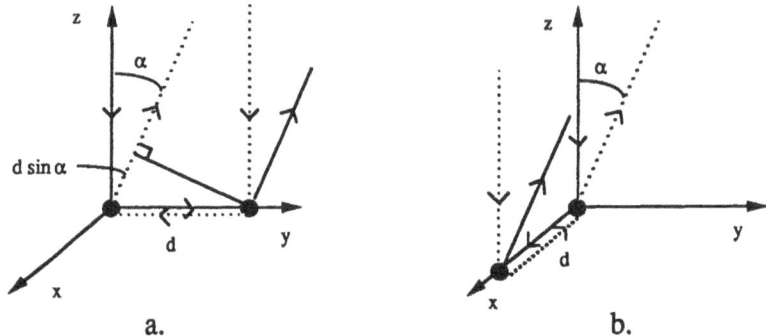

Fig. 1. Interference in second–order scattering (Muinonen 1990a). The wave components propagating in opposite directions (solid and dotted lines) interfere constructively at opposition ($\alpha = 0°$). For non–zero but small phase angles, the interference favors negative polarization: *(a)* in the yz–plane, in the scattering geometry leading to positive polarization, the interference depends on the phase difference $\delta = kd \sin \alpha$; but *(b)* the interference is always constructive in the geometry causing negative polarization.

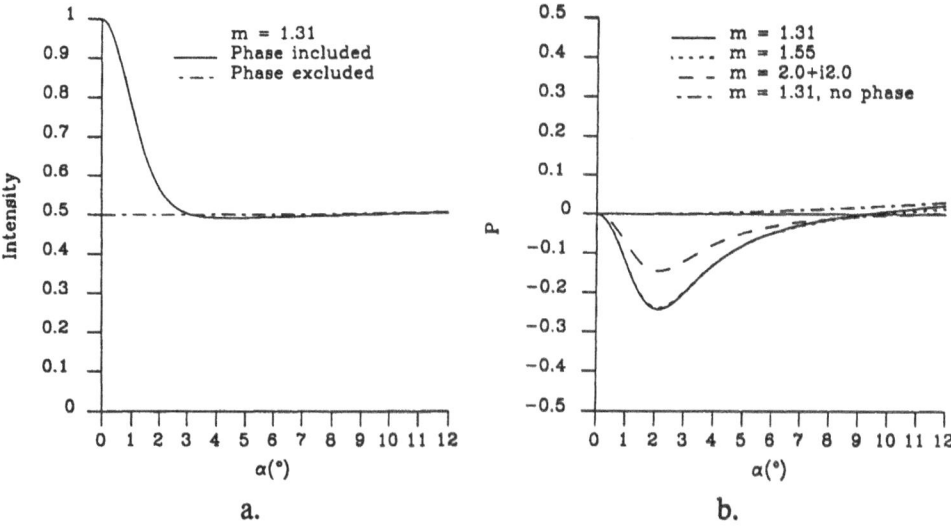

Fig. 2. Second–order reflection including and excluding the phase. Both *(a)* backward enhancement and *(b)* polarization reversal follow when the phase is included. The spherical elements (radius R) touch each other, $\kappa = 1/10k$ in an exponential size distribution $n(R) = \kappa \exp(-\kappa R)$, and m is the refractive index.

observer in the yz–plane will measure positive polarization from the geometry in Figure 1(a) and negative polarization from that in Figure 1(b). However, positive polarization undergoes a phase difference $\delta = kd \sin \alpha$, whereas the phase difference for negative polarization is zero at all phase angles (scattering centers in the xy-plane). Isotropic averaging over positions of the scattering centers will result both in an increase in brightness and in a reversal of polarization near the backward direction (at exactly zero phase angle the polarization goes to zero).

The simplest two–particle scattering problem is that of two electric dipole scatterers (Muinonen 1989). The backscattering enhancement and reversal of polarization are clearly present in second–order scattering, but due to the small scattering cross section, the phenomena do not show up in total scattering. The contribution due to multiple scattering increases when the other dipole scatterer of the previous calculation is replaced by a dielectric halfspace (Muinonen et al. 1990), and a quantitative confirmation is obtained for the backscattering peak and polarization reversal in total diffuse scattering. Finally, the second–order external reflection from two spherically curved surface elements indicates why the phase has to be included in the study of a close–packed medium (Muinonen 1990b). Figure 2 shows the results from horizontal isotropic averaging, both including and excluding the phase, for normal incidence. Neither the backscattering peak nor polarization reversal follows from calculations, in which phase is excluded.

3. Modeling Regoliths and Fluffy Particles

As shown above, coherent backscattering arises from inhomogeneities at scales of about a few microns. Spheres are known to exhibit the glory effect which, however, is due to a different interference mechanism. Both the regolith and interplanetary dust particles are believed to have sizes between 10 μm and 100 μm, which is larger than the typical range of coherence phenomena. This suggests a model in which, in the first approximation, scattering between the particles can be treated using geometric optics, although wave optics phenomena must be accounted for when calculating the single–particle phase function.

For planetary regoliths we can now assume

$$I_p(\mu_0, \mu, \psi) = \frac{1}{4} \varpi_0 P_p(\alpha) \, \mu_0 R_1(\mu_0, \mu, \psi) \, F + \mu_0 R_M(\mu_0, \mu, \psi) \, F \qquad (1)$$

for the perpendicular and parallel polarizations (subscripts $p \to \perp$ and $p \to \parallel$). In this model, μ_0 and μ are the cosines of the angles of incidence and emergence, ψ is the azimuth, α the phase angle, πF the incident solar flux density, ϖ_0 the single–particle albedo, and P_p is the single–particle phase function. The single and multiple reflection coefficients R_1 and R_M can be calculated from the geometric optics approximation as by Lumme et al. (1990), who generalized the classical radiative transfer theory, replacing the planar interface with a stochastic process. If the surface roughness tends to zero then $R_1 = S(\alpha)/(\mu_0 + \mu)$ which is the classical Lommel–Seeliger law corrected for the mutual shadowing function S.

As described earlier, we have calculated P_p for two extreme cases: when the scattering centers are very small or very large compared to the wavelength. It is, however, conceivable that the elements are on the order of 1 μm in the case of closely packed fluffy particles. Modeling this kind of situation could be done, as a first step, by the rigorous theory of two interacting spheres as formulated by Bruning and Lo (1971).

4. Discussion

At present, no valid theoretical model exists for quantitative analysis of the observations of the opposition effect and polarization reversal. We suggest that modeling could be initiated by studying dark particulate media using radiative transfer theory, including the effects of coherent multiple backscattering, mutual shadowing, and shadowing due to surface roughness.

Based on our calculations, we conclude that the width of the branch of negative polarization depends mainly on the size of the inhomogeneities, their distribution in the regolith, and on the refractive index. The observed negative branch can be ascribed to inhomogeneities on the order of 1μm. The mechanism predicts broader negative branches for larger refractive indices, in which case more energy is concentrated on low orders of scattering. This agrees well with observations.

Acknowledgment. We are grateful to Edward Bowell for valuable comments.

References

Bruning, J. H., and Y. T. Lo (1971). Multiple scattering of EM waves by spheres, parts I and II. *IEEE Trans. Ant. Prop.* **AP-19**, 378.

Lumme, K., and E. Bowell (1985). Photometric properties of zodiacal light particles. *Icarus* **62**, 54.

Lumme, K., Peltoniemi, J. I., and W. M. Irvine (1990). Diffuse reflection from a stochastically bounded, semi-infinite medium. *Trans. Theory Stat. Phys.*, in press.

Lyot, B. (1929). Recherches sur la polarisation de la lumière des planètes et de quelques substances terrestres. *Ann. Obs. Paris* **8**(1), 1.

Muinonen, K. (1989). Electromagnetic scattering by two interacting dipoles. *Proc. 1989 URSI Symp. EM Theory*, 428.

Muinonen, K. (1990a). Light scattering by inhomogeneous media: backward enhancement and reversal of linear polarization. Ph.D thesis, Report 3/1990, Observatory and Astrophysics Laboratory, University of Helsinki.

Muinonen, K. (1990b). Scattering of light by solar system dust: the coherent backscatter phenomenon. *1990 Proc. Finnish Astron. Soc.*, 12.

Muinonen, K. O., Sihvola, A. H., Lindell, I. V., and K. A. Lumme (1990). Scattering by a small object close to an interface. II: Study of backscattering. *J. Opt. Soc. Am. A*, in press.

Roosen, R. G. (1970). The Gegenschein and interplanetary dust outside the Earth's orbit. *Icarus* **13**, 184.

Seeliger, H. von (1887). Zur Theorie der Beleuchtung der grossen Planeten, insbesondere des Saturn. *Abh. Bayer. Akad. Wiss. Math. Naturwiss.*, Kl. **16**, 405.

Shkuratov, Yu. G. (1989). New mechanism of the negative polarization of light scattered by atmosphereless cosmic bodies (in Russian). *Astron. Vestnik* **23**, 2, 176.

THE OPTICAL PROPERTIES OF INTERPLANETARY DUST

P.L. LAMY, J.M. PERRIN
Laboratoire d'Astronomie Spatiale
Les Trois Lucs, 13012 Marseille, France

ABSTRACT. After briefly evaluating the observations of the Zodiacal Light and F-corona, we review the laboratory results on the light scattering by dust particles and the various theories which have been recently proposed. We then discuss the optical properties of the dust with emphasis on the phase function, the polarization, the color, the albedo and the local enhancement in the Gegenschein.

1. Optical observations of interplanetary dust and deduced properties

The present review is concerned with the light scattered by interplanetary dust that is wavelengths below approximately 3 μm; beyond, the observed brightness is dominated by thermal emission. The observations relevant to the optical properties of the dust grains are essentially the all-sky brightness and polarization encompassing the ultra-violet, the visible and the near-infrared.

No new major observations have been obtained since the previous colloquium of this series (Marseille, 1984). We therefore limit ourselves to a brief summary of the present situation.

At 1 AU, the survey of the visible brightness and polarization of the zodiacal light by Dumont and Sanchez (1975, 1976) remains the most complete and reliable source of data (see also the tabulations by Levasseur-Regourd and Dumont, 1980 and by Fechtig et al., 1981). The axisymmetric, non-spherical model of the F-corona obtained by Koutchmy and Lamy (1985) bridges nicely to the Dumont-Sanchez data along the ecliptic and meridian directions, the combination of the two sources resulting almost in an all-sky map.

As pointed out by Lamy and Perrin (1986), these results look very reasonable on one hand but are in conflict with other good-quality data on the other hand. A good example is the brightness of the anti-solar point, the center of the Gegenschein, which ranges from 150 to 250 $S_{10}(V)$. Progress in the absolute photometry and polarization are still needed and the forthcoming results from the COBE satellite will make a significant contribution in this direction. Outside the visible, the situation is far less satisfactory as we have only limited coverages in the ultra-violet - see Cebula and Feldman (1982) and Lillie (this volume) - and the near infrared-red (Leinert and Grün, 1991, for a recent review). Polarization data in these spectral ranges are especially meager.

Inside 1 AU, visible brightness and polarization measurements of limited spatial coverage have been obtained by the Helios space probes (Leinert et al., 1982). Outside 1 AU, the Pioneer spacecrafts have secure extended spatial brightness measurements out to a distance of 3 AU (Toller and Weinberg, 1985).

A.C. Levasseur-Regourd and H. Hasegawa (eds.), Origin and Evolution of Interplanetary Dust, 163–170.
© 1991 Kluwer Academic Publishers.

This observational material is classically analyzed to produce the volume scattering function $\psi(\theta)$ and its associated polarization p(θ) which characterizes the scattering phase function of a unit volume of interplanetary dust, the spectral variation of the intensity I(λ) - that is the color - and of the polarization p(λ) and the albedo. These quantities are amenable to confrontations with laboratory measurements and theoretical calculations and provide direct information on the optical properties of the grains.

Retrieving $\psi(\theta)$ and p(θ) implies a transformation from the spatial coordinates (elongation, latitude) to the scattering angle. This is the so-called inversion technique which has been pioneered by Dumont (1973). It allows to get rid of the integral over the spatial distribution and to obtain $\psi(\theta)$ and p(θ). Note that these quantities still include the integral over the size distribution. The inversion effort may be broadly divided into three main streams:

i) inversion limited to the plane of the ecliptic (Dumont and Sanchez, 1975; Leinert et al., 1976);

ii) all-sky inversion attempted by Schuerman (1979) and Buitrago et al. (1980) and successfully carried out by Lamy and Perrin (1986) who further realized the inversion at several heliocentric distances;

iii) local inversion allows to retrieve $\psi(\theta)$ at specific spatial locations with minimal uncertainty (Dumont, 1973; Dumont and Levasseur-Regourd, 1988).

Note that the all-sky inversion gives also access to the three-dimensional distribution of interplanetary dust.

2. Light scattering by dust particles

We now review the situation and progress in the field of light scatering by dust particles from the points of view of experimental measurements and theoretical treatments.

The group at Bochum University headed by the late R. Giese has been very active in experimental work, both in the visible and microwave domains, pertaining to interplanetary dust. Laser ($\lambda = 0.633$ μm) scattering measurements on single particles of diameter 30 to 80 μm performed by Weiss-Vrana (1983) have been pursued and extended by Killinger (1987). The diameter range has been increased to 20-200 μm, the laser was upgraded to allow multi-color measurements ($\lambda = 0.476$, 0.568 and 0.647 μm) and the coverage in scattering angle has been extended to 170°. This effort resulted in a wealth of high quality data which should be the basis for better understanding the scattering by complex particles. The investigation also includes grains extracted from various meteorites (Allende, Murchison) which are particularly suitable for interplanetary dust studies. An interesting finding of Killinger (1987) is the color dependence of the phase function: as expected, the diffraction lobe is colorless but a strong color effect appears at large scattering angle. This behaviour was also found by Bliek and Lamy (1988) in a totally different experiment where a jet of dust particles whose size distribution extends from 1 to 40 μm is illuminated at five different wavelengths (0.447 to 0.829 μm). The microwave facility at Bochum University has also produced significant results in the field of light scattering by complex e.g. Zerull et al. (1977). More recently Gustafson et al. (1989) has started to investigate porous aggregates of small spheres having dimensions of a few wavelengths. These authors experimentally obtained the phase function, the linear and cross polarization and also attempted to solve theoretically the problem. Another microwave facility located at the University of Florida at Gainesville has concentrated on different types of particles (e.g., ellipsoids, cylinders, cubes...) which are less appropriate to the interplanetary dust situation (Schuerman, 1980).

The field of theoretical investigations of light scattering by complex particles has been quite active in recent years. An exhaustive review is beyond the scope of the present article as the developments are further highly specialized and technical. One widespread approach which has also been actively pursued by us starts with the classical electrodynamics equation for the electric field $\mathbf{E}$ interacting with a dust praticle

$$\nabla \times (\nabla \times \mathbf{E}) - n^2 k^2 \mathbf{E} = 0 \tag{1}$$

where k is the wave number ($= 2\pi/\lambda$) and n, the complex index of refraction (a tensor in the most general case). The solution of equation (1) is given by the integral equation

$$\mathbf{E}(\mathbf{r}) = \mathbf{E}_0(\mathbf{r}) + k^2 \int_V G(\mathbf{r}\text{-}\mathbf{r'}) [n^2(\mathbf{r'}) - 1] \mathbf{E}(\mathbf{r'}) d\mathbf{r'} \tag{2}$$

where $\mathbf{E}(\mathbf{r})$ is the total field at the point $\mathbf{r}$ (inside or outside the particle), $\mathbf{E}_0(\mathbf{r})$ is the incident field at the point $\mathbf{r}$ and G is the Green function which describes the field at the point $\mathbf{r}$ resulting from the interaction of an unit incident field with an element of matter $d\mathbf{r'}$, i.e. an "elementary particle" in the framework of the classical electrodynamic theory. The integral in equation (2) extends over the volume V of the particle. Depending upon the particular case (small or large particles) and the corresponding approximations which can be introduced in the expression of the Green function, one ends up with different practical solutions. The above elements of matter $d\mathbf{r'}$ are dipoles, so the discrete dipole approximation (DDA) as first introduced by Purcell and Pennypacker (1973) is obtained. This is well adapted to small particles as the number of dipoles remain within computer capabilities. For larger grains, one may consider larger, non-dipolar, elements of matter and solve for their complex mutual interactions as performed by Grim and Greenberg (this volume). One may also recover, the so-called eikonal solution (Chiappetta, 1980). All these solutions lead to intensive calculations often requiring large, vectorial computers. The DDA is well adapted to handle inhomogeneous or porous particles as impurities are modeled by different dipoles while porosity is simply modeled by voids. An example of a highly rough and porous particle is given in Fig. 1 and compared with spheres (Mie solution) of equivalent mass or equivalent cross-section. One notes important differences, both for the phase function and the polarization. The approach of effective medium has been introduced in particular for large particles for which the DDA becomes impractical. However, we have shown (Perrin and Lamy, 1990) that this application requires very stringent conditions or leads to large uncontroled errors. So one should be excessively cautious when using it.

3. Interpretation and discussion

3.1. THE VOLUME SCATTERING FUNCTION

As introduced in section 1, the volume scattering function $\psi(\theta)$ obtained from inversion of observational data represents the phase function of a unit volume of interplanetary dust. Basically, it can be expressed by

$$\varphi(\theta,\lambda) = \frac{c}{k^2} \int F(\theta,\lambda,s)\, S(s)\, ds$$

where the integral extends over the size distribution S(s). Fig. 2 gives our nominal solution (Lamy and Perrin, 1986) characterized by a broad diffraction lobe, a shallow minimum in the interval 80-120° and a broad backward enhancement by a factor 2. We further found that the shape of $\psi(\theta)$ does not vary with heliocentric distance d while its magnitude varies as $d^{-0.3}$. This implies that the albedo increases as d decreases, a result confirmed by local inversion (Dumont and Levasseur-Regourd, 1988). It is interesting also to emphasize that the absolute value of $\psi(\theta)$ at 1 AU is fully compatible with the measured differential spatial density of interplanetary grains.

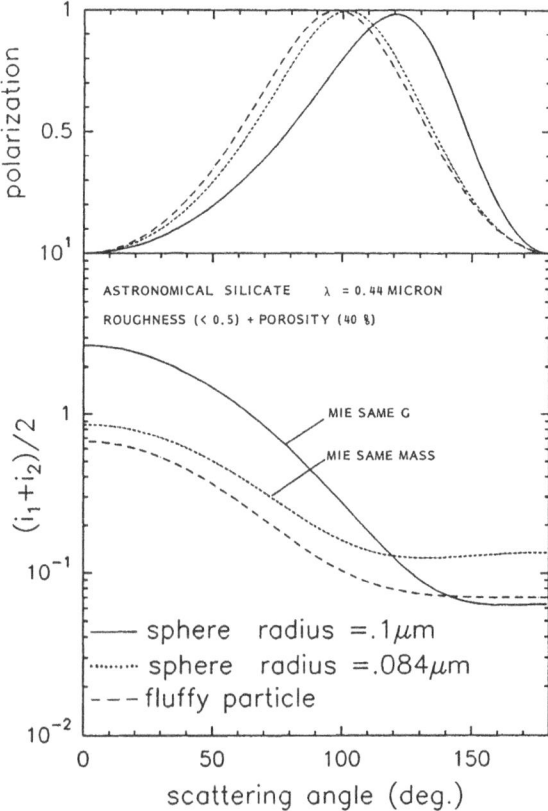

Fig. 1: The DDA result for the phase function and polarization of a highly rough and porous silicate particle of 0.1 μm radius (broken line) and comparison with the Mie result for spheres of equivalent mass (dotted line) and equivalent cross-section (solid line)

3.2. THE BACKSCATTERING SPIKE

A careful observation of the Gegenschein reveals a narrow spike (Maucherat et al., 1986) which has unfortunately escaped attention, in particular in the low resolution photometric scans. This spike has a width of approximately 1.5° to 2.5° and bears a direct ressemblance to the backward spike found in observations of several asteroids. An important point is that it is not correlated with the negative branch of polarization (see below) which extends over 15° to 20°. It may be connected with a fundamental aspect of optics known as the principle of weak localization. Among several mechanisms which may be invoked to explain this phenomenom, it seems that multiple effects (reflection, scattering) produced by the roughness of interplanetary grains is the most plausible.

3.3. THE POLARIZATION FUNCTION p(θ)

When inverting separately the two polarized components of the brightness of the zodiacal light, we retrieve the polarized components of the volume scattering function and finally, the polarization function p(θ) which is therefore also "integrated" over the size distribution. p(θ) is displayed in Fig. 2 and exhibits a broad maximum in the interval 70-100° reaching 0.29, and two negative branches: one in the backscattering domain which is not well defined since the observations themselves are not sufficiently accurate (the inversion angle is about 168°); and a second branch in the forward direction with an inversion angle of approxiamtely 22°. Overall, the polarization values decrease with decreasing heliocentric distance as $d^{0.3}$.

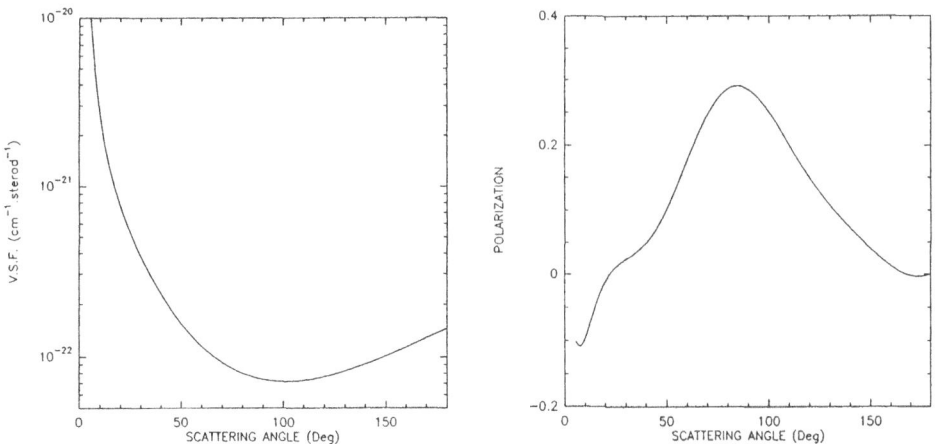

Fig. 2 The volume scattering function and its polarization at 1 AU

3.4. INTERPRETATION

As already hinted in the above sections, the behaviour of both $\psi(\theta)$ and p(θ) are best explained by **rough**, moderately absorbing dust grains. It must be emphasized that other explanations such as those obtained by the Mie theory, i;e., assuming spherical grains, require restrictive, not to say unrealistic, assumptions. Furthermore, the simultaneous presence of two negative branches of polarization (if confirmed) cannot be explained by this theory but is present in laboratory results on irregular grains (Killinger, 1987). Also, on the basis of these results, the behaviours of both $\psi(\theta)$ and p(θ) are well compatible with a mixture of Allende matrix and Murchison grains. However, this must be taken as a trend since the laboratory data are for single grains, i.e., not integrated over a size distribution. But this points to a very coherent view of the zodiacal cloud. Finally, the decrease of p(θ) and the increase of albedo as the heliocentric distance decreases are probably connected to a change in size distribution resulting from catastrophic collisions among interplanetary grains and the loss of absorbing material such as organic material.

3.5. THE COLOR OF THE ZODIACAL LIGHT

As a presentation of the observational results may be found in the review of Grün and Leinert (1991), we limit ourselves to discussing the various aspects of this question following a more detailed investigation (Perrin and Lamy, 1989). First, it must be understood that the color index of the zodiacal light deduced from brightness measurements at different wavelengths depends upon the elongation, while the color ratio for light scattered by dust grains is intrinsec since it involves the scattering cross-section. The two indices are equal only in the limit of zero elongation when, furthermore, the spatial density varies as r^{-1}. Second, the widespread view that the presence of small grains implies a blue color is incorrect as this results from the Rayleigh law which is strictly never obeyed. This is due to the variation of the refractive index with wavelength and the large values of its complex part exhibited by all materials in the ultra-violet. This is of course confirmed by a simple Mie calculation for spherical grains. Third, the color depends upon the amplitude of grain roughness with respect to wavelength: small amplitudes lead to reddening while large amplitudes lead to blueing. Finally, the very observation of the brightness of the zodiacal light as a function of elongation introduces an intrinsec color effect, either reddening or blueing; this theoretical prediction is supported by actual observations.

3.6. THE ALBEDO

Following the work of Hanner et al. (1981), the albedo Ap(θ) of a single dust grain is well defined as a function of its total intensity function i(θ) and its geometric cross-section. As a consequence, the albedo depends upon the scattering angle, the wavelength and the size of the grain as proved by laboratory measurements (Killinger, 1987; Giese et al., 1986). To derive the albedo from observed intensities which are integrated over the size distribution and along the line-of-sight certainly is an hopeless task. As we have proposed for cometary grains (Lamy et al., 1987), a reasonable approach may be to find an acceptable fit to the volume scattering function $\psi(\theta)$ - from the point of view of composition and scattering theory - and to compute the albedo in restricted size intervals, typically $\Delta s/s = 0.1$. It can be shown from theoretical considerations that the method which combines visible and infrared brightnesses and which was devised for asteroids does not apply at all to dust particles. Finally, the method relying on the slope of the negative branch of polarization calibrated for rough surfaces and used for determining the albedo of planetary objects again completely fails for dust particles (Lamy, in preparation). A sound approach to the question of the albedo of interplanetary dust is probably to measure the phase function of collected IDP.

4. Conclusion

In our opinion the situation of the observations of the zodiacal light is somewhat paradoxal. On the one hand, the set of available data, in particular from the Pioneer and Helios space probes have not been fully interpreted to reach a synthesis of the properties of interplanetary dust. On the other hand more data are needed, especially polarization data, as well as a better photometric accuracy. The forthcoming COBE data will partly fill this need but the inner zodiacal light remains poorly known. The next solar eclipse and the SOHO mission may help remedy this situation. Direct determinations of the spatial density of dust using impact detectors are also needed to help obtain the volume scattering function from inversion. The Galileo, Ulysses and Cassini missions should provide this information. A better understanding of the optics of complex dust particles has been achieved in recent years thanks to both laboratory measurements and theoretical work. Overall it appears that we now have a coherent understanding of the zodiacal light as rough grains of mostly meteoritic composition satisfy most if not all observational constraints.

References

Bliek, P. and Lamy, P. (1988) 'Laboratory measurements of light scattering by dust particles', E. Bussoletti et al. (eds.), Experiments on Cosmic Dust Analogues, Kluwer Academic Publishers, pp. 253-258.

Buitrago, J., Gomez, R. and Sanchez, F. (1983) 'The integral equation approach to the study of interplanetary dust', Planet. Space Sci. 31, 373-376.

Cebula, R.P. and Feldman, P.D. (1982) 'Ultraviolet spectroscopy of the zodiacal light', Ap.J. 263, 987-992.

Chiappetta, P. (1980) 'A new model for scattering by irregular absorbing particles', Astron. Astrophys. 83, 348.

Dumont, R. (1973) 'Phase function and polarization curve of interplanetary scatterers from zodiacal light photopolarimetry', Planet. Space Sci 21, 2149-2155.

Dumont, R. and Sanchez, F. (1975) 'Zodiacal light photopolarimetry. II Gradients along the ecliptic and the phase functions of interplanetary matter', Astron. Astrophys. 38, 405-412.

Dumont, R. and Sanchez, F. (1976) 'Zodiacal light photopolarimetry. III. All-sky survey from Teide 1964-1975 with emphasis on off-ecliptic features', Astron. Astrophys. 51, 393-399.

Dumont, R., and Levasseur-Regourd, A.C. (1988) 'Properties of interplanetary dust from infrared and optical observations', Astron. Astrophys. 191, 154-160.

Fechtig, H., Leinert, Ch. and Grün, E. (1981) 'Interplanetary dust and zodiacal light', in Landolt-Börnstein, New Series, Volume 2a, K. Schaifers and H.H. Voight (eds.),

Chapter 3.3.4., Springer Berlin-Heidelberg-New-York, pp. 228-243.

Giese, R.H., Killinger, R.T., Kneissel, B. and Zerull, R.H. (1986) 'Albedo and colour of dust grains: laboratory versus cometary results', ESA SP-250, pp. 53-57.

Gustafson, B.A.S., Zerull, R.H., Corbach, E. and Schulz, K. (1989) 'Light scattering by open-structured and filamentary dust aggregates; experiment and theory', P.M.M. Jenniskens and J.I. Hage (eds.), Fluffy structures II, University of Leiden, pp. 3-6.

Killinger, R.T. (1987) Dissertation, Ruhr-Universität, Bochum.

Koutchmy, S. and Lamy, P.L. (1985) 'The F-corona and the circum-solar dust evidences and properties', R.H. Giese and P. Lamy (eds.), Properties and interactions of interplanetary dust. D. Reidel publishing company, Dordrecht-Boston-Lancaster-Tokyo, pp. 63-74.

Lamy, P.L. and Perrin, J.M. (1986) 'Volume scattering function and space distribution of the interplanetary dust cloud', Astron. Astrophys. 163, 269-286.

Lamy, P.L., Grün, E. and Perrin, J.M. (1987) 'Comet P/Halley: implications of the mass distribution function for the photopolarimetric properties of the dust coma', Astron. Astrophys. 187, 767-773.

Leinert, C. and Grün, E. (1991) 'Interplanetary Dust', R. Schwenn and E. Marsch, Springer (eds.), Physics of the Inner Heliosphere.

Leinert, C., Link, H., Pitz, E. and Giese, R.H. (1976) 'Interpretation of a rocket photometry of the inner zodiacal light', Astron. Astrophys. 47, 221-230.

Leinert, C., Richer, I., Pitz, E. and Hanner, M. (1982) 'Helios zodiacal light measurements - a tabulated summary', Astron. Astrophys. 110, 355-357.

Levasseur-Regourd, A.C. and Dumont, R. (1980) 'Absolute photometry of zodiacal light', Astron. Astrophys. 84, 277-279.

Perrin, J.M. and Lamy, P.L. (1990) 'On the validity of effective medium theories to light extinction by inhomogeneous dust particles', Astrophys. J. 364, 146-151.

Perrin, J.M. and Lamy, P.L. (1989) 'The color of the zodiacal light and the size distribution and composition of interplanetary dust', Astron. Astrophys. 226, 288-296.

Purcell, E.M. and Pennypacker, C.R. (1973) 'Scattering and absorption of light by nonspherical dielectric grains', Astrophys. J. 186, 705-714.

Maucherat, A., Llebaria, A. and Gonin, J.C. (1986) 'A general survey of the gegenschein in blue light, Astron. Astrophys. 167, 173.

Schuerman, D.W. (1979) 'Inverting the zodiacal light brightness integral', Planet. Space Sci. 27, 551-556.

Schuerman, D.W. (1980) 'The microwave analog facility at Sunya: capabilities and current programs', D.W. Schuerman (ed.), Light Scattering by Irregular Shaped Particles, New-York, pp. 227-232.

Toller, G.N. and Weinberg, J.L. (1985) 'The change in near-ecliptic zodiacal light brightness with heliocentric distance' R.H. Giese and P. Lamy (eds.), Properties and interactions of interplanetary dust. D. Reidel publishing company, Dordrecht-Boston-Lancaster-Tokyo, pp. 63-74.

Weiss-Wrana, K. (1983) 'Optical properties of interplanetary dust: comparison with light scattering by larger meteoritic and terrestrial grains', Astron. Astrophys. 126, 240-250.

Zerull, R.H., Giese, R.H. and Weiss, K. (1977) 'Scattering functions of nonspherical dielectric and absorbing particles vs Mie theory', Appl. Optics 16, 777-778.

THE INFRARED ZODIACAL LIGHT

MARTHA S. HANNER*
Institute for Astronomy
University of Hawaii
Honolulu HI 96822

ABSTRACT. Thermal emission from interplanetary dust is the main source of diffuse radiation at λ 5-50 μm. Analysis of infrared sky maps from IRAS and ZIP lead to the result that the average optical properties of the dust change with heliocentric distance. The present uncertainties in calibration should be resolved by COBE. Existence of a dust sublimation zone at 4 solar radii awaits confirmation at the next solar eclipse.

1. Introduction

Recent surveys of the background sky at infrared wavelengths have enabled us to take a new look at the average optical properties and spatial distribution of the interplanetary dust. In fact, thermal emission from interplanetary dust is the strongest source of diffuse radiation at wavelengths 5-50 μm. Measurements at several wavelengths yield the average temperature and albedo of the grains and, by applying appropriate inversion techniques, the heliocentric gradients of these quantities. The IRAS sky survey detected structure in the zodiacal cloud--dust bands and comet trails--giving us, for the first time, an observational link to the sources of the interplanetary dust cloud (see review by Sykes, this volume).

This paper will describe the infrared observational database, discuss the dust properties derived from the infrared data, and make some recommendations for future observations and analysis.

2. Infrared Observations

Absolute calibration and separation from other diffuse radiation sources are the two major difficulties in obtaining reliable measurements of the infrared zodiacal light. Emission from the earth's atmosphere and thermal radiation from the instrument complicate the problem.

Infrared observations of point sources normally screen out foreground and background radiation by chopping--i.e., by rapidly comparing the signal from the object + background with the signal from the background alone a small angular distance away. Such a technique is inappropriate for obtaining absolute measurements of the spatially extended zodiacal dust emission. To circumvent the high foreground emission, the zodiacal emission has to be observed from high altitude (balloon, rocket) or from space, with Helium-cooled sensors. Ideally the instrument should chop against an accurately calibrated internal reference source.

*on leave from Jet Propulsion Laboratory, Pasadena CA 91109

A.C. Levasseur-Regourd and H. Hasegawa (eds.), Origin and Evolution of Interplanetary Dust, 171–178.

Early measurements from sounding rockets at large solar elongation angles were carried out by Soifer et al (1971) and Briotta (1976). Briotta obtained an 8-13 μm spectrum which showed a silicate emission feature. If confirmed, this implies that sub-μm or μm-sized silicate grains are abundant. Salama et al (1987) obtained balloon measurements of the spectral energy distribution at λ11, 19, 50, 108, 225 μm and the brightness gradient at ecliptic longitudes 10-90 deg. Because the signal was chopped against a sky position several degrees away, the measured intensities are differential.

Zodiacal emission maps at λ 11 and 20 μm and solar elongation angles 30 < ε < 75 deg were made on a July, 1974 rocket flight by Price et al (1980). The intensity at ecliptic latitude β > 30 deg was taken as the background level; this procedure will underestimate the true brightness. The absolute calibration was revised by Price et al (1982); the zodiacal light intensities in the 1980 paper should be decreased by a factor of 2.

Much more extensive maps at 15 wavelengths between 2 and 30 μm were obtained by Murdock and Price (1985) during two rocket flights in July 1980 and August 1981--the so-called ZIP experiment. The cooled radiometer included an internal absolute reference source. Pre- and post-flight calibration agreed to 5%; the absolute calibration accuracy is estimated to be 20%. Scans covered solar elongation angles 22 < ε < 180 deg and ecliptic latitudes -60 < β < 90 deg. Data within 3 deg of the galactic plane were eliminated, but otherwise no attempt at separating the zodiacal and galactic components was made. The sky coverage allowed the authors to determine a 3-dimensional model for the dust spatial distribution. The dust bands at β ~ 10 deg are evident in their plots at ε = 60 and 90 deg. Although the authors state that no spectral features were detected, it appears from their plots that a 10 μm silicate feature can not be ruled out.

Murdock and Price compared their values with previous data. The ZIP intensities are 35% higher than the (corrected) July 1974 intensities and 40% lower than Soifer et al (1971) and Briotta (1976) at ε = 103 deg . More important for modeling the dust properties, the ZIP intensities are a factor of two lower than the IRAS intensities in Hauser et al (1984) at 12 and 25 μm. The final IRAS calibration reduced the discrepancy to a factor of ~1.5 (Good 1988).

Launched in January 1983, the IRAS satellite carried out a 10 month sky survey in 4 broad spectral bandpasses centered at 12, 25, 60, and 100 μm (Neugebauer et al 1984). While the primary objective was to survey point sources and small extended objects, the diffuse sky background was measured as well. The focal plane array consisted of 62 detectors, arranged so that each point in the sky scanned across two detectors in each bandpass (Beichman et al 1988). The spacecraft was placed in a sun-synchronous 900 km polar orbit, with the orbit plane approximately perpendicular to the sun-earth line. The usual observing pattern was to scan at a fixed solar elongation angle during one orbit, then offset by 1/4 deg (half of the focal plane field of view) on each successive orbit. Scans during the first 6 months were restricted to 80 < ε < 100 deg; in the last 3 months the region 60 < ε < 120 deg was surveyed. A field near the north ecliptic pole was measured regularly to monitor the electronic baseline stability; baseline drift was < 5% per day at 12, 25 μm and < 20% per day at 60 and 100 μm. Absolute calibration of the extended emission was based on the point source absolute calibration (Beichman et al 1988).

The IRAS project prepared a zodiacal observation history file, a time-ordered listing of the background flux in each bandpass, averaged into 0.5 x 0.5 deg bins. This file contains approximately 8000 scans for studying the large- scale structure of the zodiacal emission and annual variations. No synopsis of this database has ever been published in tabular or graphical form.

Before one can study the zodiacal emission, separation of the galactic background emission has to be made. Table 1, from Boulanger and Perault (1988) compares the zodiacal and galactic emission in the 4 bandpasses. The zodiacal emission dominates at 12 and 25 μm; the galactic component can be fairly readily separated by assuming symmetry with respect to the galactic plane and the zodiacal dust symmetry plane. Separation is more difficult at 60 and 100 μm. Boulanger and Perault used the correlation between the infrared emission and the H I gas emission to identify the galactic component. They find a residual isotropic emission of ~1.2 MJy/sr at 100 μm, which could be of either solar system or galactic origin. With the present calibration uncertainties, however, it is premature to attach significance to this result.

TABLE 1. Comparison of Galactic and Zodiacal
emission in the IRAS database (MJy/sr)*

λ	Zodiacal Emission ε = 90 °		Galactic Emission	
	β=0°	β=90°	lbl < 2°	b = 90°
12 μm	40	14	6	0.05
25	85	28	10	0.08
60	38	7	30	0.2
100	10	2	130	1.0

*Boulanger and Perault (1988)

COBE, the Cosmic Background Explorer, has just completed a 10-month survey of the diffuse background sky, from an orbit similar to that of the IRAS satellite. In contrast to IRAS, which was optimized for the survey of point sources, the COBE instruments were designed to make extremely accurate measurements of the diffuse radiation, as the microwave spectrum of the 2.735 K cosmic background so eloquently demonstrates (Mather et al 1990a). The diffuse infrared background experiment (DIRBE) maps the sky simultaneously in ten bandpasses from 1-300 μm, including the four IRAS bandpasses (Gulkis et al 1990; Mather et al 1990b). Linear polarization is measured in the three short wavelength channels J(1.2 μm), K(2.2 μm), L (3.5 μm). The field of view is 0.7 x 0.7 deg and the signal is chopped between the sky and a zero-flux internal reference. The instrument views at an angle of 30 deg to the spacecraft spin axis, sampling elongations 64 < ε < 124 deg on every spin and covering half the sky every day. The complete database should allow unambiguous separation of the zodiacal and galactic emission.

A very preliminary comparison of the DIRBE fluxes with those of IRAS indicates zero-point differences of a few MJy/sr in all four bandpasses and an overestimate of the IRAS DC gain at 60 and 100 μm. The combined errors are largest at 100 μm toward faint sky regions, amounting to a factor of 3 (IRAS higher) at the ecliptic pole (Mather et al 1990b). We eagerly await the results from this experiment!

3. Models of the Zodiacal Emission

Whether viewed in visible or infrared light, the observed zodiacal light is an integral over the line of sight. The spatial distribution and the scattering or emitting properties of the dust are convolved together in this integration. At visual wavelengths, the angular scattering function of the dust is one of the unknowns complicating the line-of-sight integration. In the infrared, we may make the simplifying assumption that the thermal emission from the dust is isotropic. Instead, the temperature of the dust grains, and the variation of the temperature with heliocentric distance, have to be known or assumed. Based on the size distribution of the dust measured from space probes near 1 AU, particles 10-100 μm in size should make the major contribution to the zodiacal emission as viewed from earth (Grün et al 1985).

Two approaches have been applied to modeling the zodiacal light, in order to extract physical information about the dust. In the first approach, global models are constructed, assuming some parameters and varying others, and predictions of the integrated zodiacal light are computed for comparison with the observations. While this approach does not necessarily lead to a unique answer, one can distinguish the types of models and parameter ranges that are consistent with the data.

Over the years, numerous models for the spatial distribution of the interplanetary dust cloud have been published, based on the optical zodiacal light isophotes. A good summary is given by Giese et al (1986). The various models have been compared with the ZIP and IRAS data by Giese and Kneissel (1989) (see also Kneissel and Mann, this volume). Spatial distributions can be formulated which are compatible with both optical and infrared observations at $\epsilon \geq 60$ deg. But models which are similar at $60 < \epsilon < 120$ deg (the range covered by IRAS and DIRBE) can deviate markedly at small elongations, predicting either a minimum or a maximum dust density at small heliocentric distances. The optical data imply a maximum, while ZIP data apparently imply a minimum, although even the line of sight at $\epsilon = 22°$ samples no closer to the sun than $r = 0.34$ AU. Yet zodiacal light observations from the Helios probe at 0.3 AU show no sign of a decrease in dust density at $r \geq 0.1$ AU (Leinert et al 1981). The dust distribution is surely more complex than an idealized mathematical model, and probably consists of two or more components with different orbital distributions, size distributions, and optical properties (e.g., Kneissel and Mann, this volume; Levasseur-Regourd et al. this volume). Infrared data at small ϵ are clearly important; the answer is not so simple as to assign one dust component to explain the infrared observations and another to account for the optical scattering!

The annual variation of the ecliptic pole brightness in the IRAS database can be used to define the dust symmetry plane relative to the ecliptic near 1 AU. The inclination is found to be ~1.6 deg and the longitude of the ascending node $\Omega = 43$-77 deg (Hauser 1987). The symmetry plane derived from earth-based optical zodiacal light data is $i = 1.5 \pm 0.4$ deg and $\Omega = 96 \pm 15$ deg. (Dumont and Levasseur-Regourd 1978). In contrast, the symmetry plane measured from the Helios probe at $r < 1$ AU has $i = 3.0 \pm 0.3$ deg and $\Omega = 87 \pm 4$ deg, (Leinert et al 1980), closer to that of Venus' orbit and supporting the concept of a warped symmetry surface for the dust distribution.

The IRAS survey has revived interest in modeling the zodiacal emission (see Hauser 1987 for a review). Some models, such as that of Boulanger and Perault (1988), are strictly empirical fits to the survey scans, with the goal of accurately subtracting the "foreground" zodiacal emission from the galactic emission. Other models have been global, specifying the 3-D spatial distribution, dust temperature, and emissivity. However, the large number of parameters and the differing assumptions about grain properties make the results less than unique. Take, for example, the temperature distribution, $T(r) \propto r^{-\delta}$. Deul & Wolstencroft (1988), Rowan-Robinson et al. (1990), and Hong and Um (1987) assume $\delta = 0.5$ (blackbody), while Good (1988, 1990) solves for it, finding $\delta = 0.36$. Röser and Staude (1978), Reach (1988), and Temi et al (1989) compute $T(r)$ from Mie theory. Dirty silicate spheres or large absorbing spheres will have temperatures close to a blackbody, while small absorbing grains are considerably hotter and have a smaller radial gradient. Collected interplanetary dust particles (IDPs) have irregular structure and heterogenous composition, and their optical properties surely differ from those of homogeneous spheres. Giese et al (1978) proposed large fluffy absorbing particles to explain the observed zodiacal light polarization. Very irregular particles, even large ones (size 10-100 μm), may have wavelength-dependent cross-sections, and thus temperatures which differ from the blackbody approximation. The temperature gradient retrieved from inversion of the zodiacal emission integral yields $\delta = 0.33$, significantly less steep than that of a blackbody, as described below. Of course, the assumed temperature gradient affects the spatial distribution derived from the model.

The second approach to modeling is to make use of inversion techniques. Hong and Um (1987) developed an inversion method to recover the heliocentric gradient of the volumetric cross section, the product of the spatial density $n(r)$ and the average absorption cross section $\sigma(r, \lambda)$. Applying their method to ZIP data at several elongations, they concluded that $n(r) \sigma(r, \lambda)$ can not be represented by a single radial gradient in all directions, that the gradient is less steep than that derived from scattered light observations, and that the infrared σ (r, λ) varies with r in a manner requiring more than one dust component.

Dumont and Levasseur-Regourd (1988) have applied their "nodes of lesser uncertainty" method to IRAS and ZIP data in order to retrieve the dust temperature and albedo and their radial gradients near 1 AU. Using the 12 and 25 µm values from Hauser (1984), based on the preliminary IRAS calibration, they derive a color temperature at 1 AU, $T_o = 257$ K, and $T(r) = 257 \, r^{-0.33}$. The revised IRAS calibration would raise T_o without altering the gradient. This result is very similar to Good's (1988) global model, $T(r) = 266r^{-0.36}$. The 11 and 21 µm ZIP data yield a similar gradient but higher T_o; $T(r) = 298r^{-0.32}$.

When the method is applied to the optical data as well, the albedo at 90 deg phase angle can be retrieved; Levasseur-Regourd et al. (1990b) find $A = 0.08$ (IRAS) and $A = 0.15$ (ZIP), with A(r) $\propto r^{-0.3\pm0.1}$ near 1 AU. Levasseur-Regourd et al (1990) have shown that the polarization also depends upon distance in and perpendicular to the ecliptic plane (see Levasseur-Regourd et al., this volume, for more detailed discussion).

In summary, the most important conclusion from the modeling to date is that the optical properties of the dust (specifically the wavelength-dependent absorption cross-section and the scattering function) must change with position in the solar system. Consequently, the heliocentric gradient of the spatial density derived on the assumption of constant optical properties has to be questioned.

The revised absolute calibration and improved spectral coverage from the DIRBE experiment will make more detailed modeling possible. We have other means of analyzing the dust at 1 AU, but analyses of the dust scattering and emission remain the best means of determining average dust properties away from 1 AU.

4. Circumsolar Dust

Detecting the thermal emission from dust particles near the sun is an important means of probing their physical properties. Whereas the visible integral over the line of sight at small elongation angles is weighted towards particles near 1 AU seen in forward scattering, the thermal emission integral is weighted towards the hottest grains closest to the sun. Peterson (1963) stimulated interest in such observations when he predicted that a brightness maximum should be detectable at the edge of the sublimation zone at wavelengths of ~ 2 µm, near the peak in the Planck function for grain temperatures > 1000 K.

A distinct emission peak at 4 solar radii was recorded in coronal scans at 2.2 µm during the 1966 eclipse by both Peterson (1969) and MacQueen (1968), as well as a broad shallow feature at 3.5 R_0. The 4 R_0 peak was seen even more clearly during a 5-hour balloon flight with an infrared coronograph two months later; total radiance at 4 R_0 was a factor of 2 above the continuum (MacQueen 1968). These data have the advantage of much slower scan rate across the corona. Smaller peaks at 8.7 and 9.2 R_0 were also detected. However, balloon observations during the 1983 eclipse show only a slight inflection near 4 R_0 at 2.2 µm, although a broad peak centered at 3.8 R_0 is present at 1.65 µm (Mizutani et al 1984). Perhaps the amount of dust near the sun does vary with time; sun-grazing comets may play a role.

If silicates are a component of the circumsolar dust, then strong thermal emission is expected in the reststrahlen bands near 10 µm. Mankin et al (1974) and Lena et al (1974) detected 8-13 µm emission near the ecliptic during eclipse. A possible peak near 4 R_0 is comparable to the sky noise (Mankin et al) or to the differences between scans (Lena et al).

Since 1983, there have been great advances in infrared detectors. The coronal emission is a problem well-suited for modern, sensitive two-dimensional arrays. Experiments using such arrays are planned for the July 1991 solar eclipse at Mauna Kea (D. Hall, private communication).

The fate of dust particles near the sun is a complex function of their optical properties (Mukai and Mukai 1973, Mukai et al 1974, Lamy 1974, Schwehm & Rohde 1977, Mukai and Yamamoto 1979). Because of their low absorption at visible wavelengths, pure silicate grains can reach ~ 2 R_0 before reaching sublimation temperatures. The ratio of the repulsive force of radiation pressure to the force of gravity (β) is always < 1 for all grain sizes, and the grains remain near the sun until they sublimate completely.

Absorbing grains, on the other hand, heat to sublimation temperatures at somewhat larger solar distances. Mukai et al (1974) computed that graphite grains will concentrate in a sublimation zone near 4 R_0, creating a dust ring. As the grains sublimate and their radii shrink, β will increase and the remnant grain will be accelerated outward. Small grains from the solar direction, with speeds > 50 km/s, have been detected from space probes (Berg and Grun 1973, Zook and Berg 1975). Mukai and Yamamoto (1979) proposed a two-component model to fit both the 1-2 μm and the 10 μm observations. They compute that both graphite and p-obsidian grains will form a dust ring at 4-5 R_0, where both components sublimate rapidly. At T ~2100 K, the graphite grains produce the observed 1.6-2.2 μm peaks, while the silicate grains generate the high 10 μm flux.

Given what we have learned about the dust composition from Comet Halley and from studies of interplanetary dust, the next step towards a model for real grains near the sun should consider the progressive physical changes to the carbonaceous organic-rich grain material upon heating, and whether a high-temperature graphitic material will form, which would follow the evolutionary path described here. Moreover, the role of collisions has been neglected in this picture. Yet, Grun et al (1985) have concluded that collisions play the dominant role in the evolution of the interplanetary dust at small heliocentric distances.

5. Science Questions for the Future

This section briefly summarizes the current status and lists some of the science questions to be answered by future work.

A. *Zodiacal emission*

The zodiacal emission has now been well-observed over the year at 60 < ϵ < 120 deg from IRAS and COBE. It is important that the absolute calibration be resolved, so that an accurate spectral energy distribution can be defined. These results should then be published in tables or plots, so that those without large computers can have access to them. Future observations should concentrate on ϵ < 60 deg and ϵ > 120 deg and on improved 5-25 μm spectral resolution, to answer the following questions:

What is the infrared emission as f(ϵ, β, λ) at 2 < ϵ < 60 deg and what does this tell us about the spatial distribution and thermal propertiues of the dust at small r?

What is the spectral energy distribution as f(ϵ, β) from λ 5-25 μm? Is it consistent with our expectations based on IDP composition?

Is 10 μm emission from small silicate grains present?

What is the intensity of the dust emission at 50 < λ < 150 μm? Is there a cold, outer solar system component?

ISO and SIRTF can play an important role in answering these questions.

B. *Dust near the sun*

We can anticipate important new results with modern, sensitive array detectors during the July 1991 solar eclipse. These observations will help answer the following questions.

What is the distribution of dust near the sun? Does it form a spherical halo or is it concentrated towards a plane? (which plane?)
At what temperature(s) do grains sublimate? Is there a well-defined sublimation zone?
Is the amount of dust time-variable? Do sun-grazing comets make a significant contribution?
Is a 10 μm emission feature visible that pinpoints the location of small silicate grains?

A dust detector on a solar probe could measure the dust flux and orbital distribution and the size distribution, to help determine the relative role of collisions and sublimation in the destruction of grains near the sun (Tsurutani and Randolph, this volume).

C. *Modeling the dust emission*

Models have given us a general picture of the dust spatial distribution and optical properties. Inversion methods are a useful tool and should be applied to the DIRBE data. To progress, however, future models have to wrestle with the properties of real grains and should not ignore what has been learned about the dust from other sources. The following questions are relevant:

What physical changes to the grains can cause the albedo variation with r?
What optical/infrared properties of real grains are consistent with $T(r) \propto r^{-0.33}$?
What are the scattering and emitting properties of the mixed silicate + carbonaceous IDPs? Are they consistent with the zodiacal emission?
How will the heterogeneous IDP particles evolve when strongly heated near the sun?

Ultimately, dynamical models -- sources, sinks, orbital evolution - have to be linked to the "static" zodiacal light models.

6.   References

Beichman, C. A., Neugebauer, G., Habing, H. J., Clegg, P. E., Chester, T. J. (1988). IRAS Explanatory Supplement, NASA RP-1190, Vol. 1, 2nd. ed.
Berg, O. E. and Grün, E. (1973) *Space Research*, **13**, 1047-1055.
Boulanger, F. and Pérault, M. (1988) *Astrophys. J.*, **330**, 964-985.
Briotta, D. (1976) Ph.D. dissertation, Cornell University, Astronomy Department.
Deul, E. R. and Wolstencroft, R. D. (1988) *Astron. Astrophys.*, **196**, 277-286.
Dumont, R. and Levasseur-Regourd, A.-C. (1978) *Astron. Astrophys.*, **64**, 9-16.
Dumont, R. and Levasseur-Regourd, A.-C. (1988) *Astron. Astrophys.*, **191**, 154-160.
Giese, R. H. and Kneissel, B. (1989) *Icarus*, **81**, 369-378.
Giese, R. H., Kneissel, B. and Rittich U. (1986) *Icarus*, **68**, 395-411.
Giese, R. H., Weiss, K. Zerull, R. H., and Ono, T. (1978) *Astron. Astrophys.*, **65**, 265-272.
Good, J. C. (1988) preprint.
Good, J. C. (1990) submitted to *Astrophys. J.*
Grün, E., Zook, H. A., Fechtig, H. and Giese, R. H. (1985) *Icarus*, **62**, 244-272.
Gulkis, S., Lubin, P. M., Meyer, S. S., Silverberg, R. F. (1990) *Scientific American*, Jan. 1990, 132-139.

Hauser, M. G. (1987) in Comets to Cosmology *Lecture Notes in Physics*, **297**, 27-39.
Hauser, M. G. et al. (1984) *Astrophys. J. Lett.*, **278**, L15-L18.
Hong, S. S. and Um, I. K. (1987) *Astrophys. J.*, **320**, 928-935.
Lamy, P. L. (1974) *Astron. Astrophys.*, **35**, 197-207.
Leinert, C., Hanner, M., Richter, I., and Pitz, E. (1980). *Astron. Astrophys.*, **82**, 328-336.
Leinert, C., Richter, I., Pitz, E. and Planck, B. (1981) *Astron. Astrophys.*, **103**, 177-188.
Léna, P., Viala, Y., Hall, D. and Soufflot, A. (1974) *Astron. Astrophys.*, **37**, 81.
Levasseur-Regourd, A. C., Dumont, R. and Renard, J. B. (1990a) *Icarus*, **86**, 264-272.
Levasseur-Regourd, A. C., Renard, J. B., and Dumont, R. (1990b), these proceedings.
MacQueen, R. M. (1968) *Astrophys. J.*, **154**, 1059-1076.
Mankin, W. G., MacQueen, R. M., and Lee, R. H. (1974) *Astron. Astrophys.*, **31**, 17-21.
Mather, J. C. et al. (1990a) *Astrophys. J. Lett.*, **354**, L37-L40.
Mather, J. C. et al. (1990b) Preprint, April 1990. To appear in Observations in Earth Orbit and Beyond.
Mizutani, K., Maihara, T., Hiromoto, N., Takami, H. (1984) *Nature*, **312**, 134-136.
Mukai, T., Fechtig, H., Grün, E., Giese, R. H., and Mukai, S. (1986) *Astron. Astrophys.*, **167**, 364-370.
Mukai, T. and Mukai, S. (1973) *Publ. Astron. Soc. Japan*, **25**, 481-488.
Mukai, T., Yamamoto, T., Hasegawa, H., Fujiwara, A., and Koike, C. (1974) *Publ. Astron. Soc. Japan*, **26**, 445-458.
Mukai, T. and Yamamoto, T (1979) *Publ. Astron. Soc. Japan*, **31**, 585-595.
Murdock, T. L. and Price, S. D. (1985) *Astron. J.*, **90**, 375-386.
Neugebauer, G. et al. (1984) *Astrophys. J. Lett.*, **278**, L1-L6.
Peterson, A. W. (1963) *Astrophys. J.*, **138**, 1218-1230.
Peterson, A. W. (1969) *Astrophys. J.*, **155**, 1009-1015.
Peterson, A. W. (1971) *Bull. Am. Astron. Soc.*, 3, 500.
Price, S. D., Murdock, T. L. and Marcotte, L. P. (1982) *Astron. J.*, **87**, 131.
Price, S. D., Murdock, T. L. and Marcotte, L. P. (1980) *Astron. J.*, **85**, 765.
Reach, W. (1988) *Astrophys. J.*, **335**, 468-485.
Röser, S., and Staude, H. J. (1978) *Astron. Astrophys.*, **67**, 381-394.
Rowan-Robinson, M., Hughes, J., Vedi, K., Walker, D. W. (1990) *Mon. Not. R.A.S.*, **246**, 273-278.
Salama, A. et al. (1987) *Astron. J.*, **92**, 467-473.
Schwehm, G. and Rohde, M. (1977) *J. Geophys.*, **42**, 727-735.
Soifer, B. T., Houck, J. R. and Harwit, M. (1971) *Astrophys. J.*, **168**, L73-L78.
Temi, P., de Bernardis, P., Masi, S., Moreno, G. and Salama, A. (1989) *Astrophys. J.*, **337**, 528-535.
Zook, H. A. and Berg, O. E. (1975) *Planet. Space Sci.*, **23**, 183-203.

TEMPORAL AND SPATIAL VARIATIONS
OF THE ATMOSPHERIC DIFFUSE LIGHT

S. M. KWON[1], S. S. HONG[1], AND J. L. WEINBERG[2]
[1]*Department of Astronomy, Seoul National University, 151-742, KOREA*
[2]*Space Astronomy Laboratory, University of Florida, U.S.A.*

ABSTRACT. The Barbier's relation for the diffusely scattered airglow has been modified in such a way that it may describe, with simple changes of two parameter values, the dependence on zenith distance of the atmospheric diffuse light at any time of the night.

1. Introduction

The atmosphere related diffuse light(ADL) is composed of the directly transmitted airglow(AG) and the diffusely scattered signals of integrated starlight(IS), zodiacal light(ZL) and the AG. Conditions in the AG emitting layer are known to vary with time during a night (Tanabe 1964; Dumont 1965; cf. Roach and Gordon 1972; Tanaka *et al.* 1989). Changing configurations of the Milky Way and the ZL-cone with respect to the earth scattering atmosphere further make the ADL vary with time and sky position.

For the reduction of ZL, many methods have been devised to remove the ADL from the observed sky brightness (Weinberg 1963; Dumont 1965; Dumont and Sanchez 1975; Hong *et al.* 1985). To avoid the errors due to the time-variation, observed data were often averaged over the period extending many years. The averaging was successful in rendering overall pictures of the zodiacal dust cloud; it also masked valuable information about the detailed structures of the cloud.

2. Reduction of the ADL from the Night Sky Observations

Weinberg and Mann monitored the night sky brightness at 5080Å and 5300Å, by scanning the meridian eleven times over a night for each of the wavelengths. To characterize the dependencies of the ADL brightness on time t and zenith distance z, we analyzed this set of 22 meridian scans in the same way as done for the almucantar scans (Kwon, Hong, Weinberg and Misconi 1991, in this volume).

An example of the scans after being calibrated is shown in Fig 1 by dots; the ordinate represents the brightness in units of $S_{10}(V)_{G2V}$, and the abscissa the zenith distance in degrees. By summing up the directly transmitted contributions from resolved bright stars, IS (Toller 1981), and ZL (Levasseur-Regourd and Dumont 1980), we synthesized the profile of directly transmitted astronomical light only.

The difference between the observed (dots) and the synthetic (thin line) profiles gives the brightness of the ADL along the meridian, and is shown by the thick solid line in the figure. We drew a smooth line through the sharp spikes due to imperfect removals of the bright stars. Reading

179

A.C. Levasseur-Regourd and H. Hasegawa (eds.), Origin and Evolution of Interplanetary Dust, 179–182.

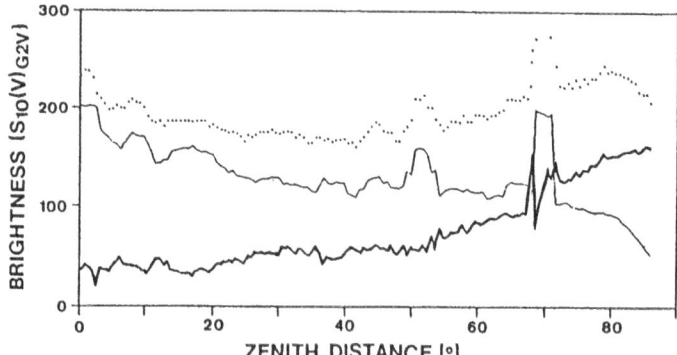

Fig 1 - A sample plot of the meridian scan simulations. Solid dots represent the observed brightness distribution and the thin solid curve represents the synthetic profile. The difference between the two curves is plotted in a thick line, which represents the ADL brightness.

the brightness off the smoothed line, we have constructed a two-dimensional (t, z) ADL table for each of the wavelengths. Figures 2a (5080Å) and 2b(5300Å) illustrate how the ADL varies with t at a few selected zenith distances.

The ADL brightness increases rather rapidly in the beginning of the night, reaches a broad maximum extending 21 to 22 hr in Hawaii standard time (HST), and declines slowly afterwards. This trend of variation is the same for the both wavelengths. The amplitude of the time-variation generally increases with increasing z, and becomes as large as 30 $S_{10}(V)_{G2V}$ at $z \simeq 80°$. This big an amplitude makes time-dependent corrections be necessary for the ADL.

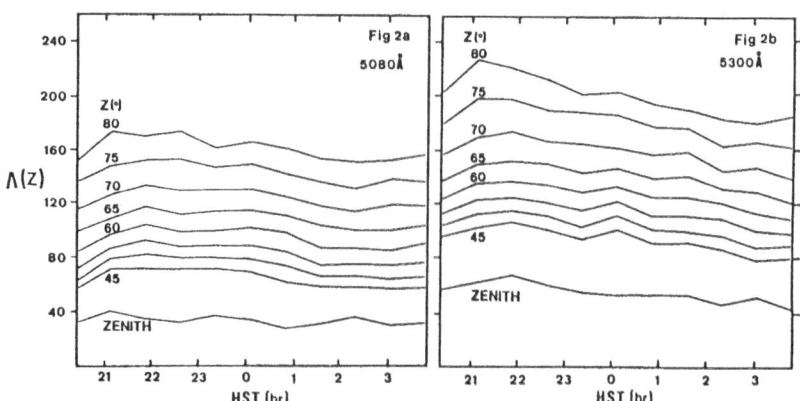

Fig 2 - Time variations of the ADL are shown for wavelength 5080Å(a), and 5300Å(b).

At a given time, brightness $\Lambda(z)$ of the ADL increases with z up to 80°, where it reaches a maximum and reverses its rising trend. We normalized the brightness profile to its zenith brightness Λ_0, which is about 40 and 50 $S_{10}(V)_{G2V}$ at 5080Å and 5300Å, respectively. A sample of the normalized profiles is shown in Fig 3 by dots. General pattern of the variation seen in the figure persists for all the 22 profiles; only maximum values are different depending on the time of observation. It would

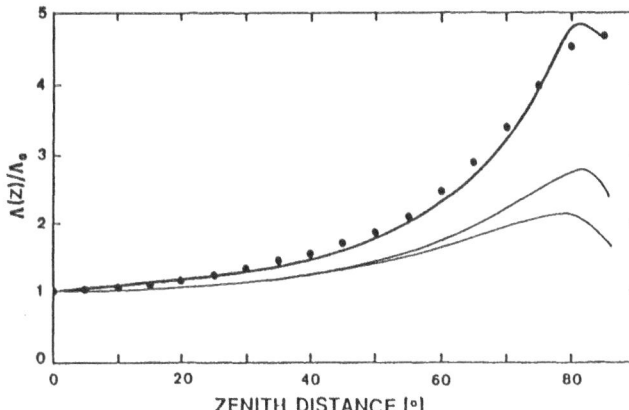

Fig 3 - Normalized distribution of the ADL brightness with zenith distance. The observed ADL is shown by the filled circles, and the non-linear least squares fit of the modified Barbier's relation to the data is represented by the thick solid line. The two thin curves are calculated from the original version of the Barbier's relation with $h = 50$ and 200 km.

then be convenient, if one could describe the normalized profiles by a simple function of z with a few parameters.

3. Parametric Representation of the Normalized ADL Profile

Barbier (1944) attempted to describe the z-dependence of the diffusely scattered AG with the function $(F_1 - F_2 \cos^2 z)\{1 - exp[-\tau_1 x]\}$, where F_1 and F_2 are parameters related to the height h of the AG emitting layer, τ_1 is the scattering optical depth at the zenith, and x is the airmass at z. When Dumont (1965) formulated his multiple height method (cf. Dumont and Sanchez 1975), he used the Barbier's relation to conceptualize the z-dependence of the diffusely scattered AG. We use this relation not because of its accuracy but because of its flexibility. The components in the ADL other than the diffusely scattered AG require the original Barbier's relation to be modified.

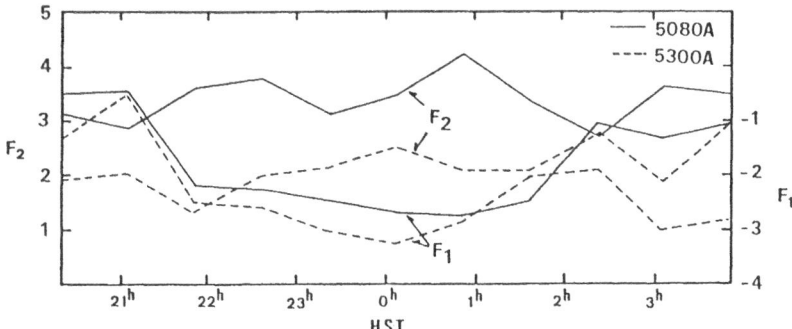

Fig 4 - The time dependent variations of the parameter F_1 and F_2 are plotted as functions of HST for the two wavelengths.

If θ denotes the angle between a line of sight and the outward normal drawn at the point where the line of sight enters the AG emitting layer, the pathlength through the layer becomes proportional to $\sec \theta$. The intensity of the AG incident upon the earth absorbing and scattering atmosphere is assumed proportional to the pathlength; $\sec \theta$ is another representation of the van Rhijn function. The directly transmitted AG may then be modeled by $\sec \theta \, exp[-\tau_0 x]$ with τ_0 being the extinction optical depth of the atmosphere at the zenith. For $h = \infty$, $F_1 = 0.5$ and $F_2 = 0$. Consequently, the diffusely scattered components of the IS and ZL are easily incorporated into the Barbier's relation by simply relaxing the definition of F_1. We thus propose $\Lambda(z)/\Lambda_0 = \sec\theta \, exp[-\tau_0 x] + (F_1 - F_2 \cos^2 z)\{1 - exp[-\tau_1 x]\}$ as an empirical function to describe the meridian profile of the normalized ADL at a fixed time.

In the analysis of night sky observations, Dumont (1965) is the first who differentiated the scattering τ_1 from the extinction τ_0 optical depth. For the night Kwon (1990) was able to determine τ_0 and τ_1 as functions of time. Therefore, the two optical depths are not free parameters.

Since θ is related to z through a known function of h and Barbier gave F_1 and F_2 as functions of h only, in a strict sense of Barbier, the trial relation we propose could be a single parameter function. The two curves in the lower part of Fig 3 illustrate the range one can have by changing h from 50 to 200 km in the original definitions of F_1 and F_2. Comparison of the curves with the observed profile clearly suggests that the meanings Barbier originally envisioned may no longer be attributed to F_1 and F_2; we thus decided to treat F_1 and F_2 as free parameters and to simply approximate θ to z, limiting the number of free parameters to only two.

With the method of non-linear least squares fit, we determined the best F_1 and F_2 values for each profile. An example of such fits is shown in Fig 3 by the thick solid line. The best parameter values are shown in Fig 4 as functions of time for the both wavelengths.

The best fit values of F_1 and F_2 turned out to be outside the ranges that are allowable from Barbier's original definitions of them. We consider our trial function as a mathematical means for simply reproducing the z-dependence of the observed ADL. The modified Barbier's relation accurately reproduces the z-dependence of the observed profile, and is flexible enough to give equally good fits for all the 22 profiles.

4. Conclusion

The brightness of ADL varies over a night to the degree that its corrections are to be made time-dependently. The Barbier's relation has been modified to describe the z-dependence of the ADL at a fixed time. We propose the modified Barbier's relation with time-dependent values of F_1 and F_2 as a practical means for effecting the time-dependent ADL corrections.

SSH and SMK were supported by the Basic Science Research Institute Program, Korean Ministry of Education. At an earlier stage, JLW and SMK were partially supported by the US AFOSR.

REFERENCES

Barbier, D. 1944, *Ann. Geophys.*, **1**, 144

Dumont, R. 1965, *Ann. d'Astrophys.*, **28**, 265

Dumont, R., and Sanchez, F. 1975, *Astr. Ap.*, **38**, 397

Hong, S. S., Misconi, N. Y., van Dijk, M. H. H., Weinberg, J. L. and Toller, G. N.1985, in *Properties and Interactions of Interplanetary Dust*, ed. R. H. Giese and P. Lamy (Dordrecht:Reidel), p.33

Kwon, S. M. 1990, *Ph. D. Thesis*, Seoul National University

Levasseur-Regourd, A. C. and Dumont, R. 1980, *Astr. Ap.*, **84**, 227

Roach, F. E. and Gordon, J. L. 1972, *The Light of the Night Sky*, (Dordrecht:Reidel)

Tanabe, H. 1964, *Publ. Astron. Soc. Japan*, **16**, 324

Tanaka, K., Miyashita, A., Takechi, A. and Tanabe, H. 1989, *Atlas of Zenith 5300Å Brightness*, National Astronomical Observatory, Japan

Toller, G. N. 1981, *Ph. D. Thesis*, State University of New York at Stony Brook

Weinberg, J. L. 1963, *Ph. D. Thesis*, University of Colorado

FINE RESOLUTION BRIGHTNESS DISTRIBUTION
OF THE VISIBLE ZODIACAL LIGHT

S. M. KWON[1,2], S. S. HONG[1], J. L. WEINBERG[2], AND N. Y. MISCONI[2]
[1] *Department of Astronomy, Seoul National University, 151-742, KOREA*
[2] *Space Astronomy Laboratory, University of Florida, U.S.A.*

ABSTRACT. Applying time-dependent corrections of the atmospheric diffuse light to the observed night sky brightness, we have determined brightness of the zodiacal light over the region $40° \leq \lambda - \lambda_\odot \leq 320°$ and $-20° \leq \beta \leq 20°$. The resulting map of equal brightness contours has an angular resolution of two degrees, and exhibits east-west and north-south asymmetries.

1. Introduction

For the purpose of reducing the zodiacal light from night sky observations, the major sources of the sky brightness may be classified (cf. Weinberg 1963) into the following four categories: the bright stars(BS) that are resolved as individual stars by a given telescope, the integrated starlight(IS) including the diffuse Galactic light, the zodiacal light(ZL) of our interest, and the airglow(AG) emitted by the earth upper atmosphere. Because of the scattering by atmospheric constituents, the telescope's field of view intercepts not only the directly transmitted light of all the four sources but also the diffusely scattered light of the latter three extended sources. Lack of accurate information about the IS distribution, difficulties in correcting for the diffusely scattered components, and uncertain nature of the AG time-variation all made it practically impossible to obtain, from single night observations of night sky, a fine resolution map of the ZL distribution. Careful analyses of the night sky observations accumulated over many seasons/years have given us a smoothed-out distribution of the ZL (Dumont 1965; Dumont and Sanchez 1975; Levasseur-Regourd and Dumont 1980), and Pioneer 10/11 space probes have actually measured the IS over almost entire sky (Toller 1981). On the basis of the space measurements and the smoothed-out ZL, a self-consistent empirical method has recently been developed for making the time-dependent corrections of the atmospheric diffuse light (Kwon, Hong and Weinberg 1991). With this method, we will analyze the single night observations of night sky brightness, and determine the ZL brightness over an extended area of sky. The resulting distribution of the ZL will be presented in the form of equal brightness contours with a 2° resolution.

2. Observational Data and Calibration

From archives of the night sky observations by Weinberg and Mann, we selected the observations over one complete night, August 21/22 1968, at Mt. Haleakala, Hawaii. The sky was scanned repeatedly over the entire 360° range of azimuth at 8 fixed zenith distances 45°(05) 80° by the photometer attached to a telescope of 1°.85-radius field of view. After each set of 8 almucantar scans, they also scanned the entire northern meridian from zenith to horizon. During the night, 11 sets of 8 almucantar and 1 meridian scans were obtained for each wavelength of 5080Å and 5300Å,

A.C. Levasseur-Regourd and H. Hasegawa (eds.), Origin and Evolution of Interplanetary Dust, 183–186.
© 1991 *Kluwer Academic Publishers.*

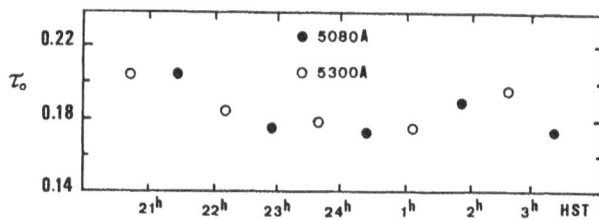

Fig 1 - Time-variation of the extinction optical depth of the atmosphere at zenith. The relative amplitude of the variation is about 15%.

making 198 scans altogether. The 22 meridian scans have provided the means for making time- and position-dependent corrections of the atmospheric diffuse light (Kwon, Hong and Weinberg 1991), and the remaining 176 almucantar scans will be analyzed in this paper. (Further details of the observations may be found from Weinberg and Mann [1967].)

Contrary to the most of the previous studies, in our scheme of data analysis, such reduction parameters as the zenith extinction optical depth τ_0 of the atmosphere, the effective field of view (FOV) of the telescope, and the calibration factor C are all simultaneously determined from the same set of observational data, from which the ZL brightness is to be reduced. This ensures us of an internal consistency of the reduction scheme, and also reduces the level of errors in the resulting ZL brightness.

Ninety three bright stars are identified with distinct peaks in the scan profiles (cf. the top curve in Fig 2). The maximum chart deflection R of a star with magnitude m_λ at wavelength λ and the calibration factor C in units of $S_{10}(V)_{G2V}$ per unit deflection are related by the relation $C \cdot R \cdot FOV = \eta \, 10^{0.4(10-m_\lambda)} \cdot exp[-\tau_0 x]$, where x is the airmass, and η, the conversion factor from $S_{10}(\lambda)$ to $S_{10}(V)_{G2V}$, is 1.20 for 5080Å and 1.05 for 5300Å. Since m_λ's for all the stars can be known by interpolating V and B–V values of the stars and x's for the stars are determined from their zenith distances at the moment of observation, the relations for a number of stars comprise a set of simultaneous equations for the two unknowns τ_0 and $C \cdot FOV$. Application of the least squares fit to the data determines the best values for τ_0 and $C \cdot FOV$.

In order to take into account for the changing atmospheric conditions, we divided the 11 sets for one wavelength into 5 sub-groups, and τ_0 is determined for each group with the value for $C \cdot FOV$ being fixed at its average for all the scans. The resulting τ_0's for the two wavelengths are plotted in Fig 1 as a function of Hawaii standard time. The means of the time-dependent τ_0's are 0.183 for 5080Å and 0.173 for 5300Å, and the relative amplitude of the time-variation amounts to about 15%. If one ignores the time-variation of τ_0, significant errors would come into the derived brightness of the ZL.

3. Eliminations of the BS, IS and Atmospheric Diffuse Light

From the General Catalogue of 33342 Stars (Roman and Warren 1985), the Nicolet Homogeneous UBV Catalogue (Nicolet 1978), the Bright Star Catalogue (Hoffleit 1982), and the U.S. Naval Observatory Photoelectric Catalogue (Blanco et al. 1968), we have collected all the positional and photometric information about the resolved stars, 8372 in total, that are brighter than 6.5 mag in the visual, and stored them in a single data-base. We search the data-base for those bright stars that come into the given FOV along the scan path, and synthesize the brightness profile due to the stars. For the IS, we also synthesize the profile by using the Pioneer observations. Sum of the two profiles is then compared with the observed profile, graphically on a computer screen. An example of the synthetic profiles is given in Fig 2 by the dashed curve. By adjusting the trial values for C and FOV until the comparison gives satisfactory results, particularly over the regions of very bright stars and the Galactic plane, we finally determine accurate values for C and FOV separately.

By subtracting the synthetic profile that is obtained with the best values of τ_0, FOV and C, from

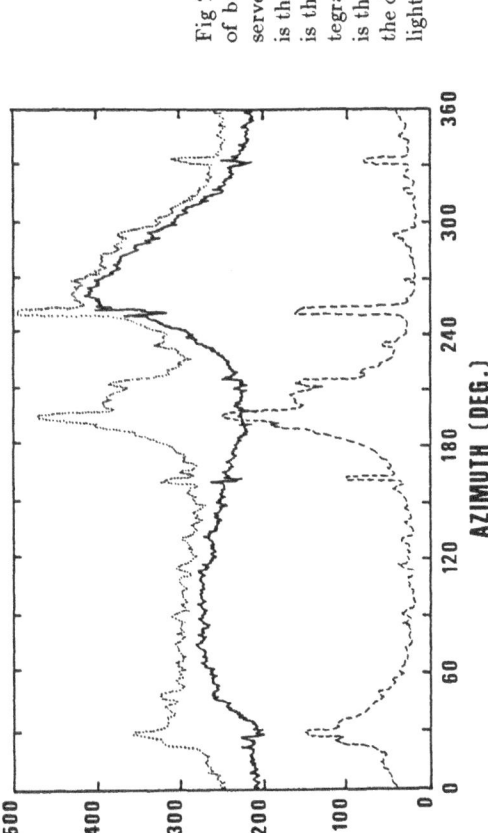

Fig 2 - Removal of the directly transmitted brightness of bright stars and integrated starlight from the observed brightness of night sky. The top curve by dots is the observed profile, and the bottom curve by dashes is the synthetic profile for the bright stars and the integrated starlight. The middle curve by thick solid line is the difference between the two profiles, and contains the directly transmitted ZL and the atmospheric diffuse light.

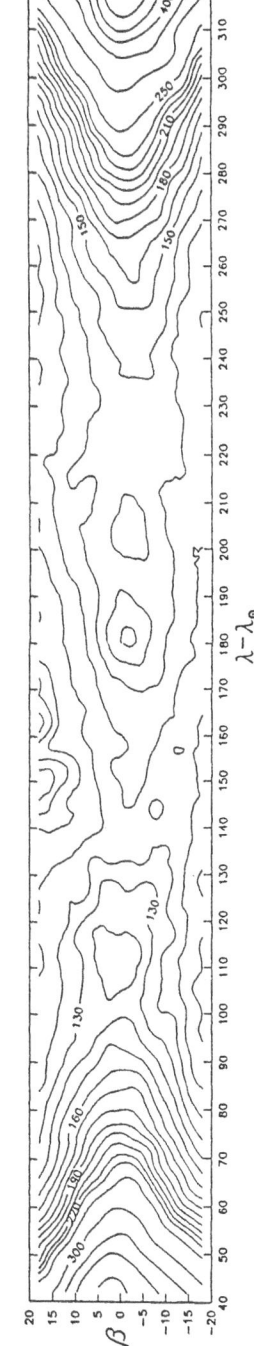

Fig 3 - Map of equal brightness contours for the ZL over an extended region of sky. The lowest contour level is 85 $S_{10}(V)_{G2V}$ and the highest level in the Gegenschein region is 160 $S_{10}(V)_{G2V}$. Please note that the contours are, purposely, not folded over the ecliptic plane, nor the line $\lambda - \lambda_\odot = 180°$.

the observed profile, we removed the directly transmitted contributions of the BS and IS from the observed brightness. The thick solid line in Fig 2 represents the profile for the residual brightness. The residual still contains, in addition to the ZL, the atmospheric diffuse light (ADL), which is composed of the directly transmitted AG and the diffusely scattered components of the IS and ZL.

The ADL brightness even at a fixed zenith distance varies with time over a night. The Barbier's relation of two parameters have been modified to describe the dependence of the ADL brightness upon zenith distance. By the use of time dependent values for the two parameters, Kwon, Hong and Weinberg (1991) are able to describe the temporal and spatial variations of the ADL with the modified relation of Barbier. Time-dependent values of the ADL are then removed from the residual brightness profiles to finally obtain the ZL brightness along the paths of almucantar scans.

4. A Fine Resolution Map of the ZL Brightness

For each wavelength, we have produced a two dimensional $(\lambda - \lambda_\odot, \beta)$ map of isophotal ZL contours with an angular resolution of two degrees on average (here λ meaning the ecliptic longitude). An internal consistency of our reduction methodology has been demonstrated by good agreements in the contour morphology and the brightness between the two maps for 5080Å and 5300Å. This study has reduced the errors in the ZL brightness from the previous level of higher than 15% to the present level of 10% or less. The two maps are stacked upon each other, and the combined map is shown in Fig 3.

The brightness distribution of the ZL clearly exhibits asymmetries between the morning side $(\lambda - \lambda_\odot > 180°)$ and evening side $(\lambda - \lambda_\odot < 180°)$, and also between the north and south of the ecliptic. The position of peak brightness in the Gegenschein is located below the ecliptic plane. These features of asymmetry are interpreted as a consequence of the disalignment of the surface of densest dust concentration with respect to the ecliptic. The ZL distribution reduced from the single night observations not only has the information about the symmetry plane(s) but also sets up the observational criteria for competing models of the zodiacal dust cloud. The ZL map presented in this study is hoped to provide a platform for probing the zodiacal cloud in three dimensions.

SSH and SMK were supported by the Basic Science Research Institute Program, Korean Ministry of Education. At an earlier stage of this study, SMK, JLW and NYM were partially supported by the US AFOSR.

REFERENCES

Blanco, V. M., Demers, S., Douglass, G. G., and Fitzerald, M. P. 1968, *Photoelectric Catalogue, Magnitudes and Colours of Stars in the UBV and $U_c BV$ Systems*, ser. 2, vol. 21 (U.S. Naval Observatory:Washigton, D.C.)

Dumont, R. 1965, *Ann. d'Astrophys.*, **28**, 265

Dumont, R., and Sanchez, F. 1975, *Astr. Ap.*, **38**, 397

Hoffleit, D. 1982, *The Bright Star Catalogue* (Yale:New Haven)

Kwon, S. M., Hong, S. S., and Weinberg, J. L. 1991, in this volume

Levasseur-Regourd, A. C. and Dumont, R. 1980, *Astr. Ap.*, **84**, 227

Nicolet, B. 1978, *Astr. Ap. Suppl.*, **34**, 1

Roman, G. N., and Warren, W. H. Jr. 1985, *Documentation for the Machine-Readable Version of the General Catalogue of 33342 Stars for the Epoch 1950*, NSSDC/WDC-A-R&S 83-07

Toller, G. N. 1981, *Ph. D. Thesis*, State University of New York at Stony Brook

Weinberg, J. L. 1963, *Ph. D. Thesis*, University of Colorado

Weinberg, J. L., and Mann, H. M. 1967, in *Proceedings of the Symposium on the Zodiacal Light and the Interplanetary Medium*, NASA SP-150, ed. J. L. Weinberg, p.3

INTERPLANETARY DUST CLOSE TO THE SUN

Ingrid Mann
Ruhr-Universität Bochum *),
D 4630 Bochum,
RFA

1. ABSTRACT

The optical and infrared brightness of the Fraunhofer-corona is produced by light scattering at the zodiacal dust particles and by their thermal emission (see Koutchmy and Lamy 1985). It is modelled within the ecliptic ($4 R_0 \leq \epsilon \leq 15 R_0$) taking into account investigations of the global zodiacal dust cloud due to remote sensing and in situ experiments. The input of near solar dust to the corona brightness is discussed.

2. SPATIAL DISTRIBUTION OF ZODIACAL DUST

The global shape of the zodiacal cloud is assumed to be mainly undisturbed from collisional effects up to the beginning of the sublimation at a distance of 5 R_0 (solar radii) close to the Sun. This zone of sublimation is based on the assumption of olivine particles representing a huge amount of silicates in interplanetary dust particles. Small components (up to 15%) of other materials give no drastic change.

3. BRIGHTNESS INTEGRALS

The brightness seen from the observer results from integration of all volume elements along the line of sight (LOS). The modelling applies a so called volume scattering function $VSF(\theta,r)$ describing the average scattering pattern of the interplanetary dust mixture at solar distance r:

$$VSF(\theta,r) = VSF(\theta,r_0) \, (r/r_0)^{-\nu^*} \quad \text{with } \nu^* = \alpha+\nu. \tag{1}$$

It includes the change of particle number density with solar distance according to a power law with with exponent $-\nu$. The included albedo of particles changes with exponent $-\alpha$. The brightness integral is described

*) New address: Max-Planck-Institut für Aeronomie D 3411 Katlenburg-Lindau

A.C. Levasseur-Regourd and H. Hasegawa (eds.), Origin and Evolution of Interplanetary Dust, 187–190.

elsewhere (cf. Giese et al. 1986). The scattering function for scattering angles $\theta < 30°$ is calculated by diffraction of compact, isotropically scattering particles, considering the size distribution of interplanetary dust at $r = 1$ AU (cf. Grün et al. 1985). Under the assumption of a geometric albedo $A_0 = 0.15$ (cf. Hanner et al. 1981) the function fits the empirical scattering function proposed by Leinert (see Weiß-Wrana 1983) for scattering angles $\theta > 30°$. For scattering angles $\theta > 30°$ the modellings are based on the latter function. Modelling of the thermal emission concerns an average temperature of particles per volume element $\langle T \rangle$ and an average emission cross section $\langle C_E \rangle$ as described elsewhere (cf. Dumont and Levasseur-Regourd 1988). The emission cross section is given as a product of emissivity $\langle E \rangle$ and average geometric cross section $\langle G \rangle$ of particles per volume element: $\langle C_E \rangle = \langle E \rangle \langle G \rangle$. The total surface of particles in a volume element is determined to amount $1.7 \ 10^{-21} \text{cm}^{-1}$ at 1 AU. This is based on the parameters of the interplanetary flux model suggested by Grün et al. (1985). The variation of the albedo results in a change of emissivity: $E = 1 - A_0 \ (r/r_0)^{-\alpha}$.

4. OPTICAL BRIGHTNESS

The modelled optical brightness is compared to observational data given by Blackwell et al. (1967) and Waldmeier (1965). The modelling refers to the brightness at $\lambda = 0.55$ μm. Further colour effects are not regarded.

	$\varepsilon = 4 \ R_0$	$\varepsilon = 8 \ R_0$	$\varepsilon = 15 \ R_0$
$\Delta Z/\%$	52	62	71

Table 1: percentage ΔZ of the corona brightness that results from near solar regions ($r < 0.2$ AU)

A good fit to the data with deviations smaller than 25% is achieved by variation of the volume scattering function with exponent $\nu^* = 1.1 \ -1.25$. Modelled brightness is rather sensitive on the distribution of particles close to the Sun. In spite of the increasing scattering function for small scattering angles, that gives rising input from particles close to the observer, the contribution of near solar particles (see table 1) is high.

5. POLARIZATION

The modelling of the polarization is based on the corona polarization data given by Blackwell et al. (1967), being less than 0.1 % at $\varepsilon = 8 \ R_0$ and going up to 0.8% at $\varepsilon = 16 \ R_0$. The polarization is modelled with respect to the relative shape of the polarization function given by Leinert (see Weiß-Wrana 1983). This function has its maximum within 80° and 85° scattering angle. That is consistent with results from S.Mukai et al., who found a move of the maximum polarization position to smaller scattering angles in case of a rough particle surface. A consistent fit to both, brightness and polarization data is achieved by variation of the polarization function with solar distance in the inner region around the Sun (r < 0.2 AU, see table 2). This decrease is much steeper than that found out for the outer dust cloud by Levasseur-Regourd (1991).

r/R_o	35	30	25	20	15	10
p_{max}	20%	13%	8.6%	4.2%	2.0%	0.6%

Table 2: maximum particle polarization p/% in relation to the solar distance r of particles.

6. INFRARED BRIGHTNESS

As well as the optical brightness the modelled thermal emission is increasing with decreasing elongation of the LOS. However in case of the thermal emission this increase is interrupted when the LOS crosses the dust free zone. Thus even thermal emission of very hot particles cannot explain the observed near infrared brightness of the F-Korona (McQueen 1968, Peterson 1967, 1969, Maihara et al. 1985). The near infrared brightness is affected by both, thermal emission and light scattering. The light scattering at dust particles has to be considered up to λ = 5 μm. The superposition of these two components leads to an edge in the near infrared brightness at the beginning of the dust free zone, that explains observational data in the near infrared. This enhancement in brightness is discussed controversy as pointed out by Tsurutani and Randolph, (this issue). The dependance of the mean temperature was assumed to be <T> (r) = T_o $(r/r_o)^{-0.5}$. A change of this dependance with solar distance, that means a change of the exponent, as discussed for the zodiacal light (cf. Dumont and Levasseur-Regourd 1988) has only modest influence in the case of the corona. As well as the change of the emissivity (see section 3) has only modest influence on the calculated brightness. The spectral variation of the calculated corona brightness in the range 1.25 ≤ λ ≤ 3.6 μm is in the range of converging observational data. The best fit, with deviations smaller than 25 % with exception of the McQueen data, was achieved for a temparature close to that of a grey body (<T_o> = 280 K).

7. DISCUSSION

Both, optical and infrared corona brightness are engraved on near solar particles and have contributions from scattered solar irradiation. Within the ecliptic the spatial distribution of optical efficient particles is mainly continued from the interplanetary dust cloud into the most inner parts. The optical brightness demands for an increase of albedo and a steep decrease of polarization for the near solar particles. Both effects may indicate a change of particle properties in the most inner dust cloud, what has to be investigated further.

8. ACKNOWLEDGEMENT

This work was inspired and encouraged by the late Professor R.H. Giese. Thanks are due to Professor H. Fechtig who enabled the continuation of this work. The author thanks Dr. B. Kneißel for the lifely discussion of the work and for his helpful advices. The work was supported by the Bundesminister für Forschung und Technologie (Contract 010S89023).

9. REFERENCES

Blackwell, D.E. Dewhirst, D.W., and Ingham M.F. 1967: The Zodiacal Light, Adv. Astron. Astrophys. 5, 1

Dumont, R. and Levasseur-Regourd, A.-C. 1988: Properties of Interplanetary Dust from Infrared and Optical Observations.I.Astron. Astrophys. 191, 154

Giese, R.H., Kneißel, B., and Rittich, U. 1986: Threedimensional Models of the Zodiacal Dust Cloud: A Comparative Study, ICARUS 68, 395

Grün, E., Zook, H.A., Fechtig, H., and Giese, R.H. 1985: Collisional Balance of the Meteoritic Complex, ICARUS 62, 244

Hanner, M.S., Giese, R.H. ,Weiss, K., and Zerull, R.H. 1981: On the Definition of Albedo, Astron. Astrophys. 104, 42

Koutchmy, S. and Lamy, P.L., 1985, in: Properties and Interactions of Interplanetary Dust (Giese R.H. Lamy P.L., eds.), Reidel, Dordrecht, 63

Levasseur-Regourd, A.C. 1991: The Zodiacal Cloud Complex, this issue

MacQueen, R.M. 1968: Infrared Observations of the Outer Solar Corona, Astrophys. J. 154, 1059

Maihara, T., Mizutani, K., Hiromoto, N., Takami, H. and Hasegawa, H. 1985, in: Properties and Interactions of Interplanetary Dust (R.H. Giese and P.L. Lamy, eds.), Reidel, Dordrecht, 63

Mukai, S., Mukai, T., and Kikuchi, S., 1991: Scattering Properties of Cometary Dust Based on Polarimetric Data, this issue

Peterson, A.W. 1969: Coronal Brightness at 2,23 µm, Astroph. J. 155, 1009

Tsurutani, B.T. and Randolph, J.E., 1991: The NASA Solar Probe Mission: Dust Sciene, this issue

Waldmeier, M. 1965: The Corona, in: Landolt-Börnstein, Gr.IV, Bd. I, Springer, Berlin, 115

Weiß-Wrana, K. 1983: Optical Properties of Interplanetary Dust, Astron. Astrophys. 126, 240-250

FUTURE OBSERVATION OF THE F-CORONA WITH THE LASCO CORONOGRAPH SPACE EXPERIMENT

P.L. LAMY[1], A. LLEBARIA[1], A. MAUCHERAT[1]
S. KOUTCHMY[2], F. GIOVANE[3]

[1] Laboratoire d'Astronomie Spatiale
Les Trois Lucs, 13012 Marseille, France
[2] Institut d'Astrophysique de Paris
[3] University of Florida

ABSTRACT. The Wide-field White Light and Spectrometric Coronograph (LASCO) to be flown on SOHO in 1995 will observe the corona from just above the limb at 1.1 out to 30 solar radii (R_0) . In addition to the fundamental problems of coronal physics (heating of the corona, acceleration of the solar wind, coronal transients), the scientific objectives incorporate the distribution and properties of dust particles including those released from sun-grazing comets, and interactions of coronal plasma with the dust.

1. Introduction

Most if not all of our knowledge of the F-corona come from ground-based, balloons and aircraft observations during total solar eclipses. The limitations of these observations are quite obvious, the available time is very short and the sky background limits the useful field-of-view to a few solar radii. Going to space has allowed to overcome the limitation imposed by the Earth's atmosphere: externally occulted coronographs have been implemented to reduce the now dominating instrumental scattered light. Surprisingly, space-born coronograph has produced no real advance in the field of interplanetary dust with the exception of Sun-grazing comets. To a large extent, the problem has to do with the poor photometric performances of the detectors used up to now (vidicons in particular). Even the elaborated Solar Maximum Mission coronograph with polarimetric capabilities suffered from this problem. As a consequence, quantitative data have seldom been obtained. The most valuable results were derived by image subtraction revealing the transitory plasma events in the corona.

The LASCO space coronograph will hopefully remedy to these shortcomings in particular thanks to its CCD cameras whose superior photometric performances are well established. It further has the capabilities of a large field-of-view (30 R_0), of performing colorimetric, polarimetric measurements as well as high resolution spectrometric profiles. All these performances should enable LASCO to make a valuable investigation of the F-corona.

LASCO is a joint project of the Naval Research Laboratory which has the overall responsability of the instrument (P.I. G. Bruckner), the Max-Planck Institute für Aeronomie,

A.C. Levasseur-Regourd and H. Hasegawa (eds.), Origin and Evolution of Interplanetary Dust, 191–194.

Laboratoire d'Astronomie Spatiale and the University of Birmingham. It is part of the scientific payload of the SOHO spacecraft which is scheduled to be launched in July 1995 and to have an operational lifetime of at least two years. The late R. Giese participated as a co-investigator in the definition of the scientific objectives of the instrument.

2. Scientific objectives

In addition to the fundamental problems of coronal physics (heating of the corona, acceleration of the solar wind, coronal transients), LASCO offers a unique opportunity to study the photopolarimetric/colorimetric properties of the F-corona (further required for the proper separation of the K-corona) and their connection to those of the Zodiacal Light, to study the dynamics of the circum-solar grains by Doppler shift and line profile measurements of a Fraunhofer line, to look for variations associated with solar coronal activity (CME's, solar flare associated shock waves, varying magnetic fields), to observe Sun-grazing comets or comets colliding with the Sun (adding a wealth of information on the dynamics of newly released dust).

The separation of the K and F components must be accurately accomplished to achieve the scientific goals. This is particularly accute in the outer corona where the K component represents only a few percent of the total brightness but is of fundamental interest for understanding the expansion of the solar wind.

Two different methods are used to achieve this separation: the spectroscopic method which works well up to 3 R_0 and the polarization method at larger distances. However, the classical approximation that the polarization of the F-corona is negligibly small no longer holds beyond about 5 R_0. Clearly, the problem must be approached from the two fronts, namely by modeling the polarization of both the K and F-corona, recalculating the polarized intensities and iterating until a satisfactory agreement is reached. A criterion of importance for the modeling of the F-corona is the continuity with the photopolarimetric properties of the Zodiacal Light.

3. Description of the instrument

The scientific goal of exploring the corona from 1.1 to 30 R_0 in a single experiment requires innovation in design of the instrument, for this has never before been attempted. Furthermore, the fundamental problem of photospheric stray light rejection, common to all coronal observations, is compounded by the enormous gradient in coronal brightness with radial distance from the Sun. The total K+F brightness varies by six orders of magnitude between 1.1 and 30 R_0. Therefore, we have chosen to break up the observing domain into three subregions, and LASCO is composed of three compact optical systems, C1, C2 and C3, each specially designed and optimized for its particular range of operation, and all held in precise coalignment through sound mechanical and thermal design (Michels et al., 1989). Characteristics of the three optical systems are shown in Table 1. The concentric fields of view allow for extensive overlapping which is essential for intercalibration of the three systems (including cross calibration on orbit), and to assist in later reconstruction of composite wide field images.

3.1. OPTICAL DESIGN OF C1 (1.1 - 3.0 R_0)

In order to image the corona with adequate spatial resolution very close to the sun's limb, an internally-occulted system is required. C1 implements this requirement with a reflective design, made possible by recent advances in mirror superpolishing technology. The objective mirror, M1, is an off-axis paraboloid. The solar image is formed at the prime focus, on a convex annular mirror, M2, analogous to the field lens in a standard refractive Lyot coronograph. This mirror provides a field stop, and also performs the function of the Lyot internal occulting disk, by removing the unwanted photospheric image from the system. The field mirror relays the light of the coronal image onto a second off-axis paraboloid, the collimating mirror, M3 identical to M1. This fully symmetric arrangement accomplishes complete cancellation of coma in the system. The collimated beam of coronal light is next sent through a narrow-bandpass tunable Fabry-Perot filter, and then into an objective which focuses the final coronal image onto a 1024 x 1024 pixel CCD image detector. Max-Planck Institute für Aeronomie has responsability for C1.

The coronal light is analyzed spectroscopically by the tunable Fabry-Perot (FP) filter which is designed by the Naval Research Laboratory. The FP is an interference (comb type passband, 0.4 Å FWHM) filter, which passes many interference orders. It works in conjunction with a broader blocking filter, with passband narrow enought to eliminate all but one selected order passed by the FP, but broad enough to allow tuning of the FP over a full free spectral range (22 Å). The filter presents the CCD detector with a monochromatic image of the entire 3 R_0 coronal field of view. Two-dimensional imaging spectroscopy of coronal line profiles is achieved by synchronizing CCD exposures with a stepwise spectral scan of the tunable filter.

3.2. OPTICAL DESIGN OF C2 (1.5 - 6 R_0) AND C3 (3 - 30 R_0)

C2 and C3 are externally occulted coronographs adapted from earlier, well developed designs (SOLWIND, SMM) and optimized for their respective domain. In this design, the external occulter completely shadows the objective lens from direct sunlight, dramatically lowering the level of instrumental stray light. The main disadvantage of external occultation is that the imaging properties of the objective lens are seriously impeded by vignetting (partial obscuration of the aperture) by the shadowing assembly (i.e. the external occulter) for object points not far removed from the occulter shadow. In other words, the instrument achieves full resolution at the outer part of its field, but relatively poor resolution near the central occulter shadow. It is considerations such as this that drive the requirement for extensive overlap of all three of the optical systems.

Laboratoire d'Astronomie Spatiale has responsability for C2 which incorporates a new multi-thread occulter and a doublet in optical contact as the objective lens.

For C3 designed by the Naval Research Laboratory, the external occulter is a triple disk assembly, with each disk sized to intercept light diffracted from the previous one. Its objective lens is a singlet, to obtain absolutely minimal scattering. For both C2 and C3, the corona is imaged onto 1024 x 1024 CCD cameras.

4. System design

LASCO consists of two boxes. The coronograph optical box (COB), realized by the University of Birmingham, houses the three optical systems and cameras. It is mounted to the spacecraft instrument pylon with isostatic mounting legs and an offset pointing mechanism, capable of correcting for unacceptably large errors in the spacecraft sunpointing system. The optical box is broken into two parts, one half contains the C1 coronograph optical train, and the other half contains the C2 and C3. A second portion of the experiment is the LASCO electronics box (LEB) which is realized by the Naval Research Laboratory. It contains microprocessors for instrument control and image processing, status and image memory, power conditioning circuitry, command and telemetry interface, etc. This box is relatively small and hard-mounted to the spacecraft.

The three telescopes are equipped with identical CCD cameras chosen because of their large dynamic range and geometric and photometric stability. Passive radiant coolers, integral to each camera package, will maintain the CCDs at the required temperatures. The three cameras are under the responsability of the Naval Research Laboratory.

TABLE 1. Characteristics of the three coronographs

	C1	C2	C3
Field of view (Annular ring)	$1.1 - 3.0 \, R_0$	$1.5 - 6.0 \, R_0$	$3 - 30 \, R_0$
Optical system	Internally Occulted	Externally Occulted	Externally Occulted
Optical Type	Mirror	Lens	Lens
Pixel(angular)	5.6 arc/sec	11.2 arc/sec	54 arc/sec
Polarization analysis:	3 linear polarizers on each coronograph		
Color Filters	FeXIV 530nm FeX 637nm CaXV 569nm Fraunhofer Orange	3 Broadband: Orange Red Blue H-alpha	3 Broadband: Orange Red Blue H-alpha

References
Michels, D.J. et al. (1989): 'LASCO - A wide-field white light and spectrometric coronograph for SOHO', ESA SP-1104, 55-62.

SYNTHETIC MAPS OF THE BRIGHTNESS AND POLARIZATION OF THE F-CORONA

Y. FANG, P.L. LAMY, A. LLEBARIA
Laboratoire d'Astronomie Spatiale
Les Trois Lucs
13012 Marseille, France

ABSTRACT. The Wide-Field Light and Spectrometric Coronograph (LASCO) to be flown on SOHO in 1995 is designed to perform accurate photopolarimetric observations of the solar corona. For simulation purpose but also to have a two-dimensional model of the F-corona, we have realized synthetic maps of its brightness and polarization.

1. Introduction

The F-corona, which represents the inner extension of the zodiacal light, is intrinsically difficult to observe because of the proximity of the Sun and of the presence of the K (electronic) corona. Carefull photopolarimetric measurements during total solar eclipses followed by an accurate separation of the F and K components are required to yield the radial variation of the brightness and polarization of the F-corona alone. Although the flattened aspect of both K and F coronae was perceived quite early, the non-sphericity of the K-corona only was analyzed and the ellipticity of the F-corona was traditionally neglected leading to classical spherical models (Van de Hulst, 1953; Saito, 1970). Recently, Koutchmy and Lamy (1985) re-analyzed various sets of observational data and produced a new, axially symmetric, non-spherical model in the form of two brightness profiles, equatorial and polar, from 1.1 to 30 solar radii (R_θ); in addition polarization values were derived up to 15 R_θ. Subsequent analysis by Baldayan (1988) have demonstrated the superiority of this model which is now generalized to produce full, two-dimensional maps of the brightness and polarization of the F-corona.

2. Procedure and observational data

A key aspect in the analysis of Koutchmy and Lamy (1985) is the ellipticity of the corona as the shape of the observed isophotes are well fitted by ellipses. The present scheme to construct the two-dimensional maps is naturally based on this property and the equatorial and polar radial profiles allow to generate a family of elliptical iso-contours. This is done separately for the two polarized brightness, radial (or parallel) and tangential (or perpendicular). As the ellipticity decreases with the elongation, the model tends to sphericity in its most inner part,

A.C. Levasseur-Regourd and H. Hasegawa (eds.), Origin and Evolution of Interplanetary Dust, 195–198.
© 1991 *Kluwer Academic Publishers.*

that is within 1.5 R_θ.

The two brightness profiles along the equatorial and polar directions of Koutchmy and Lamy (1985) were supplemented by zodiacal light data in order to insure a correct behaviour of the coronal brightness beyond 30 R_θ and to smoothly bridge to the zodiacal light. The procedure is exactly the same as that used by Lamy and Perrin (1986) except that additional data points were introduced inside 2.5 R_θ. The data points were fitted by polynomials in log-log form (6th order) to give a functional representation of the brightness.

The polarization data of the F-corona and the zodiacal light have been fully analyzed and discussed by Lamy and Perrin (1986). We presently retain the polarization profiles appearing in their Fig. 8c (equatorial = plane of symmetry ; polar = meridian plane).

The construction of two-dimensional brightness images from the observed profiles is done in two steps. We first generate a field of elliptic isophotes with adequate spatial resolution. Then, for each pixel of the desired image, we interpolate the brightness between the adjacent isophotes.

2.1. GENERATING THE FIELD OF ISOPHOTES

The kth isophote of the isophote field can be characterized by its semi-major axis a_k (given by its intersection with the equatorial axis) and its semi-minor axis b_k (given by its polar axis). Thus, the isophote field can be described by a set of pairs (a_k, b_k) with k = 1, N. For a given a_k, the corresponding b_k is read from the brightness profiles. Rather than solving the transcendal equation $B_{pol}(b_k) = B_{equ}(a_k)$ for b_k, we interpolate b_k from a table of values $B_{pol}(r_i)$ taking advantage of the monotonic variation of $B_{pol}(r)$. The total number N of isophotes is taken as four times the desired resolution (e.g., N = 1024 if a resolution of 256 pixels is selected). This is driven by the generation of the isophotes which do not intersect the equatorial and/or the polar axis and the requirement of obtaining a smooth variation of the coronal brightness.

2.2. GENERATING THE BRIGHTNESS MAP

Let $P(X_i, Y_j)$ be the center of the pixel (i, j) where X_i and Y_j are the coordinates on the equatorial and polar axis. Two adjacent isophotes (k-1) and k, can be found such that P lies in between (Fig. 1). They satisfy:

$$X_i^2/a_k^2 + Y_j^2/b_k^2 \leqslant 1 \qquad X_i^2/a^2_{k-1} + Y^2_j/b^2_{k-1} > 1$$

The brightness $B(X_i, Y_j)$ at pixel (i, j) is interpolated from these two isophotes according to the following cases.

i) Case A: For points such as P_A (Fig. 1) satisfying $1 - Y^2_j/b^2_{k-1} \geqslant 0$ the interpolation is performed in the equatorial direction (this is the most frequent situation). We can find a point $P_{k-1}(X_{k-1}, Y_j)$ on the curve (k-1), and another one $P_k(X_k, Y_j)$ on the curve k, such that

$$X_{k-1} = a_{k-1} \sqrt{(1-Y^2_j/b^2_{k-1})}, \qquad X_k = a_k \sqrt{(1 - Y^2_j/b^2_k)}$$

then, $B(X_i, Y_j) = B_{equ}(k-1) + \dfrac{X_i - X_{k-1}}{X_k - X_{k-1}} (B_{equ}(k) - B_{equ}(k-1))$

ii) Case B: For points such as P_B (Fig. 1) satisfying $\quad 1 - Y_j^2/b^2_{k-1} < 0$ the point $P_{k-1}(X_{k-1}, Y_j)$ does not exist and the interpolation must be done in the polar direction. We can find a point $P_{k-1}(X_i, Y_{k-1})$ on the curve (k-1), and another one $P_k(X_i, Y_k)$ on the curve k, such that

$$Y_{k-1} = b_{k-1} \sqrt{(1 - X^2_j/a^2_{k-1})}, \quad Y_k = b_k \sqrt{(1 - X^2_j/a^2_k)}$$

then, $B(Xi, Yj) = B_{equ}(k-1) + \dfrac{Y_i - Y_{k-1}}{Y_k - Y_{k-1}} (B_{equ}(k) - B_{equ}(k-1))$

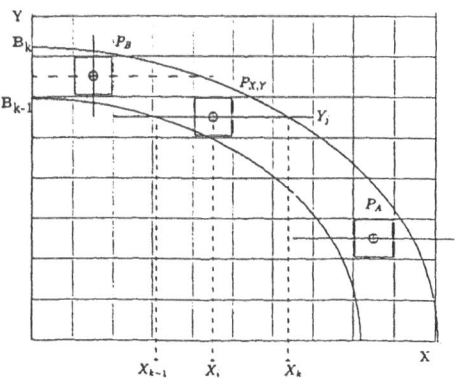

Fig. 1. Determination of the brightness at a pixel

3. Results

The symmetry of the model allows to limit the calculations to one fourth of the corona. We present an example for a field-of-view of 30 R_θ with a resolution of 256 x 256 pixels. Fig. 2 gives the isophotes of the total brightness of the F-corona, the iso-contours of its polarization and the radial profiles of these two quantities along the equatorial and polar directions. One sees that the usual assumption of zero polarization made for the purpose of separating the K and F components of the corona becomes less and less valid as the elongation increases.

We are presently generating a model of the K-corona using however a different approach as we start from the spatial density of electrons and calculate the Thompson scattering. This model incorporates an homogeneous part as well as structures (Bohlin and Garrison, 1974). We shall then study the contrast of various Fraunhofer lines for the purpose of separating the K and F-coronae. We shall also revisit the other method of separation based on polarization, relaxing the assumption of $P_F = 0$ which is not valid in the outer corona.

198

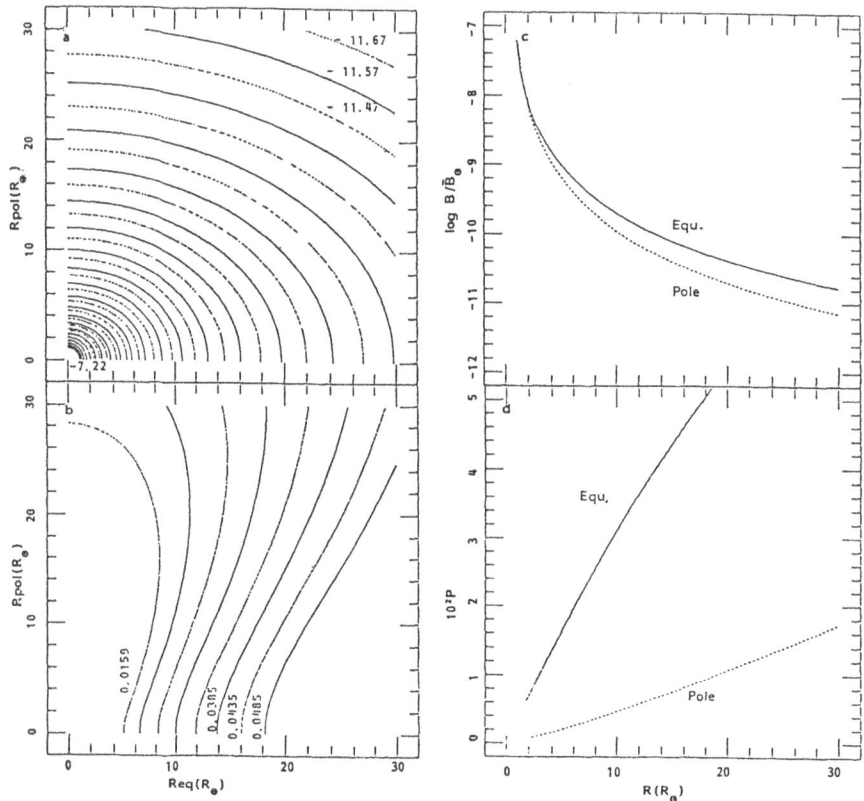

Fig. 2. The F-corona model out to 30 $R_\odot$: a) isophotes of the total brightness (some values of log B are indicated); b) isocontours of the polarization; the brightness profiles (c) and the polarization profiles (d) along the equatorial and polar directions.

References

Badalyan, O.G. (1988) The Physics of Solar Activity, Ed. Nauka, Moscow.

Bohlin, J.D. and Garrison, L.M. (1974) 'Numerical Calculation of Thompson Scattering from inhomogeneous models of the Corona', Solar Physics **38**, 165.

Koutchmy, S. and Lamy, P.L. (1985) 'The F-corona and the Circum-Solar Dust: Evidences and Properties' in Properties and Interaction of Interplanetary Dust, eds. R. Giese, P. Lamy, pp. 63-73.

Lamy, P. and Perrin, J.-M. (1986) 'Volume Scattering Function and Space Distribution of the Interplanetary Dust Cloud', Astron. Astrophys, **163**, 269.

Saito, K. (1970) 'A Non-spherical axisymmetric model of the solar K Corona of the Minimum Type', Ann. Tokyo Astron. Obs. Vol. XII, 2, pp. 53-120.

Van de Hulst, H.C. (1953) 'The Chromosphere and the Corona', in G.P. Kuiper (Ed.), The Sun, Univ. Chicago Press, Chicago, pp. 207-283.

OPTICAL PROPERTIES OF INTERPLANETARY DUST IN THE TANGENTIAL PLANE

J.B. RENARD, A.C. LEVASSEUR-REGOURD
Université PARIS 6 / Service d'Aéronomie - BP 3
91371 Verrières le Buisson, France

R. DUMONT
Observatoire de Bordeaux
33270 Floirac, France

ABSTRACT. Local intensity and emissivity, and consequently local polarization degree, temperature and albedo, can be retrieved from optical and thermal observations of zodiacal light. The local polarization degree (normalized at constant solar distance and phase angle) is found to decrease with elevation above the symmetry plane of the zodiacal cloud. The heterogeneity of the cloud, established towards the symmetry pole, is here demonstrated in the tangential plane (almost perpendicular to the ecliptic plane at 1 AU). We present a map of the local polarization degree in this plane.

Introduction

Measurements of light, either scattered or emitted by interplanetary dust, provide brightness integrals along the line of sight which extends from the observer to the outer fringe of the zodiacal cloud. Local contribution of the grains can be obtained by some local inversion. In order to perform an inversion, mathematical smooth functions (exponential, lorentzian, trigonometric, polynomial, Fourier series) which represent the elemental contributions, can be searched. The integral of these functions has to be equal to the measured brightnesses. The functions need also to satisfy various constraints because of the physical properties of the zodiacal cloud : to be monotonous, to be equal to zero at infinity, to decrease asymptotically to zero with increasing solar distance. It has been found that the curves representing the functions constrict in some nodal regions (fig.1) where the local contributions are determined with lesser uncertainties than elsewhere (Dumont, 1983; Dumont & Levasseur-Regourd, 1985, 1988; Levasseur-Regourd & Dumont, 1990).

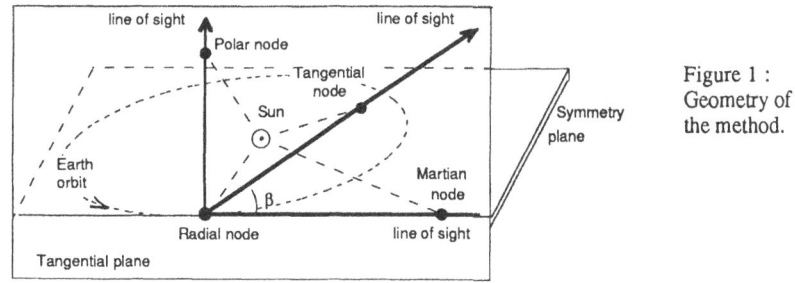

Figure 1 : Geometry of the method.

A.C. Levasseur-Regourd and H. Hasegawa (eds.), Origin and Evolution of Interplanetary Dust, 199–202.
© 1991 *Kluwer Academic Publishers.*

1. Presentation of the Method

1.1. INVERSION TECHNIQUES

In the symmetry plane of the zodiacal cloud, two integrated brightnesses are available along the same line of sight which intersects the Earth's orbit for solar elongations ε and 180° - ε. Curves which represent the functions for the local brightnesses have to focus in two nodal regions : the first one, called radial node, is located in the middle of the chord intersecting the Earth's orbit; the second one, called martian node, is located at a constant solar distance R=1.54 AU. Towards the pole of the symmetry plane, only one integrated brightness is available. So another data is required : the local intensity at α=90° and R=1AU; this result is obtained from the radial node value at an elongation of 90°. Curves have to focus in a nodal region called polar node (fig.2), at an elevation h above the symmetry plane (Levasseur-Regourd et al.,1990). Using this method for integrated intensity (2 polarized components), integrated emissivity (at 2 wavelengths), and for polarimetry and temperature results, four polar node positions are obtained with an elevation varying from 0.2 to 0.4 AU.

We present here an extension of the method previously described, which allows us to derive local values at various ecliptic latitudes from available measurements in the tangential plane (tangent to the Earth's orbit and perpendicular to the symmetry plane). Curves representing the functions have to focus in a nodal region called tangential node, at an elevation varying from 0 to 0.4 AU with ecliptic latitude β increasing from 0° to 90°. Using this method to available intensity data (by steps of 5° in β), two tangential node positions are obtained for each line of sight.

The method developed in the symmetry plane for ε=90°can also be applied in the tangential plane. One additional tangential node is derived on each line of sight for β smaller than 15°. Its elevation remains between 0 and 0.3 AU.

Various sources of data are available : Levasseur-Regourd & Dumont compilation of smoothed intensities (1980), Dumont & Sanchez polarimetric measurements (1975), Fechtig, Leinert & Grün compilation of intensity and polarimetric data (1981), and IRAS results (Hauser et al, 1984; Hauser & Vrtilek, 1988).

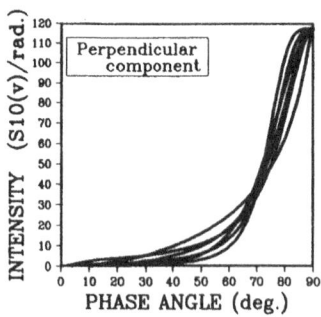 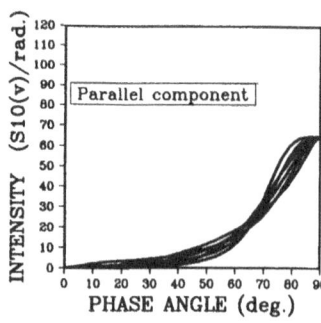

Figure 2 :
Inversion techniques for perpendicular and parallel components of intensity towards the pole. Various trigonometric and lorentzian families of functions are used.

1.2. DISCUSSION

For the lines of sight with an ecliptic latitude varying from 70° to 85°, the elevations obtained for the two positions of the node are stable (respectively 0.39 and 0.26 AU). The local values of the polarization are in excellent agreement (in average 0.08 at h=0.39 AU, α =75° and 0.16 at h=0.26 AU, α=68°), with an error lower than 5%. It is possible to compare these results with those already found towards the symmetry pole (0.08 at h=0.40

AU, α=68°, and 0.16 at h=0.27 AU, α=75°). The two sets of values are quite similar. Also, for β=0°, there is a good agreement near the martian node with the tangential plane method results and the symmetry plane method results (respectively 0.08 at R=1.51 AU, α=41.5°, and 0.06 at R=1.54 AU, α=40.4°).

Gradients of intensity and temperature with solar distance are obtained (Levasseur-Regourd, 1990), as well as a curve describing the evolution of local intensity with phase angle (Levasseur-Regourd et al., 1991). It is then possible to normalize the local contributions at constant phase angle and solar distance .

2. Towards the Symmetry Pole

Curves of integrated intensity and emissivity allow us to obtain, at the radial node position and at the four polar node positions, the values of local intensity, polarization degree, emissivity, temperature and albedo. Evolution of these values with elevation above the symmetry plane, between 0 and 0.40 AU, can be derived. Results are given at normalized phase angle and solar distance. From published IRAS data, local temperature is found to decrease significantly from 255 K in the symmetry plane to 230 K at h=0.40 AU (with an relative error of the order of 5%). Indeed, the local albedo is computed to increase from 0.08 at h=0 AU to 0.10 at h=0.40 AU (with a relative error of 5%). Simultaneously, a decrease of the local polarization degree, from 0.29 in the symmetry plane to 0.18 at 0.40 AU above it, is found with a relative error smaller than 10% (fig.3).

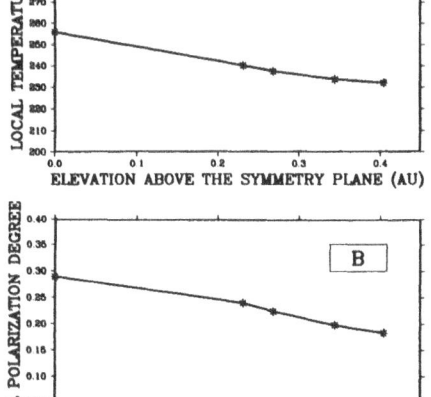

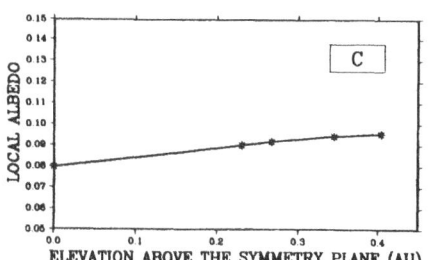

Figure 3 : Evolution with elevation towards the pole of the symmetry plane of local temperature (A), local polarization degree (B) and local albedo (C). The values are normalized at R=1 AU, α=90°.

3. In the Tangential Plane

The tangential node positions correspond to an elevation, a solar distance, a Earth's distance and a phase angle, respectively between 0 and 0.40 AU, 1.02 and 1.74 AU, 0 and 1.5 AU, 35° and 75°. The evolution of the local polarization degree, at normalized solar distance and phase angle, can be presented in the tangential plane as a function of the elevation and the Earth's distance (fig.4). The average relative error is of the order of 10%. The decrease of the polarization with elevation, previously found towards the symmetry pole, is here confirmed in the area of the tangential plane previously described. The decrease of the polarization seems to be continuous, but steeper between 0.2 and 0.4 AU than between 0 to 0.2 AU (fig.4).

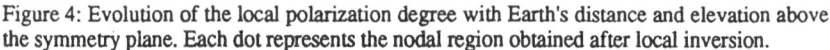
Figure 4: Evolution of the local polarization degree with Earth's distance and elevation above the symmetry plane. Each dot represents the nodal region obtained after local inversion.

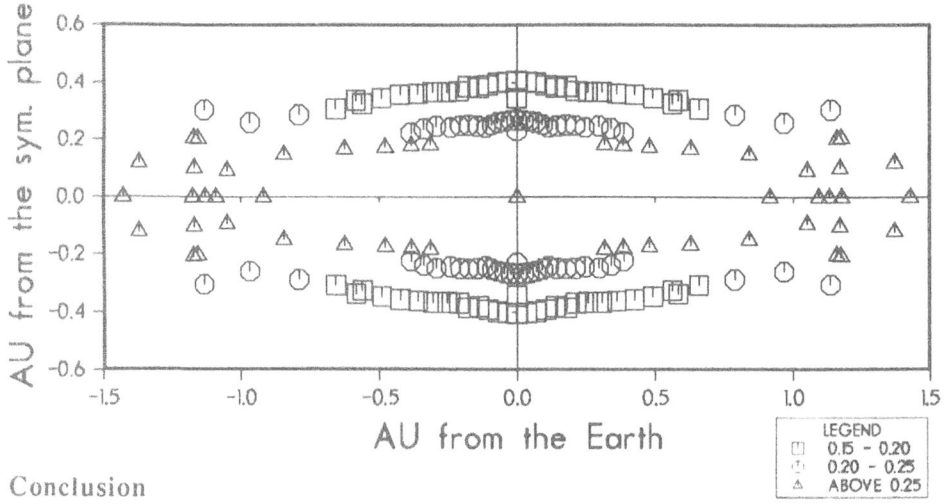

Conclusion

Local values of the optical properties of the grains, derived from measurements of integrated intensity and emissivity (lines of sight in the tangential plane), allow us to demonstrate the decrease of the local polarization degree with elevation. The evolution of the local physical properties with elevation confirms the heterogeneity of the zodiacal cloud out of its symmetry plane. The variations of the properties of the grains explain the differences between the various models (fitting visual or thermal observations) and suggest a dust population from different origins (Levasseur-Regourd et al., 1991).

References

Dumont R. (1983), *Planet. Space Sci.* 31 (12), 1381-1387.
Dumont R. & Levasseur-Regourd A.C. (1985), *Planet. Space Sci.* 33 (1), 1-9.
Dumont R. & Levasseur-Regourd A.C. (1988), *Astron. Astrophys.* 191, 154-160.
Dumont R. & Sanchez F. (1975), *Astron. Astrophys.* 38, 405-412.
Fechtig H., Leinert C., & Grün E. (1981), in K. Schaifers & H. Voigt (eds.), *Landolt-Börnstein N.S.,* 2a, Springer-Verlag, Berlin, 228-243.
Hauser M.G., Gillett, F.C., Low F.L., Gautier T.N., Beichman C.A., Neugebauer G., Aumann H.H., Baud B., Boggess N., Emerson J.P., Houck J.R., Soifer B.T. & Walker R.G. (1984), *Astrophys. J.* 278, L15-L18.
Hauser & Vrtilek (1988), Private communication.
Levasseur-Regourd A.C. (1990), in B. Battrick (ed.), *ESA SP 315, 105-111.*
Levasseur-Regourd A.C. & Dumont R. (1980), *Astron. Astrophys.,* 38, 277-279.
Levasseur-Regourd A.C. & Dumont R. (1990), *Adv. Space. Res.,* 10, 3, 163-170.
Levasseur-Regourd A.C., Dumont R., & Renard J.B. (1990), *Icarus,* 86, 264-272.
Levasseur-Regourd A.C., Renard J.B. & Dumont R. (1991), *this issue.*

SCATTERING CALCULATIONS ON THE BASIS OF THE FREDHOLM INTEGRAL EQUATION METHOD

M. MATSUMURA[1]* AND M. SEKI[2]

[1]*Astronomical Institute, Tohoku University*
Sendai 980, Japan
[2]*College of General Education, Tohoku University*
Kawauchi, Sendai 980, Japan

ABSTRACT. The Fredholm integral equation method (FIM) is one of the solutions to the scattering of electromagnetic radiation by homogeneous and isotropic ellipsoidal particles. Some numerical calculations are performed with the FIM. The results for spherical particles are compared with those by the Mie theory. It is confirmed that the agreement between them is satisfactory for all the models calculated. On the basis of the present method, we examine profiles of the absorption band around $\lambda = 10\ \mu m$ for spherical and ellipsoidal particles composed of crystalline olivine. It is found that the profile strongly depends on the shape of the particle. Even when the particle is moderately elongated (axial ratios are $2:\sqrt{2}:1$), the profile is significantly different from that for a sphere.

1. Fredholm Integral Equation Method (FIM)

The dyadic electric field E of the wave scattered by a particle with a refractive index m can be described by an integration equation

$$E(r) = I_I \exp(i k_I \cdot r) + \int_{vol} G(r, r')\gamma(r')E(r')\,dr' ,$$

(1)

where k_I is the wave number vector of the incident light and

$$I_I = J - \hat{k}_I\hat{k}_I ,$$

(2)

$$G = (J + k^{-2}\nabla\nabla)\exp(ik\,|r - r'|)/(4\pi\,|r - r'|) ,$$

(3)

$$\gamma(r) = k_0^2(m^2(r) - 1) ,$$

(4)

---

*Present address: Osaka Museum of Science, Nakanoshima 4-2-1, Kitaku, Osaka 530, Japan.

A.C. Levasseur-Regourd and H. Hasegawa (eds.), Origin and Evolution of Interplanetary Dust, 203–206.
© 1991 *Kluwer Academic Publishers.*

where J is a unit dyadic, $\hat{k}_I$ is a unit vector along k_I, and k (k_0) is the wave number within (without) the particle. When $r \to \infty$, the asymptotic form of equation (1) is

$$E(r) = I_I \exp(ik_I \cdot r) + B(k_S, k_I) \exp(ik_0 r)/r, \tag{5}$$

where k_s is the wave number vector of the scattered light, and B is the scattering amplitude given by

$$B(k_S, k_I) = (4\pi)^{-1} I_S \cdot \int_{vol} \exp(-ik_S \cdot r)\gamma(r)E(r)\,dr. \tag{6}$$

Equation (6) shows that we can evaluate numerically B provided that the field within the particle is known. The solution for E within the particle is, however, not obtained directly from equation (1), since it is an integral equation with a singular kernel. We perform a transformation such that the singularity is removed analytically. Following Holt et al (1978), we assume that the field within the particle is expressed as the spatial Fourier transform

$$E(r) = \int C(k_1) \exp(ik_1 \cdot r)\,dk_1. \tag{7}$$

Making use of Equation (7), we get a pair of the coupled Fredholm equations, which can be rewritten as matrix equations

$$B(k_S, k_I) = (4\pi)^{-1} I_S \cdot \sum_{l=1}^{n_{max}} c_l C(k_l) U(k_S, k_l), \tag{8}$$

and

$$\sum_{l=1}^{n_{max}} K(k_j, k_l) \cdot c_l C(k_l) = I_I U(k_j, k_I), \tag{9}$$

where expressions of U and K are given elsewhere (Holt et al 1978, Matsumura and Seki 1990). Once B is obtained, the calculations of the optical cross-sections are straightforward as is described in classical textbooks.

In the practical calculation with the FIM, we must choose adequate pivots k_1 ($i=1,\ldots,n_{max}$). In a spherical coordinate (θ_P, ϕ_P) which displays the directions of k_1, we first divide evenly the range of $0 < \theta < \pi$ by $2n_\theta$, i.e., intervals $\Delta\theta_P$ are equal to each other. Next, for each θ_P, we divide the range of $0 < \phi_P < 2\pi$ evenly such that the intervals $\Delta\phi_P$ are nearly equal to $\Delta\theta_P$. Practically we get $n_{max} = 2$, 12, 30, and 56 for $n_\theta = 0$, 1, 2, and 3 (when $n_\theta = 0$, the pivots are poles only).

2. Application to silicate 10 μm band

We apply the FIM to the investigation of the band profile observed near $\lambda = 10 \mu$m in many astronomical objects, which is attributed to silicate particles. The shape of the particle is assumed to be spherical or ellipsoidal (the axial ratios are 2:√2:1).

Table 1 presents values of efficiency factor for absorption Q_{abs} as a function of n_θ. The equivalent radius a_{eq} is defined as $a_{eq}=(abc)^{1/3}$ where a, b, and c are three radii of the ellipsoid, and is set to be 1.0 μm in these calculations. We confirm that the convergence for n_θ is rapid, and that the obtained values are correct enough provided that $n_\theta = 1$.

TABLE 1. Examples of calculations: Q_{abs} for $a_{eq}=1.0$ μm

λ (μm)	9.8	10.0	11.0	11.6
Re (m)	0.5109	0.6682	0.8004	2.748
Im (m)	0.6344	1.061	1.379	2.291
x_{eq} (=$k_0 a_{eq}$)	0.6410	0.6283	0.5713	0.5417
Sphere				
n_θ = 0	0.8397	1.807	2.366	0.8975
1	0.8943	1.902	2.415	0.8836
2	0.8943	1.902	2.415	0.8835
Mie	0.8943	1.902	2.415	0.8835
Ellipsoid (2:√2:1)				
n_θ = 0	0.8493	1.642	1.992	0.8934
1	0.9347	1.793	2.100	1.119
2	0.9347	1.793	2.100	1.119

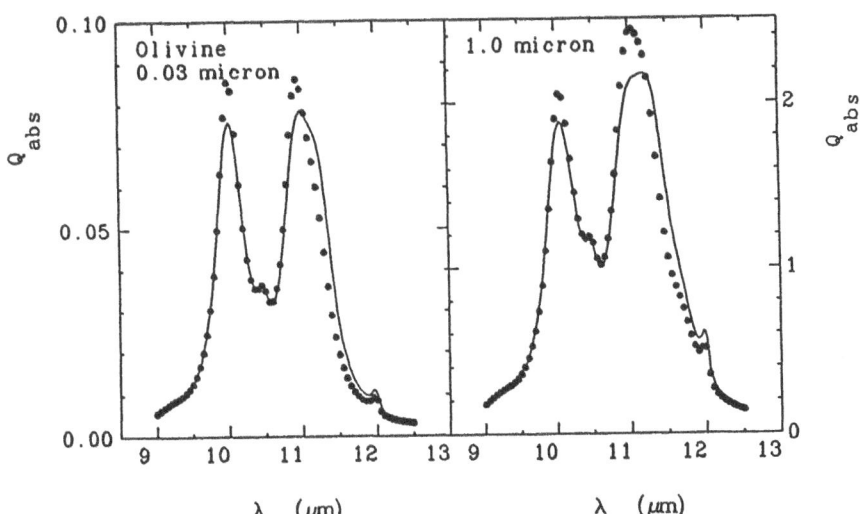

Figure 1. Band profiles for crystalline olivine. Solid lines are for ellipsoidal particles, while dots are for spherical ones. The left figure is for a_{eq} = 0.03 μm, and the right one is for a_{eq} = 1.0 μm.

Figure 1 shows the results for crystalline olivine, where optical constants are cited from Mukai and Koike (1990). We find that the profile for the ellipsoidal particle (solid line) is significantly different from that for the spherical one (dots).

The results for amorphous silicate are demonstrated in figure 2 (m is from Day 1979). The behavior of the solid line (results for ellipsoid) is closely similar to that of the dashed line (results for sphere). Thus, the effect of the shape is less significant for amorphous silicate.

Recent observations of comets show a double peak structure in 10 μm-band, and this fact is interpreted as the presence of crystalline olivine in cometary grains. The present study suggests that the effect of the shape is important to the quantitative interpretation of such observations.

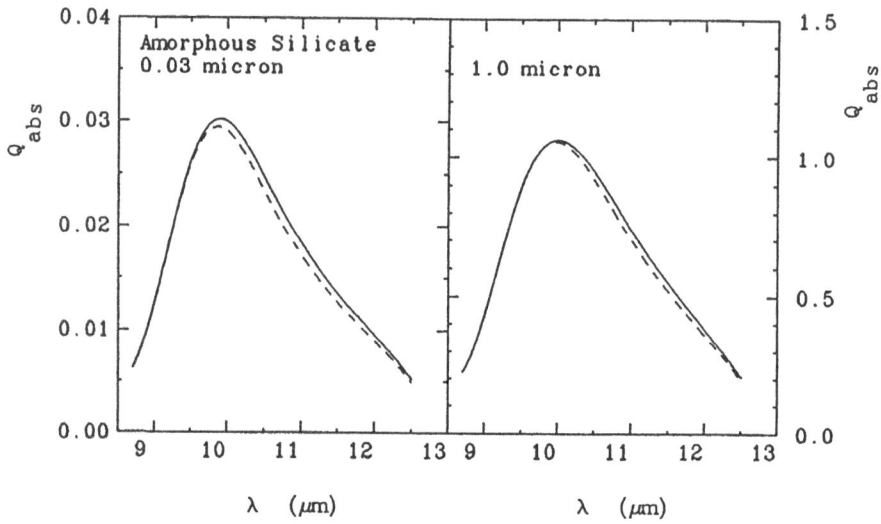

Figure 2. Same as figure 1 but for amorphous silicate. The results for sphere are shown by dashed line.

REFERENCES

Day, K.L. 1979, Ap. J., **234**, 158.
Holt, A.R., Uzunoglu, N.K., Evans, B.G. 1978, IEEE Transaction on Antennas and Propagation 26, 706.
Matsumura, M., Seki, M. 1990, Ap. Space Sci., in press.
Mukai, T., Koike, C. 1990, Icarus, in press.

THE SCATTERING MATRIX OF RANDOMLY ORIENTED
INFINITE CYLINDERS

P. STAMMES
Space Science Department   and   *Dept. of Physics & Astronomy*
ESA/ESTEC   *Free University*
P.O. Box 299   *De Boelelaan 1081*
NL-2200 AG Noordwijk   *NL-1081 HV Amsterdam*
The Netherlands   *The Netherlands*

ABSTRACT. A method is outlined to compute the scattering matrix of an ensemble of infinite cylinders in random orientation, as an approximation to the scattering matrix of an ensemble of very long but finite cylinders. Numerical checks are presented, which show that the results for infinite cylinders agree with theoretical results for very thick and very thin cylinders, as well as with numerical data for prolate spheroids and short finite cylinders.

1. Introduction

The scattering matrix of small particles is an important quantity in analyzing observations of sunlight scattered by interplanetary dust, because it completely describes the light scattering process including polarization. Often spheres have been considered as scattering models for interplanetary dust particles, since exact solutions of scattering by nonspherical particles are rare. The infinitely long circular cylinder is one of the few nonspherical shapes for which an exact analytical solution of Maxwell's equations is known. However, it poses some problems for use as a scattering model because of its infinite length. We have tried to overcome these problems in order to obtain a valid limit for a strongly elongated particle. The basic principle is that we consider a unit length of the infinite cylinder in all orientations.

In the past much research has been devoted to obtain the mathematical formulation for scattering by an infinite cylinder in arbitrary orientation with respect to the incident light (see e.g. [1–8]). In almost all previous work, however, scattering by only <u>one</u> cylinder was considered, or only the cross-sections for scattering and extinction were discussed. Here I will consider the scattering matrix of an <u>ensemble</u> of randomly oriented infinite cylinders, as an approximation to the scattering matrix of an ensemble of very long, but finite, cylinders.

2. Definition of the scattering matrix

We use the Stokes parameters I, Q, U, V to describe a beam of light. I is the intensity (in W/m²) and Q, U, V contain the polarization information of the beam. We consider scattering by an ensemble of particles which are randomly oriented. The scattering properties of the ensemble are completely contained in the 4×4 scattering matrix $\mathbf{F}(\Theta)$ (Θ is the scattering angle), which transforms the Stokes parameters of the incident beam into those of the scattered beam. For particles that have a plane of symmetry, like cylinders, the scattering

A.C. Levasseur-Regourd and H. Hasegawa (eds.), Origin and Evolution of Interplanetary Dust, 207–210.
© 1991 *Kluwer Academic Publishers.*

matrix of the ensemble has the form [2]

$$\mathbf{F}(\Theta) = \begin{pmatrix} F_{11}(\Theta) & F_{12}(\Theta) & 0 & 0 \\ F_{12}(\Theta) & F_{22}(\Theta) & 0 & 0 \\ 0 & 0 & F_{33}(\Theta) & F_{34}(\Theta) \\ 0 & 0 & -F_{34}(\Theta) & F_{44}(\Theta) \end{pmatrix}.$$

For unpolarized incident light, $F_{11}(\Theta)$ gives the intensity and $-F_{12}(\Theta)/F_{11}(\Theta)$ the degree of linear polarization of the scattered light.

3. Procedure to obtain the scattering matrix

We consider an infinitely long, homogeneous, circular cylinder with radius ρ, which is tilted with respect to the incident radiation (a plane wave). The angle of incidence, i.e. the angle between the incident beam and the cylinder normal, is α. The scattered radiation propagates along the surfaces of cones, with apex angles $\pi - 2\alpha$ (cf. [4]).

The expression for the radiation scattered by one tilted infinite cylinder, for arbitrarily polarized incident radiation, is given by, e.g., [5–8]. We have to correct this expression for the infinite length of the cylinder using a factor $\cos\alpha$, in order to approximate a very long but finite cylinder. An infinite cylinder receives the same amount of light in any position, whereas a tilted finite cylinder receives an intensity reduced by a factor $\cos\alpha$ compared to a perpendicular finite cylinder (cf. [8]).

To find the scattering matrix of an ensemble of infinite cylinders in random orientation, we have to integrate over all cylinder orientations in 3D-space. However, because of the scattering cone for one cylinder, we only have to integrate over those cylinders which are in the bisection plane of the directions of incidence and scattering. The resulting scattering matrix depends on: (i) the size parameter of the cylinder, $x = 2\pi\rho/\lambda$, where λ is the wavelength, and (ii) the refractive index m.

4. Checks

To check our numerical results for the scattering matrix of an ensemble of randomly oriented infinite cylinders, we performed the following comparisons.

(a) The scattering matrix elements obeyed four general inequalities [9].

(b) We compared with Fraunhofer diffraction, which is the limit for very thick ($x \gg 1$) cylinders. In Figure 1 $F_{11}(\Theta)$ is shown for the exact case of infinite cylinders with $x = 50$ and $m = 1.31 - 0.1i$, and for the case of Fraunhofer diffraction [10] with $x = 50$. The agreement is good at small scattering angles but degrading at larger angles, as is expected.

(c) We compared with Rayleigh-Gans scattering, which is the limit for "soft" and thin cylinders ($|m - 1| \ll 1$ and $2x|m - 1| \ll 1$). Figure 2 shows the excellent comparison between the exact result and Rayleigh-Gans theory [2] for $F_{11}(\Theta)$ in the case $x = 0.1$ and $m = 1.1$. The other matrix elements divided by F_{11} compare equally well, their shape being that of ordinary Rayleigh scattering. Here we used a correction factor $1/\cos\alpha$ in the Rayleigh-Gans theory of finite cylinders to make it applicable to infinite cylinders.

(d) We compared with computational results for short finite cylinders and prolate spheroids, obtained by the T-matrix method [11]. In Figure 3 the scattering matrix is shown for ensembles of infinite cylinders, finite cylinders and prolate spheroids, all having $m = 1.31$. The size parameter of the infinite cylinders is $x = 2.5$. As size parameter for the finite cylinders

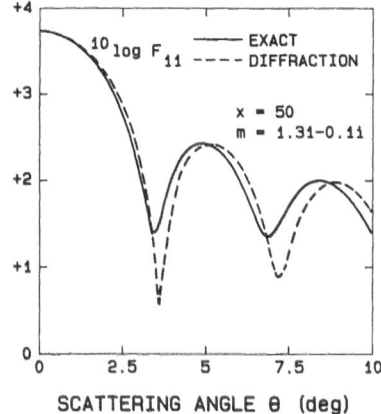

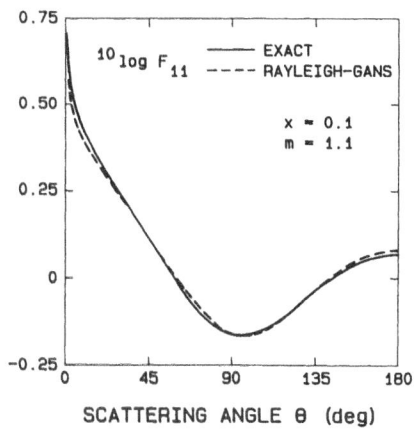

Figure 1. The scattering function $F_{11}(\Theta)$ for an ensemble of infinite cylinders in random orientation, following from exact theory and Fraunhofer diffraction.

Figure 2. $F_{11}(\Theta)$ for an ensemble of infinite cylinders, following from exact and Rayleigh-Gans theory. For the Rayleigh-Gans results we used $2\pi L/\lambda = 10^4$, where L is the cylinder length.

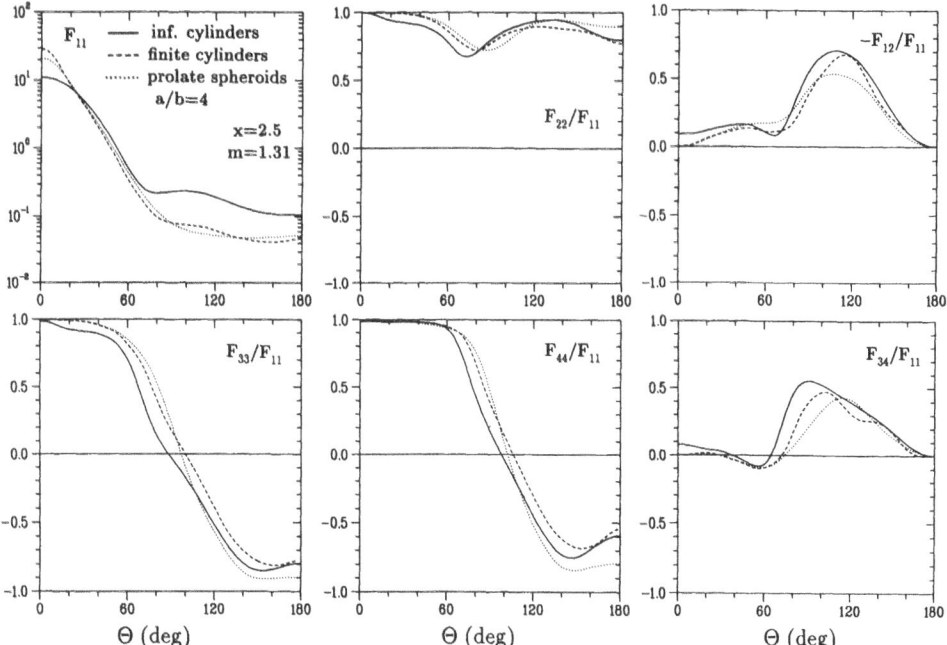

Figure 3. The scattering matrices of ensembles of infinite cylinders, finite cylinders and prolate spheroids. The finite cylinders and prolate spheroids both have axes ratio $a/b = 4$. The results for finite cylinders and spheroids have been obtained with the T-matrix method.

and prolate spheroids we take the size parameter of their semi-minor axis b, so $x = 2\pi b/\lambda$. The finite cylinders and prolate spheroids both have $x = 2.5$ and axes ratio $a/b = 4$. Almost all features in the scattering matrix elements of the finite cylinders and prolate spheroids are also found in the infinite cylinder results. If absorption is added the agreement improves.

5. Conclusions

It is possible to obtain the scattering matrix of an ensemble of randomly oriented infinite cylinders. Numerical tests have been performed to check the reliability of the results.

The presented solution for infinite cylinders is useful as an approximation to the scattering matrix of an ensemble of very long finite cylinders. Until now, it is not possible to obtain the scattering matrix of very elongated particles with other computational methods. The T-matrix method is limited to short cylinders and spheroids (axes ratio $\approx$ 4:1) and to size parameters (of the semi-major axis) of $\approx$ 15. Our infinite cylinder solution works well up to much larger (radius) size parameters of $\approx$ 100.

Some general characteristics of the scattering matrix of infinite cylinders are:
• The steepness of F_{11} is much less than for short finite cylinders due to the absence of a strong forward peak. However, details in F_{11} of short finite cylinders appearing at larger Θ also appear for infinite cylinders.
• Despite the fact that F_{11} does not agree in magnitude with short finite cylinder results, the matrix element ratios F_{ij}/F_{11} agree very well.
• The matrix element ratios $-F_{12}/F_{11}$ and F_{34}/F_{11} tend to a non-zero value for $\Theta \rightarrow 0$. This differs from the behaviour for randomly oriented finite particles. The reason is the integration over the bisection plane to obtain the scattering matrix of randomly oriented infinite cylinders. This plane acts as a plane of preference for the polarization.

Acknowledgements
I am grateful to F. Kuik for performing the T-matrix method calculations, using a program from P. Barber (Clarkson Univ., Potsdam, USA), and acknowledge helpful discussions with J.W. Hovenier. This research has been supported by ESA's Research Fellowship Programme.

References

1. Wait, J.R. (1955) Can. J. Phys. **33**, 189

2. Van de Hulst, H.C. (1957) "Light Scattering by Small Particles", J. Wiley & Sons, New York; also Dover, New York, 1981

3. Lind, A.C. and Greenberg, J.M. (1966) J. Appl. Phys. **37**, 3195

4. Kerker, M. (1969) "The Scattering of Light and Other Electromagnetic Radiation", Academic Press, New York

5. Liou, K.-N. (1972) Appl. Opt. **11**, 667

6. Cohen, A. and Acquista, C. (1982) J. Opt. Soc. Am. **72**, 531

7. Bohren, C.F. and Huffman, D.R. (1983) "Absorption and Scattering of Light by Small Particles", J. Wiley & Sons, New York

8. Barabás, M. (1987) J. Opt. Soc. Am. A, **4**, 2240

9. Hovenier, J.W., Van de Hulst, H.C. and Van der Mee, C.V.M. (1986) Astron. Astrophys. **157**, 301

10. Takano, Y. and Tanaka, M. (1980) Appl. Opt. **19**, 2781

11. Barber, P. and Yeh, C. (1975) Appl. Opt. **14**, 2864

ASTEROIDAL DUST AND THE ZODIACAL EMISSION

WILLIAM T. REACH
Astronomy Department
University of California
Berkeley, CA 94720

ABSTRACT. The contribution to the brightness of the infrared background by asteroidal dust, distinguished both by lower color temperature and 'band-pair' morphology, is determined using IRAS observations. Dust band pairs are associated with at least 7 asteroid families and groups, but very little is detected from the remainder of the asteroid belt, indicating that asteroid families and groups are the source of asteroidal dust.

1. Profile-fitting Technique

The brightness of the 25μm background away from the galactic plane is a smooth function of ecliptic latitude at 0.5° resolution, with the exception of four bumps at latitudes ±10° and ±1.4°. These features were first discovered by IRAS (Low *et al.* 1984), and it was quickly suggested that they are associated with the prominent Hirayama families (Dermott *et al.* 1984). Using spatial filtering to remove the brightness on angular scales larger than ~ 3°, Sykes (1988, 1990) showed that the inner band pair is itself split into two pairs associated with the Themis and Koronis families; the outer band pair is associated with the Eos family.

We have fit each IRAS profile of the ecliptic with a combination of a smooth (4-th order) polynomial in ecliptic latitude, a linear term in the Galactic H I column density, and 4 Gaussians intended to represent the two main asteroidal band-pairs. Although a pair of Gaussians is not the best model for a band pair (see below), they provide an excellent fit to the data and the center, amplitude, and width allow straightforward interpretation. First, it was immediately evident that the Gaussian widths of the dust bands were well resolved, with full widths at half maximum between 2 and 4°. In contrast, when viewed at higher resolution (2') in spatially-filtered maps, the inner bands are sharp and should be unresolved at low resolution (30'). Second, the ecliptic latitudes of the four Gaussians used to represent the dust bands reveal a parallactic effect that allows determination of their distance. Due to parallax, the separation between two Gaussians that constitute a band pair is larger for scans made at larger solar elongation. We measured the band-pair separation for a range of solar elongations, and found that the ±10° (Eos) band pair is 2.4 AU from the Sun, while the ±1.4° (Themis & Koronis blend) band pair is 1.7 AU from the Sun. These distances are clearly smaller than the distances (3.0 and 2.9 AU, respectively) to the Hirayama families. The sharp dust bands have been shown to be at the same distance as the Hirayama families (Sykes 1990), indicating that the broad emission is produced by dust closer to the Sun than the parent asteroid family.

A.C. Levasseur-Regourd and H. Hasegawa (eds.), Origin and Evolution of Interplanetary Dust, 211–214.
© 1991 *Kluwer Academic Publishers.*

2. Spectral Decomposition Technique

In order to determine the total contribution of asteroidal dust to the zodiacal emission (ZE), we have used the IRAS spectrum at each position in the sky to separate the infrared background into 3 components: 1) relatively hot dust near the Earth's orbit that dominates the ZE, 2) warm dust that is in the asteroid belt, and 3) cold dust in the interstellar medium. Three spectral components allow one extra degree of freedom in fitting the four IRAS wavebands at 12, 25, 60, and 100μm. The spectrum of the interplanetary components was calculated using the size distribution of particles near the Earth's orbit (Grün *et al.* 1985) and the optical properties of silicates (see Reach 1988). It was necessary to correct the color of the 'hot' component for lines of sight with different solar elongation and latitude; the corrections were most important at 12μm (in the Wien portion of the spectrum).

Maps of the three separate components of the infrared background clearly show the different morphology of the respective emitting regions: the hot component is very smooth, the warm component shows the parallel zodiacal dust bands, and the cold component shows the interstellar cirrus. A slice through the asteroidal component from the North to the South ecliptic poles at 90° solar elongation is shown in Figure 1. Although the brightness of the asteroidal component is less than 3% of the ZE brightness, the point-to-point noise is only about 0.1% of the ZE brightness. In Figure 1, the twin bumps at ±10°, representing the Eos band pair, and the central, marginally resolved bumps at ±1.4°, representing the blended Themis & Koronis band pairs, are clearly evident.

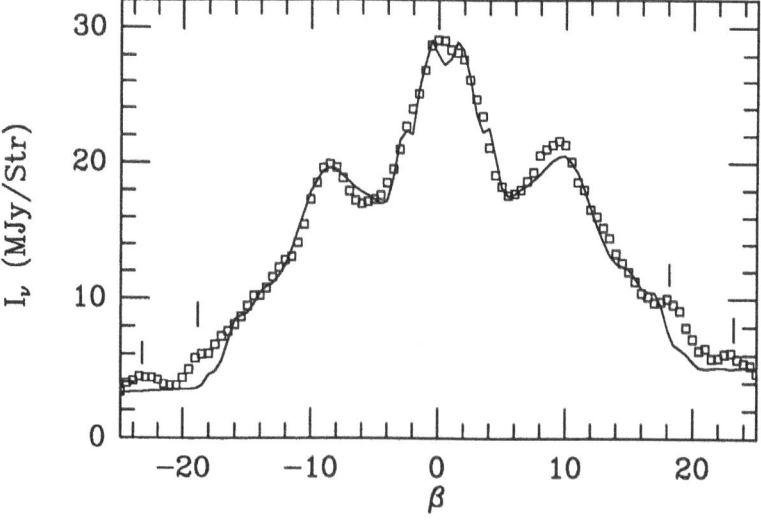

Figure 1. The 25μm surface brightness of the asteroidal component of the zodiacal emission as a function of ecliptic latitude, for solar elongation 90°. The observations (squares) are shown together with the asteroid family fit (curve). Two band pairs not associated with prominent families are indicated by vertical lines; the ±24° pair is probably associated with the Phocaea group.

Most interesting in the map of the asteroidal component of the ZE is the presence of apparent broad band pairs at a range of ecliptic latitudes. The band pairs discovered by Sykes (1988) in spatially filtered maps are confirmed, as well as new bands. Since the bands are severely blended, it is not possible to uniquely disentangle the emission of the separate bands. We have calculated the theoretical brightness profile (Sykes 1990) of band pairs associated with the main asteroid families, and used a least-squares fit to determine the amount of emission contributed by each family. The only free parameter in the fits was the total surface area of grains associated with each family. The theoretical band pair was convolved to 0.5° resolution, and the dispersion in the orbital elements was boosted for each family by factors of 1 (Themis), 2 (Eos), and 3 (all others) to match the observed band-pair separations. In addition to the asteroid family models, a Gaussian with variable amplitude, center, and width was included as a model for the non-family, main belt asteroidal dust. The Gaussian was unable to fit the observations better than the dust band models, despite a wide range of guesses for its initial conditions; we place an upper limit of 0.6 MJy/Str on the brightness of the dust associated with non-family asteroids. The least-squares fit is shown as a smooth solid line in Figure 1, and the surface areas are given in Table 1.

TABLE 1. Asteroidal dust band surface areas

Family	Surface Area (10^{19} cm^2)
Themis/Koronis/Nysa	$2.0 \pm .3$
Flora	$0.9 \pm .3$
Eos	$58. \pm 1.$
Io	$1.9 \pm .4$
Maria	$1.2 \pm .3$
Phocaea	$1.7 \pm .5$

The results in Table 1 may not represent a unique solution for the amount of emission from each asteroid family, due to blending of some of the bands. The Themis/Koronis/Nysa blend is very severe. We find that the amount of dust in the Themis and Nysa bands is greater than in the narrower Koronis band, but it is difficult to determine how much of the Themis band is actually produced by a broad component of the Koronis band. Sykes (1990) has shown that the Koronis band is much brighter than the Themis band, when viewed at high angular resolution. Some new features are evident in the residuals from the fits. We can tentatively associate the bumps at ±21° (visible at the edges of Figure 1) with the Phocaea group of asteroids, which is not considered a true family but a grouping bounded by orbital instabilities (see Valsecchi et al. 1989). If this association is real, then the dust bands are not necessarily relics of the collisions that produced true asteroid families.

3. Discussion

The predominance of asteroid families and groups as regions of active dust production indicates that the collisional environment in the asteroid belt is not uniform: family asteroids constitute some 26% of the numbered asteroids, but the actual fraction of all asteroids is likely to be closer to 4% when selection effects in asteroid searches are taken into account (Valsecchi *et al.* 1989). The 20-200μm size grains that emit the infrared radiation detected by IRAS are comminution results of relatively recent collisions, less than 10^6 yrs ago based on Poynting-Robertson lifetimes. If asteroids are indeed the source of a large fraction of the interplanetary dust that is near the Earth's orbit, it would be expected that dynamical signatures of family origin will be discovered; in the case of true family origin (as opposed to origin in the Phocaea group), chemical signatures may also be discovered.

In the ecliptic plane, the total asteroidal dust area per unit volume $\langle n\sigma \rangle = 2.7 \times 10^{-22}$ cm^2/cm^3. For comparison, near the Earth's orbit $\langle n\sigma \rangle$ is about 1.7 times (Grün *et al.* 1985) to 3 times (Reach 1988) larger. In steady state, Poynting-Robertson (PR) drag spreads the dust through the inner Solar System with the volume density increasing as inversely as the heliocentric distance— a factor of 3 decrease between the Earth's orbit and the asteroid belt. Thus the asteroidal dust bands have at least enough dust to maintain the interplanetary dust complex. The loss rate from a given family will decrease with time, as the larger, less-abundant particles (with longer PR time scales) drift inward. It is presently uncertain whether new particles are continuously being generated in asteroid families to balance the losses due to PR drag.

4. References

Dermott, S.F., Nicholson, P.D., Burns, J.A., and Houck, J.R. (1984) 'Origin of the Solar System dust bands discovered by IRAS', *Nature* **312**, 505.

Grün, E., Zook, H. A., Fechtig, H., and Giese, R.H. (1985) 'Collisional balance of the meteoritic complex', *Icarus* **62**, 244.

Hauser, M.G. (1988) 'Models for infrared emission from zodiacal dust', in A. Lawrence (ed.), *Comets to Cosmology*, Springer-Verlag, Berlin, pp. 27-39.

Low, F.J. *et al.* (1984) 'Infrared cirrus: New components of the extended infrared emission', *Ap. J.* **278**, L19.

Reach, W.T. (1988) 'Zodiacal emission I. Dust near the Earth's orbit', *Ap. J.* **335**, 468.

Reach, W.T. (1991) 'Zodiacal emission II. Dust near the Ecliptic', *Ap. J.* **387**, in press.

Reach, W.T. and Heiles, C. (1988) 'Separating the Solar System and Galactic contributions to the diffuse infrared background', in A. Lawrence (ed.), *Comets to Cosmology*, Springer-Verlag, Berlin, pp. 40-43.

Sykes, M.V. (1988) 'IRAS observations of extended zodiacal structures', *Ap. J.* **334**, L55.

Sykes, M.V. (1990) 'Zodiacal dust bands: Their relation to asteroid families', *Icarus* **84**, 267.

Valsecchi, G.B., Carusi, A., Knezevic, Z., Kresak, L., and Williams, J.G. (1989) 'Identification of asteroid families', in R.P. Binzel, T. Gehrels, and M.S. Matthews (eds.), *Asteroids II*, University of Arizona Press, Tucson, pp. 368-385.

IV

COMETARY DUST :
OBSERVATIONS AND EVOLUTION

SPECTROSCOPIC EVIDENCE OF ORGANIC MOLECULES RELEASED BY THE DUST OF HALLEY'S INNER COMA

J. CLAIREMIDI, P. ROUSSELOT, G. MOREELS
Observatoire de Besançon
BP 1615
25010 BESANCON Cedex
FRANCE

ABSTRACT. New spectroscopic arguments supporting the probable presence of organic molecules in the material released by comet Halley are deduced from the data obtained by the Vega three-channel spectrometer. Two excesses of emission on the UV side of the OH and CN bands at 305 and 383 nm are interpreted as being due to "prompt" radiation emitted by electronically excited OH and CN radicals directly produced by the photolysis of water vapor and an organic X-CN molecule. A broad-band emission progressively appears between 342 and 375 nm when the solar scattered continuum has been substracted. This emission increases approximately as the inverse of the projected distance to the nucleus. It is interpreted as a fluorescence emission of organic molecules, possibly condensed polycyclic hydrocarbons. Present observations support the hypothesis of grains coated with organic material and give arguments in favor of a probable interstellar origin for cometary dust.

1. Introduction

Two major observational facts obtained during Halley's investigation program suggest that the matter released by the comet contains organic molecules. First, a spectral signature was detected in emission in the near-infrared spectrum at 3.3 - 3.4 µm both from space (Combes et al., 1988) and from the ground (Baas et al., 1986; Danks et al., 1987). This feature is attributed to a vibration C-H band of an organic compound. Secondly, a large fraction of dust particles, approximately 30%, contain light elements (CHON particles, Kissel et al., 1986; Jessberger et al., 1988). The presence of organics in cometary grains has been suggested by Greenberg (1982) and has been argumented recently (Greenberg and Hage, 1990) on the basis of a possible similarity of composition between interstellar and cometary material.

Additional arguments supporting the probable presence of organic molecules in cometary dust may be found in the data of the Vega three-channel spectrometer. Spectra originating from gas jets (Clairemidi et al., 1990) as well as spectra obtained close to the nucleus show emission features directly produced by organic molecules or their primary photoproducts.

2. Spectroscopic data and monochromatic composite images

Between 300 and 400 nm, the main features of cometary spectra are the OH ($A^2\Sigma^+$ - $X^2\Pi_i$) (0,0) band at 309 nm, the NH ($A^3\Pi_i$ - $X^3\Sigma^-$) (0,0) band at 336 nm and the CN ($B^2\Sigma^+$ - $X^2\Sigma^+$) (0,0) band at 388 nm. The excitation mechanism is principally fluorescence excited by solar radiation. An excess of emission is detected on the UV side of the OH band at 305 nm and of the CN band at 383 nm (fig.1). The additional emission in the case of OH is probably due to the R branch of the (0,0) band. The OH ($A^2\Sigma^+$) would be produced directly by the photolysis of water vapor between 120 and 137 nm (Shafizadeh et al., 1988). In the case of CN, the excess of emission would

217

A.C. Levasseur-Regourd and H. Hasegawa (eds.), Origin and Evolution of Interplanetary Dust, 217–220.
© 1991 *Kluwer Academic Publishers.*

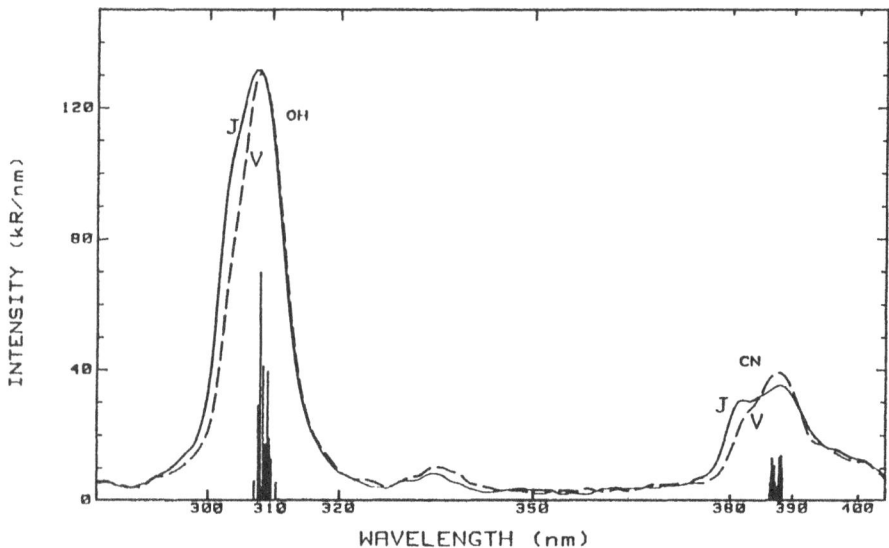

Fig. 1 Spectra in the 290-400 nm range from a jet (J) and from the valley (V); below : fluorescence spectra of OH and CN

correspond to the sequence $\Delta v=0$ of CN with $v' \leq 7$. The CN ($B^2\Sigma^+$, $v' \leq 7$) would be produced directly by the photolysis of an organic molecule of formula X-CN.

Monochromatic composite images of the prompt OH and CN emissions are displayed in Fig.2 and Fig.3. Both emissions are clearly located inside the gas jets which are present up to distances as long as 40000 km. As shown by Clairemidi et al., 1990, the gas jets are correlated with dust jets which contain a relatively important fraction of submicronic particles. They constitute a distributed source of water vapor and organic molecules detected through the OH and CN prompt emissions. It seems probable that these particles are directly connected with the CHON particles detected by Kissel et al. (1986).

3. Broad-band emission in the 342-375 nm spectral region

In the innermost region of the coma, at cometocentric distances smaller than 5000 km, a broad-band emission arises between 342 and 375 nm. This region, between the NH and CN bands, is free of emission in ground-based spectra. The intensity increases approximately as the inverse of the projected distance to the nucleus p (Fig.4 and 5). A perspective view of this increase, showing the progressive emergence of the broad band is presented in fig.6. Four peaks, at 346, 357, 363 and 373 nm may be distinguished in the band.

The 1/p intensity variation shows that the emission is due to a parent molecule fluorescence mechanism. Molecules with fluorescence in the near-UV are numerous, but their spectra are generally obtained in a solvent liquid phase. In this region, possible candidates are methanol, naphtalene, anthracene or phenanthrene, or substituted compounds. If a condensed polycyclic aromatic hydrocarbon (PAH) was the emissive species, this would support the proposition made by Léger and Puget (1984) that interstellar matter contains PAHs, since cometary and interstellar material are supposed to present some degree of similarity in composition.

4. Conclusion

Two sources of organic molecules are revealed by their spectroscopic characteristics and their spatial distribution. The first is connected with the jets which extend to distances of the order of 40000 km or more. The jets contain very small submicronic particles which release water vapor

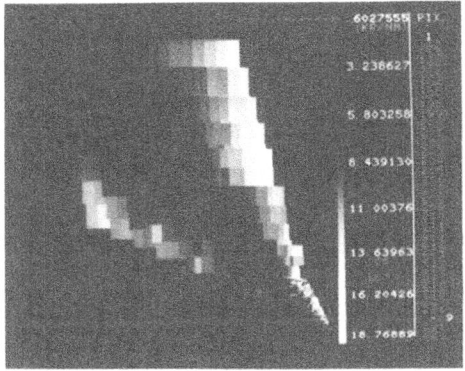

Fig. 2 Prompt emission of OH p ≤40000 km

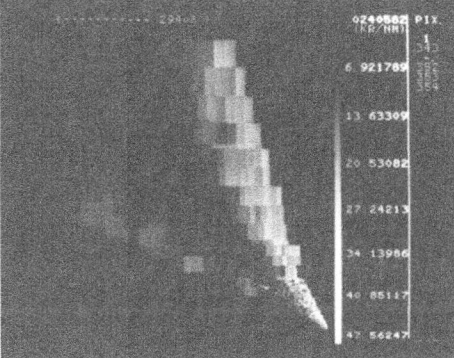

Fig. 3 Prompt emission of CN p ≤40000 km

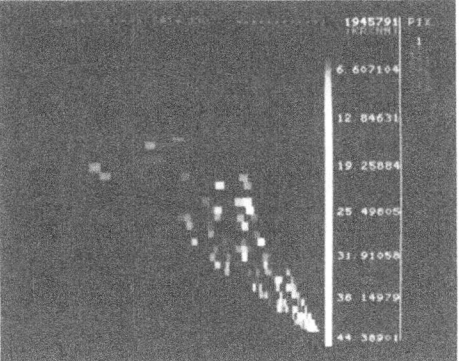

Fig. 4 Broad-band emission λ = 346 nm
Δλ = 3 nm   p ≤ 8000 km

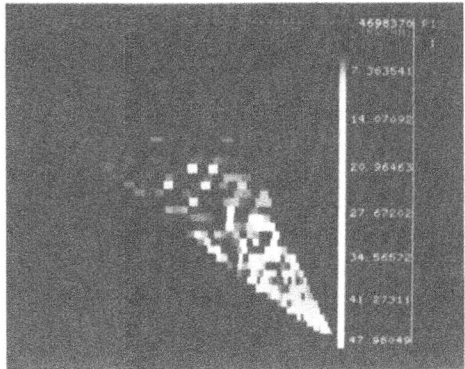

Fig. 5 Broad-band emission λ = 363 nm
Δλ = 3 nm p ≤ 8000 km

and carbon-bearing organic molecules which are detected by the prompt emission of their OH and CN photoproducts. The second is connected with the nucleus which releases organics, possibly condensed polycyclic aromatic hydrocarbons, which are detected by their broad-band fluorescence emission in the near UV. Present observations provide arguments in favor of the hypothesis of grains coated with refractory organic material (Greenberg, 1982) and in favor of an interstellar origin for cometary material (Encrenaz et al., 1989).

220

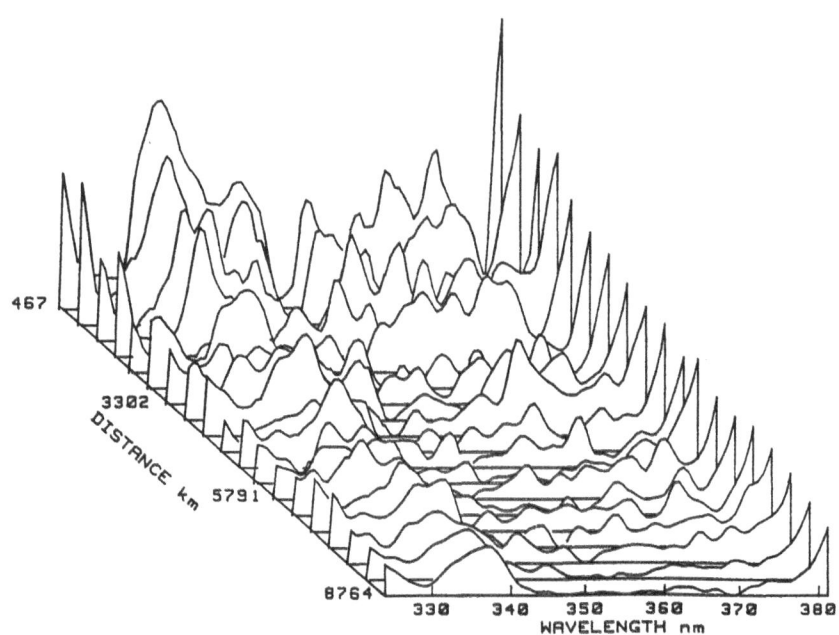

Fig. 6 Progressive emergence of a broad-band emission between 342 and 375 nm

References

Baas F., Geballe T.R., Walther D.M. (1986), Astrophys. J., 311, L97-L101
Clairemidi J., Moreels G., Krasnopolsky V.A. (1990), Astron. Astrophys.., 231, 235-240
Clairemidi J., Brandon E., Rousselot P., Moreels G., This volume, Paper P-50, 4 pages
Combes M., et al. (1988), Icarus, 76, 404-436
Danks A.C., Encrenaz T., Bouchet P., Le Bertre T., Chalabaev A. (1987), Astron. Astrophys., 184, 329-332
Encrenaz T., Knacke R.F., Crovisier J. (1989), Bull. Amer. Astron. Soc., 21, 3
Greenberg J.M. (1982) in Comets, Univ. Arizona Press, 131-163
Greenberg J.M., Hage (1990), Astrophys. J. 361, 260-274
Jessberger E.K., Christoforidis A., Kissel J. (1988), Nature, 332, 691-695
Kissel J. et al. (1986), Nature, 321, 280-282
Shafizadeh N., Rostas J., Lemaire J.L., Rostas F. (1988), Chem. Phys. Letters, 152, 11, 75-80

MODELING DUST FRAGMENTATION IN COMETS

I. KONNO and W.F. HUEBNER
Southwest Research Institute
P.O. Drawer 28510
San Antonio, TX 78228-0510
USA

ABSTRACT. We developed a 1-D hydrodynamic model of dusty gas flow with dust fragmentation in a cometary atmosphere and performed calculations for a dust-size distribution with radii $a = 10^{-4}-10$ cm and densities variable with dust size. A comparison was made with Giotto observations of dust jet intensities within 100 km of the nucleus of Comet Halley. We found that dust fragmentation cannot be solely responsible for the flattening of the dust intensity near the nucleus with respect to the $1/R$ law. We conclude that a combination of geometric effects and grain fragmentation may explain the observed intensity profiles.

1. Introduction

Dust intensity profiles obtained by the Halley Multicolour Camera (HMC) show a sharp deviation from the expected $1/R$ dependence (R is the cometocentric distance). Thomas et al. (1988), Huebner et al. (1988), and Reitsema et al. (1989) have successfully used geometric effects to explain the deviation. However, Szegö et al. (1988) found the deviation seen in profiles using Vega images to be too large to be explained by source geometry effects alone and suggested dust particle fragmentation as an alternative. From Giotto's PIA and DIDSY data, McDonnell et al. (1987) found some evidence for possible fragmentation or evaporation of grains. Vaisberg et al. (1987) obtained several pieces of evidence for an evolution (including splitting) of the grains during their motion in the cometary coma.

2. Modeling

We solve the hydrodynamic equations for a dusty gas flow in a comet coma. The fragmentation of dust particles into the next smaller size is incorporated in the model by adding source and sink terms in the continuity equations for the dust. The lifetime τ_i of dust particle with radius a_i (a_1 is the minimum size) is given by

$$\tau_i = \tau_{max}[\log(a_i/a_{max})]^{\alpha}, \tag{1}$$

where τ_{max} is the lifetime of the particle of the maximum size a_{max}. The lifetime of the smallest particles is assumed infinite. The size distribution at the nucleus surface is given by

$$\rho_i = \rho_{max}(a_i/a_{max})^{\beta}, \tag{2}$$

where ρ_i is the spatial mass density of the particles of radius a_i and ρ_{max} is the density of the particles of the maximum radius a_{max} that can be entrained by the gas. The bulk density of dust particles is a function of dust size (Lamy et al., 1987):

A.C. Levasseur-Regourd and H. Hasegawa (eds.), Origin and Evolution of Interplanetary Dust, 221-224.
© 1991 *Kluwer Academic Publishers.*

$$\rho_{bulk} = 2.2 - 1.4a/(a_0 + a) \text{ g cm}^{-3}, \quad a_0 = 10^{-4} \text{ cm.} \tag{3}$$

We perform calculations with dust particles of radius 10^{-4} cm to 10 cm with a continuous distribution approximated by 26 discrete particle sizes seperated by a logarithmic scale, so that a particle always fragments into about four particles of the next smaller size. We define the mass density by ρ, velocity by v, pressure by P, ratio of specific heats by γ, number density by n, temperature by T. The subscripts g and d refer to gas and dust, respectively.

The hydrodynamic equations for the single-fluid, inviscid, perfect gas are as follows.

$$\frac{1}{R^2}\frac{d}{dR}(R^2 \rho_g v_g) = \dot{\rho}_g, \tag{4}$$

$$\rho_g v_g \frac{dv_g}{dR} + \frac{dP_g}{dR} = -\sum_i F_{gd,i}, \tag{5}$$

$$\frac{1}{\gamma - 1}(v_g \frac{d\rho_g}{dR}) + \frac{\gamma}{\gamma - 1}\rho_g \frac{dv_g}{dR} = \dot{Q}_{photo} - \sum_i \dot{Q}_{gd,i}, \tag{6}$$

where $F_{gd,i}$ is the momentum transfer rate from the gas to the dust of size a_i, given by

$$F_{gd,i} = \frac{1}{2}C_{D,i}\pi a^2 \rho_g(v_g - v_{d,i})^2 n_{d,i}. \tag{7}$$

Here $C_{D,i}$ is the modified free molecular drag coefficient defined as follows (cf. Probstein, 1968):

$$C_{D,i} = \frac{2\sqrt{\pi}}{3}\sqrt{\frac{T_d}{T_g}} + \frac{2s_i^2 + 1}{s_i^2 \sqrt{\pi}}e^{-s_i^2} + \frac{4s_i^4 + 4s_i^2 - 1}{2s_i^3}erf(s_i), \quad \vec{s}_i = \frac{\vec{v}_g - \vec{v}_{d,i}}{\sqrt{2kT_g/m_g}}, \tag{8}$$

where k is the Boltzmann constant and m_g is the molecular mass of the gas. $\dot{Q}_{photo}$ is the rate of heating of the gas due to the photodissociation of water and $\dot{Q}_{gd,i}$ is the energy tranfer rate from the gas to the dust. At distances $R << 10^4$ km, $\dot{\rho}_g$ for H_2O is negligible, so that we set $\dot{\rho}_g = 0$ in the application in the next section.

For the dust, the following hydrodynamic equations are solved for each dust particle of size a_i.

$$\frac{1}{R^2}\frac{d}{dR}(R^2 \rho_{d,i} v_{d,i}) = -\frac{\rho_{d,i}}{\tau_i} + \frac{\rho_{d,i+1}}{\tau_{i+1}}. \tag{9}$$

$$\rho_{d,i} v_{d,i}\frac{dv_{d,i}}{dR} - \rho_{d,i}g_{comet}(R) = F_{gd,i}, \tag{10}$$

$$\rho_{d,i}v_{d,i}C_{D,i}\frac{dT_{d,i}}{dR} = \dot{Q}_{gd,i} + \dot{Q}_{rad,i}, \tag{11}$$

where $g_{comet}(R)$ is the gravitational acceleration by the comet nucleus and $\dot{Q}_{rad,i}$ is the rate of the radiation energy due to solar heating and IR cooling.

We assume for the nucleus of Comet Halley that the effective radius $R_n = 4.0$ km, the density $\rho_n = 0.5$ g cm^{-3}, and the dust-to-gas mass ratio $\chi = 1.0$. The reason that the nucleus density is lower than the average dust density is that the material in the cometary nucleus may be more porous than dust particles and contains hollow space.

3. Application: Dust Jets near the Nucleus of Comet Halley

We apply the dust fragmentation model to the problems of the dust intensity profiles near the nucleus of Comet Halley. Because cometary dust is optically thin, the observed dust intensity, which is proportional to the dust column density, should change like $1/R$. However, intensity profiles of jet-like features analyzed in the HMC images from the Giotto spacecraft (Thomas et al., 1988; Huebner et al., 1988) show strong flattening relative to a $1/R$ profile near the nucleus.

Thomas and Keller (1990) have explained the dust intensity profile using a dust fragmentation calculation. They showed that on the average, fragmentation of a particle into 2.7 particles can explain the 30% increase of $I \cdot R$ (where I is the scattered light intensity) beyond 9 km from the source.

The dust intensity profiles in Figure 1 obtained by Huebner et al. (1988) show a steep rise of $I \cdot R$ within 50 -100 km from the source. They have shown that a wide range of dust profiles can be modeled using different source geometries and obtained qualitative agreement with HMC profiles without invoking other mechanisms. Although their model fit the data very well, it assumes a constant velocity for the dust particles. Dust particles are accelerated by gas-drag force from rest to the terminal velocity of several 100 km s^{-1} to 1.0 km s^{-1}, depending on their size. Figure 2 shows an $I \cdot R$ profile obtained from a hydrodynamic calculation with the dust size distribution given by equation (10) with $\beta = 1.0$ in the case of no fragmentation. The figure shows a rapid decrease of $I \cdot R$, which is just the opposite of the observed profiles. Therefore, models for the observed dust in tensity profile must account for the dust acceleration.

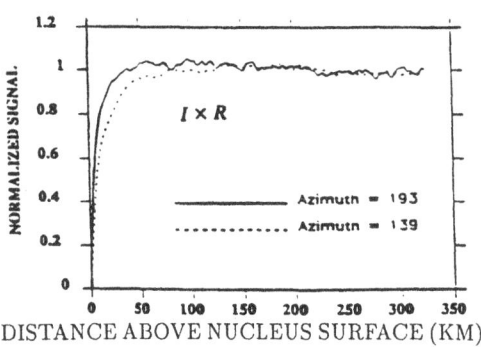

DISTANCE ABOVE NUCLEUS SURFACE (KM)

Fig. 1 - $I \cdot R$ profiles for dust obtained by the Giotto spacecraft. If the intensity $I \sim 1/R$, then $I \cdot R$ should be constant. From Huebner et al. (1988).

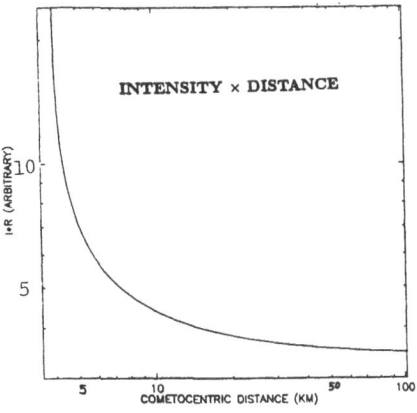

Fig. 2 - A calculated $I \cdot R$ profile for dust when there is no dust fragmentation. $I \cdot R$ is not constant because of acceleration.

We performed the hydrodynamic calculations to see if we can explain the observed dust profiles by dust fragmentation (Figure 3). The parameters for the model are $\alpha = 0$, $\beta = 1.0$, $\tau_{max} = 10$ sec. The model seems to fit the observation fairly well. The lifetime for the dust particles used for the fit, however, is much too short for particles to survive. Figure 4 shows the dust density for several sizes as a function of cometocentric distance. We see that all particles fragment into the smallest ones within 50 km, which is contradiction to observations because larger particles have been detected at much larger distances.

The fragmentation calculation by Thomas and Keller (1990) gives the lower limit to fragmentation since their profile does not include the region closer than ~ 9 km to the nucleus. We conclude that fragmentation alone cannot be responsible for the deviation of the dust intensity profile from 1km to ~ 50 km near the nucleus observed by the Giotto spacecraft. The combination of the geometric effect (Huebner et al., 1988) and dust fragmentation described here may explain the deviation.

224

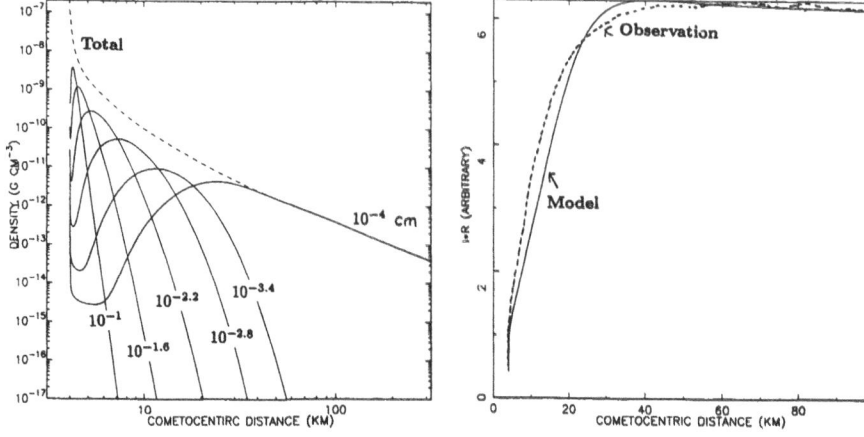

Fig. 3 - Model $I \cdot R$ profile for dust compared with the observation, the solid line in Figure 1.

Fig. 4 - Dust space density obtained for the model in Figure 3.

References

Huebner, W.F., Boice, D.C., Reitsema, H.J., Delamere, W.A., and Whipple, F.L. (1988) 'A model for intensity profiles of dust jets near the nucleus of comet Halley', *Icarus*, **76**, 78-88.

Lamy, P.L., Grün, E., and Perrin, J.M. (1987) 'Comet P/Halley: implications of the mass distribution function for the photometric properties of the dust coma', *Astron. Astrophys.* **187**, 767-773.

McDonnell, J.A.M. et al. (1987) 'The dust distribution within the inner coma of comet P/Halley 1982i: encounter by Giotto's impact detectors', *Astron. Astropys.* **187**, 719-741.

Probstein, R.F. (1968) 'The dusty gas dynamics of comet heads',*Problems of Hydrodynamics and Continuum Mechanics*, Soc. Industr. Apply. Math., 568-583.

Reitsema, H.J., Delamere, W.A., Williams, A.R., Boice, D.C., Huebner, W.F., and Whipple, F.L. (1989) 'Dust distribution in the inner coma of comet Halley: comparison with models', *Icarus*, **81**, 31-40.

Szegö, K., Tóth, I., Szatmáry, Z., Smith, B.A., Kondor, A., and Merényi (1988) 'Dust photometry in the near nucleus region of comet Halley', *preprint*, KFKI-1988-33/C

Vaisberg, O.L., Smirnov, V., Omelchenko, A., Gorn, L., and Iovlev, M. (1987) 'Spatial and mass distribution of low-mass dust particles (m< 10^{-10} g) in comet P/Halley's coma', *Astron. Astrophys.* **187**, 753-760.

Thomas, N. and Keller, H.U. (1987) 'Comet P/Halley's near nucleus jet activity', in *Symposium on the diversity and similarity of comets, ESA SP-278*, 337-342.

Thomas, N. and Keller, H.U. (1990) 'Interpretation of the inner coma observations of comet P/Halley multicolor camera', *Annales Geophysicae*, **8**, (2), 147-166.

Thomas, N., Boice, D.C., Huebner, W.F., and Keller, H.U. (1988) 'Intensity profiles of dust near extended sources on comet Halley', *Nature*, **332**, 51-52.

THE CONTRIBUTION OF LONG PERIOD COMETS TO THE INTERPLANETARY DUST CLOUD

M. FULLE
Osservatorio Astronomico
Via Tiepolo 11
I-34131 Trieste
Italy

G. CREMONESE
Osservatorio Astronomico
Vicolo dell'Osservatorio 5
I-35122 Padova
Italy

ABSTRACT. The numerical analysis of cometary dust tails (Fulle 1989) allows to obtain the mass loss rate, the size distribution, the ejection velocity and the orbital eccentricity of the dust grains ejected by the parent comets. All these physical quantities are necessary and sufficient to compute the comet mass contribution to the interplanetary dust cloud. We apply this method to three long period comets, namely C/Bennett 1970II, C/Bradfield 1987XXIX and C/Liller 1988V, and obtain that each long period comet injects in bound orbits at least half of the total produced mass. When we consider that a typical long period comet can produce more than 10^{14} g of dust along each perihelion passage, we obtain that the considered long period comets alone injected a input mass rate of about 10^6 g s^{-1} of meteoroids in bound orbits during the last 20 years, a contribution which is very close to that from all short period comets.

1. Introduction

The mass of meteoroids injected by a comet into bound orbits depends on three quantities, namely: i) the time-dependent dust mass loss rate of the comet, which gives the total mass of lost dust; ii) the time-dependent size distribution of the dust grains, which gives the percentage of the total mass released in the largest grains, i.e. in meteoroids; iii) the time and size -dependent dust ejection velocity, which gives the orbital eccentricity of the dust grains, and therefore tells us if a grain is a meteoroid or not. Dust tail analysis allows a self-consistent computation of all these quantities, thus providing a method which is applied to the dust tails of long period comets Bennett 1970II, Bradfield 1987XXIX and Liller 1988V to determine their contribution to the interplanetary dust cloud.

2. The numerical model of dust tails

The dust tail model (Fulle 1989) considers $\mathcal{N}_t \times \mathcal{N}_\mu \times \mathcal{N}_s$ sample dust grains, where $\mathcal{N}_t$ is the number of samples in the time interval of dust ejection, $\mathcal{N}_\mu$ is the number of samples in the sizes, and $\mathcal{N}_s$ is the number of grains of a fixed size uniformly distributed on a dust shell. It considers different ejection geometries for each of which the ejection of dust is restricted to a cone of half width w with its symmetry axis pointing toward the Sun. The position of each grain at the observation is derived from its keplerian motion, then projected into the photographic plane coordinate system (M,N), where M is the projected prolonged radius vector, so as to obtain the model distribution of the scattered light from the tail and the related kernel matrix A.

A.C. Levasseur-Regourd and H. Hasegawa (eds.), Origin and Evolution of Interplanetary Dust, 225–228.

The solutions are given by minimizing the functional $[AF-I]^2 + \beta[BF]^2$, where A is the kernel matrix, I is the data vector containing the dust tail surface light intensities of the N_k images sampled in $N_N \times N_M$ points, B is a regularizing matrix weighted by β, and F is the solution vector sampled in $N_t \times N_\mu$ values, from which the dust number and mass loss rates and the time dependent and time averaged size distributions can be directly computed. The dust ejection velocity v(t) is required for the computation of the matrix A, so that it must be determined by means of a trial and error procedure. The regularizing weight β tunes the constraints to our ill-posed problem: when β increases, the instability of F decreases, but so does the quality of the fit to the data. Therefore the most probable dust velocity v(t) is defined as the function giving a stable and positive vector F for a regularizing weight β as small as possible.

Table 1. The mass contribution to the zodiacal cloud of long-period comets.
$u = \partial \log v(t, d) / \partial \log d$, power index of the dust velocity size dependence. w, half width of the dust ejection cone: isotropic ejection (half width $w = \pi$), hemispherical ejections ($w = \pi/2$), and strongly anisotropic ejections ($w = \pi/4$). $\mathcal{N}_s$, $\mathcal{N}_\mu$, $\mathcal{N}_t$, dust samples on a dust shell, in the modified size and in time. N_t, N_μ, samples of the solution $F(t, 1-\mu)$ in time and in the modified size. N_M, N_N, samples of the N_k source images in the M and N directions. T, number of functions v(t) tested to find the true solution. $\mathcal{M}$, total ejected dust mass (10^{14} g) for $Ap(\alpha) = 0.06$ (Hanner et al., 1990). $\mathcal{M}_b$, percentage of the total mass $\mathcal{M}$ injected into bound orbits. S, symbol in Fig.1

u	w	$\mathcal{N}_s$	$\mathcal{N}_\mu$	$\mathcal{N}_t$	N_t	N_μ	N_k	N_M	N_N	T	$\mathcal{M}$	$\mathcal{M}_b$	S
C/1970II: 81% of the dust mass in bound orbits													
-1/6	180°	284	100	180	20	10	4	30	30	25	—	80%	O
-1/6	90°	143	100	180	20	10	4	30	30	18	—	81%	□
-1/6	45°	382	100	180	20	10	4	30	30	31	—	78%	Δ
-1/4	180°	284	100	180	20	10	4	30	30	17	—	80%	+
-1/4	90°	143	100	180	20	10	4	30	30	19	—	86%	×
-1/4	45°	382	100	180	20	10	4	30	30	31	—	81%	*
C/1987XXIX: 52% of the dust mass in bound orbits													
-1/6	180°	2578	100	180	20	10	2	30	30	10	2.1	53%	O
-1/6	90°	1285	100	180	20	10	2	30	30	15	1.8	53%	□
-1/6	45°	382	100	180	20	10	2	30	30	36	2.0	48%	Δ
-1/4	180°	2578	100	180	20	10	2	30	30	24	2.1	51%	+
-1/4	90°	1285	100	180	20	10	2	30	30	15	2.7	67%	×
-1/4	45°	382	100	180	20	10	2	30	30	17	1.6	38%	*
C/1988V: 69% of the dust mass in bound orbits													
-1/6	180°	284	100	180	20	10	3	30	30	22	0.3	48%	
-1/6	90°	143	100	180	20	10	3	30	30	17	0.4	70%	O
-1/6	45°	382	100	180	20	10	3	30	30	15	0.4	69%	□
-1/4	180°	284	100	180	20	10	3	30	30	19	0.3	47%	
-1/4	90°	143	100	180	20	10	3	30	30	20	0.5	75%	Δ
-1/4	45°	382	100	180	20	10	3	30	30	10	0.4	65%	+
-1/3	90°	143	100	180	20	10	3	30	30	30	0.5	74%	
-1/3	45°	382	100	180	20	10	3	30	30	31	0.4	66%	×
-1/2	45°	382	100	180	20	10	3	30	30	24	0.5	70%	*

The free parameters of the model are the dust bulk density and albedo (which are approximated as constant quantities in the size and phase ranges here considered), the size dependence u of the ejection velocity v, $u = \partial \log v(t, d) / \partial \log d$, where d is the dust-particle diameter, and the dust ejection anisotropy w. Since no particular value of u and w can be assumed, we show results which depend on a combination of them, so that the sensitivity of the solutions to such parameters can be directly evaluated.

3. The meteoroid mass

The quantity F allows to directly compute the dust mass loss rate and the time-dependent size distribution, and therefore also the production rate of meteoroids, the absolute values of which depend on the dust albedo and the absolute calibration of the data, which can be affected by large uncertainties. On the contrary, the dust bulk density cannot introduce any uncertainty in the dust mass loss rate, since the relation between the dust mass and the quantity F is independent of the dust density. Moreover, if we consider the percentage of the mass production rate of meteoroids with respect to the total loss rate, we obtain a quantity which is not affected by any physical uncertainty, but only by the errors of the solution F, a good estimate of which is given by the dispersion of the solution itself due to different combinations of free parameters u and w.

We apply our model to three long period comets: four images of C/Bennett 1970II (Hogner & Richter 1980), two 20/25/50 cm Schmidt images of C/Bradfield 1987XXIX obtained by A.Cimatti, and three CCD frames of C/Liller 1988V obtained by K.Jockers and coworkers. The results are summarized in Table 1, and point out that the percentage of the meteoroid mass is independent at all on the free parameters u and w, and therefore should be affected by very low uncertainties. On the contrary, it was possible to absolutely calibrate only the Bradfield and Liller images, assuming a dust albedo for the phase function of 0.06 (Hanner et al. 1990). Since such value is an upper limit for the dust albedo of this comet, and since the model can consider only finite time and size intervals, the absolute values of the produced dust mass should be considered as lower limits only.

Our results point out the high percentage of meteoroids produced by all the long period comets here considered. This fact is mainly due to the high power index of the size distributions of such comets, which are shown in Fig.1. All the power indeces are significantly higher than -4, and this fact implies that most of the mass is released in large grains, i.e. in meteoroids, the size range of which is also shown in the same figure.

Fig.1 Time averaged size distributions of C/1970II (left, power index -3.3 ± 0.1), C/1987XXIX (center, power index -3.2 ± 0.2) and C/1988V (right, power index -3.5 ± 0.1). The dust bulk density is assumed of 1 g cm^{-3} (the diameter values d depend inversely on the assumed dust bulk density). The symbols are related to Table 1. The shaded areas show the size range of the meteoroids in bound orbits.

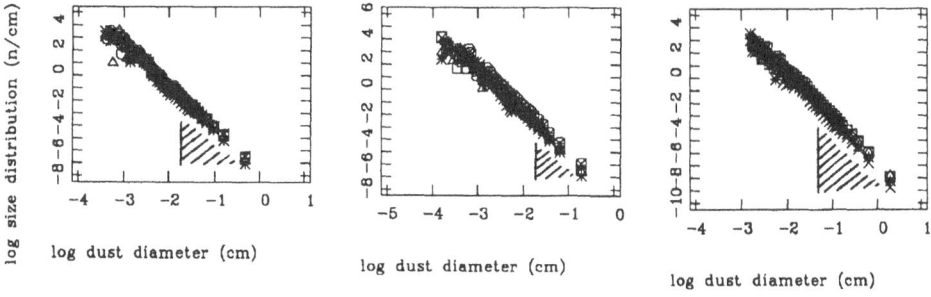

4. Discussion

When we remember that C/Bennett was by far brighter than comet Bradfield, we can conclude that in the last 20 years at least 3 long period comets contributed to the zodiacal dust cloud, with an averaged meteoroid loss of at least 10^{14} g for each comet. If we admit that such a sample has statistical significance (such a very poor statistic obviously needs further samples), we obtain from the long period comet family $\approx 10^6$ g s^{-1} of meteoroids in orbits bound in the Solar System, a value 30 times larger than the statistical estimate by Mukai (1990). Such disagreement points out that each long period comet has to be deeply analysed to deduce its effective production of meteoroids injected into bound orbits.

Our model applied to Schmidt plates (from Sekanina & Schuster 1978a,b) of short period comets Encke and D'Arrest has shown that each short period comet injects in bound orbits $\approx 5 \ 10^4$ g s^{-1} of meteoroids (Fulle 1990), an estimate very close to the results of the analysis of IRAS dust trails (Sykes 1990). Therefore the contribution of meteoroid mass from long period comets is close (probably a bit lower) to that from ≈ 100 short period comets, in agreement with the significant percentage of observed meteoric orbits which can be explained only in terms of a long period cometary source (Olsson-Steel 1990).

The long period cometary source would supply an isotropic dust cloud, because the Poynting-Robertson drag does not change the orbital inclinations. However the meteoroids from short period comets cover all the sizes observed in the related dust tails (d > 20 μm for a dust bulk density of 1 g cm^{-3}, here the sizes depend inversely on the adopted dust density), whereas only the largest grains from long period comets become meteoroids (d > 200 μm). Therefore the optical scattering from short period meteoroids is much larger than from long period ones (a factor 10 in the case of a size distribution power index of -4), so that also in the limit case of the same input mass rate from the two cometary sources, we would obtain a strong optical concentration of zodiacal dust close to the ecliptic plane.

The different distribution of orbital inclinations of the two cometary sources implies a strong correlation between ecliptical latitude and meteoroid size distribution. Very far from the ecliptic we should find no cometary meteoroids in bound orbits for d < 0.2 mm, but a significant mass for d > 0.2 mm, a bit lower than close to the ecliptical plane.

5. References

Fulle, M. (1989) Evaluation of cometary dust parameters from numerical simulations: comparison with analytical approach and role of anisotropic emissions. *Astron. Astrophys.* **217**, 283 - 297

Fulle, M. (1990) Meteoroids from short period comets. *Astron.Astrophys.* **230**, 220 - 226

Hanner, M.S., Newburn, R.L., Gehrz, R.D., Harrison, T., Ney, E.P., Hayward, T.L. (1990) The infrared spectrum of Comet Bradfield (1987s) and the silicate emission feature. *Astrophys.J.* **348**, 312 - 321

Hogner, W., Richter, N. (1980) *Isophotometric Atlas of Comets*, Par. II, Springer-Verlag

Mukai, T. (1988) Dust from the comets. *Highlights of Astronomy* **8**, 305 - 312

Olsson-Steel, D. (1990) The orbital distribution and origin of meteoroids. *Origin and evolution of interplanetary dust (126th IAU Colloquium)* Kluwer Academic Publisher

Sekanina, Z., Schuster, H.E. (1978a) Meteoroids from Periodic Comet D'Arrest. *Astron. Astrophys.* **65**, 29 - 35

Sekanina, Z., Schuster, H.E. (1978b) Dust from Periodic Comet Encke: Large Grains in Short Supply. *Astron. Astrophys.* **68**, 429 - 435

Sykes, M. (1990) Cometary and asteroidal sources of interplanetary dust. *Origin and evolution of interplanetary dust (126th IAU Colloquium)* Kluwer Academic Publisher

LONG DUST TRAILS OF SHORT PERIOD COMETS

H.U. KELLER and K. RICHTER
Max-Planck-Institut für Aeronomie
Postfach 20
3411 Katlenburg-Lindau
Germany

1. Introduction

Comets constitute an important source for the zodiacal dust cloud. Mainly large particles are contributed because the smaller particles are emitted into hyperbolic orbits relative to the sun. Radiation pressure force reduces the effective solar gravitational attraction. Information about large cometary particles can be derived from a variety of sources requiring quite different observational techniques. Many distinct meteor streams are connected to orbits of short period comets. These streams contain large dust particles that are very little influenced by radiation pressure force. In some cases such as the η Aquarids and Orionids connected to comet Halley the total mass and the age of the meteors have been derived (Hughes, 1987; Hajduk, 1987). The mass of the streams is 5 to 10 times larger than the present mass of the nucleus and their lifetime corresponds to 2000 to 3000 orbital periods. Visible meteors are typically 10^{-2} g and more of centimetre size.

Radar observations of comets coming close to the earth revealed the presence of clouds of large dust particles around cometary nuclei (Campbell et al., 1989; Harmon et al., 1989) dust particles are so numerous that their reflection can obscure the signal from the cometary nucleus itself. The observed particles had to be of the size of the radar wavelength typically around 10 cm. Little is known about the nature and physics of these clouds. Are the particles gravitationally bound? What is the total mass contained in these clouds? The radar signal are yet too noisy to extract more significant observations. Comet IRAS-Araki-Alcock was observed at a distance of only 0.033 AU. Its radar cross section corresponded to a few km^2.

The infrared satellite IRAS observed dust trails concentrated near the orbits of short period comets (Sykes et al., 1986). Dust particles were found in some cases all around the orbit, in other cases in front of and trailing the nucleus. The outflow velocities of the observed particles are small, in the range of a few metres per second, their sizes range around a few millimetres (comet P/Tempel 2 (Sykes et al., 1990)). The IRAS data have not been fully reduced and more results can be expected. Analytical tools (models of the dust density distribution) have been developed and first results are reported in the second part of the paper.

In the visible wavelength range the large particles can hardly be observed since their scattered light is masked by the much more numerous small particles in the range from submicron to several 100 μm. The observations of anomalous tails under special geometric circumstances are exceptions and yield limited information on larger particles (e. g. Richter and Keller (1988)).

229

A.C. Levasseur-Regourd and H. Hasegawa (eds.), Origin and Evolution of Interplanetary Dust, 229–234.
© 1991 *Kluwer Academic Publishers.*

The attitude of the Giotto was disturbed during its encounter with comet Halley in March 1986. The Halley Multicolour Camera (HMC) recorded several distinct dust impacts that allowed for direct determinations of lower mass limits of the particles causing the perturbations (Curdt and Keller, 1988). Calculation of the accumulated flux yields 2 additional points of measurements in the flux vs. mass diagram (see Fig. 3 of Curdt and Keller (1990) substantiating the extrapolation by McDonnell et al. (1987)). The dust mass distribution of comet Halley is dominated by large millimetre size particles and the total dust distribution is comparable to that of the gas (ratio ≥ 1). The impact events observed by HMC have been further analysed and the outflow velocities of the particles have been determined. The properties of these particles resemble in many respects the particles observed in the cometary dust trails by IRAS. Some relevant results are summarized in the following section and discussed with respect to the properties of trail particles.

1.1 PARTICLES NEAR THE NUCLEUS OF COMET HALLEY

Five clearly discernable events were selected for analysis. 2 occurred at distances around $8 \cdot 10^4$ km, 3 closer than $1.5 \cdot 10^4$ km. Further in the events became too numerous to be clearly separated. The characteristics of the 5 dust particle impacts are listed in Table I. Keplerian trajectories connecting the nucleus and the point of observation could be determined assuming that the particles had been emitted in direction towards the sun (Richter et al., 1990). Under these conditions all observed particles could have left the nucleus shortly before or after perihelion. The close in particles are about a factor of ten more massive (around 10 mg) than the ones observed earlier. Their outflow velocities, however, are considerably smaller than predicted by hydrodynamical calculations of the gas drag. Figure 1 displays solutions for the outflow velocity and ejection angle, Θ (corresponding to the zenith angle relative to the sun direction) for particle number 4. Velocities in the range from 20 to 30 km as predicted by the gas dynamic models (Gombosi et al., 1986) require ejection on the night side ($\Theta > 90°$) of the nucleus a few days before observation! Alternatively, these particles could have been ejected from the nucleus at much earlier times long before comet Halley passed its perihelion. The calculations show solutions with reasonable zenith angles (below 30°) for times hundreds of days before perihelion requiring outflow velocities below 10 m s^{-1}. These particles resemble the particles found in the dust trails observed by IRAS: same size range, similarly low outflow velocities. It is possible that HMC and the Giotto encountered these old particles that stay in the vicinity of the nucleus for a long time. In this case the larger particles would not result from the production shortly before encounter. The excess of large particles in the actual cometary dust production rate may not be as big as deduced in the papers quoted above.

If these particles were indeed members of the "trail" population they may have been liberated from the nucleus years before the Giotto encounter. When comet Halley was recovered at a heliocentric distance of more than 11 AU a weak coma was present in the images. The analyses of the dust trails of shorter period comets (typically comet P/Tempel 2) showed that the dust trails comprise particles released many orbital periods ago and along most of the cometary orbit.

1.2 PARTICLES RELEASED ALONG THE COMETARY ORBIT

Analyses of cometary tails are complicated because dust particles emitted from the nucleus at a wide variety of orbital positions contribute to the density at any one point. "Old" particles have orbited the sun many times. The positions of particles oscillate perpendicular to the orbital plane but also relative to the nucleus in the orbital plane. In addition one and the same particle (characterized by its ß value, the ratio of radiation pressure force to gravity) will reach different positions depending on the moment of its release along the orbit.

Some of these effects are illustrated in the following section. The calculations were performed for comet Tempel 2 for the time of observation when the comet passed through the

Figure 1. Solutions of trajectories connecting the nucleus and the location of impact for event No 4, characterized by the emission velocity (left ordinate) and zenith angle (right ordinate, dashed line), are displayed as function of time of release relative to the perihelion of comet Halley.

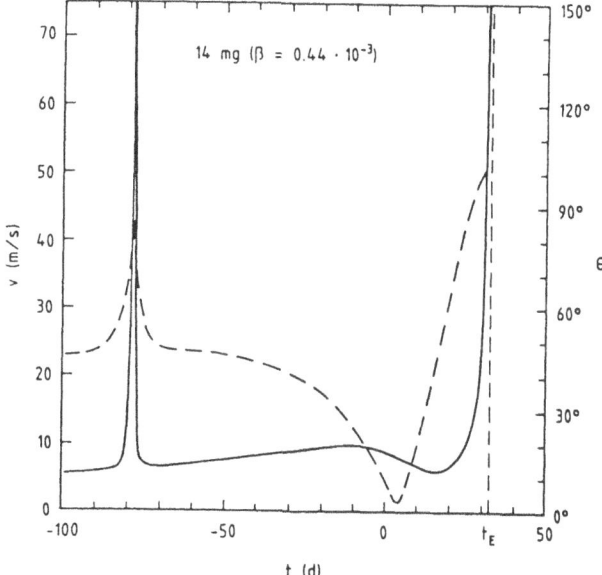

ecliptic on 22.5 July 1983. This point of time lies inside the interval of IRAS observations of the cometary trail. The orbital parameters of comet Tempel 2 are: period (5.29 a), perihelion distance (1.38 AU), and inclination of orbit (12.4°).

2. The trail width

The high resolution mode used for observations of specific targets could resolve the width of the cometary trail to about 4 arcmin (Sykes et al., 1990). A particle released from the nucleus with a velocity component $v_\perp$ perpendicular to the comet's orbital plane will pass this plane again at a the opposite nodal line at a true anomaly different by 180°. This behaviour explains, the two minima of the curve shown in Fig. 2. The comet's true anomaly at the time of observation is 36.6°. The curve depicts the distance of a particle released with $v_\perp = 1$ km s^{-1} perpendicular to the orbital plane of the comet as seen from earth and measured by its offset angle, δ. The total width of the trail would be twice this ordinate value. The actual offset angle of a particle strongly varies with time of release between 0 (nodal line) and ≈ 0.5 arcmin (for $v_\perp = 1$ m s^{-1}). The average offset angle is smaller and probably more characteristic for the width of the cometary trail. The IRAS observations limits $v_\perp$ to < 8 m s^{-1}.

2.1 ORBITAL PERIODS OF RELEASED DUST PARTICLES

The distribution of dust particles along a cometary orbit, i.e. their distances from the nucleus as a function of time is determined by the differences of the orbital periods of the particles relative to that of the nucleus. The orbital period of a released particle depends on its effective

Figure 2. The offset angle of a particle released with $v_\perp = 1$ m s^{-1} as function of true anomaly of comet P/Tempel 2 seen from earth.

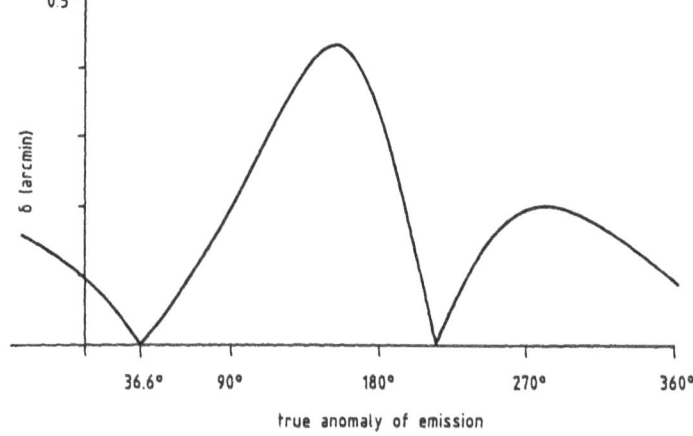

ß value and on its velocity (direction and magnitude). But it also depends on the time, i.e., orbital position, of the release. This is shown in Fig. 3 where the abscissa is the true anomaly of the particle release. Assuming that the ejection velocity of a particle is zero the relative change in orbital period, T, as a function of ß can be approximated by $\frac{1}{T}\frac{\partial T}{\partial \beta}\Big|_{\beta\,-\,0}\cdot\beta$ (left hand side of left ordinate).

Figure 3. The change of period of particles released along the orbit of comet P/Tempel 2. Partial derivations and their ratio at zero emission velocity and ß=0 are displayed as functions of true anomaly of comet P/Tempel 2. v_0 is the velocity of the cometary nucleus.

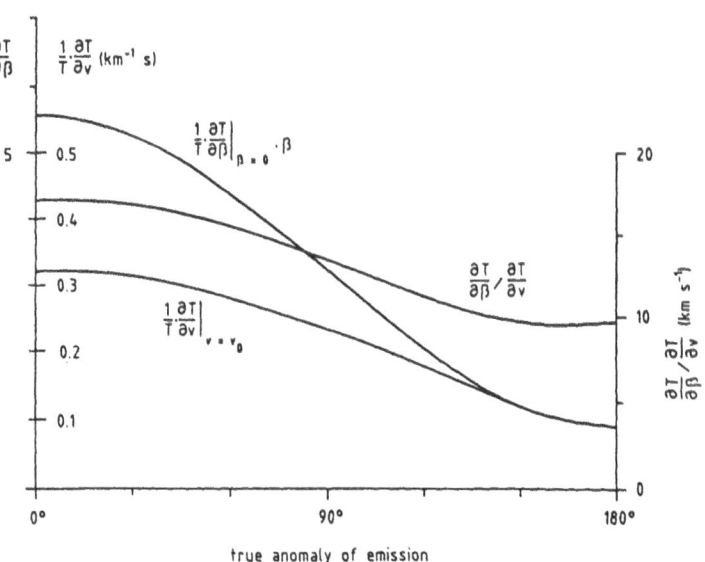

A release around perihelion (true anomaly = 0) is most effective, here the relative increase in the particle periods is 5 times larger then at aphelion.

The effects of ejection velocity can be studied by looking at the quantity $\frac{1}{T}\left.\frac{\partial T}{\partial v}\right|_{v-v_0}$, where v_0 is the orbital velocity of the comet at time of particle release. Again the effect is strongest at perihelion (left hand side, right ordinate) however, the relative variation is only little more than a factor 2. Particles released with a small velocity component in direction of the nucleus velocity will acquire orbits with a longer period and trail the cometary nucleus after some time.

The effects of outflow velocity and ß value of a particle can compensate each other. The ejection velocity Δv (into the direction opposite to the cometary motion) that compensates the effect of radiation pressure yielding $\Delta T = 0$ can be derived from $\Delta v \approx \frac{\partial T}{\partial \beta} \cdot \left(\frac{\partial T}{\partial v}\right)^{-1} \cdot \beta$. The behaviour of Δv as function of true anomaly is given as the third curve of Fig. 3 (right hand ordinate). Particles released at perihelion require the largest velocity for compensation so that they stay with the nucleus. The difference between perihelion and aphelion is not pronounced.

For a particle with $\beta = 0.5 \cdot 10^{-3}$, typical for particles impacting the Giotto spacecraft (see Table I), an ejection velocity of ≈ 9 m s^{-1} would compensate for the particles reduced effective solar attraction. This value is similar to $v_\perp$ derived from the width of the dust tail.

TABLE I

The columns represent the event number, the distance to the nucleus when the impact occurred, the median mass (1.4 times the minimum mass) of the particle, the ratio β, the minimum ejection angles, the corresponding emission times with respect to the comets perihelion, the emission velocities, and corresponding velocities derived from gas dynamic calculations (Gombosi et al., 1985).

No	Distance to nucleus [km]	m=1.4 m$_0$ [mg]	β 10^{-3}	θ_{min} [°] zenith	t [d]	v [m s^{-1}]	v [m s^{-1}] Gombosi
1	8.6·10^4	1.4	0.95	5.9	-9	40	31
2	7.6·10^4	0.7	1.20	5.2	-6	41	36
3	1.48·10^4	7	0.56	3.2	0.5	13	26
4	0.71·10^4	14	0.44	2.3	4.5	8	24
5	0.45·10^4	7	0.56	1.4	8.5	7.6	27

3. Discussion

The initial, preliminary, calculations show that particle velocities derived from the widths of cometary trails (the example is comet P/Tempel 2) are of similar magnitude as the velocities required for compensation of the reduced solar attraction of released particles. Many of these particles can stay close to the nucleus for extended times.

A release at aphelion requires a smaller compensating velocity as at perihelion. The presented calculations demonstrate the complexity of a determination of a dust particle density distribution. Questions such as the variation of production rates as function of orbital position are still open. What are the mechanism of release at large heliocentric distances? Can particles be trapped in the comet's vicinity by gravitation to form an extended source such as the clouds of large particles observed by radar? And in connection with the observations by the Halley Multicolour Camera: can the density of old (trail) particles be large enough to explain the impacts during encounter?

REFERENCES

Campbell, D.B., J.K. Harmon, and I.I. Shapiro 1989. Radar Observations of Comet Halley. *Astrophys. J.* **338**, 1094-1105.

Curdt, W. and H.U. Keller 1988. Collisions with Cometary Dust Recorded by the Giotto HMC Camera. *ESA Journal* **12**, 189-208.

Curdt, W. and H.U. Keller 1990. Large Dust Particles along the Giotto Trajectory. *Icarus* **86**, 305-313.

Gombosi, T.I., T.E. Cravens, and A.F. Nagy 1985. Time-Dependent Dusty Gasdynamical Flow Near Cometary Nuclei. *Astrophys. J.* **293**, 328-341.

Gombosi, T.I., A.F. Nagy, and T.E. Cravens 1986. Dust and neutral gas modeling of the inner atmospheres of comets. *Rev. Geophys.* **24**, 667-700.

Hajduk, A. 1987. Meteoroids from comet P/Halley. The comet's mass production and age. *Astron. Astrophys.* **187**, 925-927.

Harmon, J.K., D.B. Campbell, A.A. Hine, I.I. Shapiro, and B.G. Marsden 1989. Radar Observations of Comet Iras-Araki-Alcock (1983d). *Astrophys. J.* **338**, 1071-1093.

Hughes, D.W. 1987. P/Halley dust characteristics: a comparison between Orionid and Eta Aquarid meteor observations and those from the flyby spacecraft. *Astron. Astrophys.* **187**, 879-888.

McDonnell, J.A.M., W.M. Alexander, W.M. Burton, E. Bussoletti, G.C. Evans, S.T. Evans, J.G. Firth, R.J.L. Grard, S.F. Green, E. Grun, M.S. Hanner, D.W. Hughes, E. Igenbergs, J. Kissel, H. Kuczera, B.A. Lindblad, Y. Langevin, J.-C. Mandeville, S. Nappo, G.S.A. Pankiewicz, C.H. Perry, G.H. Schwehm, Z. Sekanina, T.J. Stevenson, R.F. Turner, U. Weishaupt, M.K. Wallis, and J.C. Zarnecki 1987. The dust distribution within the inner coma of comet P/Halley (1982i): encounter by Giotto's impact detectors. *Astron. Astrophys.* **187**, 719-741.

Richter, K. and H.U. Keller 1988. The Anomalous Dust Tail of Comet Kohoutek (1973XII) Near Perihelion. *Astron. Astrophys.* **206**, 136-142.

Richter, K., W. Curdt, and H.U. Keller 1990. Velocities of individual large dust particles ejected from comet P/Halley. *preprint*.

Sykes, M.V., L.A. Lebofsky, D.M. Hunten, and F.J. Low 1986. The Discovery of Dust Trails in the Orbits of Periodic Comets. *Science* **232**, 1115-1117.

Sykes, M.V., D.J. Lien, and R.G. Walker 1990. The Tempel 2 Dust Trail. *Icarus* **86**, 236-247.

COMETS AS A SOURCE OF INTERPLANETARY AND INTERSTELLAR GRAINS

F. Hoyle and N.C. Vickramasinghe
School of Mathematics,
University of Wales College of Cardiff,
Senghenydd Road,
Cardiff, CF2 4AG
Wales, U.K.

ABSTRACT. Properties of cometary dust with regard to bulk density, optical characteristics and sizes, derived from recent observations, are used to model scattering properties of cometary and interstellar grains. A wide range of astronomical observations are shown to be explained if cometary objects are hypothesised as a major source of dust grains in the galaxy.

1. Introduction

Although a general similarity between cometary, interplanetary and interstellar dust had been recognised for some time, more precise comparisons were made possible only after the recent explorations of Comet Halley. Resemblances that are more than casual with regard to composition, sizes or optical characteristics would point in the direction of a common origin for the three types of dust. We shall argue in the present paper that strong similarities in properties do indeed exist and that they could be interpreted to imply a causal link in the direction from comets to interstellar grains.

The connection between interplanetary dust and comets is accepted without dispute. Whipple (1967) estimated that a dust injection rate from comets amounting to $\sim 10^7 \mathrm{g}\ \mathrm{s}^{-1}$ is needed to maintain the zodiacal dust cloud. Dust particles in this cloud, with typical sizes $\sim 1\text{-}10\mu\mathrm{m}$, spiral inwards towards the sun due to Poynting-Robertson effect and the entire cloud would be drained on a timescale of $\sim 10^4$ yr. A replenishment of the cloud by the release of dust from short-period comets seems inadequate, the rate of injection falling below the requirements of $\sim 10^7 \mathrm{g}\ \mathrm{s}^{-1}$ by an order of magnitude (Rösser, 1976). A way out of the difficulty may be to suggest that a contribution arises from long-period comets by an unknown orbital mechanism (Delsemme, 1976), or from a population of dormant short-period comets with fragile dark surfaces (Hoyle and Vickramasinghe, 1986).

A fraction of the dust released from long-period comets might be expected to be expelled into the interstellar medium by the action of

235

A.C. Levasseur-Regourd and H. Hasegawa (eds.), Origin and Evolution of Interplanetary Dust, 235–240.
© 1991 *Kluwer Academic Publishers.*

radiation pressure from the Sun. However, the present day injection rate from sun-like stars would not suffice to contribute significantly to the mass of interstellar dust. Although the mass of comets currently resident in the Oort cloud is estimated at $\sim 10^{29}$g, the total mass of cometary material expelled from the entire solar system at the stage when the planets Uranus and Neptune were being accumulated could have been as high as $\sim 10^{30}$g. With an estimated 10^{11} candidate stars possessing sun-like histories, the mass of cometary dust injected into the ISM is $\sim 10^{41}$g, which is close to the total mass of the interstellar grains in the Galaxy.

2. Structure and Size Distribution of Dust

Cometary particles associated with meteor streams such as the Geminids and Giacobinids are known to be characterised by low values of the bulk density ranging from $\sim 1\text{-}10^{-2}$g cm^{-3} (Millman, 1976). Similar low values of density were also inferred from studies of photographic and radio meteors. More recently, direct sampling of cometary dust in the upper atmosphere has provided additional evidence of low density aggregates of particles with overall compositions similar to carbonaceous chondritic material (Brownlee, 1978). Furthermore, *in situ* studies of dust from Comet Halley have yielded evidence of low average values of the bulk density, ~ 0.1g cm^{-3}, with a high proportion of grains involving the elements C, H, O, N in the form of complex organic molecules (Kissell *et al.* 1986; Kissell and Krueger, 1987).

Data from the Giotto Dust Impact Detection System (DIDSY) provides important information on the mass distribution function of cometary grains (McDonnell *et al.*, 1987). The determination of a size distribution from this data demands a knowledge of both the shape of grains as well as the bulk density. Assuming a constant value of the bulk density, and a spherical shape, the main features of the particle size distribution may be represented by the power law:

$$n(a) \, da = const. \, a^{-3.6} \quad a_1 < a < a_2 \tag{1}$$

The upper and lower bounds a_2, a_1 are dependent on the precise value of the average bulk density, a parameter which is at present somewhat uncertain. It is possible that this density could vary systematically with the radius a. In view of the uncertainties that are involved here we shall leave a_1 as a free parameter to be fitted to a given observational situation, except that to within an order of magnitude we require $a_1 \sim 0.1\mu$m. The appropriate upper limit a_2 in most applications is $\sim 10^2\mu$m, corresponding to masses $\sim 1\mu$g above which the size spectrum deviates from equation (1) (McDonnell *et al.*, 1987). In problems where the upper end of the size spectrum has a dominant influence we adopt a refinement to (1) suggested by the data of Mazets *et al.* (1986). In the discussion that follows, we shall refer to such a modification as "the cometary size distribution".

3. Scattering Efficiency Factor of Cometary Dust

If a large fraction of interstellar grains have a cometary origin the extinction of starlight should be explained on the basis of a cometary distribution of grain radii and an appropriate index of refraction. We have shown earlier that a particular size distribution of hollow organic grains comprised of material that possesses an average refractive index n = 1.16 could produce a very close agreement with the visual extinction law as shown in the left hand panel of Fig. 1. The size-distribution function which is appropriate to endospore forming bacteria used for this calculation is shown in the histogram of the right hand panel. An averge refractive index n = 1.16 arises if an organic sphere of refractive index 1.4 has a vacuum cavity taking up 60 percent of the volume. The heavy curve in Fig. 1 (right hand panel) depicts the cometary distribution function (1)

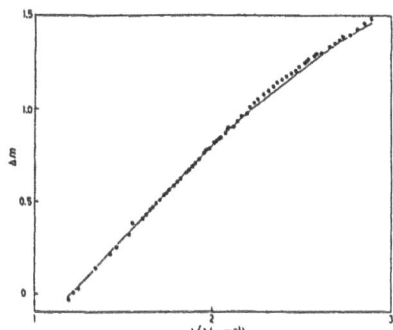 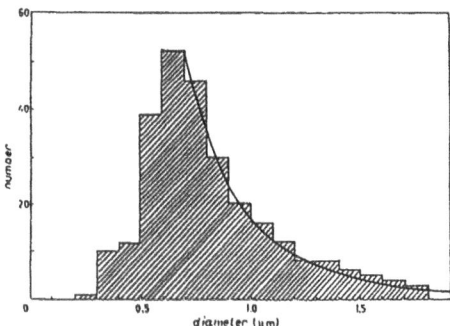

Fig. 1: (Left) Interstellar extinction data for the Cygnus region from Nandy (1964) compared with normalised extinction for a size distribution of hollow grains with average refractive index n = 1.16. (Right) Distribution of diameters of endospore-forming bacteria (histogram) compared with the observed cometary size distribution (heavy curve).

which is seen to be in remarkable agreement with the bacterial size distribution function for $a > a_1 \simeq 0.325 \mu m$. A computed extinction for this power law distribution function, assuming spheres with m = 1.16, is found to be very close to the extinction behaviour calculated in Fig. 1. The result, however, is somewhat sensitive to the assumed cut-off radius a_1, although it is insensitive to the precise value of the upper limit a_2.

4. Optical Polarization of Comet Halley

Whereas the average extinction cross-section for the size distribution given by (1) is mainly controlled by the smallest sizes, the integrated angular scattering properties tend to be dominated by larger grains. For a refractive index m = 1.4-0.05i we find that the integrated intensity of scattered radiation at a typical scattering angle ϕ = 60° is dominated by particles of radii ~ 15μm in the distribution function of cometary grains. Large sizes are similarly dominant in determining the polarization properties of the scattered light.

Measurements of the phase angle dependence of the polarization of scattered light from the dust coma of comet Halley when compared with the polarization calculated for a cometary size distribution of spherical-particles yields good agreement for the case of refractive index m = 1.385-0.032i (Mukai *et al.*, 1987; Le Borgne *et al.*, 1986). From an exhaustive set of numerical computations, Mukai *et al.* (1987) have inferred that any major departures from this best value of m cannot be permitted within the limitations of the Mie scattering model. Preliminary calculations of polarization on the basis of a hollow grain model provide a consistent explanation of the same polarization data. In this model we require a Güttler computation for hollow organic spheres with an inner to outer radius ratio a_0/a = $(0.6)^{1/3}$, where a is distributed according to (1) and m is taken to be 1.4-0.05i. The small imaginary part of m may be interpreted as being appropriate to a trace contamination with graphitic material. The agreement with the data points is found to be of a generally similar quality to that for a filled organic sphere, as indicated in the calculation of Mukai *et al.*

The wavelength dependence in the range 1900 < λ < 2900 Å of the total scattering function of dust in the coma of comet Halley was estimated by Wallis and Wickramasinghe (1990) from IUE spectra obtained on 14 March 1986. The observations implied a scattering angle of 135°. For this value of the scattering angle the normalised scattering function for a cometary size distribution of hollow organic grains has been found to fit the data.

5. Infrared Properties of Cometary and Interstellar Dust

The first announcement by Wickramasinghe and Allen (1986) of a broad emission feature centred at 3.4μm provided decisive evidence on the complex organic nature of the dust from comet Halley. This data was originally modelled in a simple way using a single grain temperature T to yield a flux

$$F_\lambda = a\tau(\lambda) \ B_\lambda(T) + \beta F_{solar}(\lambda) \tag{2}$$

where $B_\lambda(T)$ is the Planck function, $F_{solar}(\lambda)$ is the solar flux, and $\tau(\lambda)$ is the optical depth of a laboratory sample of dessicated *E. coli* (Wickramasinghe *et al.*, 1986). Our result for T = 320 K with $a:\beta$ chosen to give an optimal fit is shown in the solid curve of Fig. 2. The dashed curve shows a more realistic model computed by Wallis *et al.* (1989) using data for irradiated *E. coli* and the cometary size distribution. A flux calculation using the resulting distribution of

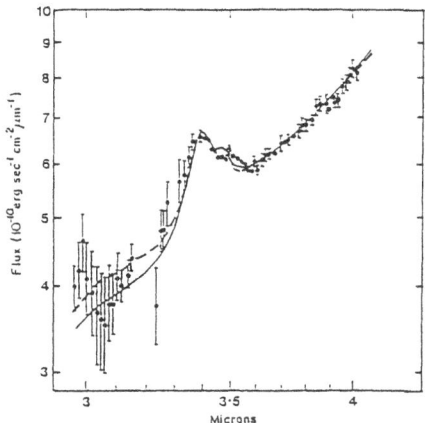

Fig. 2:   Observational data for comet Halley on 1986 March 31 compared with calculations for cometary size distribution of organic grains (dashed curve).   The solid curve is the prediction for a single grain temperature.

temperatures of the emitting grains is shown in the dashed curve of Fig. 2, for which agreement with the data is seen to be improved. Although the 2-4μm infrared spectrum of comet Halley has shown a modest extent of variation from one observed date to another, the points in Fig. 2 may be taken as representative of dust emerging in a fresh eruption from the cometary surface.   (The comet was seen to have brightened significantly on adjacent days around 1986 March 31.) An interesting feature is that the $\tau(\lambda)$ data for an organic (biotic) sample, which was earlier found to give excellent agreement to the 2-4μm absorption spectrum of the galactic centre infrared source GC-IRS7, now matches the data for freshly evaporated cometary dust.

6.   Conclusion

To conclude we note that close correspondences exist between the observed properties of cometary grains and interstellar dust over a wide range of wavelengths.   The indications are for a generic link between these grain types.   If the direction of this link is from interstellar space to comets, it is indeed surprising to find a size distribution that has in the main been unmodified throughout the entire history of the solar nebula.   Our preference is for organic (biotic) grains that are amplified within cometary objects in early solar nebula-type objects and subsequently injected into the interstellar medium.   The cosmic amplification cycle of biotic grains is depicted in Fig. 3.   Degradation products of such grains into aromatic and aliphatic hydrocarbons could account for the 3.28μm interstellar emission band and other spectral features in the mid-infrared wavelength region (Hoyle and Wickramasinghe, 1989; Wickramasinghe et al., 1990).   It is interesting to note that

240

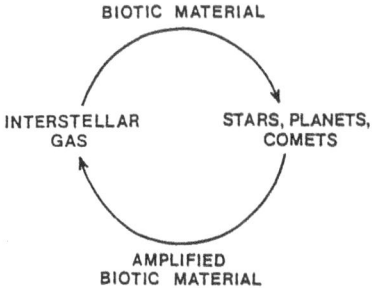

Fig. 3:   Cosmic amplification cycle of biotic grains.

emission bands near 3.3μm and 11.25μm have also been observed in spectra of comet Halley (Hoyle and Vickramasinghe, 1989;   Lynch *et al.*, 1989).

References

Delsemme, A.H., (1976) in *Lecture Notes in Physics*, eds.
  H. Elsäser and H. Fechtig (Berline: Springer Verlag)
  Vol. 48, p.314.
Hoyle, F. and Vickramasinghe, N.C. (1986).   *Earth, Moon and Planets*, 36, 289.
Hoyle, F. and Vickramasinghe, N.C. (1989).   *Proc. 22nd Eslab Symposium*, ESA SP-290, 67.
Kissell, J. (1986).   *Nature*, 321, 336.
Kissell, J. and Krueger, F.R. (1987).   *Nature*, 326, 755.
Le Borgne, J.F., Leroy, J.L. and Arnaud, J. (1986).   *20th ESLAB Symp.*, ESA SP-250, Vol. I, 571.
Lynch, D.K., Russell, R.W., Campins, H., Witteborn,
  F.C., Bregman, J.D., Rank, D.M. and Cohen, M. (1989).
  Icarus, 379.
Mazets, E.P. and 13 authors (1986).   *Nature*, 321, 276.
McDonnell, J.A.M. and 27 authors (1987).   *Astron. Astrophys.*, 187, 719.
Millman, P.M. (1976) in *Lecture Notes in Physics*, eds.
  H. Elsäser and H. Fechtig (Berlin: Springer Verlag) Vol. 48,
  p.359-372.
Mukai, T., Mukai, S. and Kikuchi, S. (1987).   *Astron. Astrophys.*, 187, 650.
Nandy, K. (1964).   *Publ. Roy. Obs. Edin.*, 3, 142 and 4, 57.
Röser, S. (1976) in *Lecture Notes in Physics*, eds. H. Elsäer and
  H. Fechtig (Berlin: Springer Verlag) Vol. 48, p.319-322.
Vallis, M.K. and Vickramasinghe, N.C. (1990) in *Evolution in Astrophysics* (ESA SP-310).
Whipple, F. (1967) in *The Zodiacal Light and the Interplanetary Medium* ed. J.L. Weinberg (NASA SP-150).
Vickramasinghe, N.C., Hoyle, F. and Al-Jabory, T. (1990).
  Astrophys. Sp. Sci., 166, 333.
Vickramasinghe, D.T., and Allen, D.A. (1986).   *Nature, 323, 44.*

CHEMICAL COMPOSITION OF AN EMANATION FROM COMETS: IDENTIFICATION OF THE 3 MICRON COMET FEATURE

A.SAKATA[1], S.WADA[2] and A.T.TOKUNAGA[3]

1. Dept. of Applied Phys. and Chem., University of Electro-Communications, Chofugaoka, Chofu, Tokyo 182, Japan

2. Dept. of Chemistry, University of Electro-Communications, Chofugaoka, Chofu, Tokyo 182, Japan

3. Institute for Astronomy, University of Hawaii, 2680 Woodlawn Dr., Honolulu, HI 96822 U.S.A.

ABSTRACT   Recent high resolution observations of comets revealed a detailed spectral shape of the 3.4 μm feature. We measured IR spectra of simple 14 hydrocarbon molecules and made "synthesized comet spectrum". Peak wavelength and spectral shape of the synthesized spectrum are well in agreement with the observed comet features.

1. Introduction

Recent high resolution observations of comet Halley (1986), comet Wilson (1987), comet Bradfield (1987), comet Brorsen-Metcalf (1989) and comet Okazaki-Levy-Rudenko (1989) showed similar features in the 3 μm region. They have a peak at 3.36–3.38 μm and small humps near 3.5 μm. Comet Brorsen-Metcalf and comet Okazaki-Levy-Rudenko showed higher feature-to-continuum ratios at 3.4 μm than other comets (Brooke et al., 1990).

We synthesized the 3.4 μm feature with IR spectra of gaseous hydrocarbon molecules. The synthesized feature is in good agreement with the observed comet features.

2. Experiment and Result

We used a JASCO-810 IR spectrophotometer to obtain IR spectra of 14 hydrocarbon molecules in a specially-designed gas cell. We made a synthesized spectrum from conbinations of each hydrocarbon spectrum, in which the highest peak is normalized as unity. The features at 3 μm of molecules are classified into three types as shown in Fig.1.

IR spectra of low-molecular-weight saturated hydrocarbons (hereafter LSHs) containing 2–6 carbon atoms, such as ethane, propane, butane, isobutane, n-pentane, and n-hexane, show a peak near 3.38 μm. A sum of LSHs spectra is shown in Fig.1 (A). Each of four saturated hydrocarbons containing of 7–10 carbons, such as n-pentane, n-octane, n-nonane, and n-decane has a very similar spectrum peaked at 3.42 μm, accompanying minor peaks at 3.38 and near 3.5 μm. An IR spectrum synthesized as a sum of spectra of these high-molecular-weight saturated hydrocarbons (hereafter HSHs) is indicated in Fig.1(B). Low molecular weight unsaturated hydrocarbons (hereafter LUHs) containing 2–4carbons, such as ethylene, propylene, butene, and butadiene, show a specific own feature. An IR

241

A.C. Levasseur-Regourd and H. Hasegawa (eds.), Origin and Evolution of Interplanetary Dust, 241–244.

242

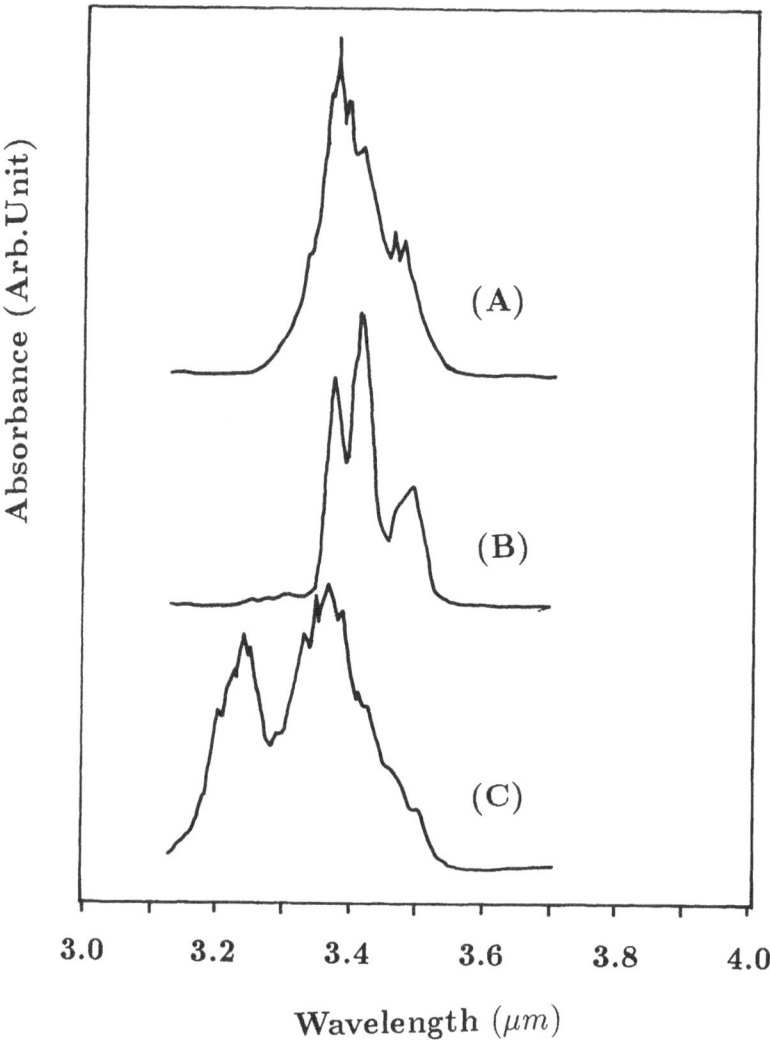

Fig.1, Three synthesized spectra, (A); from spectra of saturated hydrocarbons having 2–6 carbons, (B); from those having 7–10 carbons, and (C); from spectra of unsaturated hydrocarbons having 2–4 carbons. In the spectrum of each hydrocarbon , the main peak height is normalized as unity. Each synthesized spectrum is made from hydrocarbon spectra in the ratio of ; 1.5 ethane : 1.5 propane : 1.0 butane : 1.0 iso-butane : 0.5 pentane : 0.5 hexane in (A), 1.0 heptane : 1.0 octane : 1.0 nonane : 1.0 decane in (B), and 1.0 ethylene : 1.0 propylene : 1.0 butene : 1.0 butadiene in (C).

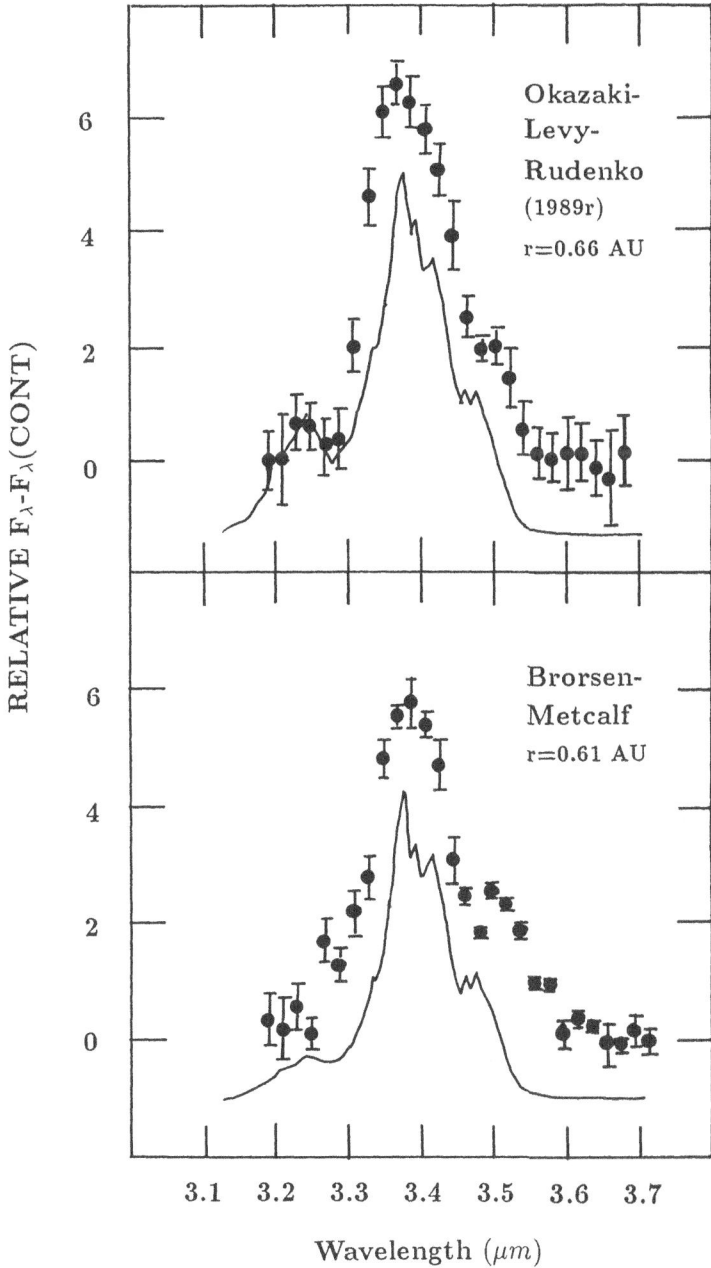

Fig.2. Comparison of "synthesized comet spectra" to the observed comet spectra. Solid lines show "synthesized comet spectra" of hydrocarbons in the ratios of 1.0(A) : 0.5(B) : 1.5(C) in an upper figure and 1.0(A) : 0.5(B) : 0.5(C) in lower figure. The symbols (A), (B), and (C) are the same used in Fig.1.

spectrum of LUHs, synthesized in a similar manner, is shown in Fig.1 (C). Its peaks are located at 3.37 and 3.23 μm.

"Synthesized comet spectra" were calculated from combination of three spectra of LSHs, HSHs, and LUHs spectra in the ratio of 2:1:1 and 2:1:2 and compared to the spectrum of comet BM and comet OLR, respectively.

3. Discussion

The intensity of the 3.4 μm feature is correlated with the water production rate (Brooke et al., 1990). These facts suggest that the 3.4 μm emitter in comets consists of a complex mixture of organic compounds (Knacke et al, 1986), which may be volatile.

We propose that the emitter in comets may be a mixture of hydrocarbon molecules.The peak position of LSHs spectra fits to comet spectra at 3.36–3.38 μm, but the width is narrower than that of the comet spectra. The addition of HSHs spectra to LSHs spectra broadens the peak in the long wavelength side, while the addition of LUHs spectra causes the broadening of the peak in the short wavelength side. The addition causes also humps near 3.5 μm and at 3.2–3.3 μm, respectively. The abundance ratios of LSHs, HSHs, and LUHs affect the shape of "synthesized comet spectra" as shown in Fig.2.

"Synthesized comet spectra" showed the similar 3.4 μm feature to those of comet OLR and comet BM. The main component of the emitter in comets may be low-molecular-weight hydrocarbons. The emitter of comet OLR contains more unsaturated hydrocarbons than that of the comet BM.

We excluded methane from our hydrocarbon mixture, because it shows a strong sharp peak at 3.32 μm. As it has a lowest boiling point amongst all hydrocarbons, it might behave differently from other hydrocarbon gases in cometary conditions.

4. Conclusion

The 3.4 μm features synthesized with IR spectra of gaseous hydrocarbons agree well with the observed comet spectra.

References

Brooke, T. Y., and Tokunaga, A. T. (1990) ' Detection of the 3.4 μm emission feature in comets Brorsen-Metcalf and Okazaki-Levy-Rudenko (1989r) and an observational summary', submitted to A. J.

Knacke, R. F., Brooke, T. Y., and Joyce, R. R. (1986) 'Observations of 3.2–3.6 micron emission features in comet Halley', *Ap. J.* **310**, L49–L53.

SPECTROPOLARIMETRY OF COMET HALLEY

N. VISVANATHAN, Z.MEGLICK
MOUNT STROMLO AND SIDING SPRING OBSERVATORIES
D.T. WICKRAMASINGHE
DEPARTMENT OF MATHEMATICS
AUSTRALIAN NATIONAL UNIVERSITY

ABSTRACT. Spectropolarimetric observations from 3800 to 7000 Å were obtained for the nucleus of comet Halley for nine nights during 1985-86. The observations were spaced over phase angle of 2 to 66 °. The continuum polarization without molecular-line contamination as well as the polarization of the molecular lines were evaluated. The plot of polarization versus the phase angle shows small negative polarization for angles less than 20°. The highest polarization measured for phase angle ≤ 66° is 19.6% .The lowest polarization of -2.0% has been measured at the phase angle ~ 10° . The variation of polarization with phase angle is nearly linear in the range 30 to 60° giving a slope of 0.4316% per degree. There is small dependence of polarization with the wavelength in the region 3800 to 6500Å. The polarization of the molecular bands CN, C_3 , $C_2(0,0)$ and C_2 (1,0) have also been evaluated for the spectra of 15th December 1985, 18th and 19th March 1986 and 8th April 1986.

1. Introduction

Comets are important astronomical objects as they are probably the most primitive objects in the solar system and have undergone little or no chemical change after their formation from the interstellar matter.Hence there is great interest in understanding the physical and chemical properties of the cometary dust - grains which may throw light into the origin and history of the interplanetary and interstellar dust. The appearance of the bright comet Halley in 1985-86 provided an unique opportunity to study the properties of the cometary grains.

Photometry and polarimetry are the two traditional techniques used to study the comets. The analysis of the degree of polarization of the solar light scattered by cometary dust with wavelength and phase angle allows in principle to determine the optical properties such as albedo, size, coarseness and composition of the grains.In the year 1985 and 1986 comet Halley has been observed for polarization at optical wavelengths by many observers (Dollfus et al. 1988) and references there in). These observations have been obtained through broad and narrow band filters.

In this paper we report the spectro-polarimetric observations of the nucleus region of the comet Halley for nine nights during 85-86 at a resolution of 6Å covering the region 3800 to 6500Å. Unlike broad - band filter polarimetry the spectropolarimetry enables us to determine the continuum polarization without molecular - line contamination. In addition the comet spectra could be used to derive the wavelength distribution of the continuum as well as the intensities of the emission lines.

2. Observations

Observations of the nucleus of the comet Halley were made on nine nights in 1985 and 1986 during which the phase angle of the comet varied from 2 to 66°. Table 1 contains the date of observation, Julian date, distance between the comet and the earth Δ , the distance between the sun and the comet R, the size of the aperture projected on to the comet S , the distance between the two apertures D, the direction of the plane of scattering ψ and the phase angle β . All observations were made with the spectropolarimeter (McLean et al 1984) in combination with the RGO spectrograph + IPCS detector at the Cassegrain focus of AAT in Sidingspring. The 250-line grating was used in combination with 25 cm camera with a resolution of 6 Å. The spectral region from 3800 to 7000Å was covered with 2000 pixel format. Two apertures of 1.34 x 2.68 arcsec with a separation of 22.8 arcsec were used. The nucleus of the comet was centered in one of the apertures.Observations were made alternately at Q and U polarimetry modes. The sky was observed

A.C. Levasseur-Regourd and H. Hasegawa (eds.), Origin and Evolution of Interplanetary Dust, 245–248.

TABLE 1. Details of the observations of the comet Halley.

Date	18 Oct 1985	12 Nov 1985	17 Nov 1985
Julian Date (day)	2446356.74 ± 0.02	2446381.69 ± 0.04	2446386.69 ± 0.02
Δ (AU)	1.4907 ± 0.0009	0.7961 ± 0.0007	0.6979 ± 0.0002
R (AU)	2.1071 ± 0.0002	1.7579 ± 0.0004	1.6879 ± 0.0002
S_A (km)	1448.7 ± 0.2	773.7 ± 0.7	678.3 ± 0.2
D_A (km)	24650 ± 15	13165 ± 12	11541 ± 4
Ψ	91°7'17" ± 8"	78°25' ± 4'	70° ± 3°
β	25°16'29" ± 9"	9°56' ± 4'	2°7' ± 2'

Date	15 Dec 1985	18 Mar 1986	19 Mar 1986
Julian Date (day)	2446414.46 ± 0.03	2446507.75 ± 0.04	2446508.75 ± 0.03
Δ (AU)	0.8192 ± 0.0007	0.8548 ± 0.0007	0.8327 ± 0.0009
R (AU)	1.2704 ± 0.0004	0.9648 ± 0.0007	0.9826 ± 0.0003
S_A (km)	796.2 ± 0.7	830.8 ± 0.4	809 ± 1
D_A (km)	13546 ± 11	14135 ± 11	13769 ± 15
Ψ	66°36'30" ± 25"	77°31'4" ± 16"	77°41'9" ± 22"
β	50°46' ± 2'	65°57'48" ± 28"	66°6'39" ± 3"

Date	8 Apr 1986	12 Apr 1986	15 May 1986
Julian Date (day)	2446528.75 ± 0.03	2446532.64 ± 0.03	2446566.40 ± 0.03
Δ (AU)	0.4270 ± 0.0003	0.4198 ± 0.0001	1.2367 ± 0.0008
R (AU)	1.2927 ± 0.0003	1.3516 ± 0.0003	1.8383 ± 0.0004
S_A (km)	415.0 ± 0.3	407.96 ± 0.04	1201 ± 1
D_A (km)	7061 ± 5	6941 ± 1	20450 ± 14
Ψ	120°52' ± 12'	153°53' ± 15'	106°20'37" ± 45"
β	39°55' ± 5'	28°44' ± 4'	31°24'0" ± 1"

separately 3° away from the nucleus. The exposures for each night varied from 1 to 2 hrs.Each night a solar spectral - type star was observed through the same instrument set - up. The polarization standards HD160529 and HD 183143 and an unpolarized star were also observed to check the performance of the equipment (Visvanathan 1965). The polarization efficiency at each pixel was evaluated by observing 100% polarised light with the same set-up.The Stokes parameters Q and U were determined and the polarization was computed for each pixel. The wavelength calibration of the spectrum was achieved through the comparison arc spectrum.

3. Polarization

The computed polarization P and the position angle of the electric vector θ as a function of wavelength for the nucleus of the comet observed on 18th March 1986 are shown in Figure 1a and 1b. The spectrum of the nucleus derived from the polarization data was divided by the spectrum of the solar spectral-type star HD186189 and is shown in Figure 1c. The 1σ error points are indicated as dotted lines.The molecular emission bands dominate the spectrum in the region from 3800 - 5600Å . Only regions with width of ~100Å around λ 3800 , 4400, 4800, 5200, 6200, 6800Å are mostly free of emission bands. Polarization reaches maximum at these wavebands. Further it can be seen that the molecular bands are also polarized but much less than the continuum. The angle θ (167°) is nearly constant with wavelength and as expected, is perpendicular to the direction of the plane of scattering , which is 77° on 18th March 86. Small deviations (~ 10°)from 167° is seen in the case of the polarized molecular bands.The polarization of the continuum at λ 3800 , 4400, 4800, 5200, 6200, 6800Å have been evaluated for 15th Dec 85, 18th Mar 86, 19th Mar 86,and 8th Apr 86 and plotted in Fig. 3. The error in P is ~ 0.5 % for values at phase angles < 40° and ~0.2 % for values at phase angles at > 40° . The wavelength dependence of polarization in the spectral region λ 3800 - 6800Å is nearly flat at phase angles < 40° . At larger phase angles P is lower at UV with a small increase at higher wavelengths. The data by other observers (Dollfus et al. 1988) indicate a similar wavelength dependence. It is interesting that our results which represents the region of ~2 seconds of arc around the

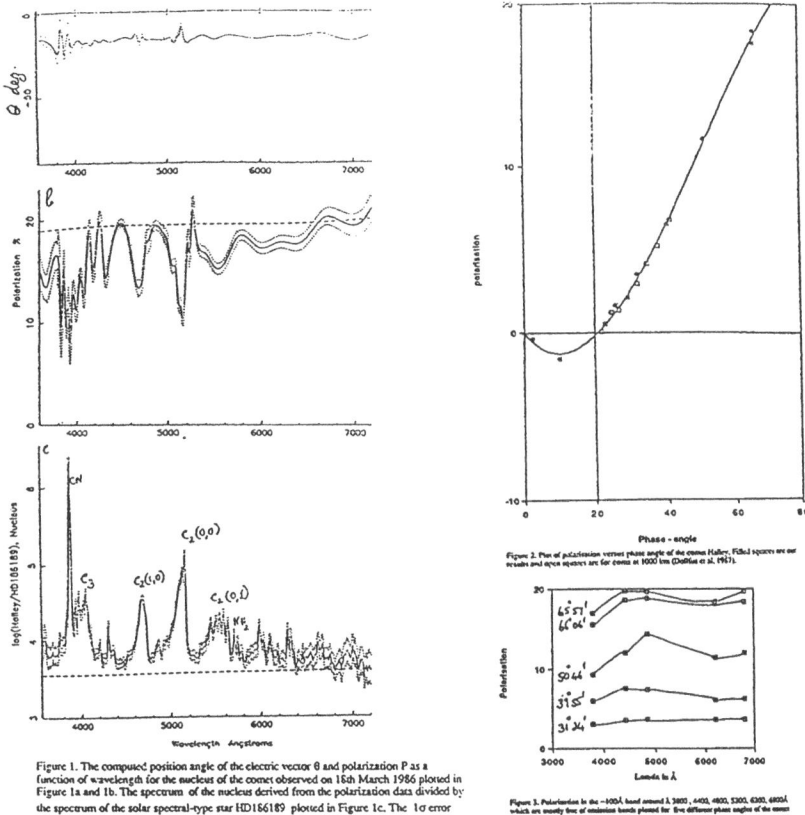

Figure 1. The computed position angle of the electric vector θ and polarization P as a function of wavelength for the nucleus of the comet observed on 16th March 1986 plotted in Figure 1a and 1b. The spectrum of the nucleus derived from the polarization data divided by the spectrum of the solar spectral-type star HD186189 plotted in Figure 1c. The 1σ error points indicated as dotted lines.Main molecular bands marked.

Figure 2. Plot of polarization versus phase angle of the comet Halley. Filled squares are our results and open squares are for comet at 1000 km (Dollfus et al. 1987).

Figure 3. Polarization in the ~100Å band around λ 3800 , 4400, 4800, 5200, 6300, 6800Å which are mostly free of emission bands plotted for five different phase angles of the comet Halley.

nucleus is in qualitative agreement with the results of other observers who used large size apertures (18-23 arcsec).

The mean polarization of the continuum of the nucleus of the comet at λ4800 and 5200 Å is computed for all the nine nights in Table 1 and plotted against the phase angle in Figure 2. It can be seen polarization becomes negative (the angle θ is equal to the direction of the plane of scattering) for phase angles less than 20°. The maximum polarization 19.6% occurs at the phase angle 66° and minimum polarization of -2.0% at ~ 10° . The relation of polarization with phase angle is nearly linear in the range 30 to 60° with a cross-over at 21° and a slope of 0.4316 ± 0.07 % per degree.Both pre-perihelion and post perihelion data form a continuous curve in Fig. 2. The data plotted in Fig. 2 represents the region around the nucleus of the comet with apertures <1000 km . Our data is in good agreement with the fresh dust polarization corrected - data at 1000 km in coma derived by Dollfus and Suchail (1987) during April 86 (Fig. 2). Though the minimum negative polarization and cross-over angle of our relation show agreement with other observers our slope of the phase-polarization relation is higher than that observed by other observers in the optical wavebands (Dollfus et al. 1988) and nearer to that observed at the IR (Brooks et al. 1987).This could be due to the fact that unlike the other data our data represents region of ~2 arcsec around the nucleus which had a star-like appearance during 1985-86. Also it is possible that the polarization - correction for molecular emission applied by other observers may be in error by a few % especially for phase angles > 30° .The polarization in the second aperture which is 22.8 arcsec away in the sunward direction is smaller than that observed for the nucleus.

4. Interpretation

Many models have been computed to interpret the polarization data of the reflected solar component from the comet. They require a mixture of rough silicate minerals and absorbing carbonaceous grains with a size distribution measured by space missions to Comet Halley (Lamy et al. 1987; Krishnasamy and Shah 1988). The negative part of the polarization - phase curve could be obtained by such a mixture. Albedo decreases with increase in size of the grains (1- 10 μm) from 0.035 to 0.02 which is in agreement with space data. This low albedo of the Comet Halley is comparable to that of the rings of Uranus and the leading hemisphere of Iapetus some of the darkest material in the solar system. The maximum polarization and the negative part of the polarization - phase curve of the Comet Halley in Figure 2 are similar than those observed in asteroids (Dollfus 1989) . This has been explained as Umov Effect which tells that the albedo A,and the maximum polarization P_{max} are related - when rough particles scatter, polarization increases with decreasing albedo. Thus again the P_{max} in Figure 2 is consistent with low albedo of the comet particles. Using this comparison between the comet and the asteroids and the laboratory measurements of several meteorite lunar and other samples Dollfus (1989) finds that the optical polarization is matched by grains far larger than wavelength, very dark with an extremely rough surface formed as aggregates of small particles.

5. References

Brooke, T.Y., Knacke, R.F. and Joyce, R.R. (1987) 'The near-infrared polarization and color of comet P/Halley ', Astron. Astrophys. **187**, 621-624.

Dollfus ,A. and Suchail J.L.(1987) 'Polarimetry of grains in the coma of P/Halley', Astron. Astrophys. **187**, 669-688.

Dollfus ,A.(1989)'Polarimetry of grains in the coma of P/Halley: II Interpretation ', Astron. Astrophys. **213**, 469-478.

Dollfus ,A.,Bastien, P., Le Borgne, J.-F., Levasseur-Regourd, A.C., and Mukai,T. (1988)'Optical polarimetry of P/Halley: synthesis of the measurements in the continuum', Astron. Astrophys. **206**, 348-356.

Krishnasamy, K.S., and Shah, G.A.(1988)'Nature of dust grains in comets', M.N.R.A.S, **233**, 573-579.

Lamy, P.L., Grun, E., and Perrin, J.M.(1987)'Comet P/Halley : implications of the mass distribution function for the photopolarimetric properties of the dust coma' , Astron. Astrophys. **187**, 767-773.

McLean, I.S., Heathcote, S.R., Paterson, M.J., Fordham, J., and Shortridge, K. (1984),'A multichannel Pockels cell spectropolarimeter for the Anglo-Australian Telescope'M.N.R.A.S, **209**,655-664.

Visvanathan, N.(1965), 'Polarization in the Galaxy and the Large Magellanic Cloud' Ph.D. Dissertation, The Australian National University, pp. 68-70.

SCATTERING PROPERTIES OF COMETARY DUST BASED ON POLARIMETRIC DATA

SONOYO MUKAI and TADASHI MUKAI[1]
Kanazawa Institute of Technology
Nonoichi, Ishikawa 921, Japan
and
SEN *KIKUCHI*
National Astronomical Observatory
Mitaka 181, Japan

ABSTRACT. Referring to the dust model in Mukai and Mukai(1990), where the scattering by large rough particles and Mie scattering by small particles are taken into account, a phase function of linear polarization of several comets is examined, especially in a region of phase angles α near a maximum polarization. A lower maximum polarization observed in comet Austin(1989c1) than those in comets West(1975n) and P/Halley leads a speculation that a mixing ratio of rough scattering to Mie scattering in comet Austin increases from a sun-comet distance r of 0.6 AU to 1.2 AU. This implies that a shortage of large particles in comet Austin occured in $r < 1$ AU.

1. INTRODUCTION

From a study of the previous cometary polarimetry(see e.g. Dollfus *et al.* 1988), we can draw a simple picture for a phase function of polarization, i.e. "the generalized phase function has a polarization maximum as high as 20-30 % at a phase angle α=90° , an inversion angle near 20° and a negative polarization branch at $\alpha < 20°$ with a polarization minimum several per cent deep"(cited from Dobrovolsky *et al.* 1986), where a phase angle is defined as a sun-comet-observer angle. Unfortunately, the reliable observations in a wide range of phase angles are very limited, especially in a region of a maximum polarization. Recently, Kikuchi *et al.*(1990) reported the visible polarization of comet Austin(1989c1) in $12° < \alpha < 110°$. These data initiate us into an analysis of the phase function of polarization near the maximum polarization.

2. POLARIZATION DATA

---

[1]present address: Dept. of Earth Sciences, Faculty of Science, Kobe University, Nada, Kobe 657, Japan

A.C. Levasseur-Regourd and H. Hasegawa (eds.), Origin and Evolution of Interplanetary Dust, 249–252.

250

Observed data of linear polarization for 3 comets in a region of maximum polarization are shown in figure 1. For reference, the data for comet P/Halley are also presented. From figure 1, we can summarize the followings: (1) A peak position of a linear polarization is at roughly a phase angle of 90°, although accuracies of the data in 90° < α < 110° are quite limited. (2) A degree of polarization in comets Austin and Kohoutek (1973f) is significantly lower in α >45° than that in comet West(1975n), even considering a wavelength-dependence and an aperture-dependence of polarization noted in Dobrovolsky et al.(1986) and Michalsky(1981). (3) On the other hand, there are no remarkable differencies among the degrees of polarization for 4 different comets in α < 45°.

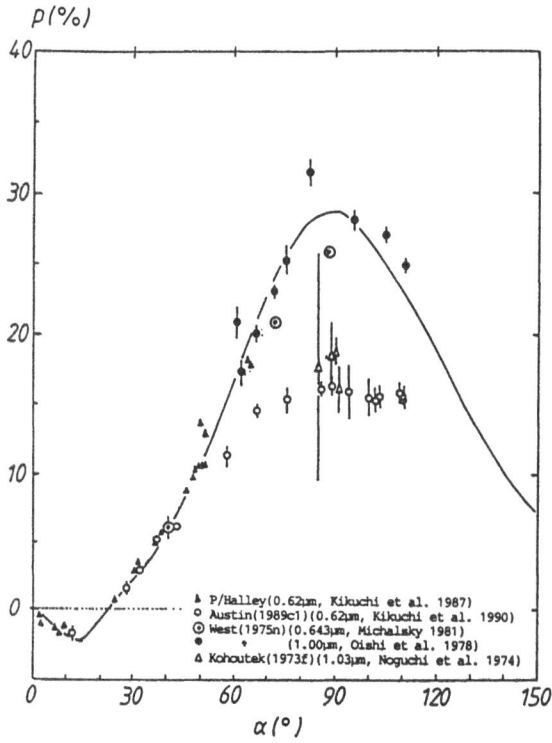

Figure 1. Polarization p(%) vs. a phase angle α. A number in unit of μm inside round brackets of each observations denotes the effective wavelength. Solid curve is computed based on a mixing model in Mukai and Mukai(1990). The uncertainty in the observed data is shown by error bars.

3. DUST MODEL

It has been shown in Mukai and Mukai(1990;to be called MM hereafter) that the scattering light by large rough particle reduces the strong negative polarization in the backward direction caused by Mie scattering alone. Simultaneously, rough scattering light increases the intensity with decreasing a phase angle(backward enhancement of intensity). These behaviours of rough scattering , accompanying with Mie scattering, lead a favorable tendency to explain the photopolarimetric data observed in the comets in the back scattering region.

The solid curve in figure 1 shows the phase function of polarization computed based on the model having the same parameters as those in MM. It looks that this model can explain the phase function observed in comets P/Halley (Kikuchi et al. 1987) and West(1975n)(Oishi et al. 1978 and Michalsky 1981) not only in the back scattering region, but also in $60° < \alpha < 110°$. However, the maximum polarization reported in comets Kohoutek(Noguchi et al. 1974) and Austin(Kikuchi et al. 1990) shows quite lower value than that expected from the model calculations.

In figure 2, the degrees of polarization computed from Mie theory alone are illustrated for spherical particles with the complex refractive index of m=1.386-k i, where k denotes the absorption coefficient of grain material. Here for the size distribution of dust grains the Halley's grain size distribution used in MM was applied. Note that in the mixing model shown as the solid curve in figure 1, we used the same dust size distribution for Mie particles and m=1.385-0.035i.

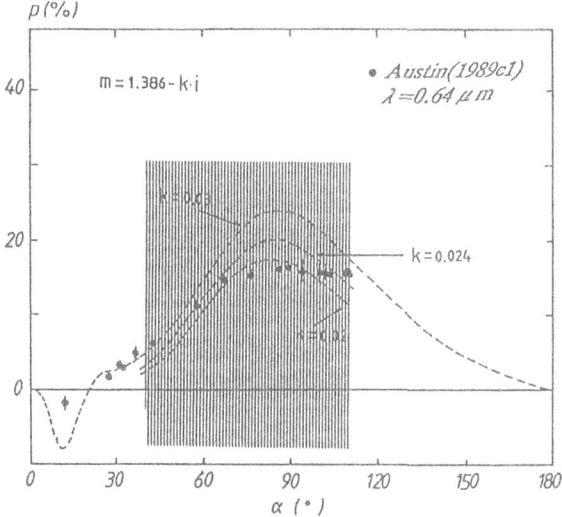

Figure 2. A phase function of polarization. Dashed curves denote the computed results based on Mie scattering alone for the grains with the Halley's dust size distribution and a complex refractive index m. A shaded area presents the region of phase angles where comet Austin(1989 c1) moved inside the Earth's orbit.

It is clear from figure 2 that the phase function computed by Mie theory alone cannot explain the observed results in the back scattering region. The maximum polarization observed near $\alpha=90°$ in figure 2, however, gives the hint that the small particles play an important role in this region of phase angles because the light coming from rough scattering by large particle moves the peak position of polarization toward the forward direction as reported in the scattering experiments (Killinger 1987). Consequently, we assume that if a mixing ratio f_R of rough scattering to total scattering decreases as the phase angle increases, the polarimetric data observed in comet Austin would be explained. Comet Austin moved inside the Earth's orbit when it took the phase angles of 60° -110°. Therefore we get a speculation that a relative abundance of large particles in the cometary coma of comet Austin was poor when this comet moved in a sun-comet distance less than 1 AU.

4. CONCLUSIONS

(1) A phase function of polarization recently observed in comet Austin(1989c1) looks very similar to that detected in comet Kohoutek(1973f), but it is significantly different from those reported in comets West(1975n) and P/Halley, even taking into account a wavelength-depencence of polarization and its aperture-dependence. (2) The phase function observed in comets West and P/Halley can be well explained by the mixing model of rough large particles and Mie scattering particles with the Halley's dust size distribution and optical constants of m=1.385- 0.035i. (3) Although a quite well agreement of polarization in four different comets of interest exists in $\alpha < 45°$, a significantly lower polarization observed in comet Austin in $45° < \alpha < 110°$ than those in comets West and P/Halley appears. This evidence strongly suggests that the dust properties in comet Austin varied with a sun-comet distance r. We speculate that in comet Austin a shortage of large rough particles occured in r less than nearly 1 AU.

References

Dobrovolsky, O.V., Kiselev, N.N., and Chernova, G.P. (1986) 'Polarimetry of comets: a review', *Earth, Moon, Planets* **34**, 189-200.
Dollfus, A., Bastien, P., Le Borgne, J.-F., Levasseur-Regourd, A.C., and Mukai, T. (1988) 'Optical polarimetry of P/Halley : synthesis of the measurements in the continuum', *Astron. Astrophys.* **206**, 348-356.
Kikuchi, S., Mikami, Y., Mukai, T., Mukai, S., and Hough, J.H. (1987) 'Polarimetry of comet P/Halley', *Astron. Astrophys.* **187**, 689-692.
Kikuchi, S., Okazaki, A., Kondo, M., and Hirata, R. (1990) 'Continuum polarization in comet Austin(1989c1)', *Proc. of the 23 ISAS Lunar and Planetary Symp.*, 39-42.
Killinger, R.T.(1987) 'Photopolarimetrische lasermessungen zur wellenlängenabhängingkeit des lightsterewerhaltens von staubpartikeln aus irdischen und kosmischen materialien' *Dissertation(der Ruhr Univ. Bochum)*
Michalsky, J.J.(1981) 'Optical polarimetry of comet West 1976VI', *Icarus* **47**, 388-396.
Mukai, S. and Mukai, T.(1990) 'Analysis of photopolarimetric data of comets at small phase angles by rough surface scattering', *Icarus* **86**, 257-263.
Noguchi, K., Sata, S., Maihara, T., Okuda, H., and Uyama, K.(1974) ' Infrared photometric and polarimetric observations of comet Kohoutek 1973f', *Icarus* **23**, 545-550.
Oishi, M., Kawara, K., Kobayashi, Y., Maihara, T., Noguchi, K., Okuda, H., and Sata, S. (1978) 'Infrared observations of comet West(1975n). I. Observational results', *Publ. Astron. Soc. Japan* **30**, 149-159.

SYNCHRONIC BAND AND ITS IMPLICATION IN THE COMETARY DUST

Jun-ichi Watanabe [1] & Kimihiko Nishioka [2]
1: *National Astronomical Observatory of Japan,*
   Mitaka, Tokyo, 181, Japan
2: *Olympus Optical Company Ltd.,*
   1-43-2, Hatagaya, Shibuya-ku, Tokyo, 181, Japan

ABSTRACT. The unique morphology of the synchronic band in the cometary dust tail is explained by finite-lifetime fragment model (Nishioka and Watanabe, 1990). However, this model needs a severe restriction on the lifetime of the dust fragments; 25-70 days (r = 1 A.U.). This implies that detailed analysis of the synchronic band may reveal physical properties of the cometary dust particles. In this paper, we suggest that the fragments in the synchronic band are relatively pure ice if they are not organic grains.

1. Introduction

The narrow striae have been observed in the dust tail of several large comets. These striae are called "synchronic bands"(SYBs hereafter). The formation of SYBs is an interesting subject in cometary dust researches. Nishioka and Watanabe(1990) studied the SYBs formation on the basis of finite lifetime fragment model(FLM hereafter), which solved two problems remained in the proposed models so far. One is why the parent particles simultaneously break up at a certain time. Although Sekanina and Farrell(1980) succeeded to simulate the morphology of SYBs with an assumption of the simultaneous fragmentation, the physical reason of this assumption is not clear. Another problem is that the observed length of SYBs was not zero even if the observed epoch was the time before the estimated fragmentation. These two problems were solved by the FLM described in the next section.

Although the FLM model succeeded to simulate the morphology of SYBs, it needs a severe restriction on the lifetime of the fragmented dust particles. This paper introduces some analyses of SYBs in several comets on the basis of the FLM, and deals with the physical properties of the cometary dust particles in the SYBs.

A.C. Levasseur-Regourd and H. Hasegawa (eds.), Origin and Evolution of Interplanetary Dust, 253–256.

2. Finite Lifetime Fragment Model

The FLM for SYBs is described schematically in figure 1. Instead of the one-time, simultaneous fragmentation, parent particles break up continuously. All the created fragments are assumed to have a finite lifetime τ_s. This lifetime is reduced to the value at the heliocentric distance of 1 A.U. This assumption means that the fragments cannot be observed when the elapsed time from the fragmentation is longer than the lifetime τ_s. The elapsed time in the solar radiation of the fragments τ_f is calculated by

$$\tau_f = \int_{t_f}^{t_{obs}} 1/\{r(\beta_f)\}^2 \, dt \tag{1}$$

where t_{obs} is the time of the observation, t_f is the fragmentation time, $r(\beta_f)$ is the heliocentric distance of the fragment, β_f the ratio of the radiation pressure to the gravitation. The fragments, of which the elapsed time τ_f is less than τ_s, survive as the scattering materials in the optical wavelength.

The merit of the FLM is that we do not need unnatural assumption of the simultaneous fragmentation of the parent particles, or of other complicated mechanisms for the dust particles acceleration proposed so far. This model explains not only the observed width of SYBs, but also the length of SYBs by using an appropriate lifetime of the fragments. The lifetime is determined to reproduce the observed morphology of SYBs using equation (1). On the basis of this model, Nishioka and Watanabe(1990) derived the lifetime of the fragments in the SYBs of three comets (33.6-43.0 days for Comet Seki-Lines 1962III; 23.0-67.1 days for Comet Mrkos 1957IX; 28.0-54.0 days for Comet West 1976VI). The lifetime of the fragments lies between 25-70 days for all SYBs. This finite lifetime of 25-70 days gives us a hint for discussing the physical properties of the dust particles.

3. Implication for the Cometary Dust in Synchronic Bands

The parent particles of SYBs must be large in size, and be fragile. It should be noted that these SYBs appeared only in the new comets in the Oort sense. Hence, the parent particles may contain more volatile materials. Then, there may be a lot of water ice grains in the fragments of SYBs. Mukai(1986) calculated the lifetime of both pure and dirty water ice grains at 1 A.U. as shown in figure 2. The lifetime of the heterogeneous grains is less than 1000sec even if the size is about 1mm. On the other hand, the lifetime of the pure ice grains is more than 1000 days. The calculated dirty grains consisted of amorphous water ice with magnetite and silicate inclusions. The volume fraction of the inclusions is 0.1. Even if the volume fraction of

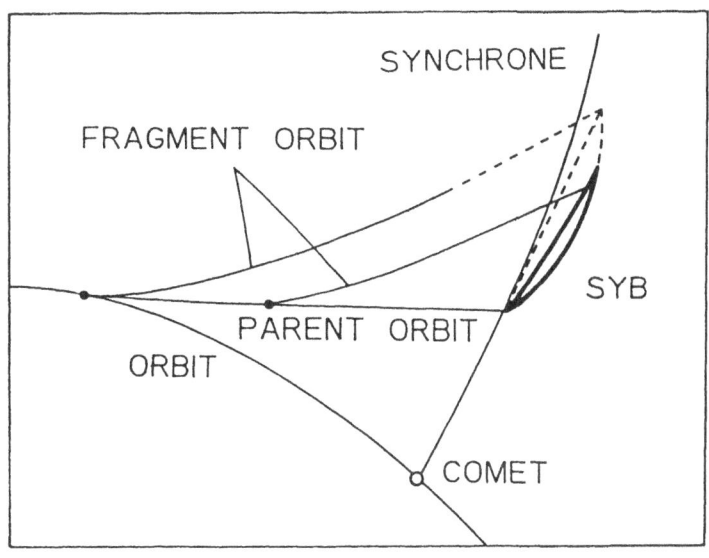

Figure 1. Formation of SYB on the basis of the finite lifetime fragment model(FLM). Parent particles ejected from the nucleus break up continuously. Fragment created earlier time vanish before the time of observation, and cannot be seen as indicated by the dashed line.

Figure 2. Lifetime of released icy grains with radius s at heliocentric distance of 1 A.U. derived by Mukai(1986). The lifetime of fragments in SYBs derived by this study is shown by two thin lines, which may suggest that they are relatively pure icy grains.

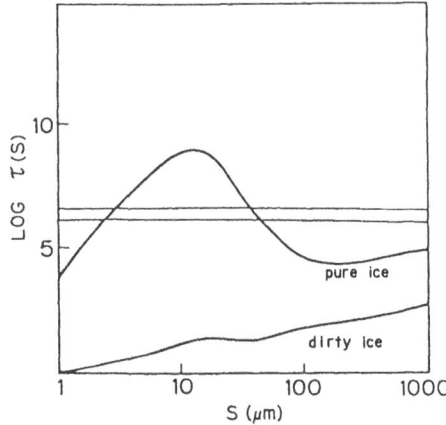

the magnetite was 0.001, Mukai(1986) indicated that the lifetime of the dirty ice would not change. However, if fraction of the inclusion of the absorbing material in the water-ice grains is less than 0.001, or if the inclusion is some kinds of organic materials, then the lifetime may be close to that of the pure ice grains. Therefore, the observed lifetime of 25-70 days may be due to the relatively pure ice grains ejected by the violent jets from the deep location in the cometary nucleus. Because the epoch of the parent particle ejection for SYBs was near the perihelion passage, these particles may have been pure materials ejected from the deep location.

Another possible candidate for the fragments in SYBs is organic grain. Although the CHON particles detected by the spacecrafts of Halley's mission may be a candidate, the lack of information on the optical properties prevented us from discussing further details. The optical constants of organic grain "tholin" are studied by Khare et al.(1984). Lamy and Perrin(1988) derived the optical properties, and concluded that the organic grains have a potential for explaining several puzzling cometary observations such as gaseous jets. Although the lifetime of the tholin itself is shorter by an order than 25-70 days, we cannot deny other organic materials as candidates for the fragments in SYBs.

References

Akabane, T. (1983) 'The secondary tail of Comet 1976VI West', Pub. Astron. Soc. Japan, **35**, 565-578.

Khare, B.N., Sagan, C., Arakawa, E.T., Suits. F., Callcott, T.A., and Williams, M.W. (1984) 'Optical constants of organic Tholins produced in a simulated Titanian atmosphere:From soft X-ray to microwave frequencies', Icarus, **60**, 127-137.

Lamy, P.L., and Perrin, J.-M. (1988) 'Optical properties of organic grains:Implications for interplanetary and cometary dust', Icarus, **76**, 100-109.

Mukai, T. (1986) 'Analysis of a dirty water-ice model for cometary dust', Astron. Astrophys., **164**, 397-407.

Nishioka, K., and Watanabe, J. (1990) 'Finite lifetime fragment model for synchronic band formation in dust tails of comets', Icarus, **87**, 403-411.

Sekanina, Z., and Farrell, J.A. (1980) 'The striated dust tail of Comet West 1976VI as a particle fragmentation phenomenon', Astron. J., **85**, 1538-1554.

ICE PARTICLE EMISSION FROM COMETARY ANALOGUES

H. KOHL and E. GRÜN
Max-Planck-Institut für Kernphysik
P.O.Box 10 39 80
W 6900 Heidelberg
Germany

ABSTRACT. Dust particles originating from comets are an important constituent of the interplanetary dust regime. In order to study the ejection mechanisms from the cometary nucleus surface simulation experiments in the laboratory have been performed. Samples consisting of water ice, carbon dioxide ice and dust grains have been studied when they are irradiated by artificial sunlight within a cooled vacuum system. It has been shown that particle emission is extremely dependent on the initial composition of the samples. For samples with a distinct amount of non-volatile, mineral particles the formation of a dust mantle and, as a consequence, rapid decrease of particle ejection has been observed.

1. Introduction

Comets are an important source for interplanetary dust particles. Heated by solar irradiation the volatile surface material sublimates and consequently emits gas, ice- and dust particles. However, it is not quite clear which proceses cause the release of such particles from the cometary surface in detail. Within classical models only dust particles smaller than a critical radius may leave the comet. Larger pebbles and rocks remain at the surface forming a non-volatile mantle.

To improve our understanding of cometary surface physics, sublimation experiments in the laboratory are performed in the comet simulation (KOSI) project (Grün et al., 1990). KOSI is an interdisciplinary and cooperative project of German, French and Israeli scientists. The scientific aim is to study the physical properties and behaviour of sublimating ice-dust-samples under interplanetary space conditions. Therefore experiments in a few meters large space simulator at Deutsche Forschungsanstalt für Luft- und Raumfahrt (DLR) in Köln, Germany, and in some other laboratories are performed.

Samples, mainly consisting of water ice, carbon dioxide ice and mineral constituents eject ice and dust particles when they are heated by irradiation of an artificial sun. In order to investigate particle emission from the sample diagnostic instrumentation like video cameras, collectors, impact and laser detectors are installed within the simulation chamber, analyzing physical properties like sizes, speed and emission distribution.

A.C. Levasseur-Regourd and H. Hasegawa (eds.), Origin and Evolution of Interplanetary Dust, 257–260.
© 1991 *Kluwer Academic Publishers.*

2. Method

The measurements of ejected particles referred to in this paper are performed by a system of ten piezoceramic impact detectors. These detectors are able to record particles hitting onto a ceramic surface. Particle momentum is transformed to an electric signal by the piezoceramic material. Sensitivity and calibration have been described by Kohl et al. (1990). Although the sensitivity of the detectors is relatively high, only particles larger than 150 to 200 μm can be detected in the KOSI experiments because of the high electric noise level within the space simulator. As the sensitivity threshold is determined both by mass and velocity of the projectiles the lower size limit for the particles is not sharp.

The physical process of momentum transfer at the detctor's surface also determines the detectibility of particles having different physical structure. Calibration experiments showed that only compact projectiles may be recorded by the impact detectors. Fluffy and weakly bound agglomerates like the clay sublimate residues (see chapter 3) are destroyed during the impact process. As a consequence, transferred momentum per unit time is much smaller and the sensitivity of the detector rapidly decreases. This means for the KOSI experiments that only particles may be recorded which contain at least a significant amount of icy components. Even mm-sized iceless agglomerates are not 'seen' by the detectors.

Size and momentum of the projectiles can be directly deduced from the signal form. Because of the time for data processing only four signals of different detectors are stored per 100 sec. Nevertheless every signal during this time exceeding a given threshold may be counted.

The detectors were installed within the vacuum chamber at a distance of 0.98 m from the sample center at different angles from -22.5 to +22.5 degrees in equal steps relative to the irradiation direction (Fig. 1). Because of the geometrical setup only particles with ejection velocities between 1.8 and 3.5 m/s may strike one of the detectors. Evaluation of particle collectors at different positions in the chamber (Thiel et al., 1990) showed that this is just the speed range of most emitted particles.

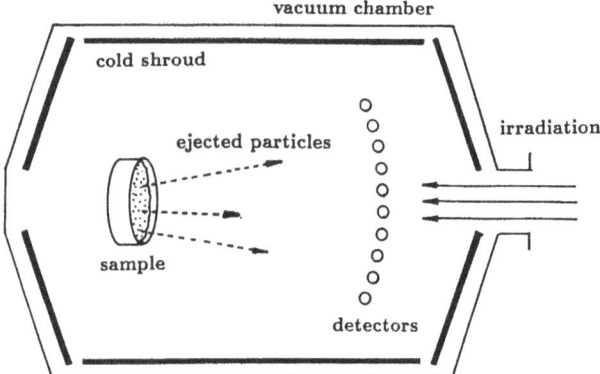

Fig. 1 Experimental setup for investigation of dust particle emission by piezoceramic detectors in the KOSI experiments.

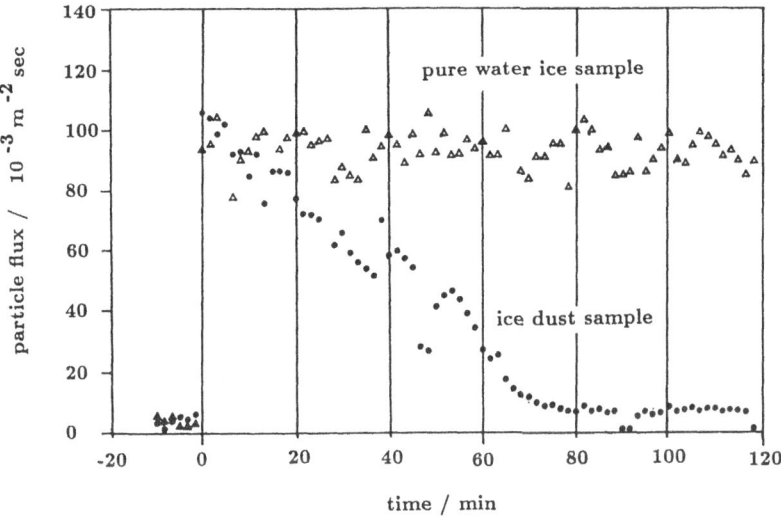

Fig. 2 Particle flux at the detectors as a function of time. At t=0 irradiation with artificial sunlight starts.

3. Simulation Experiments

In this paper we present the results of two experiment. In the first experiment a pure water ice sample was used; the second experiment was performed with a mixture of 41.6 % water ice, 15.0 % carbon dioxide ice, 7.4 % montmorillonite particles, 31.0 % olivine particles and 5.0 % fine grained carbon.

The mineral constituents in the sample show a rather different physical behaviour. Clay particles like montmorillonite which are initially mixed in the suspension with sizes in the order of a few microns tend to clump and form lattice-like agglomerates (Storrs et al., 1988). These agglomerates are already existent in the ice and keep conserved when the volatile material in the pores sublimates and disappears. Therefore sizes and forms of the agglomerates are generally determined by the droplets freezing in the liquid nitrogen. Olivine particles do not have such properties. However, they may adhere to the clay structure to a certain amount.

4. Results and Discussion

Fig. 2 shows the time development of particle flux for the first two hours of irradiation during the two experiments. For both experiments counting rates of particles increase rapidly with the start of irradiation. Particle ejection occurs on a time sclae smaller than the measuring interval of 100 seconds. In the pure water ice experiment the initial counting rate remains almost constant. This is due to free sublimation of water ice at the sample surface. In the water/carbon dioxide/mineral experiment emission activity decreases and reaches the noise level after about 60 to 70 minutes. We assume that a mantle of mineral particles has formed, probably, quenching gas and particle emission.

260

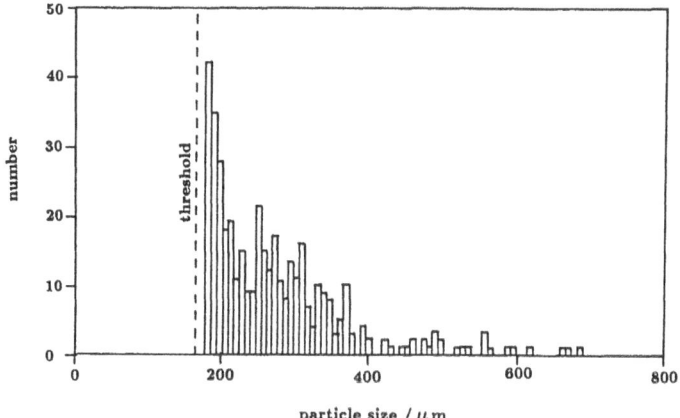

particle size / μm

Fig. 3 Size distribution of the recorded particles in the second experiment.

During the second experiment the surface temperature which was measured by the thermal emission (Heidrich, pers. comm.) reaches an equilibrium temperature of 439 K (Höppner, pers. comm.) after about one hour. This proves that a mantle of dust particles almost free of volatile material built up. The mantle causes a thermal shielding against the radiative energy input. As a consequence gas emission and particle ejection decreases.

The histogram in Fig. 3 shows the size distribution of the ejected particles as measured by the impact detectors. Particle frequency distinctly increases with decreasing diameter. This is due to the fact that particles with lower mass may be lifted more easily from the surface than large ones. The largest particles recorded are 600 to 800 μm in diameter. However, still larger ones, up to mm sizes, were found in the collectors after the experiment. An upper limit for the particle sizes, giving an estimation for a possible critical radius of emission, was not found. A significant change in the size distribution of the initial sample and the emitted particles has not been observed, as there is up to now no precise possibility to determine in situ sizes of ice-dust-particles within the sample. Changes of size distribution during the time of irradiation also could not be observed. The reason is the low number of recorded particles not allowing an exact statistical evaluation.

5. References

Grün, E., Kochan, H. and Seidensticker, K. (1990) 'Laboratory simulation, a tool for comet research', Geophys. Res. Lett., in press

Kohl, H., Grün, E., and Weishaupt, U. (1990) 'A new method for analyzing low velocity particle impacts', Planet. Space Sci. 38, 567-573

Storrs, A.D., Fanale, F.P., Saunders, R.S. and Stephens, J.B. (1988) 'The formation of filamentary sublimate residues (FSR) from mineral grains', Icarus 76, 493-512

Thiel, K., Kölzer, G., Kochan, H., Grün, E., Kohl, H., and Bremer, K. (1990) 'New results of the dust investigations in the comet simulation project KOSI', Proceedings of the 21st Lunar and Planetary Science Conference, Houston, March 1990, in press

THE DUST IN THE COMA OF COMET HALLEY

Joniek I. Hage and J. Mayo Greenberg
Laboratory Astrophysics, University of Leiden
P.O. Box 9504
2300RA Leiden
The Netherlands

ABSTRACT. The interstellar dust model of comets is numerically worked out to satisfy several basic constraints provided by observations of comet Halley and to derive the porosity of coma dust. The observational constraints are: (1) the strengths of the 3.4 μm and 9.7 μm emission bands; (2) the relative amount of silicates to organic materials; (3) the mass distribution of the dust. The results indicate that coma dust has a porosity in the range $0.93 < P < 0.975$. Preliminary calculations concerning the observed linear polarization of comet Halley are presented.

1 Introduction

The purpose of this work is to provide evidence for the model of comets by Greenberg (1985) by showing that interpretation of ground based and spacecraft observations of the coma of comet Halley may be successfully performed on the basis of this model. We present here the main ideas of recent work (Greenberg and Hage 1990; see this paper for details) and some new, preliminary results concerning the interpretation of the observed linear polarization of light scattered by the coma of comet Halley.

In terms of the present model, the dust in the coma consists of porous aggregates (figure 1a) consisting of interstellar core-mantle particles (figure 1b). Comets are assumed to be aggregates of particles as shown in figure 1c. The porosity, P, of the aggregates is defined as the relative amount of volume filled by vacuum inside the aggregate. The comet model predicts $0.6 < P < 0.83$ for comet nuclei and $0.9 < P < 0.975$ for coma dust.

2 Method

Basically, we want to show that on the basis of the present comet model, it is possible to explain the observed strengths of the 3.4 μm (Danks *et al.* 1987) and 9.7 μm (Hanner *et al.* 1987) emission bands of the coma of comet Halley, in terms of: (1) the mass ratio of silicates to organic materials as measured *in situ* (Kissel and Krueger 1987) and (2) the mass distributions of the dust, which were also measured *in situ* (McDonnell *et al.* 1989, Mazets *et al.* 1987). The shape of the 9.7 μm feature will not be considered here, although it may also be explained. The above is accomplished by going through the following scheme:

(1) Assume the porosity of the coma dust is unknown and use it as a free parameter.

(2) Calculate the thermal flux from the coma as a function of the dust porosity using: (a) the observed mass distributions; (b) the observed mass ratio in the dust of silicates:organics=2:1; (c) the interstellar dust model of comets to provide the morphology and specific chemical composition of the dust.

The model calculations to determine the thermal emission of the coma dust particles with their complicated shapes, as a function of their porosity, size and distance to the sun are described fully in Hage and Greenberg (1990) and in Greenberg and Hage (1990).

(3) Equate, if possible, the calculated results with observed values, at the wavelengths of 9.7 and 3.4 μm. If the observed values are matched, then not only are spacecraft and ground based observations tied together, but a (representative mean) value for the dust porosity is also found.

We note here that we do not calculate the continuum radiation from the coma at

261

A.C. Levasseur-Regourd and H. Hasegawa (eds.), Origin and Evolution of Interplanetary Dust, 261–264.
© 1991 *Kluwer Academic Publishers.*

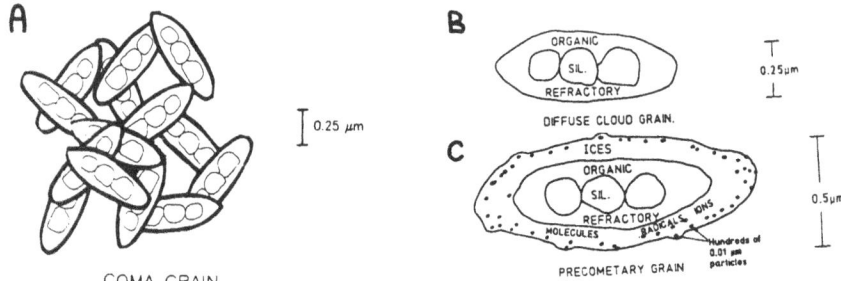

Figure 1. *(a) Schematic drawing of a coma grain according to the interstellar dust model. (b) Schematic of an interstellar dust grain which contains a core of silicates and a mantle of organic refractory material. (c) An interstellar grain after accretion of gases on its surface. These grains make up a comet.*

arbitrary wavelengths, but only the amount of emission *above the continuum* in the observed 3.4 and 9.7 μm emission bands. This restriction makes an accurate model calculation possible in spite of the fact that the mass distributions are unknown for the larger particles, because it can be shown that the contribution to the emission *above the continuum* from the large particles is negligible.

3 Results and Discussion

We have listed in table 1 the coma dust porosities which are required to match the calculated and observed emissions, on the basis of the assumptions listed in section 2. For example, the first line shows that, using a mass distribution measured by the Vega1 spacecraft to match the observed strengths of the 9.7 and 3.4 μm bands observed on March 6.85, 1986, (the 3.4 μm band strength was inferred from other observations), porosities of 0.80 and 0.90 are required, respectively. From Table 1 we can conclude that the strengths of the 3.4 and 9.7 μm bands can indeed be reproduced and that the required porosities are all rather high. The results for the 3.4 and 9.7 μm bands are slightly discrepant, but this is not surprising in view of the uncertainties in the observational data.

We summarise the conclusions from Greenberg and Hage (1990), which were based on a broader discussion: a high coma dust porosity of $0.93 \leq P \leq 0.975$ is a likely possibility which is consistent with (a) the direct results of this work, (b) comet densities as deduced independently by other workers (Sagdeev *et al.* 1988, Sekanina and Yeomans 1985) and (c) observed meteor densities (Verniani 1973). Furthermore, there are two additioanl features in the present model of comets which are *critical* in producing the observed 9.7 μm and 3.4 μm coma dust emissions: (1) The presence of organic refractory material. Without this component, it would in some cases be impossible to reproduce the amount of excess 3.4 and 9.7 μm emissions, because the coma dust would not become hot enough. (2) The presence of particles with a size similar to interstellar dust grains as basic units of the aggregates. If the mean size of the aggregate units were much larger than the size of interstellar dust grains, it would again be impossible to reproduce the amount of excess emission, because the aggregates would then neither become hot enough nor emit the emission features well.

4 The Linear Polarization of Comet Halley

Having arrived at the above results, an immediate question arises: is the observed degree of linear polarization of scattered radiation (as a function of phase angle) from the coma of comet Halley (Dollfus *et al.* 1988) also explicable in terms of very porous aggregates of interstellar dust? The main feature of the observed polarization is its negative branch at low phase angles. Ideally, a model of the dust should reproduce the following: (1) the amount of minimum polarization; (2) the phase angle at which the minimum occurs; (3) the phase angle at which the polarization goes from negative to positive values; (4) the slope of the polarization curve for larger phase angles; and (5) the wavelength dependence of these

Spacecraft	date	P(9.7)	P(3.4)
Vega1	March 6.85	0.80	0.90
Vega2	March 12.8	0.90	0.96
Giotto	March 12.8	0.97	0.98
Giotto	March 13.75	0.84	0.95

Table 1. *Required porosities of the coma dust. P(9.7) is the porosity required to match the strength of the 9.7 μm emission band, and P(3.4) is the result for the 3.4 μm band. The date refers to the time of the infrared observations from Earth.*

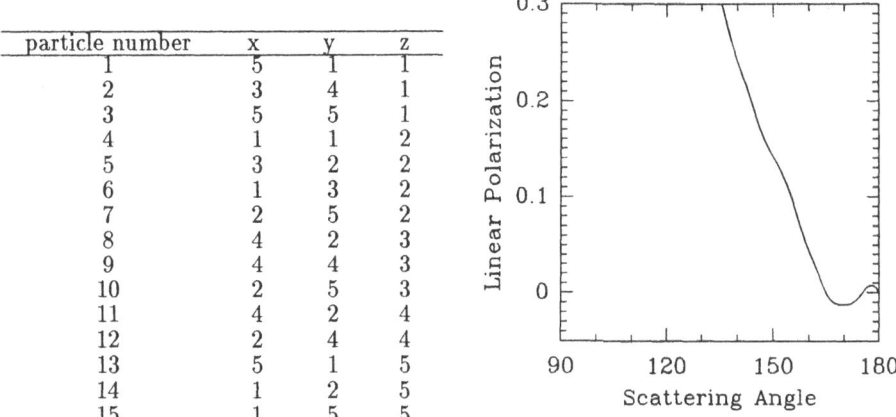

particle number	x	y	z
1	5	1	1
2	3	4	1
3	5	5	1
4	1	1	2
5	3	2	2
6	1	3	2
7	2	5	2
8	4	2	3
9	4	4	3
10	2	5	3
11	4	2	4
12	2	4	4
13	5	1	5
14	1	2	5
15	1	5	5

Table 2 *(Left). Positions of the inclusions in the model aggregate in rectangular cartesian coordinates.* **Figure 2** *(Right). Linear polarization as a function of scattering angle produced by the model.*

quantities.

It seems very likely that the same physical mechanism which produces negative polarization for two dipoles (Muinonen 1990) could work also for porous aggregates of interstellar dust. The main question is: what are the physical properties an aggregate should have in order to produce the observed polarization behaviour? For example, what should be its size and porosity; what should be the refractive index and size of the inclusions? Furthermore, are these requirements consistent with the assumptions of the present comet model?

We have started preliminary calculations to investigate the polarization produced by porous aggregates of interstellar dust. We use a numerical method (Hage and Greenberg 1990) which has been experimentally verified for porous particles (Hage, Greenberg and Wang 1990). The first step in these calculations is to build a model to represent the most important features of a porous aggregate. As one of the first attempts we have built a model representative of 15 interstellar dust grains. The model consists of 15 identical cubes with a size parameter of about 1.2, which are randomly distributed in a $5 \times 5 \times 5$ cubic lattice, simulating a structure with a porosity of $P = 1 - 15/125 = 0.88$ (see table 2). The refractive index was chosen as $m = 1.6$ and the scattering was averaged over 225 orientations of the model with respect to the scattering plane (furthermore $k = 0.9396$ and $\epsilon = 10^{-5}$; these parameters are defined in Hage and Greenberg 1990). The resulting linear polarization is shown in figure 2 as a function of the scattering angle (which corresponds to $180° -$phase angle).

Figure 2 shows that the model does indeed produce negative linear polarization at low phase angles. One notes that the calculated curve does not reproduce the observed polarization curve exactly, but this is not to be expected, one reason being that the observed polarization is produced by a large range of particles with different sizes.

The convergence of the results in figure 2 is not yet rigorously established. The reason is that the polarization at high scattering angles turned out to be highly variable as a function of scatterer orientation, so that averaging over very many orientations is required to acquire a reliable mean. A problem associated with these numerical calculations is that they take a considerable amount of computer time, even on a supercomputer. We are still in the process of solving this problem. We believe, however, that the negative branch of the polarization in figure 2 is real, because: (1) Other calculations with different parameters consistently show the negative polarization; (2) we understand the physical mechanism producing negative polarization. However, the numerical values of the polarization presented in figure 2 are expected to have a non-negligible error at scattering angles close to 180°.

Our present conclusions are, based on computations similar to those described above: (1) a high porosity is necessary to obtain the negative polarization, consistent with the results derived in section 3; (2) the particles in the aggregate must have a size parameter $x > 1$ (for moderate refractive indices), otherwise there is no negative polarization. This is consistent with the physical mechanism producing the negative polarization, which is a second order effect. If the constituent particles are too small, they do not scatter enough radiation to produce a second order effect. This also leads to the prediction that the negative polarization should vanish in observations further toward the infrared. (3) Maximum polarization should occur at 90 degrees phase angle.

References

Danks *et al.* "The Spectrum of comet P/Halley from 3.0 to 4.0 μm", 1987, *Astr. Ap.* **184**, 329–332.

Dollfus A. *et al*,"Optical polarimetry of P/Halley: synthesis of the measurements in the continuum", 1988, *Astr. Ap.*, **206**, 348– 356.

Greenberg, J. M., 1985,"The Chemical and Physical Evolution of Interstellar Dust", *Physica Scripta*, **T11**, 14–26.

Greenberg, J. M. and Hage, J. I., 1990,"From Interstellar Dust to Comets", *Ap. J.* **361**, 260–274.

Hage, J. I. and Greenberg, J. M., 1990,"A Model for the Optical Properties of Porous Grains", *Ap. J.* **361**, 251–259.

Hage, J. I., Greenberg, J. M. and Wang, R. T., 1990,"Scattering by Arbitrarily Shaped Particles: Theory and Experiment", accepted by Apl. Opt.

Hanner, *et al.*, 1987, "Infrared emission from P/Halley's dust coma during March 1986", *Astr. Ap.*, **187**, 653–660.

Kissel, J. and Krueger, F. R., 1987,"The organic component in dust from comet Halley as measured by the PUMA mass spectrometer on board VEGA1", *Nature*, **326**, 755–760.

Mazets E. P. *et al.*, 1987,"Dust in comet P/Halley from Vega observations", *Astr. Ap.*, **187**, 699–706.

McDonnell, *et al.*, 1989, "The in-situ cometary particulate size distribution measured for one comet: P/Halley", in *Workshop on Analysis of Returned Comet Nucleus Samples, Milpitas, California*, in the press.

Muinonen, K., 1990, "Light scattering by inhomogeneous media", thesis, University of Helsinki.

Sagdeev, R. Z., Elyasberg, P. E. and Moroz, V. I., 1988, "Is the nucleus of comet Halley a low density body?", *Nature*, **331**, 240- -242.

Sekanina, Z. and Yeomans, D. K., 1985, "Orbital motion, nucleus precession, and splitting of periodic comet Brooks 2", *A. J.*, **90**, 2335–2352.

Verniani F., 1973, *J. Geophys. Res.*, **78**, 8429.

POLYOXYMETHYLENE IN COMETARY DUST: LABORATORY TESTS

D.C. BOICE, D.W. NAEGELI, and W.F. HUEBNER
Southwest Research Institute
P.O. Drawer 28510
San Antonio, TX 78228-0510
USA

ABSTRACT. We have investigated the stability of gas-phase formaldehyde oligomers and its implications for cometary science. Our experiments indicate that when a formaldehyde-methanol solution is vaporized in the mass spectrometer, high molecular mass POM species (45, 61, 75, 91, 105, 121, 135, 151, 165 amu) exist in the gas phase in the temperature range 473 K to 773 K. These laboratory results complement our previous experiments using a formaldehyde-water solution and indicate that formaldehyde oligomers are stable in the gas phase up to at least 6 monomeric units in length. Methanol is important in the end-capping process of the oligomers, leading to increased stability and a richer mass spectrum when compared to the formaldehyde-water solution. The results are consistent with mass spectra obtained by the *Giotto* PICCA instrument exhibiting alternating 14-16 amu mass peaks.

1. Introduction

One of the surprises of the recent spacecraft encounters with Comet Halley was the discovery of the "CHON" particles. While the exact nature of the organic constituents of CHON remains uncertain, polyoxymethylene (POM) is likely to be an important constituent. We have previously proposed a schematic comet dust particle composed of silicate grains, carbonaceous material, and complex organic molecules including formaldehyde oligomers (Huebner et al., 1987). The affinity of these oligomers for silicate surfaces binds the grains and holds the particle together. As the particles drift away from the nucleus, solar heating causes them to fragment on time scales related to the decomposition properties of the oligomers. This has also been suggested by Boenhardt et al. (1990). In addition, POM and other semivolatile compounds are slowly released from the grains and contribute to the distributed sources of CO, H_2CO, CH_n ($n=1,2$), and other coma species. This grain picture is consistent with many observations of Comet Halley (including the mass spectra obtained with the PICCA instrument) as pointed out by Huebner and Boice (1989) and Boice et al. (1990).

2. Laboratory Experiments and Results

Experimental work on the identification of vapor phase products of polyoxymethylene has been carried out using a mass spectrometer equipped with a variable temperature sample probe. A preliminary description of the laboratory study and first results using dry paraformaldehyde (solid polyoxymethylene) and a formaldehyde-water solution have been given by Boice et al. (1989). A brief summary is given below.

It is well known that the principal product of paraformaldehyde decomposition at room temperature and above is formaldehyde gas. Previous studies (Walker, 1964) have shown that when paraformaldehyde is heated to 423 K, the evolved gas consists mainly of formaldehyde monomer.

A.C. Levasseur-Regourd and H. Hasegawa (eds.), Origin and Evolution of Interplanetary Dust, 265–268.

However, recent results obtained by Moller and Jackson (1989) indicate that polymeric forms of formaldehyde are present and stable in the gas phase at temperatures as low as 300 K. Our initial work (Boice et al., 1989) extended this result to higher temperatures and masses, and implied the existence of an oligomer, 5 monomeric units in length. POM fragments were also produced by Moore and Tanabe (1990) during 700 KeV proton sputtering and by Mahaffy (1989) with 4 KeV Cs+ bombardment. These studies, performed by a variety of methods, all yield results consistent with the stability of ionic POM fragments in the gas phase, with a characteristic pattern separating mass peaks of alternating 14-16 amu.

In the present experiments, we choose to investigate formaldehyde-methanol solutions for two reasons. Recent work by Busca et al. (1987) shows that formaldehyde, in the form of dioxymethylene absorbed on metal oxide surfaces, undergoes a disproportionation reaction forming formate and methoxide groups, probably via a Cannizzaro-type mechanism. It is possible that the Cannizzaro-type reaction produces some formic acid and methanol from formaldehyde adsorbed on the surface of comet dust particles. Additionally, methanol provides (CH_3)-groups for end-capping polyoxymethylene, a likely circumstance in comet dust rich in carbonaceous material.

The mass spectra were obtained using a Finnigan model 3300 mass spectrometer with a CDS model 120 pyroprobe. The pyroprobe has a platinum ribbon to hold the sample and can be heated to various temperatures in a few milliseconds. In the actual experiments, the samples were heated to temperatures ranging from 473 K to 773 K. The samples consisted of a solution of 30% formaldehyde in methanol, prepared by refluxing a mixture of paraformaldehyde and methanol in a thick-walled glass vessel and heating to 373 K for about 30 hours.

In our previous study using the formaldehyde-water solution, mass peaks were observed at 47, 61, 77, 91, 107, 121, and 137 amu. While these results gave the characteristic spacing of the mass peaks observed in the PICCA spectra, the peaks at 47, 77, and 107 were shifted by 2 amu to higher masses. Furthermore, the experimental mass spectra were sparce, with peaks centered at the above masses and very little in between.

For the formaldehyde-methanol solution, mass peaks were observed at 45, 61, 75, 91, 105, 121, 135, 151, and 165 amu as shown in Figure 1. Probable identification of the ions corresponding to these masses (as well as those of the formaldehyde-water solution) are listed in Table 1. This sequence of mass peaks precisely matches that observed in the PICCA spectrum reported by Korth et al. (1986) and Mitchell et al. (1987). The mass distribution shows considerably more detail than that of the formaldehyde-water mixture, with many smaller peaks surrounding the major ones. We note that all of the structure in the mass spectrum results from molecules containing only combinations of H, C, and O. A change in relative peak intensities is also apparent, with higher mass species being relatively more abundant in the formaldehyde-methanol solution. It is preliminary to claim that the observed mass spectrum has been reproduced, but continued analysis of the laboratory data will give likely candidates for some of the satellite peaks that may account for the broadening of the observed PICCA mass peaks.

In our previous study, high concentrations of water were found to be important in the end-capping process that stabilizes these species. This suggests that the proton affinity of POM-related oligomers is rather high as all mass peaks are consistent with protonated species. End-capping by methyl groups also promotes stability of the resulting molecule. These experiments were repeated with a Gas Chromatograph (model 5890)/Mass Selective Detector (model 5970) instrument to support the above results. A relatively neutral column of about 5 meters in length was used. The results obtained with the formaldehyde-water mixture were inconclusive as none of the higher-mass peaks were found in the spectra. It was suggested that the high-mass species either precipitated onto the column wall or decomposed during the long journey through the column. In contrast, gas chromatography of the formaldehyde-methanol solution yielded several high-mass peaks for neutral species roughly corresponding to those in the ionic mass spectra. Our preliminary conclusion is that end-capping polyoxymethylene chains with methanol-related groups leads to relatively stable neutral gas-phase species.

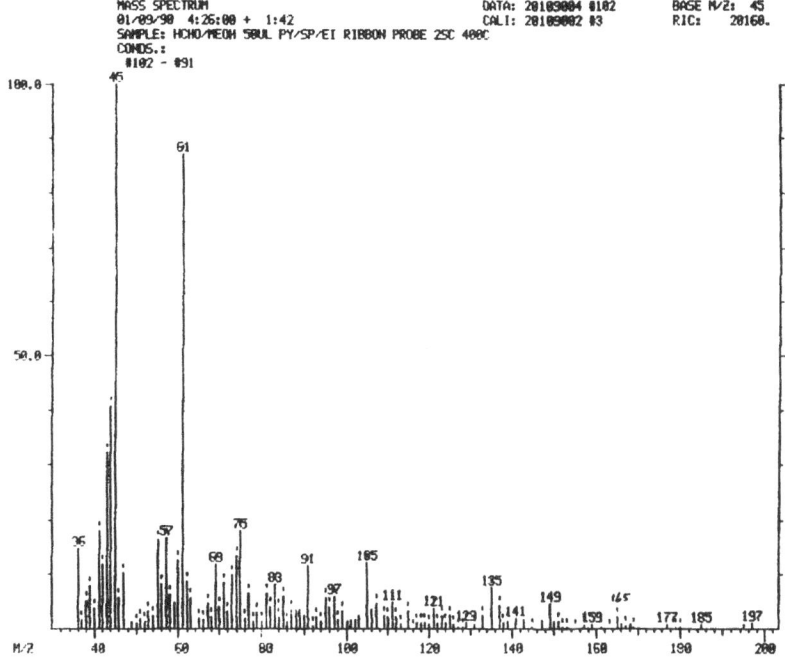

Figure 1. Mass spectrum of the formaldehyde-methanol solution heated to 673 K and using a 20 eV electron source in the mass spectrometer. Note the mass peaks at 45, 61, 75, 91, 105, 121, 135, 151, and 165 amu, and the many satellite peaks surrounding the major peaks.

TABLE 1. Mass spectra species identification

Mass	Probable POM-related species	Designation of ion fragment
45	$CH_3\text{-}O\text{-}CH_2^+$	Methyl ether
47	$HO\text{-}CH\text{-}OH^+$	Methylene glycol
61	$(CH_2\text{-}O)_2\text{-}H^+$	POM dimer glycol
	$H\text{-}(CH_2\text{-}O)_2^+$	POM dimer methyl ether
75	$CH_3\text{-}(O\text{-}CH_2)_2^+$	POM fragment methyl ether
77	$HO\text{-}CH_2\text{-}O\text{-}CH\text{-}OH^+$	Protonated POM fragment
91	$(CH_2\text{-}O)_3\text{-}H^+$	POM trimer glycol
	$H\text{-}(CH_2\text{-}O)_3^+$	POM trimer methyl ether
105	$CH_3\text{-}(O\text{-}CH_2)_3^+$	POM fragment methyl ether
107	$HO\text{-}(CH_2\text{-}O)_2\text{-}CH\text{-}OH^+$	Protonated POM fragment
121	$(CH_2\text{-}O)_4\text{-}H^+$	POM tetramer glycol
	$H\text{-}(CH_2\text{-}O)_4^+$	POM tetramer methyl ether
135	$CH_3\text{-}(O\text{-}CH_2)_4^+$	POM fragment methyl ether
137	$HO\text{-}(CH_2\text{-}O)_3\text{-}CH\text{-}OH^+$	Protonated POM fragment
151	$H\text{-}(CH_2\text{-}O)_5^+$	POM pentamer methyl ether
165	$CH_3\text{-}(O\text{-}CH_2)_5^+$	POM fragment methyl ether

3. Summary

The results of this research indicate that ionic POM-related oligomers are stable in the gas phase, up to at least 6 monomeric units in length. Methanol is important in the end-capping process for the stability of the oligomers, and it appears to lead to species that are more stable than those formed in formaldehyde-water mixtures.

The mass spectra of the formaldehyde-methanol solution are consistent with the maxima of the mass distributions obtained by the PICCA instrument in Comet Halley. Futhermore, these mass spectra show considerable detail with many small peaks surrounding the major ones. Continued analysis of these data will give likely candidates for some of the satellite peaks that may account for the broad PICCA mass peaks.

The presence of POM as one constituent in the complex organic mixture associated with cometary dust is consistent with many observations of comets and with several laboratory studies.

Acknowledgements

This study was supported by the Internal Research Program at Southwest Research Institute and the NASA Planetary Atmospheres Division.

References

Boenhardt, H., Fechtig, H., and Vanysek, V. (1990) 'The possible role of organic polymers in the structure and fragmentation of dust in the coma of comet P/Halley,' *Astron. Astrophys.* **231**, 543-547.

Boice, D.C., Naegeli, D.W., and Huebner, W.F. (1989) 'Physico-chemical properties of formaldehyde-ice-dust mixtures,' *Proceedings of an International Workshop on Physics and Mechanics of Cometary Materials*, **ESA SP-302**, 83-88.

Boice, D.C., Huebner, W.F., Sablik, M.J., and Konno, I. (1990) 'Distributed coma sources and the CH_4/CO ratio in comet Halley,' *Geophys. Res. Lett.* **17**, in press.

Busca, G., Lamotte, J., Lavalley, J.-C., and Lorenzelli, V. (1987) 'FT-IR study of the adsorption and transformation of formaldehyde on oxide surfaces,' *J. Am. Chem. Soc.* **109**, 5197-5202.

Huebner, W.F. and Boice, D.C. (1989) 'Polymers in comet comae,' in J.H. Waite, Jr., J.L. Burch, and R.L. Moore (eds.), *AGU Monograph* **54**, pp. 453-456.

Huebner, W.F., Boice, D.C., Sharp, C.M., Korth, A., Lin, R.P., Mitchell, D.L., and Reme, H. (1987) 'Evidence for first polymer in Comet Halley: Polyoxymethylene,' *Symposium on the Diversity and Similarity of Comets*, **ESA SP-278**, 163-168.

Korth, A. *et al.* (1986) 'Mass spectra of heavy ions near comet Halley,' *Nature* **321**, 335-336.

Mahaffey, P. (1989) 'SIMS of polyoxymethylene and questions of cometary dust composition,' preprint.

Mitchell, D.L. *et al.* (1987) 'Evidence for chain molecules enriched in carbon, hydrogen, and oxygen in comet Halley,' *Science* **237**, 626- 628.

Moller, G. and Jackson, W.M. (1989) 'Laboratory studies of polyoxymethylene: application to comets,' *Icarus*, in press.

Moore, M.H. and Tanabe, T. (1990) 'Mass spectra of sputtered polyoxymethylene – implications for comets,' *Astrophys. J. Lett.*, in press.

Walker, J.F. (1964) *Formaldehyde*, Reinhold Publishers, New York.

ON THE ANTI-TAIL OF COMET BRADFIELD (1987XXIX)

HIROKI AKISAWA, TAKUMA OKA
and KEN SUGAWARA
8-13-607, Shiroyama 3-chome,
Odawara-shi, Kanagawa-ken,
250 JAPAN

ABSTRACT. The anti-tail of comet Bradfield 1987XXIX ($= 1987s$) was observed when the earth passed through the orbital plane of this comet on December 20, 1987. The time variation of this phenomenon was monitored continuously by Japanese amateur astronomers. Analyzing these photographs by using the Bessel-Bredkhin theory, we deal with the dust particles in the anti-tail.

1. Observations

Many photographs which contributed to "Gekkan-Tenmon (Monthly Astronomical magazine in Japan)" by 10 amateur astronomers were used for this work. The anti-tail was shown on twelve photographs as listed in Table 1. These observations were carried out from December 14 through December 24. It should be noted that the sharp edge of one side in the anti-tail was recognized on eight photographs among them.

Table 1. Observational data

No.	UT Date 1987 Dec.	Exposure (min.)	Observer	Instrument	Emulsion	Δ	r	phase
1	14.40	15	M.Tsumura	300mm F2.8	TP2415	0.834	1.097	59.5
2	18.43	15	M.Tsumura	300mm F2.8	TP2415	0.844	1.140	57.1
3	18.45	13	K.Iwakami	528mm F3.3	TP2415	0.844	1.140	57.1
4	18.50	10	M.Yamashita	300mm F2.8	TP2415	0.844	1.140	57.1
5	19.45	2.5	K.Nishioka	400mm F2.5	X-ray	0.847	1.151	56.5
6	20.40	10	S.Kashiwagi	800mm F4	GX3200	0.851	1.163	55.9
7	21.41	15	Y.Kushida	200mm F4	Tmax400	0.855	1.174	55.3
8	21.42	19	S.Maeda	768mm F4.8	Try-X	0.855	1.174	55.3
9	22.39	8	K.Nagashima	400mm F4	TP2415	0.861	1.186	54.7
10	22.44	3	K.Yoshioka	350mm F2.8	Tmax400	0.861	1.187	54.7
11	23.42	15	M.Tsumura	300mm F2.8	TP2415	0.867	1.198	54.1
12	24.49	10	M.Ohkura	300mm F2.8	TP2415	0.873	1.209	53.6

A.C. Levasseur-Regourd and H. Hasegawa (eds.), Origin and Evolution of Interplanetary Dust, 269–272.
© 1991 *Kluwer Academic Publishers.*

2. Measurements

Two or three combinations of comparison stars were selected on each photograph. The position angle of the anti-tail was measured by using these selected S.A.O. stars. The error of these measurements is about ±1 or less degrees.

Fig. 1 shows the observed position angle of the anti-tail versus UT date. Clearly, the positional angle of the anti-tail turned over about the solar direction from south to north side at the time when the earth passed through the orbital plane of this comet. Moreover, the direction of sharp-edge side in the anti-tail changed from north to south side. This phenomenon indicates that we, on the earth, observed the anti-tail from both sides within this short period.

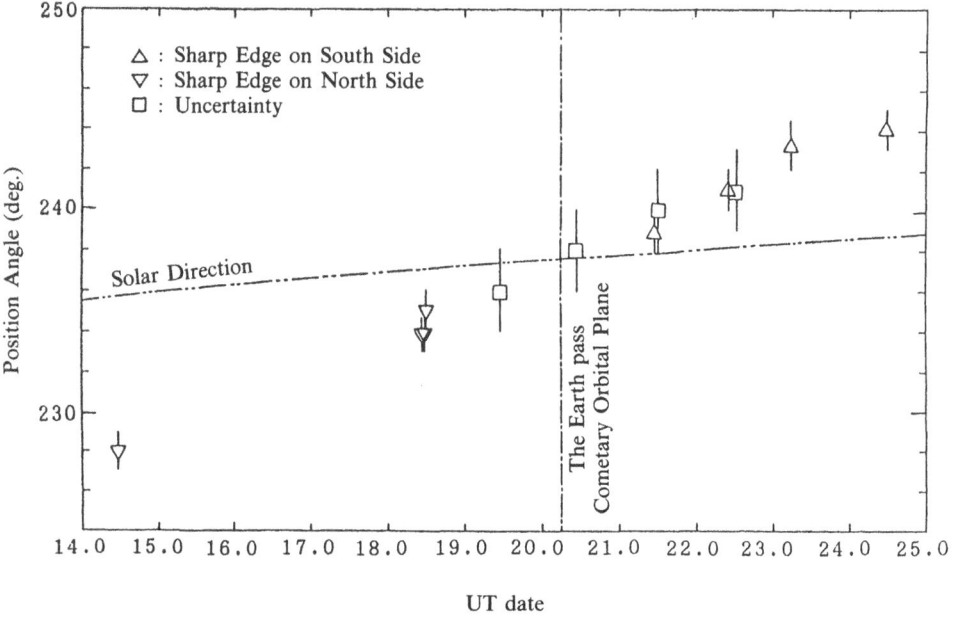

Figure 1. Time variation of the solar direction and the observed position angle of the anti-tail of comet Bradfield 1987s in December 1987.

3. Analysis

Some examples of the comparison between the observed anti-tail and the calculated synchrones using Bessel-Bredkhin dynamical theory are shown in fig. 2. The relation of the synchrones to the observed sharp-edge in the anti-tail with the variable inclination between the earth and the orbital plane of this comet is also shown in fig. 3. Careful measurements of the sharp edge in photographs on Dec. 14.40 and Dec. 24.49 give these position angles as 228 and 244 ±1 degrees.

Fig. 4 shows the measured position angle at the time of observation (top), the ratio of the radiation pressure to the gravitation β at 15 arc min from the nucleus at the time of observation (center), and the heliocentric distance at the time of ejection (bottom) versus the time of dust

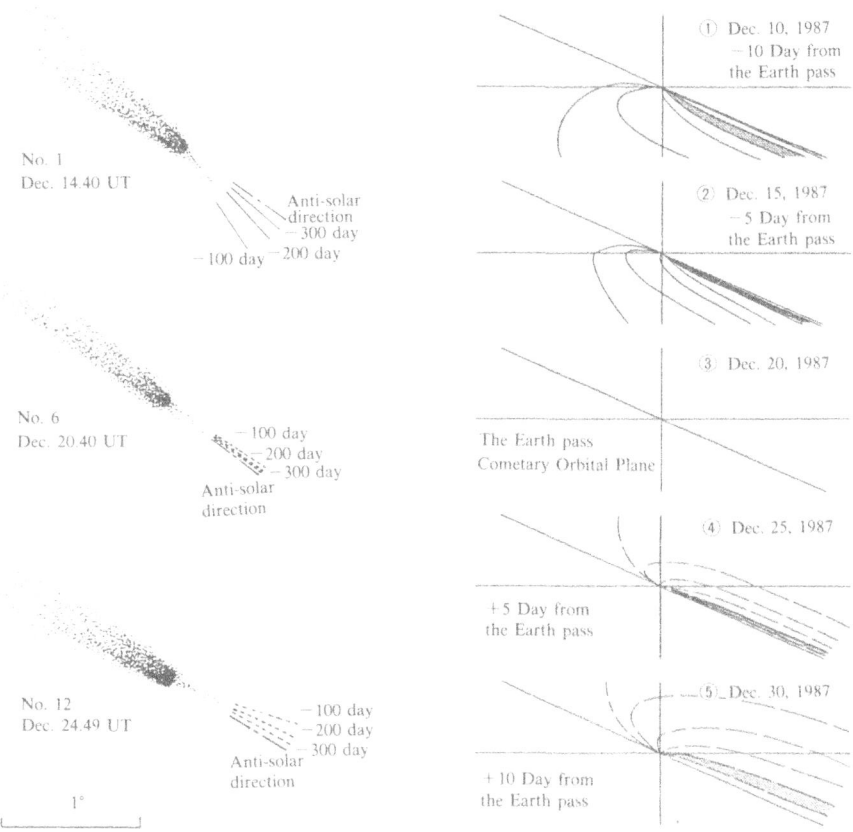

Figure 2. Comparison of the observed anti-tail with the calculated synchrones.

Figure 3. Schematic view of the edge variation in the anti-tail along with calculated synchrones.

particle ejection (synchrones). The solid curves refer to the computed synchrones, and the x axis refers to the ejection time from the nucleus. The little marks near the y axis give the observed quantities, and the two + marks on the curves give the limits on the observed quantities, including the uncertainty. Judging from this figure, we concluded that the dust particles were ejected about -200 days before the perihelion passage.

We determined β_{max} on the synchrone of -200 day. The lower limit of the dust size was calculated as

$$\beta = \frac{1.74 \times 10^{-4} Q_{pr}}{\rho \ d}$$

where Q_{pr} is the scattering efficiency of the particle; for the large particles under consideration, $Q_{pr} \leqq 1$. ρ and d are density and the diameter of the particle, respectively.

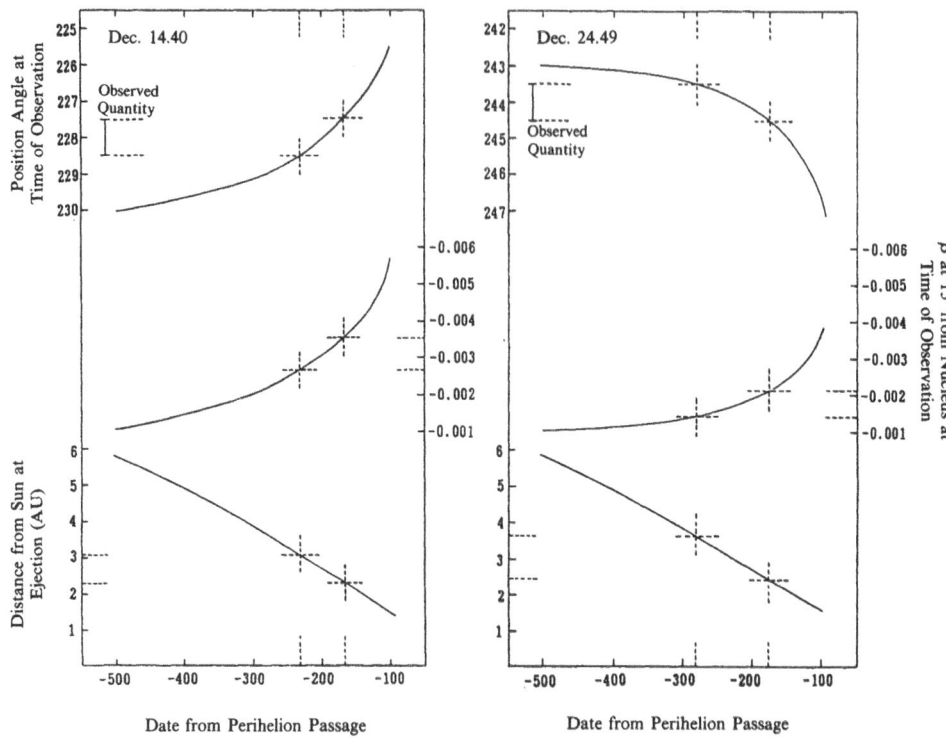

Figure 4. The relation between the position angle and fitted synchrones (upper), β at 15 arc min from the nucleus (middle) and heliocentric distance at ejection (bottom).

4. Conclusions

We found the following results.

(1) The anti-tail direction and the direction of the sharp-edge side were turned over at the time when the earth passed through orbital plane. The sharp-edge always existed in the side toward the solar direction where the old synchrones were crowded in narrow space.

(2) The dust size on 15 arc min from nucleus are estimated as $0.4 - 0.3$ mm on Dec. 14.40, and as $0.8 - 0.5$ mm on Dec. 24.49 observations. The estimated size might be uncertain because the synchrone curves were too crowded to be distinguished. In eiter case, the conclusion on the size of dust in anti-tail is approximately order of 10^{-1} mm. The diameter of dust particles derived from the β_{max} in the syndyne curves is > 0.2 mm, which is also consistent with the above estimated values.

(3) The dust particles in the anti-tail were ejected at -170 to -280 days before the perihelion passage, i.e. the heliocentric distance was about $2 - 3$ AU. Therefore, the life-time of the dust particles in the anti-tail is approximately $> 207 - 328$ days.

FORMATION MECHANISMS OF THE SPLIT TAIL OF COMET BRADFIELD 1987XXIX

Ken Sugawara
Atsugi City Children's Science Center
1-1-3, Nakachou, Atsugi-shi, Kanagawa 243, Japan
    and
Jun-ichi Watanabe
National Astronomical Observatory
Mitaka, Tokyo, 181, Japan

ABSTRACT. Comet Bradfield 1987XXIX showed a well-developed split tail, which is a dust tail feature divided into two branches by a dark gap. Similar structures have been observed in several comets. The split tail of comet Seki-Lines is one of them; some authors tried to explain it by rapid dust vaporization or self-regulation model. However, these models can not explain in the case of comet Bradfield 1987s. It was suggested in our previous works that the trend of variation of position angle of the dark gap coincides well with that of the plasma tail or projected radial vector. We discuss the formation mechanisms of the split tail from the view point of dust-plasma interactions.

1. Introduction

Comet Bradfield 1987XXIX became a naked-eye object during a few months around the perihelion passage. Many photographs also show the existence of dark gap(hereafter DG) in the dust tail. The dust tail was divided into two branches by DG. This structure has been called a split tail, which has been observed in several comets. However, comet Seki-Lines is the only example analyzed for formation mechanisms(Jambor, 1973). In this paper, the morphological characteristics of this structure are reported, and probable formation mechanisms are discussed.

2. Observational Material and Analysis

Photographs of comet Bradfield taken from late Sep.1987 to late Feb.1988 by many amateur astronomers in Japan were examined in order to know the morphology of the dust tail. We found DG in most of the photographs taken after late Oct. 1987. These structures are dust tail features, because the DG exists on the plates taken in red spectral region(Watanabe et al., 1989).

For quantitative analysis of the morphology, the dust tail was fitted by calculated synchrone and syndyne curves with the Bessel-Bredikhin model, which is a mechanical theory based on the orbital motion of dust particles. Figs. 1 and 2 show the time variation of the position angle

273

A.C. Levasseur-Regourd and H. Hasegawa (eds.), Origin and Evolution of Interplanetary Dust, 273–276.
© 1991 Kluwer Academic Publishers.

for DG, plasma tail and synchrone curves(the locus of particles emitted
from the nucleus at the same time), respectively. These show the
location of DG near the plasma tail. The variation of position angle of
DG coincides well with that of the plasma tail or projected radial
vector rather than that of synchrones.

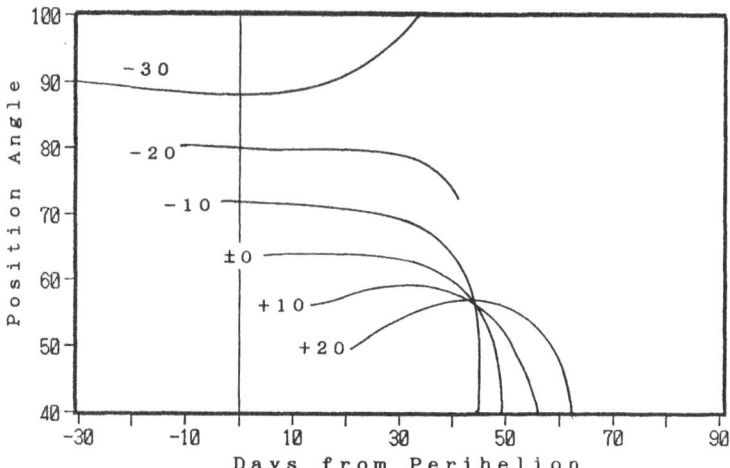

Fig. 1. Time (measured as the elapsed time from periherion passage)
variation of the position angle(measured counterclockwise from the
North) of dark gap(dots or vertical lines), plasma tail(cross marks),
and projected radial vector(solid line).

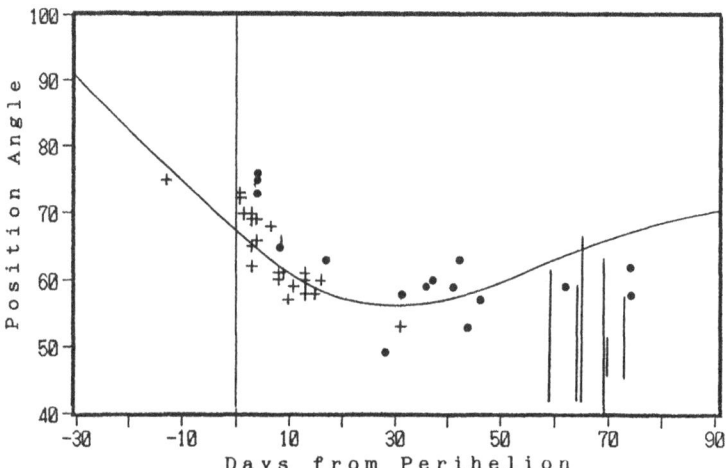

Fig. 2. Time variation of the position angle of initial directions of
synchrones. Noted numbers correspond to respective times (interval to
perihelion passage) of ejection.

3. Discussion

3.1. RAPID VAPORIZATION MODEL OF DUST PARTICLES

The position of the DG in Seki-Lines coincided with the synchrone for
dust particles emitted at perihelion(Jabor, 1973). Jambor suggested that
evaporation of dust emitted at perihelion(q=0.03 a.u.) could be the
cause of the DG, although Sekanina(1976) noted that it was impossible to
explain a DG in this way, unless the particles emitted at perihelion
were different in composition(i.e. more volatile) than the particles in
the dust tail which were emitted earlier. At the perihelion distance of
comet Bradfield (q=0.8 a.u.), however, the dust vaporization will be
negligible, and the rapid vaporization model can be ruled out.

3.2. SELF-REGULATION MODEL

The self-regulation model suggested by Sekanina(1976) is as follows. A
sharp peak in the dust production rate near perihelion caused a high
opacity for solar radiation, reducing the solar energy impinging on the
cometary nucleus.
  If the optical depth of the cometary coma suddenly increases by an
outburst, a lot of dust particles are required. Then a clear synchrone
feature should appear in the dust tail. However, corresponding
synchrone structure was not observed. This model also requires that the
direction of DG coincides with the synchrone curve at perihelion,
contrary to the observational data. Therefore, it is difficult to
explain the formation of the split tail of this comet by the
self-regulation model.

3.3. RELATION BETWEEN PLASMA TAIL AND DUST TAIL

The direction of DG coincides with that of the projected radial
vector(or plasma tail) rather than that of synchrone around perihelion.
It is known that an electric current flows along the tail axis at any
time. Thus, there is a possibility that DG was formed by electromagnetic
effects. Two models are following.

3.3.1. *Sweeping Model*. The broad tail extending from the head consists
of relatively small dust particles, on which the repulsive force by the
solar radiation is large. The electrostatic charging effect is also
larger with smaller particles. Interaction of the cometary dust with
the solar wind was studied by several authors(e.g. Ip et al. ; 1985).
They concluded that the Lorentz force for the charged dust particle was
important. The magnetosphere of comets has been revealed by in-situ
observations, which have confirmed the existence of strong magnetic and
electronic fields in the vicinity of the plasma tail axis. Hence the
Lorenz force on dust particles would be more important near the axis
than far from the tail axis. The Lorentz force could rapidly accelerate
the dust particles in the current sheet, creating a void which is seen
as a dark gap.

3.3.2. *Electrostatic Disruption Model*. Rhee(1976) showed that fluffy dust particles of micron or submicron sizes will be blown up if their surface potential reached a certain value. Following Ip(1984), the critical size for electrostatic disruption of spherical dust particles is

$$R_o (\mu m) = 9.4 \mid \Phi \mid / F_t^{1/2} \qquad (1)$$

where R_o is critical size, F_t is tensile strength of dust particle, Φ is surface potential[V]. Following Burns et al.(1979), we convert R_o into β (ratio of the solar gravitation to solar radiation pressure) in the case of the Mie scattering efficeincy for radiation pressure is 1 and the density of dust particle(assumed as sphelical) is 1[g/cm^3]. Following Rhee (1976), we assumed $F_t = 10^4$[dyne/cm^2]. If $\beta = 0.5$, dust particles disrupt into submicron sizes when $\mid \Phi \mid \sim 10$. The β value of DG derived from comparing Bessel-Bredikhin theory with observations is approximately < 0.5. Judging from model calculations(Hill and Mendis;1980), and previous discussion, it is suggested that DG formed by electrostatic disruption if the surface potential Φ reached sufficiently high value.

In order to conclude what is most important in the effects of electrostatic disruption, sweeping, or combination of them, more work would be necessary.

Acknowledgements
We express our thanks to many Japanese amateur astronomers, especially Messrs. M. Tsumura, S. Mizuno, K. Osada and S. Yamane, H. Akisawa, T. Oka, and H. Hasegawa. We are also thankful to the editors of Gekkan-Tenmon, of Tenmon-Gaido and of Sky Watcher for their permission to measure the published and non-publishd photographs.

References
Burns, J.A., Lamy, P.L., and Soter, S.(1979)' Radiation Forces on Small Particles in the Solar System.' ICARUS, 40, 1-48.

Hill, J.R. and Mendis,D.A.(1980)' On the Origin of Striae in Cometary Dust Tails.' Astrophys. J., 242, 395-401.

Ip., W.-H.,(1984)'Comet-Solar Wind Interactions:A Dusty Point of View.' Adv.Space Res. Vol.4, No.9, 239-247.

Ip., W.-H., Kimura, H., and Liu, C.-P.(1985)' Interaction of the Cometary Dust with the Solar Wind and Cometary plasma.' in Properties and Interactions of Interplanetary Dust, edt. R.H.Giese and P.Lamy(D. Reidel), 325-328.

Jambor, B.J.(1973)' The Split Tail of Comet Seki-Lines.' Astrophys. J., 185, 727-734.

Rhee,J.W.(1976)' Electrostatic Disruption of Lunar Dust Particles.' in Interplanetary Dust and Zodiacal Light,p.238

Sekanina, Z.(1976) ' Progress in Our Understanding of Cometary Dust Tails.' in The Study of Comets, pt.2, 893-939.

Watanabe, J., Aoki, T., Taniguchi, Y., and Tarusawa, K. (1989) 'Photographic Observations of Comet Bradfield 1987XXIX at the Kiso Observatory.' Publ. Natl. Astron. Obs. Japan, 7, 71-84.

SPATIAL DISTRIBUTION AND COLOR OF DUST IN HALLEY'S INNER COMA

J. CLAIREMIDI, E. BRANDON, P. ROUSSELOT, G. MOREELS
Observatoire de Besançon
BP 1615
25010 BESANCON Cedex
FRANCE

ABSTRACT. Composite images of the intensity of solar radiation scattered by dust in Halley's coma are constructed by using the three-channel spectra obtained during the approach phase of the Vega 2 spacecraft. They cover a sector centered on the nucleus that has a radius of 40000 km and an angular extent of 50°. A radial plot of dust-scattered intensity shows that it varies as the inverse of impact parameter p where p is smaller than 3200 km or higher than 7000 km. In the intermediate 3200-7000 km distance range, the intensity varies as $p^{-1.52}$. At longer distances, two jets are present with a contrast comparable to the gas jets which appear in the OH and CN images.

The color of dust shows a slight excess of near-UV radiation in a diffuse region between 10000 and 30000 km which appears to be connected with the two jets. In the region called "valley", between the jets at distances p > 25000 km, the dust-scattered intensity shows an excess of red. The color, expressed as the ratio of intensities at 377, 482 and 607 nm is interpreted in terms of Mie theory. It is suggested that the dust particles progressively differentiate. A proportionally more important population of small submicronic grains appears at p > 8000 km. This population seems to correlate with the jets.

1. Introduction

During the observation campaign of Halley's comet, two types of instruments were used to determine the physical parameters of cometary dust: impact particle counters and optical instruments. The particle counters and analysers on board the Vega and Giotto spacecrafts provided two main results; i) the mass spectrum of dust extends to much lower masses than previously expected (Mazets et al., 1987; McDonnell et al., 1987); ii) a large proportion of grains contains a relatively high content of light elements, C, H, O and N (Jessberger et al., 1988). The optical instruments of space missions consisted of video cameras (Keller et al., 1986; Sagdeev et al., 1986), a photopolarimeter (Levasseur-Regourd et al., 1986), an IR spectrometer (Combes et al., 1988) and a near-UV and visible spectrometer (Moreels et al., 1987). Earth and satellite-based optical instruments provided a large amount of observations in addition to the data gathered in space. Polarimetric measurements were used to precisely determine the complex index of Halley's dust (Mukai et al., 1987) and to compare cometary and interplanetary dust particles (Levasseur Regourd et al., 1990). Ground-based images in narrow-band spectral ranges, when processed with image-enhancement techniques, showed spiral structures identified as gaseous jets (A'Hearn et al., 1986; Hoban et al., 1989; Suzuki et al., 1990) .

Optical observations can be used to infer physical properties of dust grains, mainly their size. In the present study, three wavelengths : 377, 482 and 607 nm were chosen to investigate the spatial distribution and the size range of dust. The data consist of the spectra provided by Vega 2 three-channel spectrometer on March 9, 1986 during the approach session.

A.C. Levasseur-Regourd and H. Hasegawa (eds.), Origin and Evolution of Interplanetary Dust, 277–280.
© 1991 *Kluwer Academic Publishers.*

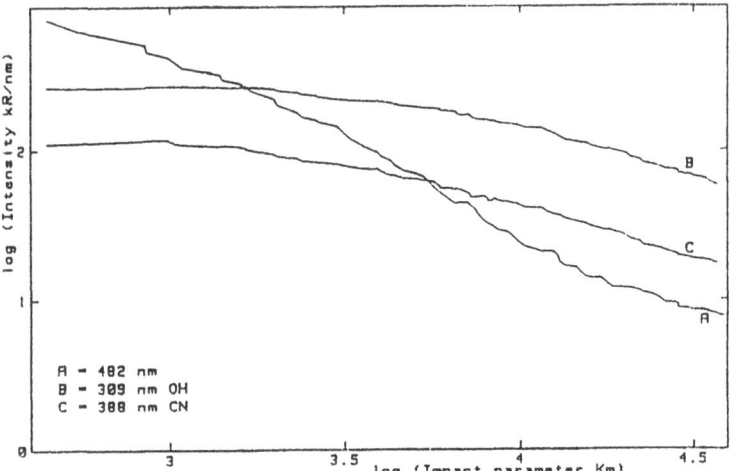

Fig.1 Radial distribution of dust-scattered intensity at 482 nm. The log-log plot shows that the slope is -1 for p < 3200km and p > 7000 km. The slope is -1.52 in the intermediate 3200-7000 km range. The OH and CN intensities, corrected of the solar-scattered continuum have also been plotted

2. Spatial distribution and color of dust in the inner coma

A cometary spectrum presents atomic and molecular features superposed on a solar-scattered continuum which is very intense close to the nucleus. The dust-scattered component was measured at three wavelengths in the near-UV, the visual and the red parts of the spectrum. The intensity at 482 nm is plotted as a function of the projected distance p in Fig.1 with a logarithmic scale. The graph has a slope equal to -1 when p < 3200 km or > 7000 km. In the intermediate 3200-7000 km range, the slope is equal to -1.52. The OH and CN intensities at 309 and 388 nm, corrected of the dust continuum component are also plotted in Fig.1. Both molecular emissions show a typical daughter-molecule radial decrease.

The three-channel spectrometer had a scanning capacity which permits to construct composite spectral charts at selected wavelengths. The explored region is a triangle with a summit located at the nucleus. The maximum radial distance is 40000 km. A composite image of dust intensity is presented in Fig.2. It may be compared with a similar image (Fig.3) obtained in the CN emission at 388 nm in a bandpass of 1.8 nm (Clairemidi et al., 1990). The spatial distributions show two differences; i) the dust intensity decreases as 1/p and ii) the jets appear comparatively more important in the CN emission. In fact, the jets are also present in dust-scattered intensity. A circular cut at p = 25000 km shows that the contrast at 482 nm between the vertical jet and the valley is equal to 3

The color of dust is represented by the intensity ratios I_{377} / I_{482} (Fig.4) and I_{607} / I_{482} (Fig.5). An excess of near-UV scattered light appears between 10000 and 30000 km mainly at the location of the jets. An excess of red intensity appears at distances p > 25000 km principally in the valley region. The color of dust-scattered radiation is a parameter which can provide useful information about the particle size when used in conjunction with Mie theory.

3. Mie theory calculations. Discussion

A Mie calculation of the scattering quantum efficiency Q_{sca} was performed following the scheme given by Eaton (1984). This calculation applies to spherical particles and does not take into account the actual shape of cometary particles which are highly porous (Greenberg and Hage, 1990). However, useful information about the size should be obtained from the knowledge of wavelength where scattering is maximum. Fig.5 shows the ratios of scattering efficiencies Qsca

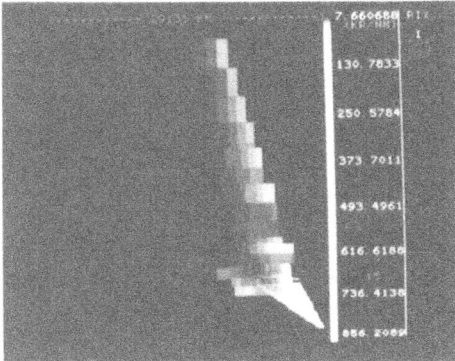

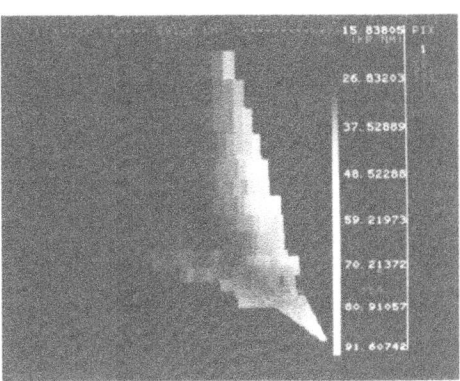

Fig.2 Intensity of dust continuum at 482 nm   Fig.3 Intensity of CN emission at 308 nm

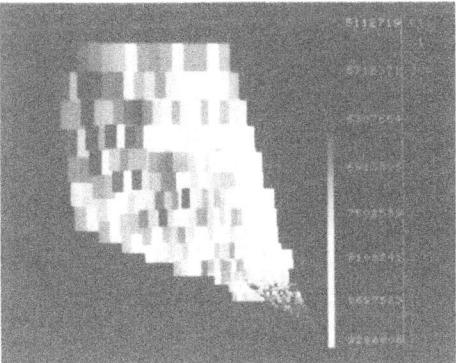

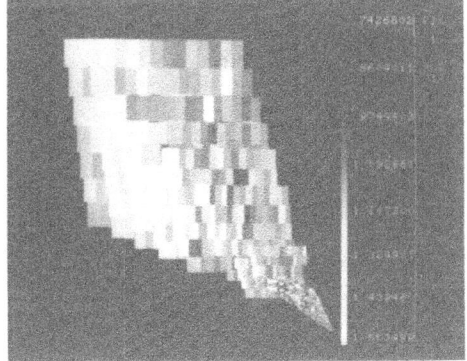

Fig.4 Intensity ratio I 377 / I 482   Fig.5 Intensity ratio I 607 / I 482

377 / 482 and Qsca 607 / 482. An excess of near-UV radiation is obtained with a particle radius a < 0.4 μm. An excess of red radiation is obtained with 0.4 μm < a < 0.8 μm. When the particle radius is > 0.8 μm, the color effect is less pronounced and disappears when a > 2 μm. An excess of color in the near-UV and visible range, according to Mie theory, shows that solar radiation is diffracted by tiny submicronic particles. These particles originate partly from the jets which probably constitute a diffuse source (Eberhardt et al., 1987) and partly from the fragmentation of particles of a larger diameter. The fragmentation process very close to the nucleus has been studied in detail by Thomas and Keller (1990). At larger distances, present results suggest that the fragmentation is a noticeable mechanism in the 3200-7000 km range. The fact that the dust scattered intensity varies as $p^{-1.52}$ in this region is in good agreement with the Didsy-Hope results of Giotto (Nappo et al., 1989) which mention a local r^{-s} density variation with s = 2.9 for Didsy and s = 2.6 for Hope.

4. Conclusion

The spatial distribution of dust measured by the solar-scattered continuum shows a p^{-s} decrease. The exponent s equals 1, except in a belt approximately located between 3200 and 7000 km from the nucleus where s = 1.52. The jets observed in the major molecular emissions are also present in the dust distribution. The color of continuum shows an excess of near-UV radiation between 8000 and 30000 km correlated with the jets. It shows an excess of red emission at distances higher than

280

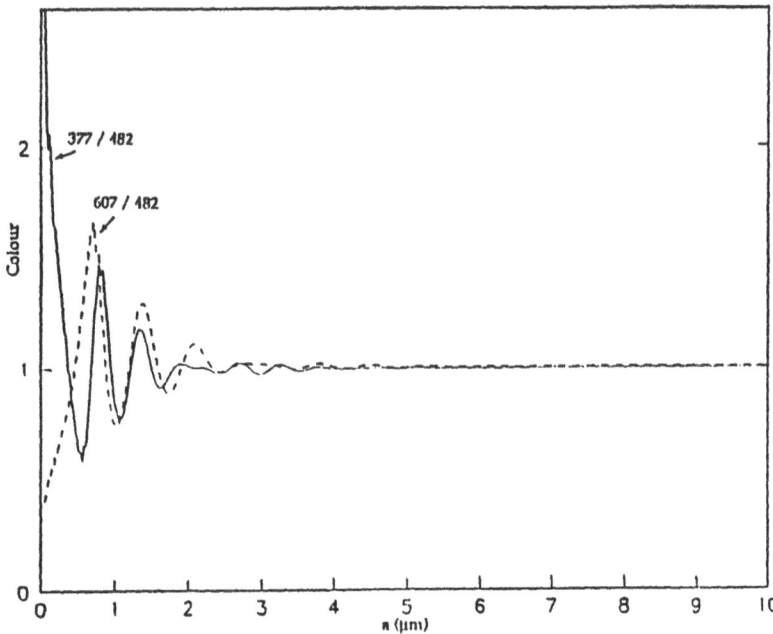

Fig.6 Ratios of scattering efficiencies Qsca 377 / 482 (and Qsca 607 / 482)

25000 km. A Mie theory calculation shows that an excess of near-UV is due to particles having the radius a < 0.4 μm and an excess of red is due to 0.4 < a < 0.8 μm. It is suggested that the submicronic particles responsible for this emission could be produced by fragmentation in the 3200 - 7000 km range and could also originate from the jets which, in this case, constitute a diffuse source.

References

A'Hearn et al. (1986). Nature 324, 649-651
Clairemidi J., Moreels G., Krasnopolsky V.A., (1990). Icarus, 86, 115-128
Combes et al. (1988). Icarus, 76, 404-436
Eaton N. (1984). Vistas in Astronomy 27, 111-129.
Eberhardt P. et al. (1987). Astron. Astrophys. 187, 481-484
Greenberg J.M., Hage J.I., (1990). Ap. J. 361, 1, 260-274
Hoban S., A'Hearn M.F., Birch P.V., Martin R. (1989). Icarus, 79, 145-158
Jessberger E. K. et al. (1988). Nature 332, 691-695
Keller H. U. et al., (1986). Nature 321, 320-326
Levasseur-Regourd A. C. et al. (1986). Nature 321, 341-344
Levasseur-Regourd A. C., Dumont R., Renard J.B., (1990). Icarus, 86, 264-272
Mazets et al. (1987). Astron. Astrophys. 187, 699-706
McDonnell J.A.M. et al. (1987). Astron. Astrophys. 187, 719-741
Moreels et al. (1987). Astron. Astrophys. 187, 551-559
Mukai T., Mukai S., Kikuchi S., (1987). Astron. Astrophys. 187, 650-652
Nappo S. et al. (1989). Asteroids, comets, meteors III, University Uppsala, 397-400
Thomas N., Keller H.U. (1990). Ann. Geophys. 8,2, 147-166
Sagdeev R. Z. et al. (1986). Nature 321, 262-266
Suzuki B., Kurihara H., Watanabe H.,(1990), Publ. Astr. Soc. Japan (Letters), submitted

PENETRATION OF HYPERVELOCITY PROJECTILES INTO LOW DENSITY MATERIALS

A. FUJIWARA, T. KADONO, and A. NAKAMURA
Department of Physics,
Kyoto University,
Kyoto 606 Japan

T. ISHIBASHI[+], N. FUJII
Department of Earth Science,
Kobe University,
Kobe 657 Japan

ABSTRACT.   Spherical nylon projectiles of 7mm diameter and up-to 4km/s velocities were penetrated into three types of targets; aluminum multisheet stacks, foamed polystyrene, and 1-atm air. Penetration depth, recovery rate of the projectiles were determined as a function of projectile velocity and target density, and a new type of dust collector is proposed.

1. INTRODUCTION

Observation of comet Halley by space probes brought many useful informations on the chemical abundances of the comet. However, determination of the molecular species of the constituent of the cometary particles will be possible after the sample of the cometary dust particles are intactly captured and returned back to the ground-based laboratory. An ISAS/NASA mission plan is proposed to capture the cometary dust particles intactly when the space vehicles fly through the comet coma(Veverka et al.,1989). In the mission, the relative velocity of dust particles and the space vehicle should be made as low as possible to avoid the damage of the dust particles during the capture. However, the planned velocity is still as high as 10-km/s. In order to develop a device to collect dust particles intactly, we are continuing the penetration experiments using projectiles launched by a two-stage light-gas gun at Kyoto University and various kind of low density materials as the catcher materials. Here we present the results of the experiments using (1) multi-sheet stack, (2) foamed polystyrene, and (3) 1-atm air.

+) Now at Japan DEC Computer Co.

A.C. Levasseur-Regourd and H. Hasegawa (eds.), Origin and Evolution of Interplanetary Dust, 281–284.
© 1991 *Kluwer Academic Publishers.*

2. ALUMINUM MULTI-SHEET STACK

Spherical nylon projectiles of 7mm diameter and 0.213g mass were launched into aluminum multi-sheet stacks. A number of aluminum sheets, each 20cm x 20cm were set with some spacings. In this system, the materials are distributed in a localized state, but as we make the sheet thickness thinner and sheet number larger, it gradually tends to the system equivalent with a target of uniformly distributed low density

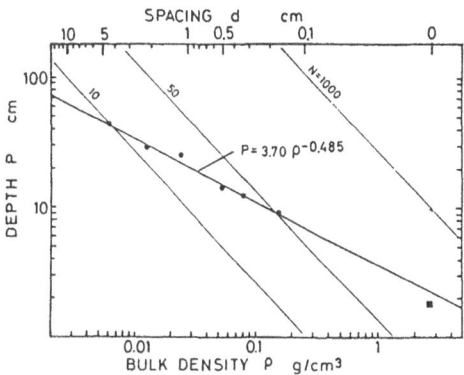 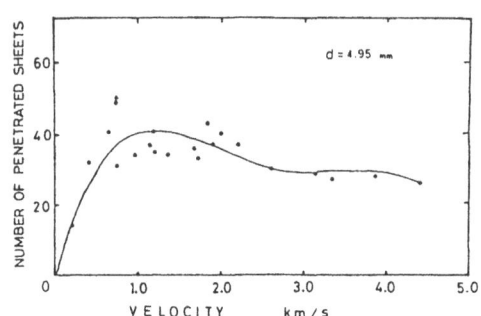

Figure 1. Penetration depth vs bulk density and spacing. velocity 3.85km/s. N is sheet number.

Figure 2. Number of penetrated sheets vs projectile velocity. spacing 4.95mm.

material. Two series of experiments were performed (Fujiwara and Kadono, 1990); one was carried out at a constant projectile velocity and varying the spacing of the sheets from 0cm to 4.4cm, and another was by varying the projectile velocity and keeping the spacing at a constant value of 5mm.

RESULTS. The penetration depth vs bulk density or spacing is shown in Fig.1, where the bulk density is defined as the total mass of the aluminum sheets divided by the total volume of the stack including the space between the sheets. The figure shows the penetration depth depends on the -0.485th power of the bulk density. Fechtig et al.(1980) report that the depth of the crater formed on the bulk target of normal density depends on the density as the power of -0.5. It is surprising that the target density dependence of the penetration depth on the target density both for the bulk target and the sheet stacks is essentially the same in spite of the apparent difference of the penetration mechanism and the definition of the density. The number of penetrated sheets at a constant spacing vs projectile velocity is shown in Fig.2. It has a broad peak around 1km/s. The recovered projectile mass normalized by the original mass as a function of the projectile velocity decreases monotonically to about 0.6 at the velocity of about 3km/s. At the velocity higher than 3.3km/s, no fragments of the projectile could be found.

3. PENETRATION INTO FOAMED POLYSTYRENE

The same projectiles were launched into foamed polystyrene block targets (Ishibashi et al., 1990). Commercially sold foamed polystyrene has a bulk density of $0.011g/cm^3$ (hereafter called L-type target), and it was uniaxially compressed to get other two kinds of densities; one is 0.033 to 0.040 g/cm^3 (M-type) and another 0.069 to $0.079g/cm^3$ (H-type).

RESULTS. The penetration depth is found to be expressed by $\rho^{-1.2}$ (where ρ is the target density). In Fig.3, the penetration depth is shown against the impact velocities. As in the case of the aluminum sheet stack, a broad peak is seen as found by Tsou et al.(1986), and it shifts toward higher velocities as the target density becomes lower. The

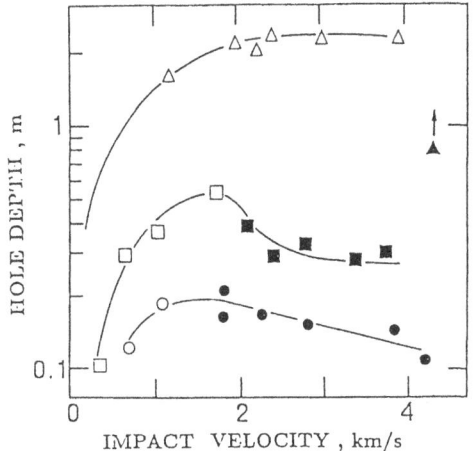

Figure 3. Penetration depth vs impact velocity for foamd poly-styrene. target density (g/cm^3): △ 0.011, □ 0.037, ○ 0.074. Filled marks denote the projectile is destroyed.

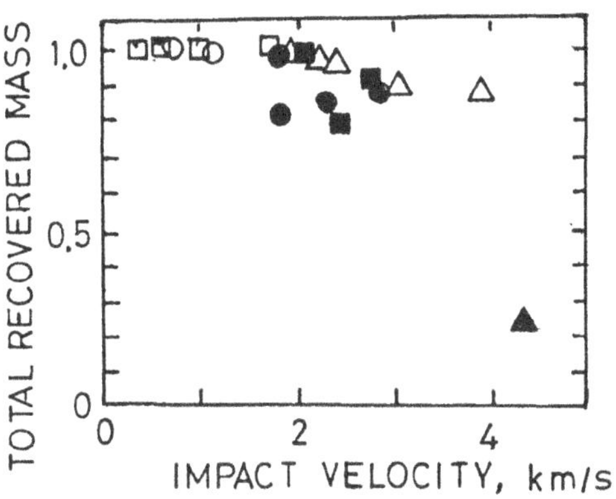

Figure 4. Recovered mass of the projectile in foamed polystyrene. Meaning of the marks is same as in Fig.3.

recovery rate is much better for L-type styrofoam(Fig.4). The projectiles were recovered without fragmentation up to 4km/s. By extrapolating the velocity at which the fragmentation of the projectile starts as a function of target densities, we found that, if we want to capture the (nylon) projectile of velocity about 10km/s without any fragmentation, we must prepare materials of density $0.001g/cm^3$. This density corresponds to that of 1atm air. Therefore, we further proceeded to make experiments using air targets.

4. PENETRATION INTO 1-ATM AIR

The projectiles were launched to test the deceleration efficiency in the open air. The projectiles thrust into the open air after perforating a very thin plastics film which separates the vacuumed chamber and the open air. During the flight in the air, the times when the projectile cut successively the very thin grids put at three points along the path were measured. By repeating these measurements with varying initial projectile velocities, we obtained velocity vs flight distance relation. Extrapolating this curve toward higher velocities, it is found that about 6 m is needed to decelerate the 7mm nylon projectile from 10km/s to 4km/s.

5. CONCLUSION

The system proposed for capturing the dust particles consists of two stages; the first is a gas cell, and the second is a soft material such as foamed polystyrene or something like that. The particle with velocity of 10km/s decelerates to about 4km/s in the first stage, and finally trapped in the second stage. The total length of the system will be in the practical range for the submillimeter dust particles instead of the 7mm-sized projectiles used in the present experiments.

REFERENCES

Fechig, H., K. Nagel, and N. Pailer (1980) 'Collisional Processes of Iron and Steel Projectiles on Targets of Different Densities', in I. Halliday and B. A. McIntosh (eds.), Solid Particles in the Solar System, D. Reidel Publishers, Boston, pp. 357-364.

Fujiwara, A., and T. Kadono (1990) 'Penetration of Hypervelocity Projectiles into Aluminum Multisheet Stacks', Jap. J. Appl. Phys. 29, 1620-1624.

Ishibash, T., A. Fujiwara, and N. Fujii (1990), 'Penetration of Hypervelocity Projectiles into Foamed Polystyrene', Jap. J. Appl. Phys. in press.

Tsou, P., D. E. Brownlee, and A. L. Albee (1986), 'Comet Coma Sample Return', ESA spec. Publ. ESA SP-250 3, 237-24.

Veverka, J., Y. Langevin, R. Farquhar, and M. Fulchignoni (1989) 'Spacecraft Exploration of Asteroids: The 1988 Perspective' in R. P. Binzel, T. Gehrels, and M. S. Matthews (eds.), Asteroids II, The University of Arizona Press, Tucson, pp. 970-993.

POLARIMETRIC PROPERTIES OF HALLEY'S DUST

A. K. Sen, M. R. Deshpande, and U. C. Joshi
Physical Research Laboratory,
Navrangpura, Ahmedabad 380 009, India

Abstract: Comet P/Halley was observed polarimetrically for seven nights in IHW and other continuum filters, during the pre and post perihelion passages. The polarimetric observations have been combined with the observations taken by other investigators, to get a complete picture of phase angle and wavelength dependence of polarization of comet P/Halley. Assuming Mie type scattering by cometary grains, the observed polarization data were fitted for a set of complex refractive indices which are (1.387, .032), (1.375, .040) and (1.374, .052) at .365, .484 and .684μm respectively.

1 Introduction: Linear polarization measurements of comet P/Halley in the continuum were made by several investigators during it s recent apparition (Bastien et al., 1986; Brooke et al., 1987; Dollfus and Suchail, 1987; Kikuchi et al., 1987; Lamy et al., 1987; Le Borgne et al., 1987b; Sen et al., 1988 etc). Polarization occurring due to the molecular fluorescence emission also has been studied by some investigators (Le Borgne et al.,1987a; Sen et al., 1989). The observed value of continuum polarization is a function of (1)incident wavelength (2)phase angle (3)shape and size of the cometary dust particles and (4)the composition of dust particles in terms of complex refractive index (n,k). The cometary grains are irregularly shaped and since the scattering computations for irregularly shaped grains are rather complicated, we make the simple assumption that the cometary particles are spherical in shape and proceed in this paper to study the dust properties of comet P/Halley using Mie theory.

2 Observations: Observations were made with the 1 m telescope of the V. B. Observatory, Kavalur, India on Dec 9 and 10, 1985 and Mar 17, 18 and 19, 1986, with entrance aperture diameter of 60 and 24 respectively, through the IHW continuum filters (centered at .365, .484 and .684 μm). On Mar 14 and 15 observations were taken through several other non IHW continuum filters (centered at .342, .442, .526, .575 and .641 μm, FWHM .005μm), with aperture 15". A photopolarimeter (Deshpande et al., 1985) was used at the Cassegrain focus, for polarimetric measurements.

3 Results and Discussions: In figure 1 and 2 the polarization observed through non-IHW and IHW filters are plotted respectively. The observed features are:

A.C. Levasseur-Regourd and H. Hasegawa (eds.), Origin and Evolution of Interplanetary Dust, 285–288.
© 1991 Kluwer Academic Publishers.

286

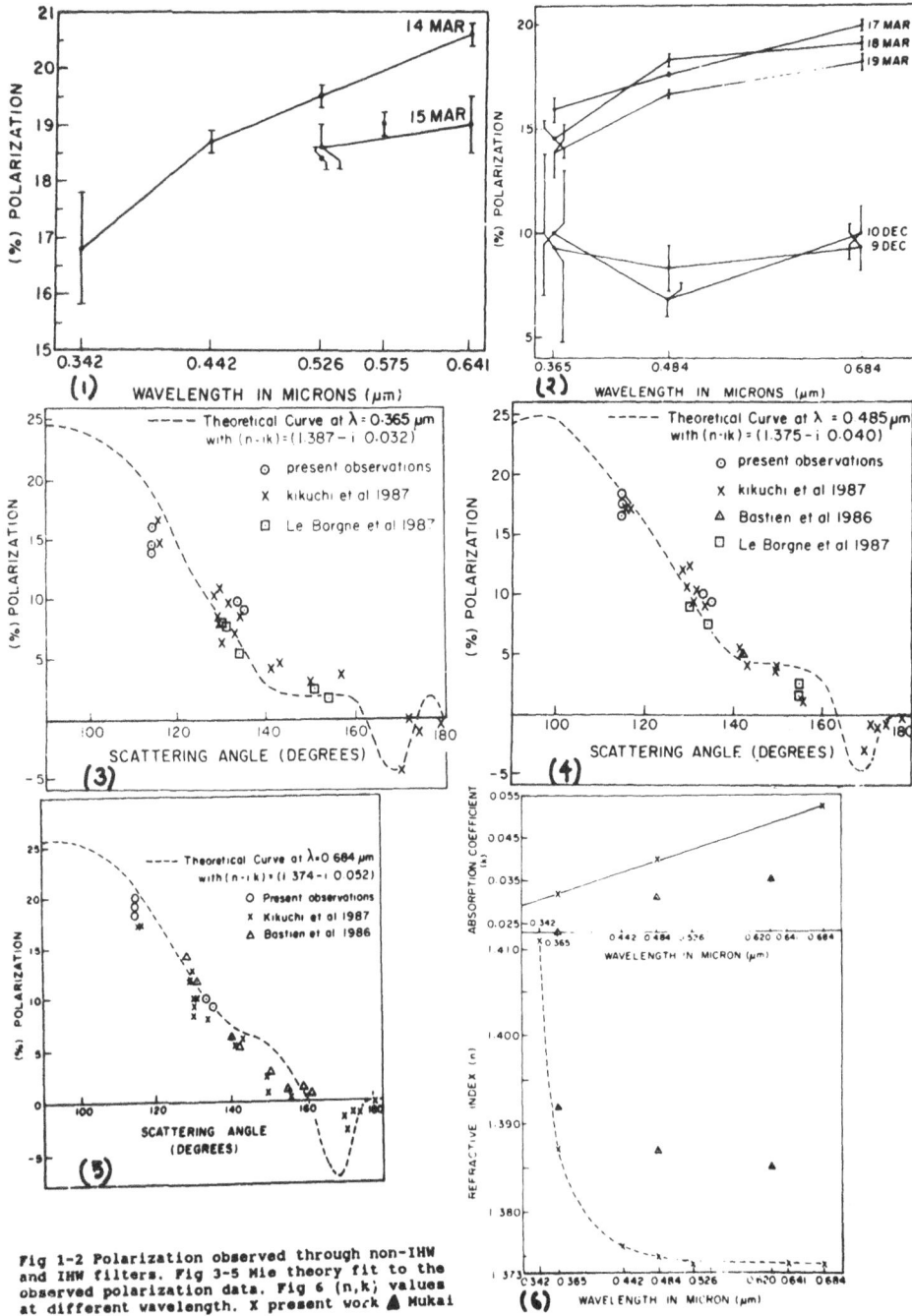

Fig 1-2 Polarization observed through non-IHW and IHW filters. Fig 3-5 Mie theory fit to the observed polarization data. Fig 6 (n,k) values at different wavelength. X present work ▲ Mukai et al (1987).

(1)At large phase angles polarization increases with wavelength, where at low phase angles no dependence is seen. As has been reported by other authors (Kikuchi et al., 1987; more general review by Dollfus et al., 1988) comet P/Halley's polarization showed a clear increase with the wavelength for higher phase angles, but no such dependence was seen for smaller phase angles. Brooke et al.(1987) confirmed this trend from their IR polarimetric observations also. (2)To see whether the polarization changes with the size of the entrance aperture, we have repeated the observation at .526μm with 24" on the night of Mar 15 (Pl. see fig 2). The polarization does not seem to change, within the errors, when we change the aperture from 15" to 24". Bastien et al.(1986) have found by changing the aperture from 4" to 18", a general trend for the polarization to decrease as the aperture increases (barring some exceptions). Kikuchi et al.(1987) changed the aperture from 13" to 33", and found no systematic dependence of polarization on aperture size.

4 Cometary grain properties: In order to study the cometary grain properties, we combined the polarization values reported by other authors, at wavelength .365, .484 and .684 μm and plotted them in figures 3,4, and 5 respectively. The dust mass detectors on board Vega and Giotto spacecraft, have found the dust mass distribution functions of comet P/Halley. Assuming that the grains are spherical in size (with density $\rho \sim 1$ gm/c.c) the following dust size distribution functions are those obtained by Mukai et al. (1987) from the in situ measurements (Mazets et al., 1987)

$$n(s) \sim s^{-\alpha} \quad \text{where } \alpha = 2, 2.75, \text{ and } 3.4 \qquad (1)$$

for s < 0.62; 0.62 < s < 6.2; and 6.2 < s μm respectively, where s = radius of the grain in μm and n(s)= number of grains with radius s. From Krishna Swamy and Shah (1988) we adopt the upper and lower limit of particle size to be .001 and 20 μm, meaningful for optical polarimetric work. Having the grain size distribution fixed, we explored a wide range of (n,k) values to calculate the expected values of polarization using Mie theory, which will match with the observed values at different phase angles. For this we have employed the method of least square and minimized the sum of square of deviations between observed and expected polarization values. We introduce a quantity σ^2, equal to the above sum divided by the number of data points. Thus at σ_{min} we get the following best fit values of (n,k).

$$(n,k) \quad \text{is} \quad (1.387, .032) \text{ at } .365 \text{ μm with } \sigma = 2.9$$
$$(1.375, .040) \text{ at } .484 \text{ μm with } \sigma = 1.6$$
$$(1.374, .052) \text{ at } .684 \text{ μm with } \sigma = 2.4.$$

As can be seen from figure 6, the k values can be fitted into a straight line 'k = 0.062 λ + 0.009' by the method of least square. The dependence of 'n' on λ seems to be nonlinear and does come out clearly only from three data points. However, from the k-λ straight line, we interpolate the k values at the wavelengths .342, .442, .526 and .641μm and keeping these k values fixed, we find out the n values which can generate the polarization values as close as possible to the observed polarization values. As a result we get more data points along the n-λ curve (in figure 6), which now help us to see the n-λ dependence in more detail. For a comparison we have also plotted n and k values obtained by Mukai et al. (1987) in a similar work. There seems to a discrepancy between the two sets of (n,k) values. It is not clear whether Mukai et al. (1987) also followed the method of least square for the selection of (n,k) values, otherwise the discrepancy between the two sets of (n,k) values may be resulted. It should also be noted that Mukai et al. (1987) have fitted the theoretical polarization curve, on the polarization values which they have observed. But in the present case we have tried to include all the polarization values available in the literature till now. As a result with more number of data points, we expect to get more accurate values of (n,k) which can fit to the observed data.

Brooke et al. (1987) have found that a two-component grain model explains better their IR polarization data. By extrapolating the (n,k) values towards the IR wavelength side (using figure 6) we found that, in a single component grain model, it fails to explain the IR polarization. While doing these calculations one should keep in mind that the dust properties of comet may change with heliocentric distance and also they may be different for different parts of the comet (Sen et al., 1990). Further it is most probable that the grains have a density somewhat smaller than 1 gm/cc and assuming a smaller value say, ρ=.7 gm/cc, we find that the size limits .62 and 6.2 in relation (1) only change to .7 and 7. Such a change in grain size limits does not appreciably change the best fit values of (n,k). Also we know that the grains are far from being spherical, but at present any treatment on nonspherical grains is beyond the scope of this paper. Assuming that the nonspherical grains don't produce polarizations significantly different from that produced by the spherical ones, we conclude that, the dust distribution obtained by Mazets et al. (1987) explains the observed polarization of comet P/Halley for particular kind of grains discussed above.

References

Bastien, P., Menard, F. and Nadeau, R. 1986, Mon. Not. R. Astron. Soc., **223**, 827

Brooke, T. Y., Knacke, R. F. and Joyce, R. R. 1987, Astron. Astrophys. **187**, 621

Deshpande, M. R., Joshi, U. C., Kulshra, A. K., Banshidhar., Vadher, N. M., Mazumdar, H. S., Pradhan, S. N., Shah, C. R. 1985, Bull. Astron. Soc. India, **13**, 157.

Dollfus, A., Bastien P., Le Borgne, J.-F., Levasseur-Regourd, A. C., Mukai, T. 1988, Astron. Astrophys. **206**, 348.

Dollfus, A., and Suchail, J. L. 1987, Astron. Astrophys. **187**, 669

Kikuchi, S., Mikami, Y., Mukai, T., Mukai, S., Hough, J. H. 1987, Astron. Astrophys. **187**, 689

Krishna Swamy, K. S. and Shah, G. A. 1988, Mon. Not. R. Astron. Soc. **233**, 573

Lamy, P. L., Grun, E., Perrin, J. M. 1987, Astron. Astrophys. **187**, 767

Le Borgne, J. F., Leroy, J. L., Arnaud, J. 1987a, Astron. Astrophys. **173**, 180

Le Borgne, J. F., Leroy, J. L., Arnaud, J. 1987b, Astron. Astrophys. **187**, 526

Mazets, E. P., Sagdeev, R. Z., Aptekar, R. L., Golenetskii, S. V., Guryan, Yu. A., Dyachkov, A. V., Ilyinskii, V. N., Panov, V. N., Petrov, G. G., Savvin, A. V., Sokolov, I. A., Frederiks, D. D., Khavenson, N. G., Shapiro, V. D., Shevchenko, V. I. 1987, Astron. Astrophys. **187**, 699

Mukai, T., Mukai, S. and Kikuchi S. 1987, Astron. Astrophys. **187**, 650

Sen, A. K., Joshi, U. C., Deshpande, M. R., Kulshretha, A. K., Babu, G. S. D., Shylaja, B. S. 1988, Astron. Astrophys. **204**, 317

Sen, A. K., Joshi, U. C., Deshpande, M. R. 1989, Astron. Astrophys. **217**, 307

Sen, A. K., Joshi, U. C., Deshpande, M. R., Debi Prasad, C. 1990, ICARUS **86**, 248

V

METEOROIDS AND METEOR STREAMS

THE ORBITAL DISTRIBUTION AND ORIGIN OF METEOROIDS

DUNCAN STEEL
Department of Physics and Mathematical Physics
The University of Adelaide
G.P.O. Box 498, Adelaide, SA 5001
Australia

ABSTRACT. Approximately 68,000 orbits of meteoroids, ranging from sizes of 10 cm and more down to microgram masses, are now available through the IAU Meteor Data Center. These orbits were measured in surveys based in the U.S.S.R., the U.S.A., Canada, Somalia, and Australia, using photographic, radar and television techniques; the data represent our best knowledge of the orbital distributions of smaller solid bodies in the solar system. It is found that quite different distributions result in different mass regimes, with implications for the origin and evolution of these particles: for example the larger bodies, observed as fireballs, are associated with meteorites in coming from the region of the asteroid belt with low-inclination orbits, whereas the smaller meteoroids have more comet-like orbits. There is also evidence for several meteoroid streams associated with specific Apollo asteroids. The data may additionally be viewed as a suitable source function in investigations of the production of interplanetary dust from the fragmentation of larger meteoroids in mutual collisions. However, inspection of the data raises many questions: for instance there seem to be many meteoroids on small retrograde paths, but no possible parent objects are known to exist on such orbits.

1. Introduction

Over the past decades, many different observational programs have been set up in which meteor orbits have been determined using a variety of techniques: starting in the 1930's using cameras with rotating shutters initially, and later, starting in the late 1940's but not reaching real fruition until the 1960's, using radar techniques which allow 24 hour coverage independent of weather conditions. In more recent years TV apparatus and other electronic detectors have been used (Jones and Sarma, 1985; Hawkes and Jones, 1986, and papers cited therein), and modern computing techniques have allowed the development of more sophisticated radars (Steel and Baggaley, 1985). Whilst some lists of individual orbits have been published (*e.g.* Jones and Sarma, 1985), in some cases not all of the orbital elements have been given (*e.g.* Lebedinets *et al*, 1981, 1982) and in other cases only the results of stream searches have been published (*e.g.* Sekanina, 1976). In any case, for some surveys many thousands of orbits have been determined so that

291

A.C. Levasseur-Regourd and H. Hasegawa (eds.), Origin and Evolution of Interplanetary Dust, 291–298.
© 1991 *Kluwer Academic Publishers.*

the data are required by other researchers in a machine-readable form rather than on paper, if later independent analysis is to be feasible.

In order that this data resource might be fully exploited, a Meteor Data Center was set up at the Lund Observatory in Sweden with sponsorship from the IAU, under the direction of B.-A. Lindblad. All available meteor orbit data have been collected together, converted into a common format, checked for consistency of the elements and the given velocities and radiants, and the missing elements (such as the perihelion distance, the argument of perihelion, or the nodal longitude) have been filled in for each orbit; this task was largely carried out by the author whilst on an ESA Fellowship in Lund. The full data set, which is still being added to as other data (such as fireball orbits from the European Network) become available, may be obtained by any interested researcher. The presently-available data are as follows, a brief description also having been given by Lindblad (1987):

Survey name	Country	Number
Harvard graphical photographic meteors	USA	2529
Harvard precise photographic meteors	USA	1245
Prairie Network fireballs	USA	336
Meteorite Orbit and Recovery Program fireballs	Canada	218
Soviet photographic meteors	USSR	1111
Television meteors	Canada	531
Harvard radar meteors 1961-65	USA	19327
Harvard radar meteors 1968-69	USA	19818
Obninsk radar meteors	USSR	9358
Mogadisho (Somalia) radar meteors	USSR	5328
Kharkov radar meteors	USSR	5317
Adelaide radar meteors 1960-61	Australia	2092
Adelaide radar meteors 1968-69	Australia	1667

Since some Harvard photographic meteors were reduced both precisely and graphically, the total number of independent meteor orbits available from the above is about 68,000. The data are of variable quality, and in checking the orbits it was found that in some surveys 10% or more were clearly in error; for example one particular element not fitting with the others, but doing so if a typographical error in one digit were corrected. Additionally, the different surveys pertain to rather differently sized meteoroids: the photographic surveys detect mostly the meteors from particles which were originally of mass 1 mg and larger (sizes rather greater than 1 mm) whilst the radar surveys (plus the TV meteors) are dominated by meteoroids smaller than this; for the Harvard and the Kharkov radar surveys, which had the lowest limiting magnitudes, masses down to micrograms.

2. Usefulness of these data

Several points can be made concerning the distribution of meteoroids in the inner solar system, and what such information can tell us about their origin and evolution. The data can be used for specific searches for streams, and also their association with certain parent objects. For example, Olsson-Steel (1988) used the Adelaide orbits to show that there are several Apollo-type asteroids which apparently have spawned meteoroid streams, so that it is not purely comets which generate meteoroids (unless the asteroids are defunct cometary cores). In addition, one of the major sources of the zodiacal dust is the catastrophic destruction of meteoroids (sizes 100 μm - 10 cm) in impacts with smaller particles (mainly 10 μm - 100 μm) so as to re-generate this cloud as it decays under other effects (*e.g.* Grun *et al*, 1985). In order to attack this problem, the regions where dust is generated needs to be known, along with the collisions velocities and other parameters. The meteoroid data described herein provide a suitable source function for such investigations since the spatial densities and thus the collision frequencies can be calculated (*e.g.* Kneissel and Giese, 1987).

3. New meteor orbits

Whilst there is a wealth of meteor orbit data available, it is also necessary to collect new orbits for a number of reasons, for instance searching for newly-appeared showers, or checking on the variation of stream parameters as they evolve. There are several amateur groups world-wide which are engaged in photographic observations which lead to excellent meteor orbits, notably in northern Europe and in Japan (Lindblad, this volume).

In addition it is notable that comparatively few southern hemisphere orbits have been collected; the Adelaide radar surveys, rendering less than 4,000 orbits in total, have been the only programs based in the southern hemisphere, although the Soviet expedition to Mogadisho also resulted in many southern radiants being measured. It is thus noteworthy that a new radar survey has begun in New Zealand under the direction of W.J.Baggaley. The equipment and techniques were described by Steel and Baggaley (1985). Routine data collection began in November 1989, with two days data collection occurring every month. Since approximately 2,000 orbits are determined in each 24-hour run, this survey has already resulted in more individual meteor orbits than the Harvard radar program listed above.

4. Example of meteor orbit data

As an example of the sorts of data which are available, the Obninsk radar survey (Lebedinets *et al*, 1981, 1982) is chosen here. This survey had a limiting radar magnitude of +7.5. The reasons for this choice are: (i) There is a large number of individual orbits available; (ii) The meteoroid sizes here are approximately 1 mm, this being about the mid-point between the larger photographic meteors and the smaller particles detected in the Harvard and Kharkov surveys; (iii) There is clear evidence of several streams in the data; and (iv) The data have been found to be of very high quality, with comparatively few errors. The Adelaide orbits are of similar quality, and some element distribution plots have been presented by Olsson-Steel (1988). Here a more extensive set will be given for the Obninsk data, similar sets for all other surveys listed above being given by Steel and Lindblad (1991).

In Figure 1 we show the distributions of the orbital elements a, e, q, Q, i and ω, where the symbols have their usual meanings. The number on each plot

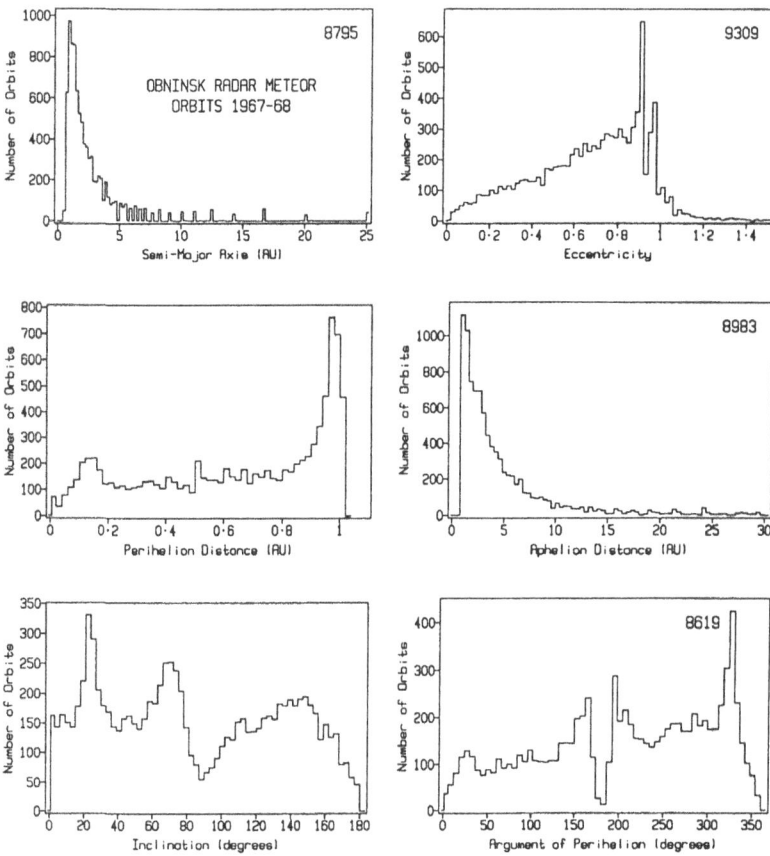

Figure 1. Distributions of orbital elements a, e, q, Q, i and ω from the Obninsk radar meteor survey (Lebedinets *et al*, 1981, 1982).

indicates the total number of meteors contributing; for example for a, of the 9358 orbits in the entire survey only 8795 had $a \leq 25$ AU. For ω it was necessary to calculate the value from the other elements, and this was not possible in some cases. The plot for a clearly shows discrete larger values; this is because the semi-major axes were stored as the orbital energy $(1/a)$ and this was only given to two decimal places. We note that the a and Q plots indicate that these meteoroids are in quite small orbits, with aphelia well within Jupiter: these are not, in this sense, by any means 'cometary'. However, they are largely on high-eccentricity orbits: the plot for e shows a general increase, with structure at $e > 0.8$ being due to the presence of various streams. In particular the Geminids have $e \cong 0.9$, and these are further evidenced by the bump at $q \cong 0.15$ AU and at $i \cong 23°$. We note also the apparent existence of some hyperbolic orbits, the reality of which has been a matter of some debate in the past (see *e.g.* Jones and Sarma, 1985). Unlike many other surveys, but similarly to the Harvard radar data, the inclination distribution does not show a preponderance of low-i orbits, rather many high-i and retrograde orbits being seen. This may be due to various selection effects which depend upon the velocities of the meteors, and the radiants compared to the antenna patterns. Thus there are few orbits with $i \cong 90°$, since these would all have very northerly radiants and would additionally need to have $\omega \cong 0°$, and these are not seen in the ω plot; similarly there are few orbits with $\omega \cong 180°$ since these would tend to emanate from deep southerly radiants. Another selection effect is due to the collision probability with the Earth: although the q-plot shows a strong peak at $q \cong 1$ AU this is due to such orbits having much higher collision probabilities, so that this distribution does not represent the true distribution in space.

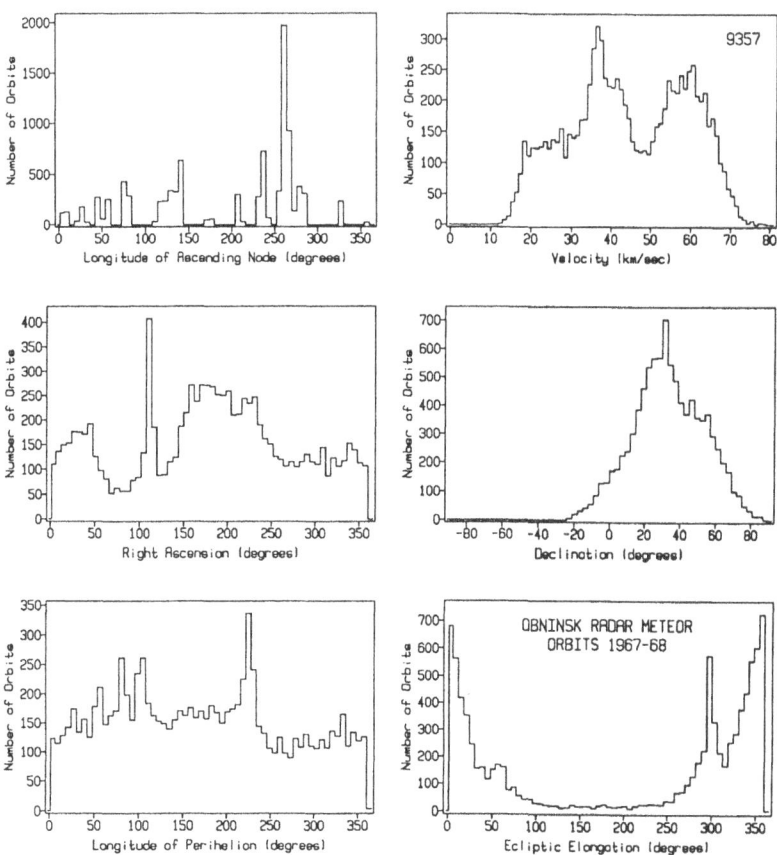

Figure 2. Distributions of various parameters of the Obninsk radar meteors.

In Figure 2 we show distributions for various other parameters of interest. The longitude of the ascending node (Ω) plot is very patchy, this being due to the distinct times of year in which observations were carried out and the number of meteors detected at each time; as pointed out below, these struck the Earth at their descending nodes. The peak near $\Omega = 265°$ is due to the Geminids, this shower also being apparent in the other plots. The velocity distribution shows twin peaks, as seen in other surveys, but here there are comparatively few low-velocity meteors: the peak at V ≅ 60 km/sec is usually much less prominent. If there were no showers, and if the data collection were continuous and not patchy, then the RA distribution would be expected to be flat, and this is clearly not the case. As expected the Declination plot peaks near the zenithal value for this observation site. The Longitude of Perihelion is a useful parameter since it conveniently indicates the presence of showers amongst sporadic data, as here. Finally the plot of the Ecliptic Elongation (ε) of the radiant is similar to that seen in other surveys, with peaks being evidenced at $\varepsilon \cong 0°$ (Apex source), 60° (Helion source) and 300° (Anti-helion source), although the second of these is quite muted.

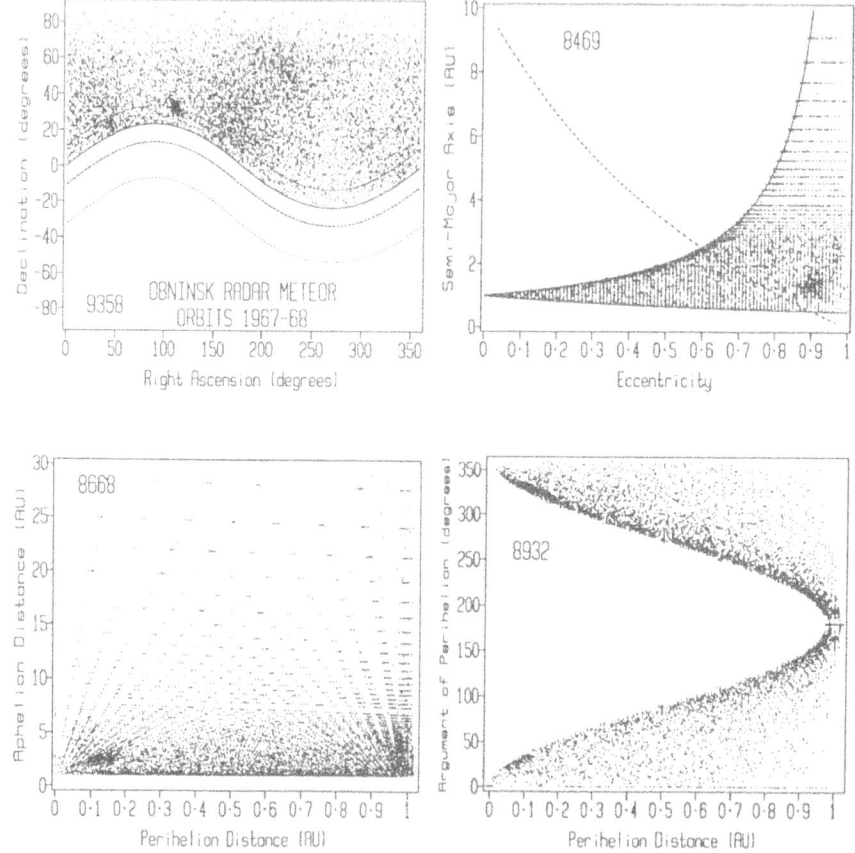

Figure 3. Radiant and orbital element distributions for the Obninsk data.

Figures 1 and 2 were just number distribution plots. We now present plots in which two parameters are plotted against each other in order to see what sort of information may be gained in this way. In Figure 3 we plot Declination versus RA, *a* versus *e*, *Q* versus *q*, and ω versus *q*. The Obninsk data which were published contain only radiants north of the ecliptic (with one exception, given in error) so that the first of these plots has no radiants from the south; this also means that these meteors were all observed at their descending node. There are several concentrations in this plot, due to showers, and in particular the Geminids stand out. For the *a* versus *e* plot three curved lines are shown. The two solid lines delineate the region for which Earth-intercept is possible (*i.e.* $q < 1.0167$ AU and $Q > 0.9833$ AU); some orbits lie just beyond this region due to round-off errors. The dashed line shows Whipple's K-Criterion, orbits lying above it being classed as being 'cometary', those below it being 'asteroidal'. There is a more-or-less equal split, this being a useful indicator of the origin of these bodies. The Geminids are seen at ($a \cong 1.3$ AU, $e \cong 0.9$) in this plot, and in the *Q* versus *q* plot at ($q \cong 0.15$AU, $Q \cong 2.5$ AU). In that plot it is notable that many of the orbits with $Q < 5$ AU have $q > 0.9$ AU. In the final plot of Figure 3 the form of the scatter is due to these orbits being observed at the descending node only; for the ascending node a chevron opposing that plotted would be formed.

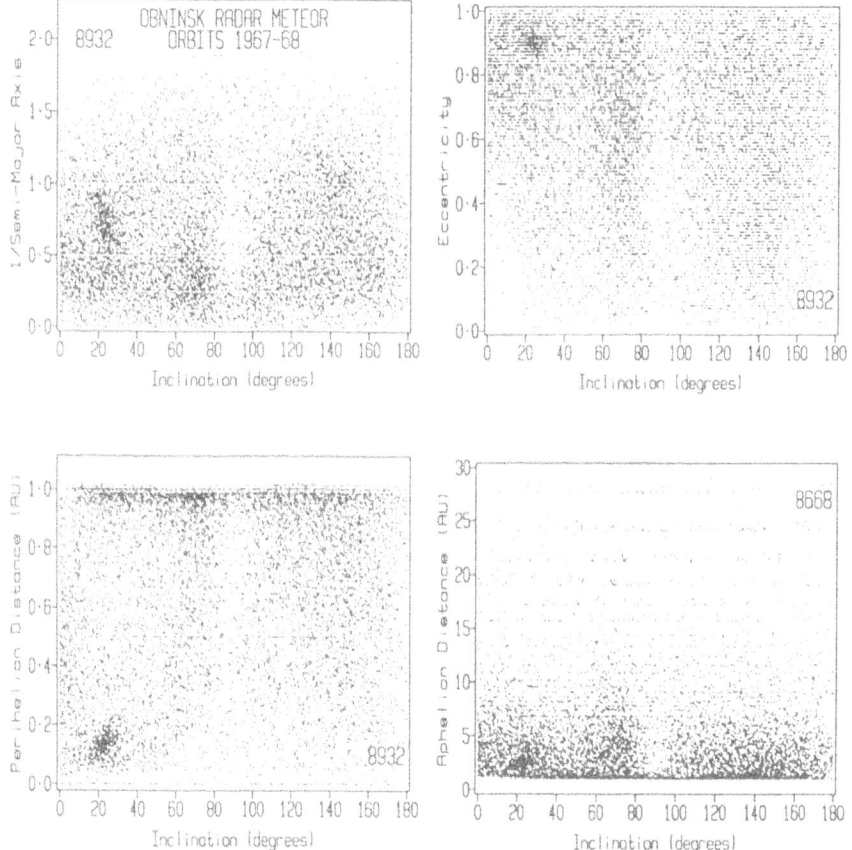

Figure 4. $(1/a)$, e, q and Q plotted against i for the Obninsk radar orbits.

In Figure 4 we plot $(1/a)$, e, q and Q in turn against the inclination, i. The Geminids in particular are prominent in each plot. In the $(1/a)$ plot we note that there are many meteoroids in retrograde Aten-type $(a < 1$ AU$)$ orbits, for which no parent objects are known: these seem to be spread over all value of q and e. Similar distributions are seen in other surveys so that the existence of such meteoroid orbits is certain. The question which then needs to be asked is: Were these meteoroids released onto such orbits by some as-yet undiscovered class of parent, or did they evolve onto these orbits under the influence of planetary perturbations/radiative effects/other influences? This dissimilarity to the known parent objects in the inner solar system is emphasized by the Q versus i plot, which shows no preponderance of retrograde meteors with large aphelia (*i.e.* these orbits are *not* like long-period comets, or P/Halley, P/Swift-Tuttle and the other known retrograde intermediate- or short-period comets).

Acknowledgements: This work was supported by the Australian Research Council and The University of Adelaide.

References

Grun, E., Zook, H.A., Fechtig, H. and Giese, R.H. (1985). 'Collisional balance of the meteoroitic complex', *Icarus*, **62**, 244-272.

Jones, J. and Sarma, T. (1985). 'Double-station observations of 454 TV meteors. II. Orbits', *Bull. Astron. Inst. Czechoslov.*, **36**, 103-115.

Hawkes, R.L. and Jones, J. (1988). 'Electro-optical meteor observation techniques and results', Q. Jl. Roy. Astron. Soc., **27**, 569-589.

Kneissel, B. and Giese, R.H. (1987). 'The dynamics of the zodiacal dust cloud on account of optical and infrared observations', *Publ. Astron. Inst. Czechoslov. Acad. Sci.*, **No. 67, vol. 2**, 241-243.

Lebedinets, V.N., Korpusov, V.N. and Manokhina, A.V. (1981, 1982), 'Radio Meteor Investigations in Obninsk: Catalogue of Orbits', Materials of the World Data Center B, Moscow, U.S.S.R.

Lindblad, B.-A. (1987). 'The IAU Meteor Data Centre in Lund', *Publ. Astron. Inst. Czechoslov. Acad. Sci.*, **No. 67, vol. 2**, 201-204.

Olsson-Steel, D. (1988). 'Identification of meteoroid streams associated with Apollo asteroids in the Adelaide radar orbit surveys', *Icarus*, **75**, 64-96.

Sekanina, Z. (1976). 'Statistical model of meteor streams. IV. A study of radio streams from the synoptic year', *Icarus*, **27**, 265-321.

Steel, D.I. and Baggaley, W.J. (1985). 'Meteoroid orbits determined by southern hemisphere radar', in *Properties and Interactions of Interplanetary Dust, IAU Colloquium No. 85* (eds. R.H.Giese and Ph.Lamy), D.Reidel, Dordrecht, Holland, 299-303.

Steel, D.I. and Lindblad, B.-A. (1991). 'The meteor orbit database', *Space Science Rev.* (submitted).

A STUDY OF METEOR ORBITS OBTAINED IN JAPAN

B.A. Lindblad
Lund Observatory
Box 43, S-221 00 Lund
Sweden

ABSTRACT. A list of 325 two-station photographic meteor orbits obtained with 35 mm cameras has recently been published by Japanese amateur groups. The present study analyzes the data and concludes that the orbits are of high quality and very useful for scientific purposes.

Introduction

The first double station observations of meteors in Japan were made at the Tokyo Observatory in the 1950:s (Hirose, Kako and Tomita 1950, Hirose and Tomita 1950, 1955). In the 1960:s several amateur groups started to make double station observations of meteors using equatorially mounted 35 mm cameras equipped with rotating shutters. Exact timing of the meteor was accomplished by visual observers. The resulting orbits were originally published in various reports of the Nippon Meteor Society (Ochai 1984, Ochai, Ohtsuka and Sekiguchi 1989). A summary list containing 325 radiants and orbits, in the following denoted NMS data, has recently appeared in the scientific literature (Koseki 1990, Koseki, Sekiguchi and Ohtsuka 1990). The author become aware of these observations a few years ago and has used the orbital data in two recent publications (Lindblad 1990a, b). The purpose of the present paper is to search the NMS data set for meteor streams and to compare the mean stream orbits with those of other investigators. Various data checks are also applied to the NMS data sample.

Meteor Stream Search

In the present study a computer program based on the D-criterion of Southworth and Hawkins (1965) is used for identifying meteor streams. The computer program searches for similar orbits and groups them into streams at a specified level of orbital similarity D_s. The appropriate level D_s to be used in a computer search depends on sample size (Lindblad 1971a, b). For a sample of 325 orbits the appropriate rejection level is $D_s = 0.20$. The computer search at this level identified 14 streams with two or more members in the NMS sample; comprising in all 269 stream members. Streams with only one member in the sample are not recognized by the program. It follows that at least 83% of the listed NMS meteors are in streams. The results of the stream search (which entirely confirmed the shower classifications previously made by Koseki et al.) are

A.C. Levasseur-Regourd and H. Hasegawa (eds.), Origin and Evolution of Interplanetary Dust, 299–302.
© 1991 *Kluwer Academic Publishers.*

Table 1. Mean orbital elements of meteor streams (35 mm cameras)

Name	N	α	δ	V_G	q	a	e	i	ω	Ω
α Capric.	4	310	-23	19	0.727	2.682	0.729	2.8	70.8	317.4
N χ Orion.	3	87	24	30	0.425	2.492	0.830	0.1	284.7	259.7
S χ Orion.	3	79	18	26	0.572	2.647	0.785	4.1	87.8	80.1
Taurids	3	51	13	32	0.343	2.327	0.852	6.5	115.0	39.9
Geminids	100	113	33	35	0.149	1.246	0.878	23.2	324.8	261.3
κ Cygnids	4	285	42	23	0.955	3.378	0.717	29.1	207.9	139.7
Monoc.	5	102	7	43	0.202	11.223	0.981	35.3	127.4	80.8
Quadr.	24	231	49	43	0.983	2.964	0.668	71.8	169.0	282.5
Lyrids	2	272	34	47	0.913	13.934	0.935	78.8	215.9	31.3
Perseids	79	46	57	59	0.940	7.770	0.877	113.0	148.6	139.2
Hyp. Pers.	17	-	-	-	0.955	12.484	1.224	115.4	154.5	139.0
Orionids	6	95	15	65	0.556	5.099	0.892	163.2	85.6	27.8
Eta Aq.	13	335	-2	65	0.590	8.354	0.933	163.9	95.1	44.0
Leonids	3	154	21	72	0.985	19.531	0.950	163.5	170.7	235.6

Note. Mean semi major axis is calculated as a = $1/\overline{1/a}$, other mean elements are arithmetic means.

Table 2. Mean orbital elements of meteor streams in search in IAU data

Name	N	α	δ	V_G	q	a	e	i	ω	Ω
α Capric.	18	307	-9	25	0.576	2.386	0.758	7.3	270.2	126.9
N χ Orion.	4	103	26	29	0.463	1.995	0.768	4.1	282.7	274.2
S χ Orion.	-									
Taurids	140	47	15	31	0.370	2.048	0.821	3.1	112.3	37.8
Geminids	84	111	32	37	0.140	1.389	0.899	23.5	324.3	260.2
κ Cygnids	7	289	55	27	0.985	4.257	0.769	38.2	199.8	147.7
Monoc.	2	102	8	44	0.183	65.789	0.998	37.1	129.1	80.3
Quadr.	14	230	49	43	0.978	3.028	0.677	71.9	171.1	282.4
Lyrids	5	273	34	48	0.920	21.552	0.957	79.4	214.4	31.9
Perseids	182	45	58	60	0.947	25.641	0.964	113.1	150.2	138.3
Hyp. Pers.	4	43	57	68	0.971	-	1.690	117.6	159.1	137.8
Orionids	27	94	16	68	0.575	11.487	0.951	164.3	82.7	28.2
Eta Aq.	-									
Leonids	7	152	22	72	0.985	14.265	0.931	162.3	172.1	234.5

summarized in Table 1. Table 1 lists the numbers of members in each stream, mean values of radiant, geocentric and heliocentric velocity and orbital elements (referred to the mean equinox of 1950). For comparison the corresponding mean values obtained by the author in a computer search in a sample of 1827 precisely reduced photographic orbits from various professional programs are shown in Table 2 (Lindblad 1971c, 1987). Only streams represented in the NMS data are listed in Table 2.

Since the NMS sample included many long-period meteor streams the mean semi major axis is computed in Tables 1 and 2 as a = $1/\overline{1/a}$. Mean radiant and geocentric velocity are computed by the program. All other orbital elements are arithmetic means. Inspection of the data shows that there is good agreement between the two sets. With the exception of the streams identified by the author as α Capricornids and κ Cygnids (which are very complex and the identification of which are doubtful), the differences in radiant position are of the order of 1 degree. Agreement in the orbital elements is also very good. This shows that the NMS data set is very useful for statistical studies. It is interesting to note that several minor meteor streams (S χ Orionids, Monocerotids and Eta Aquarids) are better represented in the Japanese amateur sample than in the much larger sample obtained by professional astronomers! The main difference between the two data sets is that the professional data in general show slightly higher values of velocity V_G and semi-major axis (Table 3). The discrepancy is most likely caused by a neglect of the correction for atmospheric deceleration in most of the NMS data.

Table 3. Derived values of semi-major axis

Stream	a_{NMS} (a.u.)	a_{1827} (a.u.)
Geminids	1.248	1.389
Perseids	7.770	25.641
Quadrantids	2.964	3.028

Checks of Consistency

Various routine checks of the consistency of the orbital elements of individual orbits have been made. These checks are based on the following equations:

$$D1 = q - a(1 - e) \tag{1}$$
$$D2 = R - q(1 + e)/1 + e \cos \omega \tag{2}$$
$$D3 = 1/a - (2/R - c \cdot V_h^2) \tag{3}$$

where R is the radius vector of the Earth's orbit in a.u., V_h is the heliocentric velocity of the meteor in km/s, c is a conversion factor and q, a, e and ω are the meteoroid's orbital elements. If the data are consistent, then D1, D2 and D3 $\approx$ 0. Checks in the NMS data based on eq. (1) always gave D1 = 0. The data sample was next divided into four subsets as shown in Table 4 and checks based on eqs. (2) and (3) were made. Since there is some overlap between the subset designations of Koseki et al. the subsets were selected in the order TN, * and +.

Table 4.

Data sample subsets	Designation in Koseki et al.	Total no.	s.d. of $\Delta(1/a)$	No. of approx. solutions R = 1
Tokyo network	TN	16	0.0021	0
Re-reduced orbits	*	46	0.0098	2
Amended orbits	+	86	0.0076	3
Remaining orbits		177	0.0197	74

Here $\Delta(1/a)$ = D3 is the difference between the published value of 1/a and the value computed from the heliocentric velocity. The s.d. of this value in each subset is a measure of the errors in the semi major axis. Table 4 shows that the re-reduced data sets marked TN, *, or + in Koseki (1990) and Koseki et al. (1990) are practically error free. Of the remaining 177 orbits about half exhibit minor discrepancies in eqs. (2) and (3). A further study showed that the orbit calculation had been based on the approximation $R = 1$, i.e. a circular earth orbit. This approximation is sometimes used in studies of radio meteors, but it should be avoided in precise photographic meteor studies. The assumption $R = 1$ implys that the errors in the orbital elements will vary slightly with date, being largest at the Earth's aphelion and perihelion points. It would be desirable to recompute these orbits. However, for most of these objects the films and original measurements are not available any more. We have therefore instead introduced two different quality classes for the data, depending on whether or not the exact value of R was used.

Conclusion

The NMS meteor data is of high quality and is very useful for statistical studies. This new data set represents a significant increase in the total number of precisely reduced photographic orbits available to the meteor scientist. The sample will be included in the official IAU meteor data center file.

Acknowledgements

Travel grants from Kungl. Fysiografiska Sällskapet, the Swedish Natural Sciences Research Council and the Scandinavia-Japan Sasagawa Foundation are gratefully acknowledged. The author is indebted to M. Koseki and K. Ohtsuka for valuable comments.

References

Hirose, H., Kaho, S. and Tomita, K. (1950) Tokyo Astron. Bull., Second Ser., No. 29, pp. 207-215.
Hirose, H. and Tomita, K. (1950) Proc. Japan. Acad., 26, pp. 23-28 (Tokyo repr. 66).
Hirose, H. and Tomita, K. (1955) Tokyo Astron. Bull., Second Ser., No. 77, pp. 755-762.
Koseki, M. (1990) in Lagerkvist, C.-I., Rickman, H., Lindblad, B.A. and Lindgren, M. (eds.) Asteroids, Comets, Meteors III, 543.
Koseki, M., Sekiguchi, T. and Ohtsuka, K. (1990) in (eds) Lagerkvist, C.-I., Rickman, H., Lindblad, B.A. and Lindgren, M., Asteroids, Comets, Meteors III, pp. 547-550.
Lindblad, B.A. (1971a) Smithson. Contr. Astrophys. 12, pp. 1-13.
Lindblad, B.A. (1971b) Smithson. Contr. Astrophys. 12, pp. 14-24.
Lindblad, B.A. (1971c) Meteor Streams, Space Res. 11, pp. 287-297.
Lindblad, B.A. (1987) Physics and Orbits of Meteoroids, in Fulchignoni, M. and Kresak, L. (eds.) The Evolution of the Small Bodies of the Solar System, pp. 229-251.
Lindblad, B.A. (1990a) in (eds) Lagerkvist, C.I., Rickman, H., Lindblad, B.A. and Lindgren, M., Asteroids, Comets, Meteors III, pp. 551-553.
Lindblad, B.A. (1990b) Bull Astron. Inst. Czechosl., 41, pp. 193-200.
Ochiai, T. (1984) The Friend of Stars, 30, 59-62 (in Japanese).
Ochiai, T., Ohtsuka, K. and Sekiguchi, T. (1989) Mimeographed report, Nippon Meteor Soc.
Southworth, R. and Hawkins, G. (1963) Smithson. Contr. Astrophys., 7, pp. 261-285.

THE MICROMETEOROID IN THE UPPER ATMOSPHERE

F.KAMIJO
Department of Astronomy
University of Tokyo
Bunkyo-ku, Tokyo 113
Japan

ABSTRACT. The temperature and the radius variation of micrometeoroids in the thermosphere and the mesosphere are calculated theoretically. If the radius and the initial velocity are 100 μm and 30 km/sec respectively, the evaporation height and the velocity coincide almost exactly with those of the Capricornids and the Virginids from the meteor stream observation.

Moreover, it is shown that the not evaporated debris till the end of the sublimation may become spherules in the bottom of deep sea; and that fluffy micrometeoroids (10μm size) floating in the stratosphere are also consistent with our calculation.

The recondensation and the coagulation of the evaporated gas molecules from the meteoroid are also calculated, and it is shown that these secondary particles are very small and few.

1. Introduction

There are two kinds of small meteorites.

1) The spherules which are found in the bottom of the deep sea Their shape is spherical and the typical size is about 100 μm.

2) The micrometeoroids which are collected in the stratosphere by airplane. Their shape is not spherical and its typical character is that it is fluffy. The typical size is about 10 μm.

We calculate the origin of these two types by a simple model.

2. The Method of Calculation

The assumptions of the calculation are:

1) The kinetic energy of the atmospheric molecules which collide the meteorite is converted entirely to heat up the collided meteorite.

2) The absorbed heat is changed to the elevation of temperature and the radiation loss (Stefan-Boltzmann) of the meteorite.

3) If the temperature of the meteorite becomes higher than the sublimation temperature, the absorbed heat is changed to the evapo-

A.C. Levasseur-Regourd and H. Hasegawa (eds.), Origin and Evolution of Interplanetary Dust, 303–306.
© 1991 *Kluwer Academic Publishers.*

ration heat and the radiation loss, while the temperature of the meteorite becomes constant.

3. The Result of Calculation

Three examples are plotted in Fig.1. The upper three curves are the
variation of temperature (abscissa) of three meteorites against the
hight from the ground level (ordinate). The initial velocity v and
the initial radius r are v =30 km/sec, r =10 μ m; 30 km/sec,100 μ m;
and 20 km/sec, 100 μ m; respectively.
     The most typical case (30 km/sec, 100 μ m) is shown as the second
curve. The temperature begins to become gradually higher at the height
of 120 km and reaches the sublimation temperature at 97.2 km.
     After this point variation of the radius in percent of the in-
itial radius (abscissa) is plotted against the height (ordinate) in
the left lower part of Fig.1.
     Physical constants are assumed, corresponding to silicates, as
follows: specific heat C= 10^7 erg/K /g ,sublimation heat L= $6*10^{10}$ erg
/g and sublimation temperature is 2100 K.
     To explain the left-lower part of Fig.1, which is rather compli-
cated, we take as an example 30 km/sec, 100 μ m particle. As mentioned
above, the temperature becomes high enough to reach the sublimation
temperature at Z=97.2 km. From this point, the heat obtained by the
collision with the atmospheric gas molecules is converted into
sublimation heat and temperature becomes constant. The radius becomes
smaller and smaller by evaporation. It is plotted in the second
curve of the left-lower part of Fig.1.
     Namely, 100 % (r = 100 μ m) at Z=97.2 km,
             50 % (r = 50 μ m) at Z=86.2 km.
The radiation loss becomes gradually larger, compared with the
sublimation loss,and at last at Z=81.3 km, where r = 8.3 μ m,the radiation loss is large enough to be equal to the energy gain. Thus the
sublimation stops there.
     In other words,the meteorite is observed as a meteor at 97.2 km >
Z > 81.3 km and radius of the debris : r = 8.3 μ m.
     Two other cases are also plotted as the first and the third
curves in Fig.1, and final r = 4.0 μ m and 24 μ m ,respectively.

4. Secondary Particle

"Secondary particle" means the recondensation of the evaporated
vapour. As the surrounding temperature is very low, the evaporated
vapour condenses into grains soon after the evaporation.
     We have calculated several cases, but to avoid to integrate the
complicated equation of coagulation numerically, we adopt "similarity
solution" (Onaka et al. 1978).
     The result of an example (30 km/sec, 100 μ m) at the height of 90
km is shown in Fig.2. The ordinate is number ratio (log scale) of the
recondensed grains to the evaporated gas molecules,and the abscissa is
radius of the grain(on the top:number of atoms included in the grain).

τ (0.1 sec) is time duration after the evaporation. It is clearly seen, that 10 sec after the evaporation, radius is about 100 A and the ratio of the number density to the gas molecule is 10^{-10}

We have to conclude that the secondary particles are very few and small compared with the primary meteorite by this calculation, in which the meteorite is very small (micrometeoroid). If the meteorite is large and fast enough, secondary particles would be important.

5. Conclusion

1) The comparison with the meteor stream observation from the ground is very interesting. Simultaneous observations from several points on the ground can determine the height and the velocity of a meteor stream:

    a) Virginids   96 km >Z> 78 km, 30.8 km/sec

    b) Capricornids   96 km >Z> 86 km, 25.6 km/sec

These observations coincide almost exactly with our (30 km/sec, 100 μm) particles.

2) It is guessed that the spherule's origin is "debris" of our calculation. The trouble is that spherules, collected in the deep sea bottom, have about 100 μm size. It is conjectured that "sampling effect" is rather essential. If we can collect also small ones this inconsistency may be solved.

3) If the meteorite is (20 km/sec, 10 μm), heating is not much enough and it cannot reach the sublimation temperature, even if it is not fluffy. If the particle is fluffy, the heating is much smaller and it is quite natural that it is floating in the stratosphere in original shape.

4) The secondary particles are very few and small.

Reference

Onaka,T.,Nakada,Y.and Kanijo,F. : Experiment on the Clustering of Fine Particles, Astrophys.& Space Sci.65,103(1978).

Figure captions are written precisely in the text.

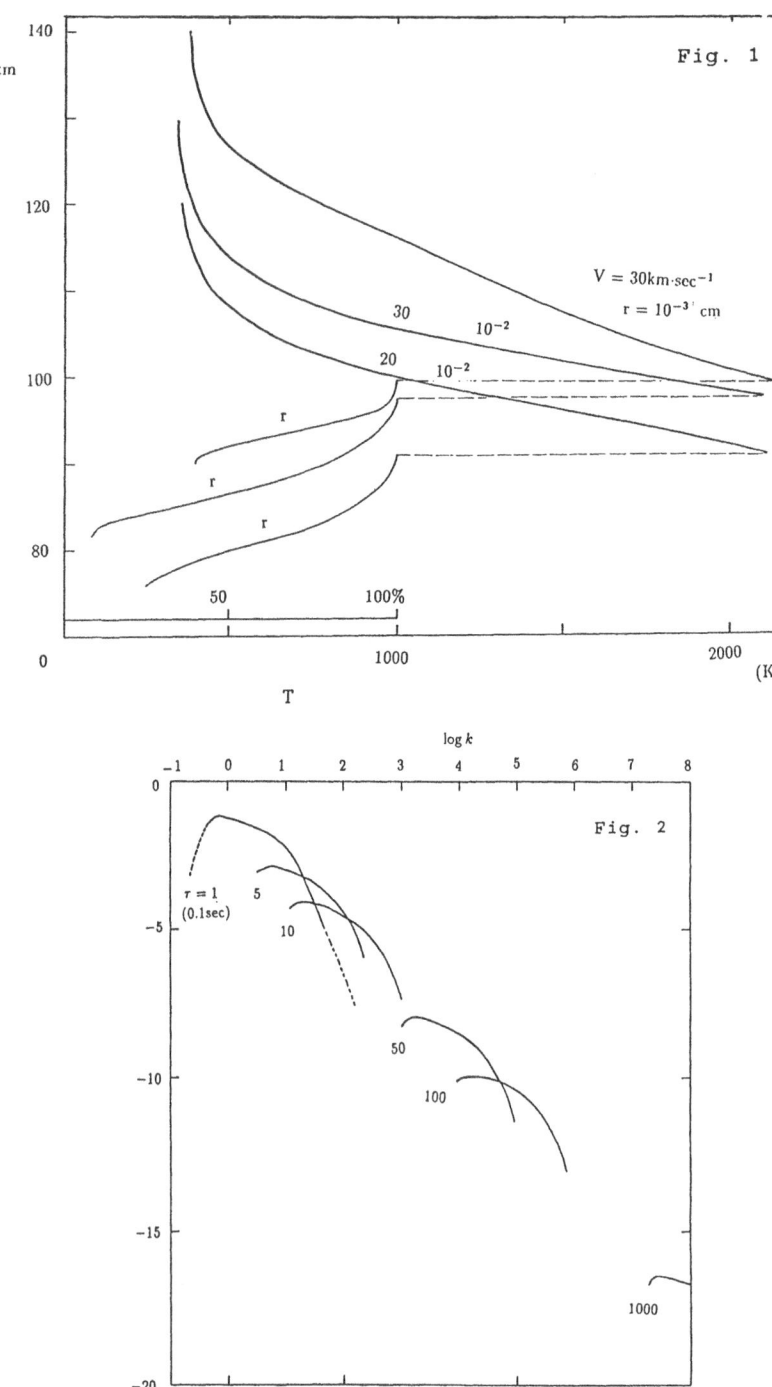

Fig. 1

$V = 30 \text{km} \cdot \text{sec}^{-1}$

$r = 10^{-3} \text{ cm}$

Fig. 2

AN ESTIMATION OF METEOROID FLUX AT OUTER MARTIAN SPACE FOR STEADY METEOR STREAMS

Ko NAGASAWA
Earthquake Research Institute, The University of Tokyo,
No. 1-1, Yayoi 1-chome, Bunkyo-ku, Tokyo,
JAPAN

ABSTRACT. This work was done to estimate meteoroid fluxes of steady meteor streams at places far away from the sun, for example, beyond Martian orbit. For the estimation, a Normal Distribution Model is introduced. This model assumes that the flow of meteoroids is steady and the meteoroids are distributed in the form of a normal distribution around the orbit of their parent comet. Keplerian motion of each meteoroid is also assumed. Some parameters, necessary for determining a practical model, can be obtained through observations. The model will give the overall structure of meteor streams and their fluxes can be calculated for any points on the orbits of their parent comets. Calculations show that fluxes of streams sharply decrease as the solar distance increases and that meteoroids scatter chiefly along the orbital planes of the parent comets. The model seems to be useful to deduce rough structure of meteor streams.

1. INTRODUCTION

We have chances to see active meteor showers such as Perseids, Orionids and Geminids several times in a year. Some of these showers seem to show nearly the same activity every year and the flow of the meteoroids can be considered to be nearly steady in the solar system.

We can, however, observe the flow of meteoroids only on a line along the orbit of the earth. As a result, it is difficult to comprehend the whole structure of a meteor stream.

Direct measurements of the particle flux have often been tried using space probes. However, these measurements are restricted to particles of much smaller size. It is almost impossible for space probes to observe meteoroids of visual meteor size because their particle density in space is extremely low.

Then, in order to estimate meteoroid flux at places far away from the sun, in other words, to determine whole structure of a meteor stream, it is necessary to assume certain models for the distribution of meteoroid.

We introduce here a "Normal Distribution Model" for steady meteor streams. This model is based on the idea that meteoroid particles of a stream distribute nearly in the form of a normal distribution around the orbit of their parent comet. Though this simple assumption may differ from the real one, this model still gives some aspects to the meteoroid flux of steady streams in the solar system.

A.C. Levasseur-Regourd and H. Hasegawa (eds.), Origin and Evolution of Interplanetary Dust, 307–310.
© 1991 *Kluwer Academic Publishers.*

2. DETERMINATION OF MODEL PARAMETERS

In order to specify the Normal Distribution Model, we have to define local coordinate systems. Take an arbitrary point P on the orbit of a parent comet and define a right-handed rectangular coordinate system (u, v, w) with P as its origin. Here, we take the positive direction of v-axis in the direction of the comet's motion and that of the u-axis in the outward direction of the comet's orbit in the orbital plane. Then, the u-v plane coincides with the orbital plane of the comet. In this way, we can define a local coordinate system for any points on the orbit of the comet. Since the u-w plane is always perpendicular to the flow of meteoroids, the number of meteoroids which pass through the plane per unit time gives the total flux of the stream. The positions where meteoroids pass through the plane give the distribution of the stream at P.

Let P be the nearest point to the orbit of the earth, then a Normal Distribution Model is defined as follows:

(1) Each particle in the stream has Keplerian orbit.

(2) The flow of all meteoroids in the stream is steady.

(3) The positions where meteoroids pass through the u-w plane per unit time are two dimensional normal distribution with rexpect to (u.w) coordinates.

For applying the model to an actual meteor stream, total flux N and variances and covariance s_{uu}, s_{ww} and s_{uw} have to be determined as the model parameters. Assuming that the orbits of the parent comet and of the earth are straight and the meteoroid flow is uniform within the region where we observe the meteor shower, we can theoretically give the change in the meteor rate along the earth's orbit using the Normal Distribution Model. On the other hand, we can determine the earth's position F where the shower is most active, the maximum flux M and the variance S for the duration of the shower from observations. Since the point F generally differs from E, the nearest point to the comet's orbit, we can calculate the distance L between E and F. We call M, S and L observed parameters.

Model parameters can be calculated by comparing the observed parameters to those obtained from the theory. The model parameters are given as follows:

$$s_{uu} = \alpha \, (\, \alpha W - \gamma U \,) \, S \, / \, (\, \gamma L + U \,)$$
$$s_{ww} = -\gamma \, (\, \alpha W - \gamma U \,) \, S \, / \, (\, \alpha L + W \,)$$
$$N = 2\pi S M \sqrt{ \frac{ \alpha \gamma \, (\, \alpha W - \gamma U \,)^2 }{ (\, \alpha L + U \,) \, (\, \gamma L + W \,) } } \exp \left[- \frac{ (\, \alpha L + U \,) \, (\, \gamma L + W \,) }{ 2 \alpha \gamma S } \right]$$

where (α, β, γ) are the direction cosines of the earth's orbit at E and (U, 0, W) are the coordinates of E. Covariance s_{uw} is assumed to be zero because it is impossible to determine four model parameters from three observed parameters.

3. FLUX CALCULATION

In order to calculate a stream flux for an arbitrary point on the orbit of its parent comet, it is necessary to trace the entire path of each meteoroid in the stream. The Normal Distribution Model enables us to do the calculation.

This calculation corresponds, in short, to a transformation of the model parameters from a local system to another one. This includes, however, lengthy calculations and it is impossible to give all equation used. We show here only outline of the procedure.

Take a meteoroid particle which pass through the u-w plane at positions (u, 0, w) with velocity components $(\dot{u}, \dot{V}+\dot{v}, \dot{w})$. $\dot{V}$ is the velocity of the parent comet. We assume u, w, $\dot{u}$, $\dot{v}$ and $\dot{w}$ are small quantities and their quardratic terms can be neglected. Let the orbital elements of the comet be $(a, e, \Omega, i, \omega)$ and those of the particle be $(a+\Delta a, e+\Delta e, \Omega+\Delta\Omega, i+\Delta i, \omega+\Delta\omega)$. Then, Δa, Δe, $\Delta\Omega$, Δi and $\Delta\omega$ can be expressed in linear forms of u, w, $\dot{u}$, $\dot{v}$ and $\dot{w}$. This means that if Δa, Δe, , Δi and $\Delta\omega$ are given the position coordinates (u', 0, w') and velocity components $(\dot{u}'$, $\dot{V}'+\dot{v}'$, $\dot{w}')$ can be calculated for any other local system at P'. This also means that we can calculate model parameters for any local system and that the meteoroid flux and the particle density are easily obtained from them.

4. RESULTS

We made the calculations for some permanent meteor streams. Though this shows us various characteristics of the meteor streams, some results are given in Table 1. In the calculations, orbital elements of the parent comets are taken from Marsden (1989) and those of the earth from Sidelman et al. (1974). Since the velocity distribution of each stream is difficult to obtain, we calculated it from the spread of the radiant. The spreads are from Millman (1963) and Koseki et al. (1990). Here, we assume the distribution of (u, v, w) are isotropic and uniform for every point on the u-w plane. Observed parameters are from Cook (1973).

Table1. Total flux, mean flux and particle density for three meteor streams at 1AU and 5AU from the sun.

		η-Aquarids	Perseids	Lyrids	
	Total Flux	1.2×10^{13}	7.6×10^{10}	7.4×10^{9}	hour^{-1}
1AU	Flux	1.1×10^{3}	67	63	$(10^5 km^2 hour)^{-1}$
	Density	73	4.5	4.1	$(10^9 km^3)^{-1}$
5AU	Flux	68	0.26	8.6×10^{-3}	$(10^5 km^2 hour)^{-1}$
	Density	11	4.1×10^{-2}	1.3×10^{-3}	$(10^9 km^3)^{-1}$

5. DISCUSSION AND CONCLUSIONS

The largest problem in this estimation is to determine whether the Normal Distribution Model actually approximates the real distribution of meteoroids. For checking this, it is necessary to observe the flux in space. This seems to be difficult in the present stage of techniques. However, the following points are clarified about the structure of meteor streams through our calculations.

(1) The flux of meteoroids in a meteor stream shows extreme decrease as the solar distance increases.

(2) The meteoroids spread out mainly in the direction of orbital plane of heir parent comet.

(3) An erroneous choice in the radiant spread, that is, a wrong assumption in the velocity distribution near the earth, doesn't introduce serious error in the flux of meteoroid at places far away from the sun.

(4) The determination of model parameters doesn't always give reasonable values in the case when the minimum distance between the orbits of the earth and the comet is large.

REFERENCES

Cook, A.F. (1973), 'A working list of meteor streams' in C.L. Hemenway, P.M. Millman and A.F. Cook(eds.) Evolutionary and Physical Properties of Meteoroids, NASA SP-319, pp.183-191.

Koseki, M., Sekiguchi, H. and Ohtsuka, K. (1990), 'Photographic Meteor Observations in Japan' in C.-I. Lagerkvist, H. Rickman, B.A. Lindblad and M. Lindgren (eds.) Asteroids, Comets, Meteors III, Uppsala, SWEDEN, pp.547-550.

Marsden, B.G. (1989), in Catalogue of Cometary Orbits, 6th edition, IAU Central Bureau for Astronomical Telegrams, Minor Planets Center.

Millman, P.M. and McKinley, D.W.R. (1963), 'Meteors' in B.M. Middlehurst and G.P. Kuiper (eds.) The Moon , Meteorites and Comets, The Solar System IV, The University of Chicago Press, U.S.A., pp.674-773.

Sidelmann, P.K., Doggett, L.E. and DeLuccia, M.R. (1974) 'Mean Elements of Principal Planets', A.J.**79**, pp.57-60.

THE IAU METEOR DATA CENTER IN LUND

B.A. Lindblad
Lund Observatory
Box 43, S-221 00 Lund
Sweden

ABSTRACT.   The purpose of the IAU Meteor Data Center in Lund is to archive, document and disseminate   information on meteoroid orbits. At present some 6 000 photographic double-station orbits and 60 000 radio determined orbits are archived.

1. Introduction

Information on photographic meteoroid orbits is widely scattered in the scientific literature and often in publications with a limited circulation. Information about (individual) radio meteoroid orbits has only been available on internal observatory listings or tapes. In the absence of key scientific personal much of this information was in the 1970's lost. A major effort has in the last few years been made to retrieve this data. At the 1976 IAU General Assembly it was proposed by Commission 22 that a meteor data center be established at the Lund Observatory, Sweden, for the archiving of meteor observations by photographic and radio techniques. The decision was confirmed by the 1982 IAU General Assembly. The archived data are two-station photographic and vidicon orbits or multi-station radio orbits. For a preliminary description of the data see Lindblad (1987a). In the present paper only the photographic data will be discussed.
  For most photographic meteors both orbital and geophysical data are available. The first file record contains the identification no., time of appearance, orbital elements, mass and shower classification; the second, geophysical, record contains identification no. and time plus earth encounter data such as magnitude, heights, radiant coordinates and velocities. An information pamphlet is available on request.

2. Photographic Orbits

The Harvard Super-Schmidt program in New Mexico, which operated from February 1952 to January 1959, recorded some 6 000 doubly photographed meteors. About 3 200 Super-Schmidt orbits have been reduced to date. About 2 300 meteoroid orbits have been obtained in various other photographic programs in the USA, Canada, Czechoslovakia, the USSR, Japan and elsewhere. These programs have in the literature rather arbitrarily been referred to as "small camera" programs. Table 1 summarizes the photographic orbit catalogues which presently are included in the IAU file. For references and a more detailed discussion of the data see Lindblad (1971, 1987a, 1987b).
  The largest number of orbits has been obtained in the Harvard photographic meteor

311

A.C. Levasseur-Regourd and H. Hasegawa (eds.), Origin and Evolution of Interplanetary Dust, 311–314.
© 1991 *Kluwer Academic Publishers.*

program. In the reduction of the Harvard Super-Schmidt data the same meteor was often measured by several investigators using different techniques. Hence, one should note that there is considerable overlap between the various Harvard catalogues.

The second largest contribution comes from the USSR stations in Dushanbe, Kiev and Odessa. A comprehensive catalogue of the Odessa data has recently appeared (Kramer, Shestaka and Markina, 1986). The USSR data have been collected over several decades and thus represent a valuable random sample - in contrast to the published Super-Schmidt data which were mainly obtained in the period 1952-54.

A photographic meteor program in New Mexico was operated in 1974-1977 by the New Mexico State University (NMSU) and the NASA Langley Research Center. This program produced 45 double-station meteor orbits (Harvey and Coffey 1976, Tedesco and Harvey 1976). For 25 of these meteors accurate timing was available. The main emphasize was to obtain simultaneous spectral - and orbital information on meteors.

A recent addition to the photographic file is 285 orbits obtained in the Czechoslovakian meteor program 1953-85. I am indebted to Dr. Z. Ceplecha and the Director of the Ondrejov Observatory for kind permission to include this data. The Czechoslovakian data are of high precision and they represent a random sample collected over many years.

Orbits obtained in three major fireball/meteorite recovery programs: Prairie Network, MORP and the European Network are included in the IAU file. The author is indebted to Drs. R. McCrosky, I. Halliday and Z. Ceplecha for kind permission to include unpublished material.

Some 450 photographic meteoroid orbits have been obtained in various amateur programs. See reports by Betlem (1985), Betlem and de Lignie (1990), Ochiai (1984, 1985), Koseki (1990) and Koseki, Sekiguchi and Ohtsuka (1990). The meteors have mostly been recorded with short focus 35 mm cameras. Studies by the present author (see these proceedings) indicate that the precision of the data is fully adequate for scientific applications. These orbits are presently being included in the IAU file.

3. Accuracy of Catalogue Data

It is difficult to assess the quality of the orbital data obtained at a particular station or by a particular investigator. An investigator may select only the very best photographic images for reduction, or study a random sample of the data, or analyze the available data in full. In fireball-meteorite-recovery programs the emphasis is on reducing photographic trails of meteors with low terminal heights. Fireballs from well known meteor showers are often not reduced. In the early Harvard studies the time of appearance of the meteor was not recorded, and the mid-exposure time was used with resulting loss of accuracy. Some investigators give a measure of the relative accuracy of each orbit. This index is included in our records. When no index of relative accuracy is given by the original investigator, various other measures of orbital accuracy have been introduced. For a discussion see Lindblad (1973). The early small-camera orbits were reduced using simple desk calculators, in which case computational errors are not uncommon. Some inconsistencies or misprints in the published data have been corrected after correspondence with the original investigators. At the Data Center the photographic meteoroid orbits are routinely checked for internal consistency. (For details see another paper by the author in this volume).These checks revealed some inconsistencies in the orbital elements. The author is corresponding with the original investigators in order to clarify these matters. An independent study of the errors in the photographic orbital data has been made by Koseki (1986). A preliminary list of errors from this study can be supplied on request.

Table 1. List of photographic meteor orbit catalogues

Station		Years	No. of orbits		Authors
Harvard (Mass.)		1936-52	139	(144)	Whipple
" "		1951-52	27		Whipple unpubl.
" (New Mex.)		1952-54	413		Jacchia and Whipple
" " "		1952-54	313	(359)	Hawkins and Southworth
" " "		1956-59	353		Posen and McCrosky
" " "		1956-59	253		McCrosky and Shao, unpubl.
" " "		1952-54	1801	(2529)	McCrosky and Posen (Graph. red.)
NMSU (New Mex.)		1975-76	12	*	Harvey and Tedesco
" "		1976-77	13	*	Drummond, Hill and Beebe
Prairie Network		1963-75	334	(336)	McCrosky, Shao and Posen
MORP "		1971-84	218	*	Halliday, Griffin & Blackwell (and unpubl.)
Dushanbe	1	1940-55	73		Katasev
"	2	1957-59	181		Babadjanov and Kramer
"	3	1960-63	72		" " "
"	4	1964	77		Babadjanov et al.
"	5	1965-66	15	(18)	Babadjanov and Getman
"	6	1968-77	44		Babadjanov et al.
"	7	1965-67	20	*	Babadjanov and Getman
Odessa	1	1957-59	133		Babadjanov and Kramer
"	2	1960-61	92		" " "
"	3	1961-65	122	(124)	Kramer and Markina
"	4	1962-72	50		" " "
"	5	1973-83	62	*	Kramer et al.
Kiev	1	1957-66	100		Benyukh et al.
"	2	1967-76	70	*	Sherbaum et al.
Ondrejov	1	1955-59	109		Ceplecha et al. (and unpubl.)
"	2	1947-89	176		Ceplecha et al. (and unpubl.)
NMS (Japan)		1964-89	325		Koseki, Sekiguchi and Ohtsuka

* An asterisk indicates that geophysical (encounter) data is partly or entirely missing. A number in parentheses gives the total number of orbits listed in a catalogue (including overlapping catalogue data and/or later rejected orbits). For detailed references see Lindblad 1987a.

314

Acknowledgements

The author is indebted to the scientists/institutions mentioned above and to the IAU for financial support. Grants from Kungl. Fys. Sällskapet, the Swedish Natural Science Research Council and the Scandinavia-Japan Sasagawa Foundation are acknowledged.

References

Betlem, H. (1985) Radiant, J. Dutch Meteor Soc., 7, 73.
Betlem, H. and de Lignie, M.C. (1990) in Lagerkvist, C.-I., Rickman, H., Lindblad, B.A. and Lindgren, M. (eds.) Asteroids, Comets, Meteors III, 505.
Harvey, G.A. and Cuffey, J. (1976) Contr. Obs. NMSU, 1, 166.
Koseki, M. (1986) J. Brit. Astron. Assoc., 96, 232.
Koseki, M. (1990) in Lagerkvist, C.-I., Rickman, H., Lindblad, B.A. and Lindgren, M. (eds.) Asteroids, Comets, Meteors III, 543.
Koseki, M., Sekiguchi, T. and Ohtsuka, K. (1990) in Lagerkvist, C.-I., Rickman, H., Lindblad, B.A. and Lindgren, M. (eds.) Asteroids, Comets, Meteors III, 547.
Kramer, E.N., Shestaka, I.S. and Markina, A.K. (1986) Meteor Orbits from Photographic Observations 1957-1983, Materials of the WDC B, Moscow.
Lindblad, B.A. (1971) Space Res., 11, 286.
Lindblad, B.A. (1973) The Distribution of 1/a in Photographic Meteor Orbits, In Hemenway, C.L., Millman, P.M. and Cook, A.F., Evolutionary and Physical Properties of Meteoroids, NASA SP-319, Washington, D.C.
Lindblad, B.A. (1987a) The IAU Meteor Data Center in Lund, in Ceplecha, Z. and Pecina, P., Interplanetary Matter, Publ. Czech. Acad. Sc., No. 67.
Lindblad, B.A. (1987b) Physics and Orbits of Meteoroids, in Fulchignoni, M. and Kresak, L., The Evolution of the Small Bodies of the Solar System, 229, North-Holland, Amsterdam.
Ochiai, T. (1984) The Friend of Stars, No. 30, 59 (in Japanese).
Ochiai. T. (1985) Werkgroepnieuws, 13, 88.
Tedesco, E.F. and Harvey, G.A. (1976) Astron. J., 81, 1010.

η LYRID METEOR STREAM ASSOCIATED WITH COMET IRAS-ARAKI-ALCOCK, 1983 VII

K. Ohtsuka
Tokyo Meteor Network
1-27-5 Daisawa, Setagaya-ku,
Tokyo 155, JAPAN

ABSTRACT. The probable association of Comet IRAS-Araki-Alcock with the
η Lyrid meteor stream is suggested, and the possible relation of the
radio meteor shower on 1983 May 10 is also discussed.

1. INTRODUCTION

Long-period Comet IRAS-Araki-Alcock 1983Ⅶ (hereafter, IAA) has closely
encountered with the Earth up to the distance of 0.03 AU, and has passed
through the descending node, of which distance to the Earth's orbit is
0.006 AU, in May 1983. Drummond (1983a) predicted an apparition of meteor shower associated with IAA on May 10.1 UT, 1983, and several meteor
observations were carried out around the theoretical maximum.
     The orbital elements of IAA are given in "Catalogue of cometary
orbits" (Marsden 1989). The original, osculating and future reciprocal
semimajor axes are +0.010466, +0.009972 and +0.010572, respectively.
This means that IAA has been belonging to the solar system, revolving
around the Sun at least two times with a period of ≈1000 yr. The periodicity of IAA may be also supported by some physical properties found
from ultraviolet, optical and infrared observations (e.g. Festou 1983;
Hanner et al. 1985; Festou et al. 1987; Watanabe 1987). These facts suggest that an associated meteor stream will be formed along the cometary
orbit.

2. η LYRIDS: PHOTOGRAPHIC METEORS ASSOCIATED WITH IAA

In the present work, five meteors probably associated with IAA are selected out among some 5800 orbits in the published photographic meteor
catalogues. For comparing the date, radiant and orbital elements of individual meteors with those of IAA, the D-criteria were applied, as
listed in Table 1. D_{SH} and D' denote the D-criterion of Southworth and
Hawkins (1963), and of Drummond (1981), respectively. The orbital elements of IAA computed by Marsden (1989), and the theoretical IAA meteor
radiant computed by Hasegawa (1990), are also given in Table 1.
     It should be noted in Table 1 that the orbital elements of IAA and

315

A.C. Levasseur-Regourd and H. Hasegawa (eds.), Origin and Evolution of Interplanetary Dust, 315–318.
© 1991 Kluwer Academic Publishers.

Table 1. Orbits of photographic η Lyrid (a) meteors and Comet IRAS-Araki-Alcock, 1983Ⅶ (equinox B1950.0)

No.	Date 1980+ UT	R.P. α_{1950} δ	V_G km/s	e	q AU	$1/a$ AU^{-1}	i 1950	ω 1950	Ω 1950	D_{SH}	D'
7598 1)	53- 5- 9.36	288° +43°	44.6	1.00	1.00	-0.0043	75°	194°	48.4	0.039	0.014
203 2)	61- 5- 9.96	289.2 +42.5	46.1	1.085	0.996	-0.0855	76.6	193.3	48.9	0.112	0.050
12068 1)	54- 5-10.39	290 +43	46.3	1.12	1.00	-0.1149	77	191	49.1	0.149	0.066
8441 3)	56- 5-11.42	287.7 +44.7	44.4	1.083	0.996	-0.0840	72.9	193.5	50.6	0.103	0.049
640322 4)	64- 5-11.77	290.1 +43.3	44.9	1.021	0.998	-0.0209	75.1	193.1	50.9	0.064	0.026
Mean	5-10	289.0 +43.3	45.3	1.062	0.998	-0.0619	75.3	193.0	49.6	0.093	0.041
S.D.		±1.0 ±0.8	±0.8	±0.044	±0.002	±0.0421	±1.4	±1.0	±1.0	±0.039	±0.019
1983Ⅶ 6)	5- 9	288.0 +44.0	43.8	0.990	0.991	+0.0100	73.2	192.8	48.4	0.006*	

1) McCrosky and Posen (1961). 2) Kramer and Markina (1966). 3) McCrosky and Shao (1969).
4) Babadzhanov et al. (1968). 5) Marsden (1989), a radiant prediction is from Hasegawa (1990).
* minimum distance of orbits between IAA - Earth.

five meteors are considerably similar to each other. Individual D-value is remarkably small in the ranges of $D_{SH} \lesssim 0.15$ and $D' \lesssim 0.07$; therefore, this ensures the existence of the meteor stream associated with IAA. As a matter of fact, Terentjeva (1968) has already found the same meteor stream: this is Stream No. 203 in her catalogue, designated as η Lyrids, segregated from the three groups of (a), (b) and (c). The five meteors found out in the present work are quite identical with the (a) group of η Lyrids, because Terentjeva gives the mean orbit of two photographic meteors, Nos. 7598 and 12068, which are listed in Table 1. In the present study three photographic members of the η Lyrids (a) are newly discovered. Moreover, two other photographic meteors, of which D-values are in the association range of $D_{SH} \lesssim 0.25$ or $D' \lesssim 0.105$, are also found out: they are Nos. 11994 and 7686 in the catalogue of McCrosky and Posen (1961). They might be members of the η Lyrids (b) or (c). Although their orbits are somewhat different, they are still possibly associated with IAA judging from the D-criteria.

3. METEOR OBSERVATIONS IN 1983

According to Drummond's (1983a) prediction of the IAA meteor shower on May 10, series of radar, visual and photographic observations were carried out in a few days around May 10 (Millman and Cook 1983; Clifton 1983; Drummond 1983b). However, these observations did not give any positive results upon the shower activity.

On the other hand, Shimoda and Ono (1983) detected enhanced forward-scattering echoes due to an activity of some meteor shower by using FM radio. The shower peaked on May 9.9 UT, and the activity duration was less than two hours. The maximum rate was about 80/hour, which was almost the same scale of η Aquarids. This rate was twice as the background rate at the same time in the previous and following day. The solar longitude $\lambda_\odot$ at the maximum corresponds to 48.3°(B1950.0), which was close to the theoretical $\lambda_\odot$ of 48.4°; therefore, this shower activity is possibly related with Comet IAA. Nevertheless, the negative result was obtained by the radar observations at Ottawa in the same time (Millman and Cook 1983), which was caused by the situation that radiant point was below or near horizon in that place.

4. DISCUSSION

It is no wonder that the association between a long-period comet and meteors is confirmed. Several cases have ever been reported, e.g. Comet 1861 I ($a \simeq 56$ AU) and Lyrids, Comet 1911 II ($a \simeq 185$ AU) and Aurigids, Comet 1739 (assumed $e = 1$) and Leo Minorids, and recently, Comet 1987 III ($a \simeq 180$ AU) and ε Geminids (Olsson-Steel and Lindblad 1987; Olsson-Steel 1987; Hasegawa 1990), and Comet Mellish 1917 I ($a \simeq 28$ AU) with relatively long period and Monocerotids (Ohtsuka 1988; Lindblad and Olsson-Steel 1990). Therefore, the association between IAA ($a \simeq 100$ AU) and η Lyrids would be an additional example.

The five meteors listed in Table 1 were observed 19-30 yr earlier than the time at the latest periherion passage of IAA, i.e. May 21 ET, 1983. Let us consider when these meteoroids were released from the IAA nucleus. We assume that these were ejected from the nucleus towards anti-IAA motion with a velocity of 1-2 $m \cdot s^{-1}$ near the IAA periherion in the previous return. We neglect the effect of gravitational and non-gravitational perturbations. Assuming the period of IAA as 1000 yr, then, we can obtain the result that these meteoroids should return 15-30 yr preceding to the comet; it is approximately consistent with the meteor observations. According to the radar observation of IAA at Arecibo, it was implied that large particles, of which radius corresponds in the order of 2-3 cm, exist within the dust-cloud around the IAA nucleus. Thier ejection velocity from the nucleus was estimated as 2.7 $m \cdot s^{-1}$ (Harmon et al. 1989). This value is consistent with that of above assumed. If such a large particle will enter into the Earth's atmosphere, the meteor should brighten up to be a fireball, which would be well-observable by using small cameras. Such a bright meteor has actually been observed at Odessa (Kramer and Markina 1966): this is Meteor No. 203 in Table 1. Kramer and Markina reduced the initial mass of meteoroid m_∞ as 14 gram. If the bulk density for the meteoroid is assumed to be ~ 0.5 $g \cdot cm^{-3}$, the radius will be ~ 2 cm. Therefore, some of the five meteoroids may be ejected from the IAA nucleus in the previous return.

Although the orbital eccentricity of IAA is smaller than 1, those of five meteors given in Table 1 are larger than 1. It may be accounted that these hyperbolic orbits were caused by observational errors, especially by poor determinations of the no-atmospheric velocities. Therefore, those hyperbolical characters are far from realistic.

Now we discuss the FM radio meteor shower in May 1983, observed by Shimoda and Ono (1983). Considering the facts that the theoretical and observed $\lambda_\odot$ were close to each other, and that the duration of meteor shower activity was very short, we can judge that the meteoroid particles should be concentrated to the cometary orbital plane. It means that the stream is at early stage in its formation. The elapsed time would have been short since released from the comet. This stream might consist of small particles (mass of $10^{-2} \sim 10^{-4}$ gram) because large echoes due to the visual-size meteoroids could have scarcely been received (Shimoda and Ono 1983). IAA has passed through the descending node on May 12.1, two days later than the Earth, where the distance of orbits between IAA and the Earth is only 0.006 AU, and the difference of mean anomalies between IAA and the FM radio meteors is only 0.002°! In spite of such an excellent condition for a meteor shower production, there was no strong shower. This means that the large-dust cloud near IAA was much sparser than the cases of Leonids and Draconids. This may be connected with the fact that infrared and radar observations for the dust production rate indicate IAA as a dust-poor comet (Hanner et al. 1985; Harmon et al. 1989).

ACKNOWLEDGEMENTS

The author is grateful to Dr. J. Watanabe and Prof. I. Hasegawa, for their helpful discussions and suggestions.

REFERENCES

Babadzhanov P.B., Getman T.I., Zausayev A.F. and Karaselnikova S.A. (1968) Byull. Inst. Astrofiz. No. 49, 3.
Clifton S. (1983) IAU Circ. No. 3811.
Drummond J.D. (1981) Icarus 45, 545.
Drummond J.D. (1983a) IAU Circ. No. 3801.
Drummond J.D. (1983b) IAU Circ. No. 3817.
Festou M.C. (1983) IAU Circ. No. 3802.
Festou M.C., Encrenaz T., Boisson C., Pedersen H. and Tarenghi M. (1987) Astron. Astrophys. 174, 299.
Hasegawa I. (1990) Publ. Astron. Soc. Japan 42, 175.
Harmon J.K., Campbell D.B., Hine A.A., Shapiro I.I. and Marsden B.G. (1989) Astrophys. J. 338, 1071.
Hanner M.S., Aitken D.K., Knacke R., McCorkle S., Roche P.F. and Tokunaga A.T. (1985) Icarus 62, 97.
Kramer E.N. and Markina A.K. (1966) Probl. Kosm. Fiz. 1, 21.
Lindblad B.A. and Olsson-Steel D. (1990) Bull. Astron. Inst. Czechosl. 41, 193.
Marsden B.G. (1989) Catalogue of Cometary Orbits, 6th Ed., SAO, MA.
McCrosky R.E. and Posen A. (1961) Smithson. Contr. Astrophys. 4, 15.
McCrosky R.E. and Shao C.-Y. (1969) Meteor Res. Program, Semiannual Tech. Rep. No. 7.
Millman P.M. and Cook A.F. (1983) IAU Circ. No. 3811.
Ohtsuka K. (1988) The Heavens 69, 199 (in Japanese).
Olsson-Steel D. (1987) Mon. Not. Roy. Astron. Soc. 228, 23p.
Olsson-Steel D. and Lindblad B.A. (1987) IAU Circ. No. 4414.
Shimoda C. and Ono K. (1983) FM Radio Obs. Rep. No. 87 (in Japanese).
Southworth R.B. and Hawkins G.S. (1963) Smithson. Contr. Astrophys. 7, 261.
Terentjeva A.K. (1968) in "Physics and Dynamics of Meteors", L. Kresák and P.M. Millman (eds.), D.Reidel, Dordrecht, p.408.
Watanabe J. (1987) Publ. Astron. Soc. Japan 39, 485.

THE ANNUAL VARIATION OF RADIO METEOR ECHOES OBSERVED FROM 1981 TO 1985

KAZUHIRO SUZUKI
Science Laboratory
Miya Fisheries High School
Miya cho, Gamagori, Aichi
443 Japan

ABSTRACT.   The radio observations of meteor echoes using FM broadcasting were carried out from January, 1981 to June, 1985.   The annual variation of meteor rates in the 5 years was obtained from the observations made from 25h through 29h in local time (16h-20h UT).
     Some features in this annual variation can be explained by the occurrence of the major meteor showers.   After the effects of these major showers are removed there remains a significant annual variation which shows higher rates in the winter half of the year (October to March).   This seems to be caused by the annual variation of non-shower meteors radiating from the apex region.

1. INTRODUCTION

In Japan, meteor radio observations using FM broadcasting have been carried out since 1970.   The method of observation is to count the appearances of meteors by utilizing the fact that meteor ionized columns reflect VHF waves (Suzuki 1976).   The observations from 25h through 29h local time (LT) have been continued with the same obser- vational equipment since 1981.   The time (from 25h through 29h LT) was selected for the following reasons: (1) The radio condition at midnight is undisturbed. (2) There are a lot of data observed optically. (3) Meteor trails radiating from the apex source are obtained mainly in the period 25h-29h LT.

2. APPARATUS AND METHODS OF OBSERVATION

If we have an ordinary audio FM tuner, we can observe the meteor echoes. There are nearly 200 FM stations in Japan, and they are broadcasting in the frequency bands of 76-90 MHz.   Tuning an audio FM tuner to a distant FM radio station which is not usually received, we can count the number

A.C. Levasseur-Regourd and H. Hasegawa (eds.), Origin and Evolution of Interplanetary Dust, 319–322.
© 1991 Kluwer Academic Publishers.

of meteor occurrences as the momentary enhancements of FM broadcasting reception.

The aerial used for this observation is a 5-element Yagi antenna. It is directed to the zenith, and is elevated 5 m above the ground. The echo receiver consists of 2 FM tuners. One of these is used for a noise canseller. The transmitting station is the FM Tokyo (frequency:80.0 MHz, peak power: 10 kW), which is located in Tokyo (long.139.7° E, lat.35.7° N). On the other hand, the receiving station is located at Toyokawa site (long.137.3° E, lat.34.8° N). The transmitting station is about 240 km east-northeast of our receiving site in Toyokawa.

The ground waves of the FM Tokyo usually cannot be received at night. Therefore, tuning the receiver to the FM Tokyo (80.0 MHz), the meteor echoes are recorded on the chart of a pen-recorder as a signal increase. The recorded echoes are divided into 5 classes according to the received power, and counted respectively.

3. RESULTS OF OBSERVATION

From January, 1981 through June, 1985, the observation of meteor echoes were carried out for 4 hours (25h through 29h LT) every day. Echoes whose received power was larger than 0.5 mV were recorded, and the number of hourly meteor echoes was counted. As a result of simultaneous observations of a TV camera equipped with Image-Intensifier and this

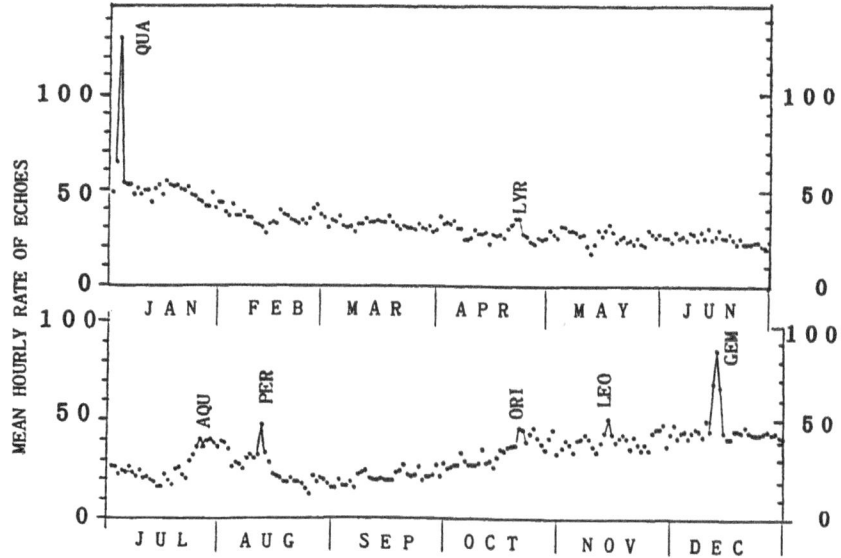

FIGURE 1. The daily mean hourly rates of meteor echoes observed during 25h-29h LT in the 5 years (1981 to 1985).

radio method, it was ascertained that meteors brighter than about +6 visual absolute magnitude were observed by our method. Fig. 1 shows the annual variation of the mean hourly rates of radio meteor echoes observed during 1981 to 1985.

4. MAJOR METEOR SHOWERS AND THE ANNUAL VARIATION

Several sharp or smooth peaks caused by the activities of meteor showers are shown in Fig. 1. These are major meteor showers observed every year. As the observational period is limited to only 4 hours a day, not all meteor showers can be found in Fig. 1. The three letter marks indicate the following meteor showers: Quadrantids in January, Lyrids in April, δ Aquarids in July, γ Perseids in August, Orionids in October, Leonids in November and α Geminids in December. Their dates of activities, normal and maximum hourly rates are given in Table 1.

Even if the effects of major meteor showers are removed, there seems to remain a significant annual variation. In order to confirm this, mean monthly hourly rates, excluding the periods of shower activities shown in Table 1, are calculated and plotted in Fig. 2. Of course, it is possible that the echoes belonging to several minor showers are included, therefore, some influences of showers may remain in the annual variation in Fig. 2. The annual variation revealed from our observations shows higher rates in the winter half of the year (October to March).

TABLE 1. The major showers observed by our method during 25h- 29h LT in 1981-1985.

Major Shower	Duration (UT)	Normal Hourly Rate	Max. (UT)	Maximum Hourly Rate	
Quadrantids (QUA)	Jan. 2-4	82	Jan. 3	120 —	180
Lyrids (LYR)	Apr. 21-23	34	Apr. 22	30 —	50
δ Aquarids (AQR)	July 26-30	39	July 29	40 —	50
γ Perseids (PER)	Aug. 11-13	37	Aug. 12	40 —	60
Orionids (ORI)	Oct. 21-25	44	Oct. 22	40 —	60
Leonids (LEO)	Nov. 16-18	42	Nov. 17	40 —	50
α Geminids (GEM)	Dec. 10-14	63	Dec. 13	70 —	120

5. DISCUSSIONS AND CONCLUSION

The annual variation of visual or radio meteor rates has been investi- gated by a number of meteor scientists. A proposed model of the radiant distribution of non-shower meteors has three main point sources: apex, helion, and antihelion. Stohl (1968) derived the annual variation of non-shower meteors radiating from each source. According to his result,

322

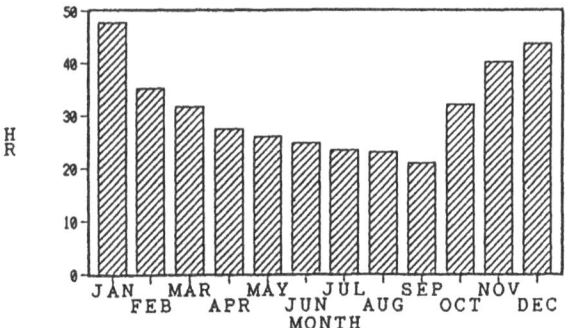

FIGURE 2. The annual variation excluding major meteor showers.

the annual variation for the apex source has its maximum in December, and its minimum in April.

Because of our geometrical relation between the transmitting station and the receiving site, meteor echoes radiated from the apex source are received effectively at 25h-29h LT. In this observation period, echoes radiated from the helion source cannot be obtained as it is below the horizon, and equally those from the antihelion source can hardly be obtained as it has just crossed the meridian. Therefore, the annual variation shown in Fig. 2 is supposed to be caused mainly by the non-shower meteors radiating from the apex region.

Hasegawa (1989) investigated a number of cometary orbits which intersect the earth's orbit, and he pointed out that more cometary orbits encounter the earth in the last quarter of the year. The annual variation observed by our method approximately coincides with this. That is to say, the annual variation of non-shower meteors radiating from the apex region may be caused by the comets coming near the earth.

ACKNOWLEDGEMENTS

The author is greatly indebted to Dr. Ichiro Hasegawa for his valuable suggestions and comments.

REFERENCES

Suzuki, K. et al. (1976) `Recording meteor echoes by FM radio', Sky and Telescope 5, 359-362.

McKinley, D. W. R. (1961) Meteor Science and Engineering (McGraw-Hill Book Company, Inc., New York), 112-120.

Stohl, J. (1968) `Seasonal variation in the radiant distribution of meteors', in Physics and Dynamics of Meteors, 298-303.

Hasegawa, I. (1989) in the Proceedings of Comets and Interstellar matter, 11- 13 (in Japanese)

LIFETIME OF METEOR STREAMS ASSOCIATED WITH COMET HALLEY

M. HAJDUKOVA
Department of Astronomy
and Astrophysics,
Comenius University,
842 15 Bratislava,
Czechoslovakia

A. HAJDUK
Astronomical Institute,
Slovak Academy of Sciences,
842 28 Bratislava,
Czechoslovakia

ABSTRACT. Critical examination of the orbital parameters of particles ejected from comet Halley rejects the low age hypotheses for meteor showers associ- ated with the comet. The diffusion of the orbits of large particles is too slow for explaining the observed structural features of the stream. The mass-loss process as derived from space observations compared with the mass of the stream of particles deduced from flux data lead to comet lifetimes of the order of 10^5 years.

1. INTRODUCTION

After the perihelion passage of Halley's comet in 1986 series of articles occurred discussing the past orbital history of this comet and its meteor streams. Weissman (1987) has concluded in agreement with Hajduk (1985) and Hughes (1985) that the age of the comet in the inner solar system can be placed between 1800 and 2500 revolutions ago. Olsson-Steel (1987) studying the probability of the encounter of the comet with Jupiter came to similar conclusion that there is less than 50 % chance that during 2×10^5 years there have been no close planetary encounters which have substantially altered the orbit of the comet. On the other hand McIntosh and Jones (1988) and Jones et al. (1989) modelling the orbital motion of test particles suggest that structural features observed in meteor streams associated with comet Halley may be explained as a consequence of particle diffusion in much shorter time corresponding to about 300 revolutions. Carusi et al. (1987) analysing Halley type comets in general, indicate these active lifetimes as mostly longer than 300 revolutions. For P/Halley they do not exclude capture by Jupiter 150 revolutions ago. Numerical simulations by Chirikov and Vecheslavov (1989) on the other hand show that the sojourn time of comet Halley within the solar system crucially depends on weak nongravitational forces acting upon the comet near the Sun and that it may reach values of 10^5 revolutions or 10^7 years. The present paper, based on the mass-loss rates determined by spaceprobes and on the derived parameters of the comet and its stream sets limits for the possible orbital and physical history of the comet and of the stream.

A.C. Levasseur-Regourd and H. Hasegawa (eds.), Origin and Evolution of Interplanetary Dust, 323–326.
© 1991 *Kluwer Academic Publishers.*

2. STREAM STRUCTURES AS AGEING CRITERION

The cloud of particles released from a comet during its perihelion passage creates, in time, a narrow band along the comet orbit. The non-homogeneous initial velocity distribution of ejecta, the spectrum of perturbations along the orbit with rapid motion of nodes, ($\Delta\Omega = 1$ deg/rev) can create concentrations of particle orbits in a broad belt. McIntosh and Jones (1988) and Jones et al. (1989) suggest that stream filaments are caused by relatively fresh ejecta, requiring only tens of revolutions. Similarly one may conclude that many such filaments should exist within the whole broad stream as a consequence of the variation of orbital parameters of the comet. If we suppose that the observed maxima in meteor rates correspond to the increased particle flux within such filaments, then the existence of these features along the broad interval of solar longitudes, especially larger than $209°$, $47°$ in both showers respectively, cannot be explained by ejecta from the present libration cycle of the comet orbit, starting with $\omega = 47°$, or $\Omega = -18°$ respectively, as defined by Kozai (1979): Firstly, because the nodes of the comet orbit are too far from the mentioned positions on the Earth's orbit, and secondly, because the ejection velocities, considered in paper by McIntosh and Jones (1988) are overestimated for a given mass. They are based on Whipple's formula (Whipple 1951) but contradict to numerous observations (Hajduk and Hajduková 1989). However, it does mean that the observed activity peaks within the showers may well be caused by the superposition of short-term and long-term processes of the spread ejecta. This also explains the gradual shifts in solar longitude of sec-ondary peaks in consecutive returns of the shower and the relatively smooth average activity from superposed returns. We can conclude, there-fore, that filamentary stream structure does not necessarily correspond to a young age. The main criterion of ageing remains the total width of the stream (not to be confused with the shower width in solar longitude, as it depends on the part of the stream intersecting Earth's orbit), which can be used as a measure for the slow gradual orbital diffusion process of the particles supplied in the series of perihelion passages of the comet. The width of the P/Halley streams, extending over 14 degrees in solar longitude, with observable particle masses well over 10^{-5} kg requires at least a few libration cycles of the comet, which im-plies that the age of the stream is of the order of 10^{5} revolutions. The age of the parent body may be determined in different ways. Hajduk (1985) and Hughes (1985) obtained the same age for P/Halley from differ-ent parameters, although both took into account the comet's mass produc-tion rate and its change with time. However, as shown by Kresák (1985) and Hughes (1988), both the time scale and irregularity of the systemat-ic decrease of comet brightness and the instrumental and selection ef- fects involved make it difficult to determine the long-term smooth de- pendence of the comet's mass production on the decreasing comet's size.

3. COMETARY DUST-LOSS RATE AND THE PARTICLE FLUX WITHIN THE STREAM AS AGEING CRITERIA

On the basis of data from space probes and ground based observations we have constructed, in Figure 1, limits for different possibilities of the evolution of the comet and its stream of particles. The total mass-loss rate of the comet from different spaceborne experiments give a mean value of about 1.5×10^{11} kg in the last apparition (Whipple 1986). Adopting the critical remarks to the analyses of the gas/dust ratios with the substantial contribution of large dust particles (Crifo 1987, Hajduk 1987) the mass of particles forming the stream does not differ substantially from the mass-loss ratio. The mass-loss process of the comet is, of course, changing considerably with the comet's age. When we postulate that the mass production of the comet changes proportionally to its surface $(R_0/R)^2$ we obtain the dependence of the mass-loss on the number of revolutions. In Figure 1 is the mass-loss curve of cometary dust ΔM_{cd} constructed for Kozai's cycles (Kozai 1979)

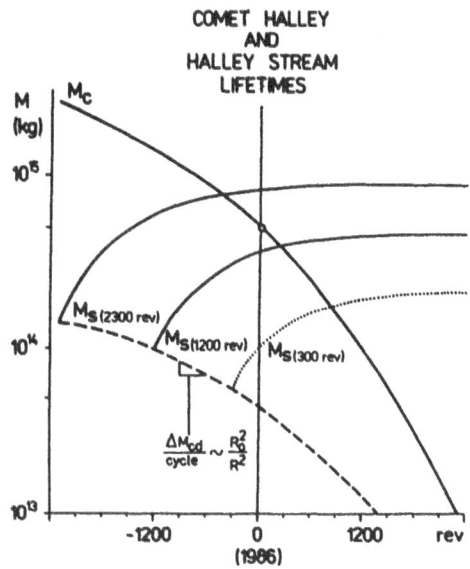

COMET HALLEY
AND
HALLEY STREAM
LIFETIMES

Figure 1.
Evolutionary diagram for the mass of the comet (M) and its stream (M) for different capture times. Dotted curve corresponds to the mass-loss of cometary dust in 300 revolution cycles.

of 300 revolutions. In similar way is expressed the change of the mass of the comet. The present mass is uncertain due to very different views concerning the density of the nucleus. We take here the mass of the nucleus $M = 5 \times 10^{14}$ kg. The same value was derived from meteor flux data (Hajduk 1982) for the present mass of the whole Halley stream. The mass of the stream M depends, naturally, on the time of the beginning of the stream forma-tion. The M curves in Figure 1 show the mass cumulated in the stream from the beginnings of this process 2300, 1200 and 300 revolutions ago respectively. Because of some uncertainty in the derived values we can slightly shift these evolutionary curves; however, the shift of the be-ginning of the stream forming process (which is equivalent to the time of the capture of the comet in the inner solar system) over 2300 revolu-tions or below 1200 revolutions appears problematic, as unprobably high or unprobably low values of the particle flux and mass-loss rates would be required. This argument, in connection with that one, following from the formation of the total width of the stream, makes the statements

like "Halley's comet is quite young" (Maddox 1989) more than question-
able.

REFERENCES

Carusi, A., Ľ. Kresák, E. Perozzi, and G.B. Valsecchi (1987). Long-term
    resonances and orbital evolution of Halley-type comets. 10th Europ.
    Reg. Astron. Meeting 2, 29-32.
Chirikov, B.V. and V.V. Vecheslavov (1989). Chaotic dynamics of comet
    Halley. Astron. Astrophys. 221, 146-154.
Crifo, J.F. (1987). Are cometary dust mass loss rates deduced from
    optical emissions reliable? 10th Europ. Reg. Astron. Meeting 2, 59-66.
Hajduk, A. (1982). The total mass and structure of the meteor stream
    associated with comet Halley. Sun and Planetary System (eds. W. Fricke
    and G. Teleki), D. Reidel; 335-336.
Hajduk, A. (1985). The past orbit of comet Halley and its meteor stream.
    IAU Coll. No. 83, 399-404, Reidel, Dordrecht.
Hajduk, A. (1987). Dust production of comet Halley with account of large
    particles contribution. 10th Europ. Reg. Astron. Meeting 2, 177-178.
Hajduk, A. and M. Hajduková (1989). From the formation of meteor streams
    up to their decay. Asteroids, Comets, Meteors III , Uppsala Univ.,
    pp. 531-534.
Hughes, D.W. (1985). The size, mass, mass-loss and age of Halley's
    comet. Mon. Not. R. Astron. Soc. 213, 103-109.
Hughes, D.W. (1988). Cometary magnitude distribution and the ratio
    between the numbers of long- and short-period comets. Icarus 73,
    149-162.
Jones, J., B.A. McIntosh and R.L. Hawkes (1989). Mon. Not. R. Astron.
    Soc. 238, 179- .
Kozai, Y. (1979). Secular perturbations of asteroids and comets. In:
    Dynamic of the Solar System, IAU Symp. No. 81, 231-237, Reidel,
    Dordrecht..
Kresák, Ľ. (1985). The aging and lifetimes of comets. In: Dynamics of
    Comets: Their Origin and Evolution, IAU Coll. No. 83, 279-302, Reidel,
    Dordrecht.
Maddox, J. (1989). Halley's comet is quite young. Nature 339, 95.
McIntosh, B.A. and J. Jones (1988). The Halley comet meteor stream.
    Mon. Not. R. Astron. Soc. 235, 673-694.
Olsson-Steel, D. (1987). The dynamical lifetime of comet P/Halley.
    Astron. Astrophys. 187, 909-912.
Weissman, P.R. (1987). How typical is Halley's comet? ESA SP-278, 31-36.
Whipple, F.L. (1951). A comet model. Astrophys. J. 113, 464.
Whipple, F.L. (1986). The cometary nucleus: current concepts.
    ESA SP-250, II, 281-288.

THE TAURID COMPLEX: GIANT COMET ORIGIN?

D.I.STEEL[1,2], D.J.ASHER[2] and S.V.M.CLUBE[2]
(1) Department of Physics and Mathematical Physics
University of Adelaide, G.P.O.Box 498, Adelaide, SA 5001, Australia
(2) Department of Physics, University of Oxford
Keble Road, Oxford, OX1 3RH, United Kingdom

ABSTRACT. The formation and evolution of the Taurid Complex of interplanetary objects is modelled on the basis of the parent being a giant comet which entered the inner solar system some time in the past 10,000-20,000 years. The orbital element distributions for the presently-observed meteor showers are discussed in terms of how these can constrain any model for the origin of the overall complex. As a baseline model we present results from the numerical integrations of fictitious meteoroids released from a comet over ten millenia, this comet having initial elements similar to those derived from a backwards integration of P/Encke. Large relative velocities at perihelion, above those feasible in conventional ejection scenarios, are necessary; we ascribe these to jetting of organics and other volatiles soon after release. Such a model gives a good first-order fit to the observed orbits, although additional processes (cometary splitting or asteroidal collisions) appear also to be necessary to explain the Taurids.

1.Introduction

The Taurid Complex (hereafter TC) of interplanetary objects consists of several night-time (optical and radar) and daytime (radar) meteor showers, P/Comet Encke and possibly other comets, and also several Apollo-type asteroids. The TC may well be the major source of the zodiacal dust cloud and also a large fraction of the broad sporadic meteoroid complex; it also appears to contain many other large objects which have orbits intersected by the Earth during the last few days of June each year, such as the Tunguska object. We believe that there is strong evidence to link the TC with climatic variations and catastrophic events which have occurred over the past few millennia (Clube and Asher, 1990), the root cause of which is the decay of the remnant components of a giant comet which entered a small-perihelion, short-period orbit within the last few tens of thousand years. This time-scale is supported by other recent modelling of the TC (Babadzhanov *et al*, 1990). In order to understand past events within this framework, and to make predictions as to the future influence of the TC upon the terrestrial environment, it is necessary to explain observed TC data in terms of a consistent model of the evolution of the objects spawned by the giant comet (smaller comets, asteroids, meteoroids). In this paper we consider a large number of the meteoroid orbits which give us most of our information on the orbital distribution of the TC. We then perform numerical integrations of the orbits of fictitious meteoroids released from a model comet over ten millenia, to see whether a fit to the observed data is attainable; a more detailed analysis is given elsewhere (Steel, Asher and Clube, 1991; hereafter SAC).

A.C. Levasseur-Regourd and H. Hasegawa (eds.), Origin and Evolution of Interplanetary Dust, 327–330.
© 1991 *Kluwer Academic Publishers.*

2. Observed Taurid meteors compared with model integrations

We selected from the many thousands of orbits available from the IAU Meteor Data Center 313 meteors which we believe to be members of the TC. Of these 170 were optically-recorded, 143 radar-recorded, and of these 143, 56 were from the daytime showers around June. The selection was on the basis of radiants, times of occurrence and orbital similarity, using similar but distinct selection criteria to those used by Stohl and Porubcan (1990); for details see SAC. In Figure 1 we plot the orbital elements a, e, q and i (referred to Jupiter, which controls the orbital evolution of these bodies) against the nodal longitude Ω for the case of the optical Northern Taurids; 53 meteors contribute. We find, as did Stohl and Porubcan, evidence for a trend in the different elements with Ω, but we also find that a and e appear to be correlated. This counts against the simple collision model of Whipple and Hamid (1952), since the orbits produced in a collision in the asteroid belt would have their aphelion distances near-constant and hence a and e anticorrelated. Although it is possible that such collisions have contributed to the stream formation and evolution, the question arises as to whether the spread in q, the main reason for Whipple and Hamid's adoption of a collision model, can be explained solely by planetary perturbations.

Our provisional model is that of a large comet with small perihelion distance which entered the inner solar system about 20,000 years ago, producing meteoroids over some 10,000 years which have long since dispersed into the zodiacal cloud, and of subsequent activity within the last 10,000 years giving rise to the present appearance of the showers. The parent object is loosely based upon a backwards integration of P/Encke, and 9,000 years ago had $a = 2.2$ AU, $e = 0.89$. The ejections are taken to occur at perihelion with velocities of -2.0, -1.5, -1.0, -0.5, -0.3, -0.1, +0.1 and +0.3 km/sec along the direction of motion of the parent at that point; this means that the ejections are strongly non-isotropic, but unless the semi-major axis is substantially different from the parent the orbital evolution is very similar, so there is little point in integrating objects ejected in other directions at perihelion. In addition, particles with larger positive velocities (resulting in $a > 2.5$ AU) are rapidly displaced from the TC by the effect of Jupiter, so there is no point in including these in the integrations either. The velocities and overall time-scale (less than 10,000 years) were ultimately chosen so as to fit the observed (a, e) plots. For each of these eight velocities we consider nine separate ejections spaced about 1000 years apart from 9000 to 1000 years ago. These 72 orbits were then integrated forward to the present, taking into account the effect of the planets Jupiter, Saturn, the Earth and Venus. In Figure 2 we have plotted orbital element distributions for the 26 particles in this model which would have intersected the Earth's orbit within the last 4,000 years under conditions such that they would be classified as Northern Taurids; this Figure should therefore be compared with Figure 1. We note the following:

(a) A sufficiently large range in Ω has been produced by the model integrations to fit the duration of the meteor shower.

(b) Because of the choice of relative velocities there is a sufficient spread in semi-major axes, but the observational data show some larger ($a > 1.5$ AU) orbits at the onset of the shower (near $\Omega = 180$ degrees) not seen in the model. It may be that our initial Ω for the parent was about ten degrees too high, and a better fit may be obtained in our later integrations by making such a correction.

(c) There is too small a scatter in the e and q plots produced by the model; more precisely, we do not see $q > 0.4$ AU. This may be due in part to the choice of the positive relative velocities: +0.1 km/sec results in the particle being injected into the 7:2 resonance with Jupiter, similarly ejection at +0.3 km/sec places a particle into the 3:1

resonance, and there is a tendency for particles in resonances not to show the increase in *q* (*i.e.* decrease in *e)* that most particles demonstrate.

(d) In Figure 2 the particles at Ω < 205 degrees result from one cycle in ω of the parent, whilst those at Ω > 205 degrees result from the next cycle, the parent's ω having increased by 360 degrees. This may explain the existence of the two groupings seen in the observed Northern Taurids (see SAC).

(e) For the inclinations the fit to the data is reasonable although again there is not as much scatter as in the observational data.

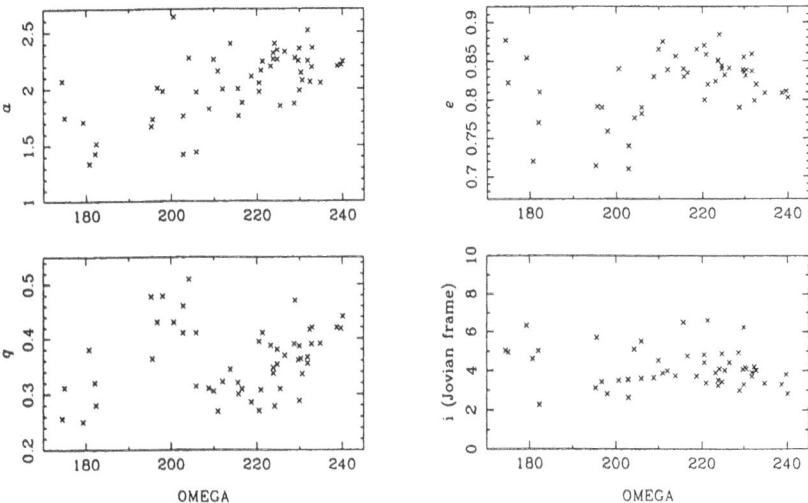

Figure 1. Orbital element distributions for 53 Northern Taurids observed in various optical meteor surveys.

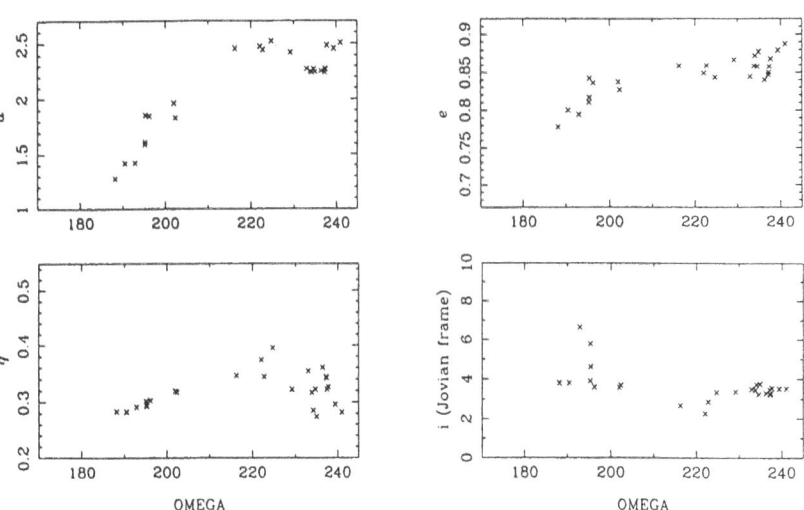

Figure 2. Model orbital element distributions for 26 meteoroids derived from a giant comet after integrations of their orbits spanning ten millenia.

3. Discussion

On the basis of these preliminary integrations we believe that the provisional model, that of a giant comet which entered the inner solar system some time within the last 20,000-10,000 years and was captured into a short-period orbit with small perihelion distance (0.15 - 0.25 AU), is a suitable baseline for explaining the gross phenomena observed in the TC. One drawback is that large relative velocities are required, and gravitational perturbations cannot produce these since they leave the semi-major axis largely invariant. Ejections from the cometary nucleus, with the velocity dispersion due to meteoroids being swept out by the expanding coma also cannot give such high ejection velocities: even for a comet nucleus larger than 100 km with a very small perihelion distance (q < 0.1 AU) the velocities tend to be below 0.1 km/sec for 1mm-10cm meteoroids. We are thus forced to look for an alternative explanation: this is not a problem unique to the Taurids since both the Geminids and the Perseids have velocity dispersions that cannot be explained in terms of either ejection velocities from the parent or planetary perturbations. One possibility is that the dispersion is due to large amounts of organic and other volatile materials known, since the advent of the spacecraft missions to P/Halley, to exist in solid particles ejected from cometary nuclei. The jetting effect as these constituents rapidly evaporate would cause a severe alteration in the velocity of both small and large objects relative to the cometary nucleus.

In our model we do not exclude the possibility *inter alia* that the original comet may have been split, by collisions or otherwise, to render two or more discrete sources for meteoroid ejection: these sources may be seen now as P/Encke, several Apollo-type asteroids, and quite likely many other large bodies which are yet to be discovered, including possibly the main fragment (Clube and Asher, 1990). Since the majority of the ejecta from a comet, especially the larger masses, would remain in orbits close to that of the parent(s), we find from our integrations that there are epochs of large influx to the Earth which last for several hundred years and are spaced by several thousand years. If there were several large fragments of the original giant comet which split to form secondary sources, then such periods of enhanced meteoric activity with large body (Tunguska-sized) impacts may be interwoven in time. These are thought to be evidenced in the historical record (Clube and Napier, 1990).

Acknowledgements: This work was supported by the Science and Engineering Research Council and the Australian Research Council.

References

Babadzhanov, P.B., Obrubov, Yu.V. and Makhmudov, N. (1990). 'Encke meteor showers', *Astron.Vestn.*, **24**, 18-28.

Clube, S.V.M. and Asher, D.J. (1990). 'The evolution of proto-Encke: Dust bands, close encounters and climatic modulations', in *Asteroids, Comets, Meteors III* (eds. C.-I. Lagerkvist, H. Rickman, B.A. Lindblad and M. Lindgren), University of Uppsala Press, Sweden, 275-280.

Clube, S.V.M. and Napier, W.M. (1990). *The Cosmic Winter*, Basil Blackwells, Oxford, U.K.

Steel, D.I., Asher, D.J. and Clube, S.V.M. (1991). 'The structure and evolution of the Taurid complex', *Mon. Not. Roy. Astron. Soc.* (submitted).

Stohl, J. and Porubcan, V. (1990). 'Structure of the Taurid meteor complex', in *Asteroids, Comets, Meteors III* (eds. C.-I. Lagerkvist, H. Rickman, B.A. Lindblad and M. Lindgren), University of Uppsala Press, Sweden, 571-574.

Whipple, F.L. and Hamid, S.E. (1952). 'On the origin of the Taurid meteor streams', *Helwan Obs. Bull.* No. **41**, 1-30.

MASS DISTRIBUTION AND BULK DENSITY DISTRIBUTION OF INTERPLANETARY DUST

A. HAJDUK
Astronomical Institute
Slovak Academy of Sciences
842 28 Bratislava
Czechoslovakia

ABSTRACT. Mass distribution of the interplanetary dust is reexamined taking into account bulk density distribution of the dust and larger particles. It can be shown that the mass index of particles depends on the evolutionary stage of the population and changes along the mass scale. The flattening of the mass distribution at the higher mass range may explain the problem of the equilibrium between the source and sink of the interplanetary dust.

1. MASS DISTRIBUTION OF PARTICLE POPULATIONS

Classical models of meteor stream formation suggested that old showers have lower mass index (s) values because Poynting-Robertson drag and solar radiation pressure gradually eliminate small particles from the stream. It was argued similarly that large particles should dominate in the inner side of the stream. However, these theoretical conclusions have not been confirmed by observations, at least for particles with masses $m \geq 10^{-6}$ kg, corresponding to the visual and radar range of detection. As shown recently by Šimek (1987), in contrary, showers without an active parent body (e.g. Geminids or Quadrantids) are characterized by higher values of the mass index, implying a lower proportion of larger particles. This is in agreement with the results of spaceprobes, indicating the dominant role of large particles in P/Halley mass production (McDonnell et al., 1987; Hajduk, 1987a) supported also by the improved dust/gas ratios for comet Halley, showing much higher dust contribution (Crifo, 1987; Hajduk, 1987b) than previously reported. Moreover, as shown in the same papers, the mass index is not constant over the mass scale and clearly indicates the superposition of two populations of particles with considerably different mass distribution. This result obtained from spaceborne experiments coincides totaly with results of radar meteor observations showing two separate levels of the mass index for the Halley showers, with values s = 1.8 and s = 2.2 (Hajduková et al., 1987). As it is seen in Figure 1 (A) different particle flux distribution corresponds to these two populations of particles within the shower. Ceplecha (1987) has classified different

331

A.C. Levasseur-Regourd and H. Hasegawa (eds.), Origin and Evolution of Interplanetary Dust, 331–334.
© 1991 *Kluwer Academic Publishers.*

types of cometary material, having different bulk densities, not
necessarily originating in different comets. However the bimodal nature
of particle size distributions has been reported also from infrared and
optical observations of different comets (Liu and Kimura, 1985).

The question arises, whether the coincidence of the same particle
flux and the same abundance of particles with different bulk densities
(ρ = 0.75 and 2.0) at the mass range between 10^{-5} kg – 10^{-4} kg is
accidental or dependent. (See Figures 1 A and 1 B. Figure 1 B is based
on Ceplecha's data quoted above.) The mass distribution of meteor
showers, in general, has a maximum mass contribution between the limits
of 10^{-5} – 10^{-4} kg. Fig. 1 C is constructed for the Halley showers. The
coincidence with the flux curves crossing is, of course, not accidental;
it shows the meaning of particle populations in the stream: the derived
mass contribution of particles of different mass categories is very
sensitive to the value of the mass index. We will deal here with the
integrated mass index S, as defined by Millman (1970) On the balance
near the critical value of S = 1 depends the mass contribution to the
stream. In Figure 1 D it is shown that the observed values of S for the
showers (SH) (Hajduková et al., 1987) are between the values derived for
the cometary halo (H) from space observations (McDonnell et al., 1987;
Hajduk and Kapišinský, 1987) and for the large particles from fireball
(F) observations (Rendtel and Knöfel, 1989). The differences in the mass
distribution correspond clearly to the age of these populations, as it
was shown by Hajduk (1989). As a consequence of a mixture of old and new
particle populations, different mass distribution may be observed,
depending on the combination of stream filaments met by Earth in a
particular return of the shower. Hence we can conclude that showers with
an active parent body cannot be characterized by a single mass index
value. However, the separation of populations corresponding to the
stream structures dynamically bound to possible ejection times, coould
be used for the determination of the oldest fraction with the highest
value of the mass index.

2. EROSION OF LARGE PARTICLES

Classical theories of the stream formation supposed a quick elimination
of small particles from the stream by the action Poynting-Robertson
effect and solar radiation pressure. The formula of Wyatt and Whipple
(1950) gives the age of the spiralling particle depending on the
particle radius and density and on the orbital parameters. However, the
physical erosion processes change drastically along the particle mass
scale. As shown by Kapišinský (1984), direct light pressure, solar wind
corpuscular pressure, Poynting-Robertson effect and other processes have
much less effect on the lifetime of particles with m > 10^{-12} kg than the
destructive processes. The greatest influence on the larger particles is
the effect of impact erosion and of corpuscular sputtering. These two
effects dominate in the range of meteoroid size particles (Kapišinský,
1987). (See Figure 1 E). Grün (1987) considers collisional fragmentation
to be the dominating process for particles with masses m $\geq 10^{-8}$ kg. The
maximum mass contribution in meteor streams comes from $10^{-5 \pm 1}$ kg and the

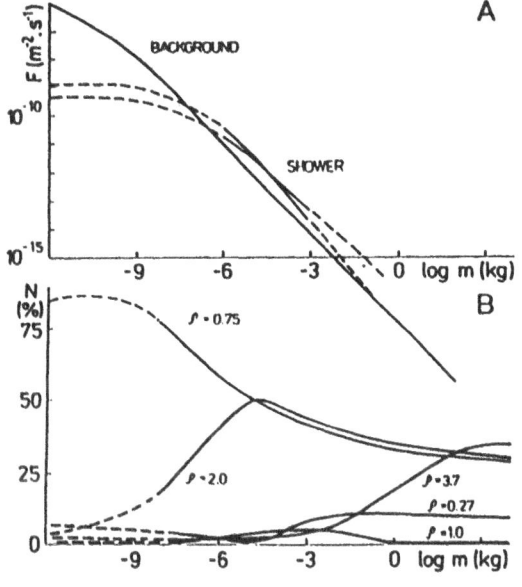

A
B

Figure 1
A: Particle flux to mass relation for shower and back-
ground.
B: Particle bulk density distribution along the mass scale.
C: Mass distribution in meteor
shower (Halley showers data).
D: Integrated mass index S for
the halo (H) population
(space probes data), Halley
showers (SH) (radar observations) and fireballs (F). S
denotes the mean values of S.
E. Lifetimes of particles due to
combined impact erosion (IE)
with corpuscular sputtering
(CS) compared with Poynting-
Robertson (PR) effect.

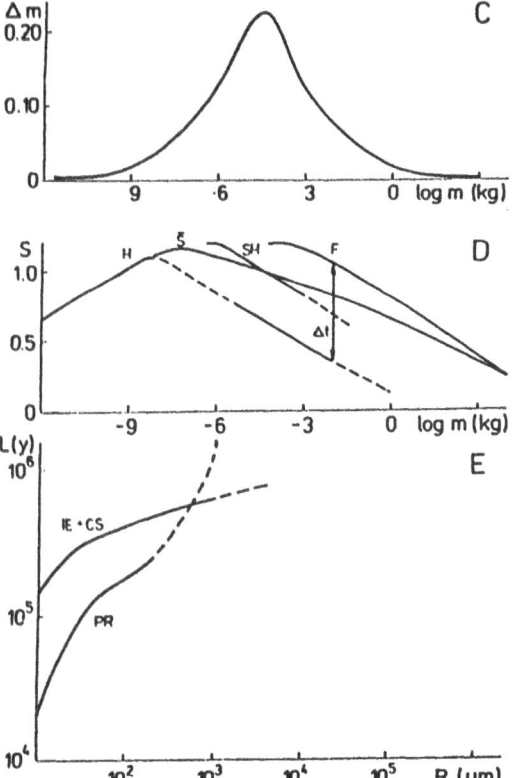

C
D
E

maximum mass contribution of
back ground meteoroids is at
$m = 10^{-8+1}$ kg (Millman, 1975).
This corresponds to the change
of the slope of the flux for the
background particles (Fig. 1 A).
The erosion is, of course, more
rapid for low density particles.
Packing forces, produced by an
anisotropic sublimation of
mantle material of grains at the
surface of fluffy particles,
shift the grains towards the
center, reduce the voids and increase the mass density of such
particles with time (Mukai and
Fechtig, 1983). Sublimation, at
least on the orbits with $q \leq 7$
AU, is also the main process by
which water-ice grains lose
their ice parts (Mukai et al.,
1989). Hence, by determining the
shift of the maximum mass contribution of particles in different meteor showers, the relative age of these showers can be
determined. Showers with a decreasing supply of larger particles will show a shift of the

maximum mass contribution towards smaller masses. This effect, verifiable from the magnitude distributions of meteors, can then be applied as an independent ageing scale of streams.

References:

Ceplecha, Z. (1987) 'Numbers and masses of different populations of sporadic meteoroids from photographic and television records, in Proc. 10th Europ. Regional Astronomy Meeting of the IAU Praha 2, 211-215.

Crifo, J. F. (1987) 'Are cometary dust mass loss rates deduced from optical emissions reliable?' in Proc. 10th Europ. Regional Astronomy Meeting of the IAU Praha 2, pp. 59-66.

Grün, E. (1987) 'Dynamics of interplanetary dust' in Proc. 10th Europ. Regional Astronomy Meeting Praha 2, pp. 177-178.

Hajduk, A. (1989) 'Evolution of cometary debris: physical aspects', IAU Coll. 116: Comets in the Post Halley Era, Bamberg, in press.

Hajduk, A. and Kapišinský, I. (1987) 'The evolution of the mass distribution of cometary particles', in Symp. on the Diversity and Similarity of Comets, Brussels, ESA SP-278, pp. 441-444.

Hajduková, M., Hajduk, A., Cevolani, G., Formiggini, C. (1987) 'The P/Halley meteor showers in 1985-1986' Astron. Astrophys. 187, 919-920.

Kapišinský, I. (1984) 'Nongravitational effects affecting small meteoroids in interplanetary space', Contrib. Astron. Obs. Skalnaté Pleso 12, 99-111.

Kapišinský, I. (1987) 'Double erosion of dust particles', Bull. Astron. Czechosl. 38, 7-12.

Liu, C. P. and Kimura, H. (1985) 'On the bimodal nature of the particle-size-distribution function of cometary dust', Marseille Symp. 9-12 July 1984, Properties and interactionof interplanetary dust, pp. 279-282 (IAU Colloq. No. 85)

McDonnell, J. A. M., et al. (1987) 'The dust distribution within the inter coma of comet P/Halley 1982i: encounter by Giotto's impact detectors', Astron. Astrophys. 187, 719-741.

Millman P. M. (1970) 'Meteor showers and interplanetary dust', Space Research X, 260-265.

Millman, P. M. (1975) 'Dust in the Solar System', The Dusty Universe, McGraw-Hill, New York, 185-209

Mukai, T. and Fechtig, H. (1983) 'Packing effect of fluffy particles', Planet. Space Sci. 31, 655-658.

Mukai, T., Fechtig, H., Grün, E., and Giese, R. H. (1989) 'Icy particles from comets' Icarus 80, 254-266.

Rendtel, j and Knöfel, A. (1989) 'Analysis of annual and diurnal variation of fireball rates and population index of fireballs from different compilations of visual observations', Bull. Astron. Inst. Czechosl. 40, 53-62.

Šimek, M. (1987) 'Dynamics and evolution of the structure of five meteor streams', Bull. Astron. Inst. Czechosl. 38, 80-91.

Wyatt, S. P. and Whipple, F. L. (1950) 'The Poynting-Robertson effect on meteor orbits', Astrophys. J. 111, 134-141.

THE INTERNATIONAL METEOR ORGANIZATION

M. Gyssens   A. Knöfel   J. Rendtel   P. Roggemans
Heerbaan 74   *A.-Fischer-Ring 96* *Gontardstraße 11* *Pijnboomstraat 25*
2530 Boechout *O-1580 Potsdam*   *O-1570 Potsdam*   *2800 Mechelen*
Belgium   *Germany*   *Germany*   *Belgium*

ABSTRACT. Founded in 1988, the International Meteor Organization (IMO) is an international scientific non-profit association with members all over the world. The IMO was created in response to an ever growing need for international cooperation of amateur work. As such, the main objectives of the IMO are to encourage, support and coordinate meteor observing, to improve the quality of amateur observations, to make global analyses of observations received world-wide, to develop contacts between amateurs and professionals, and to disseminate the observations and results obtained to other amateurs and to the professional community.

1. Organizational Structure

The International Meteor Organization (IMO) is an international, scientific non-profit association. It is composed of individual members, rather than corporate or incorporate societies or groups; the latter apply widely varying standards to their work and are therefore unattractive for implementing international scientific observing programs. At present, the membership of the IMO consists of about 200 meteor workers from 31 countries.

The prime decision-making body within the IMO is the General Assembly which consists of all voting members of the organization. Since amateurs usually do not have funds at their disposal for long-distance traveling to meetings, all decisions are made by written vote. The daily management of the IMO is in the hands of the Council. Council members, including the President of the IMO, are elected by the General Assembly. The present Council consists of 16 amateurs and professionals from all over the world.

2. Objectives

As stipulated in its Constitution, the IMO has set itself the following objectives:

- to affirm the need for studying meteors and related phenomena;
- to promote a global perspective towards the study of meteors and related phenomena, by striving towards common standards and research programs and by developing the international spirit among meteor workers;
- to encourage and develop the study of meteors and related phenomena by amateurs and professionals, both on practical and theoretical levels.
- to encourage and develop international cooperation, among as well as between amateur and professional meteor workers;

335

A.C. Levasseur-Regourd and H. Hasegawa (eds.), Origin and Evolution of Interplanetary Dust, 335–338.
© 1991 *Kluwer Academic Publishers.*

- to centralize and distribute scientific data on meteors and related phenomena in order to guarantee both their conservation and their accessibility by amateurs and professionals;
- to provide assistance to individual meteor workers as well as groups for the organization of scientific activities.

In order to achieve its goals, the IMO publishes an international scientific journal in English called "WGN", makes available various scientific publications, proposes international standards for observing and reporting data, initiates and/or organizes observing programs, collects and distributes observational data, and organizes, provides assistance in the organization of, or participates in: conferences, lectures, exhibitions, observing camps, etc.

3. The IMO and amateur-professional cooperation

One of the main objectives of the IMO listed above is to provide the framework and facilities for future professional-amateur cooperation.

Most IMO amateur members have a long standing experience as a meteor observer. Moreover, many of them also have an academic degree and thus are amateurs only in the sense that they are not paid for the work they do. In addition the IMO has the support of several professional meteor workers. This rather exceptional composition of its membership allows the organization to assess the reliability of observations by individuals or groups of amateurs and to achieve reliable results. The IMO has thus realized the basic conditions for a durable amateur-professional cooperation.

In order to further advance the quality of amateur meteor work, a set of handbooks is currently being published in which the IMO observing standards are described in great detail, and, at the same time, a lot background information is given. To improve the interpretation of observational results, the IMO takes great efforts in providing easier access for amateurs to the meteor literature. In this respect, a list with over 8000 references covering the period 1794–1987 has already been published by Roggemans (1988). An update of this extensive literature survey is currently under preparation.

For the dissemination of its observations and results, the IMO publishes an international scientific journal called "WGN". WGN consists of two series: an Observational Report Series and a Bimonthly Journal. The Observational Report Series contains the collected observational results obtained worldwide using a standard methodology and should therefore become the key source of amateur observations for the professional community. The Bimonthly Journal serves as the principal interface between amateurs and professional astronomers. In addition, the IMO's annual conferences bring together amateurs and professionals, and thus provide a forum for discussion and the preparation of joint ventures.

4. Scientific structure

In order to coordinate the observations as well as the analyses and the publication of the results, several commissions were established within the IMO. The various commissions specialize in a specific observing technique. In addition, a common annual observing calendar is distributed to publishers of amateur astronomical periodicals with the aim of drawing the attention of as many amateurs as possible worldwide to the international observing efforts and obtaining statistically significant coverage of meteor stream appearances.

Below, we present the commissions that are presently in operation.

4.1. VISUAL COMMISSION

A program for the Visual Commission was discussed and established at the International Meteor Conference in 1989. One major goal is to derive data concerning the structure of meteor streams (particle flux density, spatial number density, mass index) as well as the sporadic background. In order to achieve these goals, appropriate observing techniques are recommended. The program proposes different methods for major and minor streams.

All data of visual observations are stored in the Visual Meteor DataBase (VMDB). This database serves as the central archive and, at the same time, allows a detailed analysis of the data. The VMDB is the only global source of amateur data available for the study of meteor stream structure.

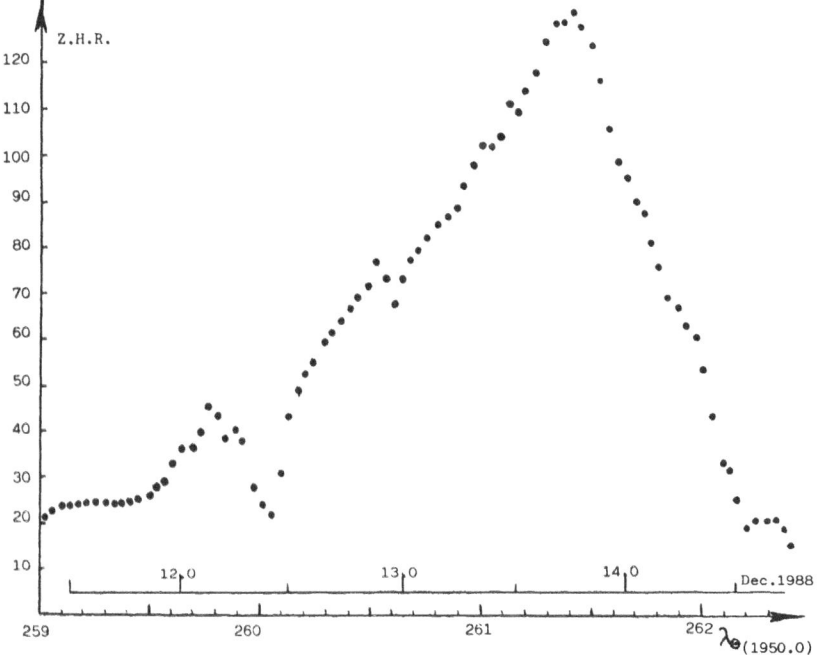

Figure 1. Activity profile of the 1988 Geminids, derived from IMO observations in the VMDB.

In order to demonstrate the effectiveness of the VMDB, we present as an example a summary of the analysis of the 1988 Geminid observations. The global effort to observe this shower yielded 14 193 Geminid meteors. The averaged zenithal hourly rates were plotted in an activity profile, shown in Figure 1 (Roggemans (1989)). The skew activity profile obtained is in very good agreement with results from Šimek and McIntosh (1989). It shows a gradual increase in activity from 20 Geminids per hour on December 12.0 to about 120 Geminids per hour on December 14.0. The peak occurs at solar longitude $\lambda_\odot = 261°38$ (1950.0) and covers 14 hours. The activity profile is furthermore characterized by a steep decrease immediately following the maximum peak: the Geminid activity dies out in about 24 hours. The asymmetric concentration of particles in the meteor stream can be explained as a consequence of the process of the meteor stream build-up (Fox et al. (1983)). The role of the assumed parent body, asteroid (3200) Phaeton, is not yet fully understood, as shown by Hunt et al. (1985).

4.2. PHOTOGRAPHIC COMMISSION

The IMO's Photographic Meteor DataBase (PMDB) contains some 1200 processed meteors. The data on these meteors allow radiant position and size determination. For the meteors photographed from at least two stations trajectory and heliocentric orbit could be derived. This long term project has to go on for many more years in order to accumulate sufficient information on the spatial distribution of meteor orbits in the Solar System.

4.3. FIREBALL DATA CENTER

Fireballs are striking and rare phenomena. They occur when meteoroids of reasonable size enter the atmosphere of the Earth. In the case of double or multiple station observations, it is possible to calculate the trajectory and the heliocentric orbit of such objects. This is of great interest, especially if a meteorite fell. Then, its relation to other minor bodies in the Solar System can be derived. Most data necessary for such calculations are provided by camera networks. Visual observations and reports from eye-witnesses may add important information such as the precise time of the event, fragmentation, and in particular accompanying sound phenomena. Eye-witness reports may be especially helpful in case a meteorite has to be searched. Visual and, if possible, photographic observations of persistent trains may also give additional hints on the material entering the atmosphere.

All available data on fireballs are included into a database of the IMO's FIreball DAta Center (FIDAC). The FIDAC database allows analyses of fireball rates and magnitude data (mass index) for rather large particles.

4.4. TELESCOPIC COMMISSION

Telescopic observations allow analyses of positions and structure of radiants. Furthermore, the magnitude range allows one to derive characteristics, such as the mass index, towards smaller particles than observed visually. A detailed program was also established at the International Meteor Conference in 1989.

4.5. RADIO COMMISSION

Radio observations are independent of clouds, daylight and other influences as city lights and so on. The radio program (not using radar but reflections of radio signals from commercial broadcasting stations) encourages people to observe on a more continuous basis. In this way, it is possible to obtain a complete activity profile down to fainter, poorly studied magnitude classes (+10). Studies of the relationship between the visual meteor appearance, their brightness, and the duration of the received echo signal are also needed.

5. References

Fox K., Williams I.P. and Hughes D.W. (1983), "The Rate Profile of the Geminid Meteor Shower", Mon. Not. R.A.S. 205, 1155–1169.

Hunt J., Williams I.P. and Fox K. (1985), "Planetary Perturbations on the Geminid Meteor Stream", Mon. Not. R.A.S. 217, 535–538.

Roggemans, P. (1988), Bibliographic Catalogue of Meteors 1794–1987, IMO.

Roggemans, P. (1989), "The Geminid Meteor Stream in 1988", WGN 17, 229–239.

Šimek M., McIntosh B.A. (1989), "Geminid Meteor Stream Activity as a Function of Particle Size", Bull. Astr. Inst. Czechosl. 40, 288–298.

VI

CIRCUMPLANETARY DUST
COLLISIONAL AND ELECTROSTATIC PROCESSES

PHYSICAL PROCESSES ON CIRCUMPLANETARY DUST

JOSEPH A. BURNS
Departments of Theoretical and Applied
Mechanics and of Astronomy
Cornell University
Ithaca NY 14853 USA

ABSTRACT. The life cycles of grains in circumplanetary space are governed by various physical processes that alter sizes and modify orbits. Lifetimes are quite short, perhaps 10^2-10^4 years for typical circumplanetary grains of 1 micron radius. Thus particles must be continually supplied to the circumplanetary complex, probably by the grinding down of larger parent bodies in collisions. Dust is eroded gradually through sublimation and through sputtering by the magnetospheric plasma but also is catastrophically destroyed through hypervelocity impacts with interplanetary micrometeoroids. Orbits evolve through momentum transfer (light drag, plasma or Coulomb drag, and atmospheric drag), and through resonant gravitational and electromagnetic forces. Plasma drag is generally the most effective evolution mechanism, with the possible exceptions of exospheric drag at Uranus and of electromagnetic schemes for some conditions. Since grains become charged (with typical electric potentials of a few volts), they undergo associated orbital perturbations: variable electromagnetic forces can cause the systematic drain of energy (orbital collapse) or, at specific (resonant) orbital locations can force large orbital inclinations/eccentricities. Solar radiation induces a periodic orbital eccentricity that can reach substantial values for 1 micron particles distant from the giant planets.

1. Introduction; The Circumplanetary Environment

Besides the interplanetary motes that are the subject of most articles in this book, dust is also found in faint ring systems about all of the giant planets (Burns *et al.* 1984, Ockert *et al.* 1987, Smith *et al.*1989, Esposito *et al.* 1991, Showalter *et al.* 1991, Showalter 1991) and exists at some level about the Earth and probably Mars (Soter 1971). Most of this material is located within several planetary radii of the parent planet concentrated near its equator. The typical circumplanetary particle has a size measured in microns. Even though the spatial density of circumplanetary dust is enormous compared to that of interplanetary grains, circumplanetary particles may still generally be considered to be isolated with mutual collisions being unimportant. The various physical processes that act on circumplanetary grains have been reviewed by Burns *et al.*(1979,1980, 1984), Grün (1984), Mendis *et al.* (1984), Mignard (1984), Schaffer (1989) and Goertz (1989).

These tenuous disks of material are of appreciable interest to planetary scientists for several reasons. First, they provide a testing ground for processes that may be obscured by the opacity of dense rings or that may be smoothed out by the frequent collisions occurring therein. Second, dust grains may regionally coat and color satellites embedded within the faint rings (cf. Cheng *et al.*1986). Third, the largest of the dust particles may prove hazardous to orbiting spacecraft such as Galileo and Cassini (Cuzzi *et al.* 1989, Burns *et al.* 1989).

A.C. Levasseur-Regourd and H. Hasegawa (eds.), Origin and Evolution of Interplanetary Dust, 341–348.
© 1991 *Kluwer Academic Publishers.*

Lastly, circumplanetary particles exhibit interesting dynamical behavior.

The discussion below will first describe the environment in which these particles reside, then will consider the electrical charge that develops on the grains, will tabulate the magnitude of the accelerations suffered by the typical grain, will list how the circumplanetary complex might be supplied and lost, and finally will depict a few orbital histories of grains.

Circumplanetary dust is affected by the planet's gravitational and magnetic fields, and is continually bombarded by the local magnetospheric plasma. The planet's gravity dominates the orbital motion of micron-sized particles so that orbits are nearly Keplerian but, owing to various perturbing forces (some of which are described below), do undergo gradual evolution. The giant planets have crudely dipolar magnetic fields (Stevenson 1983), much like Earth's, although the fields of Uranus and Neptune are highly tilted and displaced from their planet's centers. Each magnetosphere is filled by a temporally and spatially variable plasma; the constitutive ions differ from system to system, with protons, oxygen and sulfur prominent at Jupiter, protons and oxygen dominant at Saturn, and protons predominant at Earth, Uranus and Neptune. Because the magnetospheric plasma is tied to the planet's rotating magnetic field and, due to its approximately infinite conductivity, cannot sustain an electric field in its *own reference frame*, a "corotational" electric field $E=\Omega Br/c$ where Ω is the planet's angular spin rate, r is the radial vector to the particle, B is the local magnetic field and c is the speed of light) is present in *inertial* space (see Burns and Schaffer 1989). The electromagnetic and plasma environments for those planets with little or no intrinsic magnetic field are fixed by the solar wind's flow past the planet (see Horanyi *et al.* 1990).

Particles are also subject to the Sun's radiation, which produces a force (Burns *et al.* 1979, Mignard 1984) and affects the particle's charge, and to a continual rain of micrometeoroids, which can catastrophically shatter circumplanetary dust (Burns *et al.* 1980).

2. Electrical Charging

Surfaces in space build up electrical charges because they are bathed in a plasma and because they eject electrons through the solar photoelectric effect. Various schemes have been developed to compute the surface potential that will form under ambient conditions (Whipple 1981, Schaffer 1989). Typically the resultant potential is negative since, in the usual magnetospheric plasma, the temperatures of the electrons and the ions are nearly the same (owing to equipartition of energy), meaning that the electrons move much faster and are thus able to reach the surface first. The potential builds to the point where the current flow in ambient electrons and ions plus photo-ejected particles is zero. As a rough rule of thumb, a typical potential (expressed in volts) for an isolated particle in a hydrogen plasma is a factor of 2.5 times the plasma temperature (in electron volts); for a micron-sized grain, this corresponds to an excess charge N of several thousand electrons. Figure 1 gives some examples. Timescales for acquiring this charge depend on the composition and density of the local plasma as well as the grain's size since these properties determine the frequency at which charged particles strike the grain; for micron dust about the giant planets, charging timescales are tens of seconds to hours whereas orbits take tens of hours to complete.

The charge on any particular grain varies for several reasons. First the charging process is fundamentally stochastic because any plasma has a distribution of speeds and a charge's ability to gain access to the grain will depend on its speed. Typical variations are about $N^{1/2}/2$ (Schaffer 1989), a factor of two less than the usual statistical process because, once a grain's charge drifts away from the equilibrium value, that grain preferentially attracts charges of the appropriate sign so as to return toward equilibrium. Second, dust is likely to sample different plasma compositions, temperatures, and densities as the planet is orbited. Similarly, the plasma flow onto the grain depends partly on the dust's speed relative to the mean flow of the plasma (which approximately corotates with the planet's magnetic field) and the dust's speed varies systematically along any orbit but a circular equatorial one (Burns and Schaffer 1989, Northrop *et al.* 1989). Finally the particle's charge will be periodically modulated whenever the grain moves across the planet's shadow; this variation is only appreciable at Jupiter or closer to the Sun for elsewhere the total photoelectron current flow is negligible.

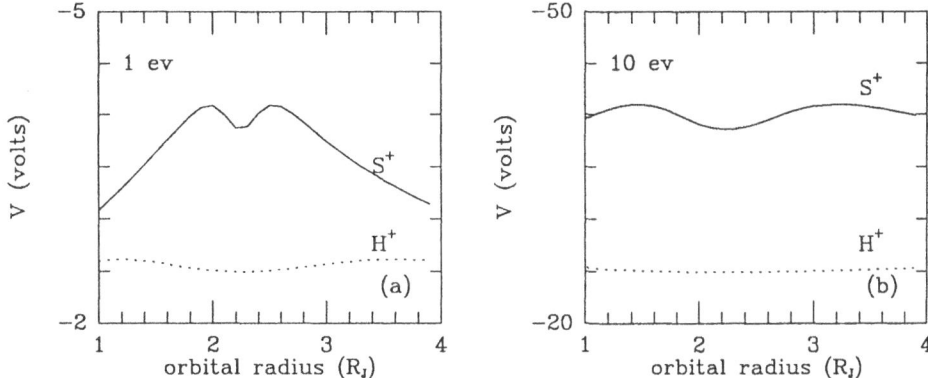

Figure 1. Equilibrium potentials vs. orbital radius in Jupiter's magnetosphere for two uniform density plasmas: a.) a S^+ or H^+ plasma with a temperature T of 1 ev; b.) T = 10 ev. The minima near the plots' centers occur at the synchronous orbit where Ω=n (from Schaffer 1989).

The above discussion assumes that the particles are isolated in space. O. Havnes, C. K. Goertz, G. E. Morfill, E. Whipple and their colleagues have addressed the circumstances under which a layer of dust might operate collectively to ward off the surrounding plasma or a planetary ring might develop vertical structure to support the electrostatic potential (Goertz 1989). For most tenuous planetary rings, with the unlikely exception of Uranus where a Coulomb lattice might develop, this complication of shielding is unimportant.

Electrical effects have also been invoked to modify particle sizes (see Burns *et al.* 1980, Mendis *et al.* 1984) and to produce Saturn's spokes (Grün *et al.* 1984, Goertz 1989).

3. Accelerations; Loss and Supply

Accelerations provide only a rough idea of the long-term significance of various forces since orbital evolution depends on the phasing of the perturbations (see, e.g., Burns 1976). Nevertheless in Table 1 we simply estimate the largest accelerations that act on 1 micron particles placed in the equatorial planes of the giant planets at r=1.8R, where R is the planetary radius. The acceleration due to planetary oblateness (J_2 being the gravity field's oblateness coefficient) is approximately $(3/2)J_2(R/r)^2$ *times* the local gravitational acceleration g. The solar radiation pressure acceleration is directed away from the Sun and is $(3SQ_{pr})/(4\rho csd^2)$, where S is the solar constant, Q_{pr} is the radiation pressure coefficient, ρ is the particle's density and s its radius, and d is the planet's distance from the Sun in AU (Burns *et al.* 1979). The Lorentz force felt by a charged grain moving relative to the planet's B field (assumed to be a dipole: $b_{1,0}R^3/r^3$) is $3\Phi(n-\Omega)b_{1,0}R^3/(4\pi\rho cGMs^2)$ *times* g, where n is the particle's orbital rate, Φ is the surface potential, G is the gravitational constant and M is the planet's mass (Schaffer and Burns 1987). Note that forces, such as drags, that depend on surface area πs^2 have accelerations varying as s^{-1} whereas electromagnetic accelerations differ as s^{-2} since surface potentials are generally constant for a specific plasma. This of course means that nongravitational accelerations, especially electromagnetic ones, become more important on small particles. The tabulated numbers for dust in the rings of Jupiter and Saturn come from Burns *et al.* (1984) while those of Uranus are taken from Esposito *et al.* (1991).

TABLE 1. Orbital Accelerations and Life/Death Processes for 1 µm Dust at 1.8 R

Accelerations (in cm/sec^2)	Jupiter	Saturn	Uranus	Neptune
Gravity ($\sim r^{-2}$)	800	350	275	350
Oblateness ($\sim r^{-4}$)	5	3	0.5	0.6
Radiation Pressure ($\sim s^{-1}$)	0.1	0.03	0.01	0.003
Electromagnetic ($\sim s^{-2}$)	1	0.05	0.03	0.02

Lifetimes (in years)	Jupiter	Saturn	Uranus	Neptune
Orbital Evolution				
Poynting-Robertson ($\sim s$)	10^5	$\leq 10^6$	10^6	10^6–10^7
Plasma Drag($\sim s$)	$2 \times 10^{2\pm1}$	$10^{5\pm1}$	$10^{5\pm1}$	$10^{6\pm2}$
Exospheric Drag ($\sim sr^{1/2}e^{-3/4}$)	—	—	$10^{2\pm1}$	10^5–10^6
Electromagnetic Variations ($\sim s^2 r^?$)	?	?	?	?
Erosion				
Sublimation ($\sim s$)	10^7	10^{17}	$\ll\infty$	$<\infty$
Sputtering ($\sim s$)	$10^{3\pm1}$	$10^{3\pm1}$	$10^{5\pm2}$	$10^{7\pm2}$
Micrometeoroid Shattering($\sim s^{-2}$)	$10^{5\pm1}$	$10^{6\pm1}$	$10^{6\pm2}$	$10^{6\pm2}$

Many processes act to eliminate small grains. Orbits evolve via radiation drag, plasma drag, atmospheric drag and electromagnetic effects; meanwhile tiny grains may be destroyed—but also be born—in catastrophic collisions, or may sublimate or sputter away. Fine debris may be sloughed off moonlets in mutual collisions (Cuzzi and Burns 1988, Colwell and Esposito 1990) or following energetic impacts by interplanetary meteoroids (Burns *et al.* 1980)—but debris may also reaccumulate on such parents and may thereby be temporarily lost from the complex. Minute grains may be launched by volcanoes or geysers onto circumplanetary orbits similar to that of the satellite. Particles may also condense directly from the local gas. Burns *et al.* (1984), Grün *et al.* (1984) and Esposito *et al.* (1991) summarize many of these processes and provide the timescales listed in Table 1.

4. Orbital Histories

The long-term orbital evolution of circumplanetary dust grains under various combinations of the perturbations described above has been an active research subject over the last few years. Most papers have merely demonstrated the efficacy of particular mechanisms for evolving dust. Here, because of limited space, we will restrict ourselves to simply showing the nature of some of the mechanisms; the original papers should be consulted for details of the calculations. As yet, no comprehensive treatment of these processes is available; nevertheless it is clear that such processes are important in transporting circumplanetary grains.

Most of these mechanisms rely on resonant forces; such periodic forces can produce long-term consequences if they are accompanied by a phase delay. Periodic Lorentz forces occur owing to motion through a spatially variable magnetic field (Fig. 2) or a periodic charge (Figs. 4 and 6). Since elliptical orbits always precess (e.g., due to J_2), any evolutions that rely on orbital alignment always are temporary (Figs. 5 and 6).

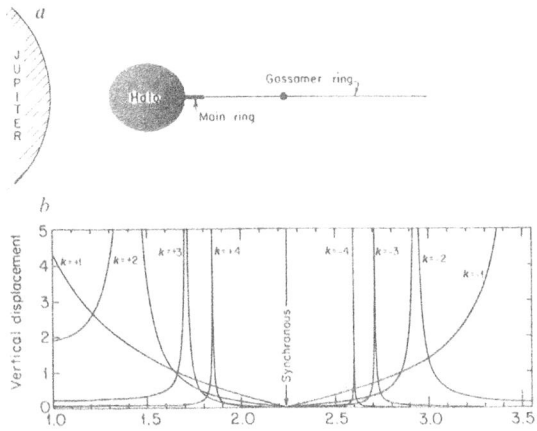

Figure 2. Jovian ring boundaries, plotted vs. planetary radii, are compared with positions of some low-order Lorentz resonances (LR). LR occur wherever the particle's orbital rate n is commensurate with the rate $k(\Omega-n)$ at which the kth component of the planet's magnetic field is traversed; LR peak at a doubly infinite set of planetary radii. At these positions the Lorentz force, which in general has both vertical and horizontal components, contains a frequency that resonates with n. Vertical responses to low-order Lorentz force components across the ring region are shown in arbitrary units. Jupiter's vertically extended halo appears to be confined by the k=2 and the k=3 LR (Burns *et al.* 1985).

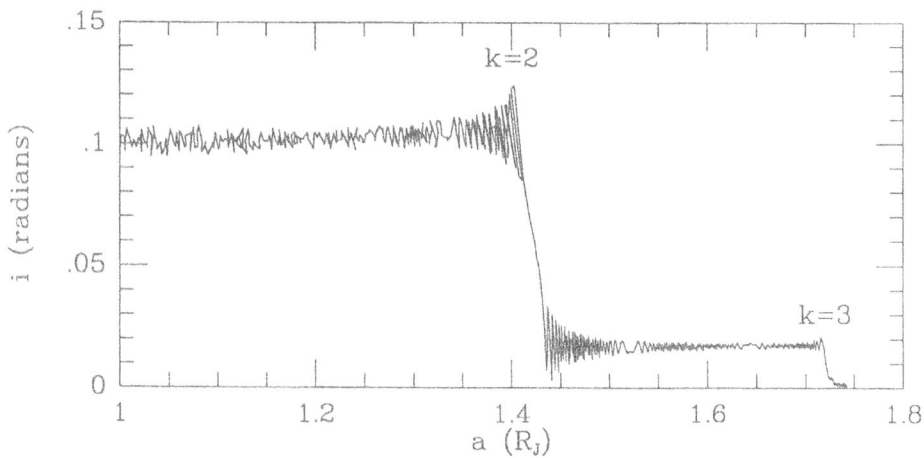

Figure 3. Passage through the k=3 and k=2 inner LR of Jupiter for a 1 μm, −1 volt grain that drifts inward radially at rate $\dot{a} \approx -0.2$ m/sec. As shown here, when particles move through Lorentz resonances, their inclinations (and eccentricities) undergo large jumps which persist after resonance passage if passage does not occur too rapidly. Dust from Jupiter's main ring may thus cause the halo. Similar phenomena occur at gravitational resonances. (From Schaffer 1989).

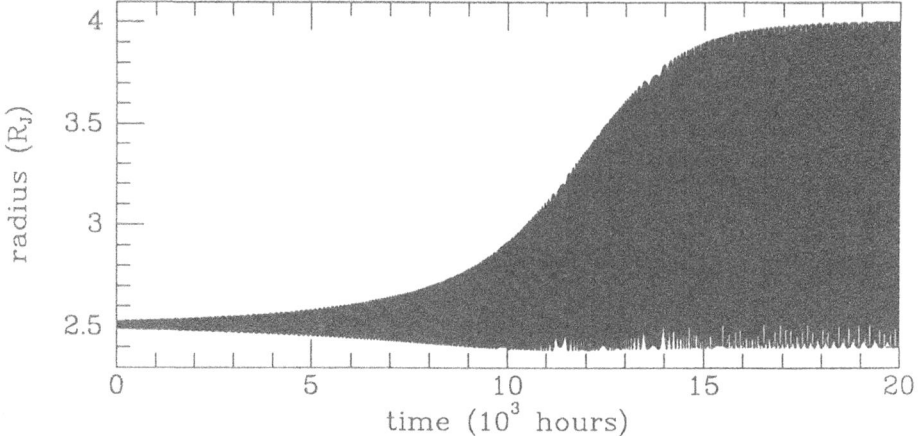

Figure 4. The evolution of a grain due to *resonant charge variation*. Grains that move along elliptical orbits experience charge variations that, if properly timed, can cause work to be done **by** (or **to**) the corotational electric field when integrated over a complete orbit; this work occurs because of a time delay in the grain's charging and is ultimately drawn from the orbit's energy (Burns 1976). For this plot, the grain was launched on a circular equatorial path at 2.54 R_J but, because of the electric charge it developed, its orbit quickly becomes elliptic, allowing this orbital evolution mechanism to proceed. (From Burns and Schaffer 1989).

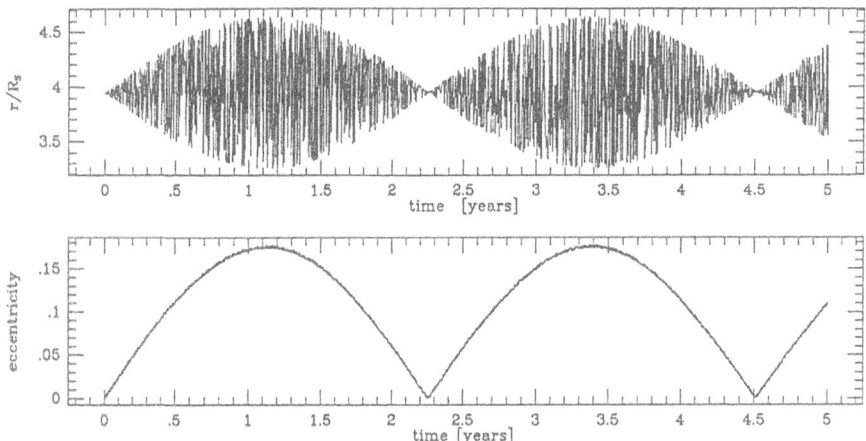

Figure 5. The effect of *solar radiation pressure*. Circumplanetary particles that start on prograde circular paths immediately develop an increasingly elongated orbit, initially pointing in the direction of the planet's orbital motion about the Sun. However, since the eccentricity growth per orbital period depends on the orientation of the elongated orbit relative to the solar force and since elliptical orbits precess (at roughly a constant rate) due to either the planet's oblateness, the corotational electric field or the planet's orbital motion about the Sun, the eccentricity growth will eventually be halted as the particle's orbital line of apses passes through the line directed toward the Sun. After this point the orbit's eccentricity gradually decreases until the orbit becomes circular once again at which time the cycle repeats. (From Burns and Horanyi 1990). Since energy is conserved, the orbit size does not change.

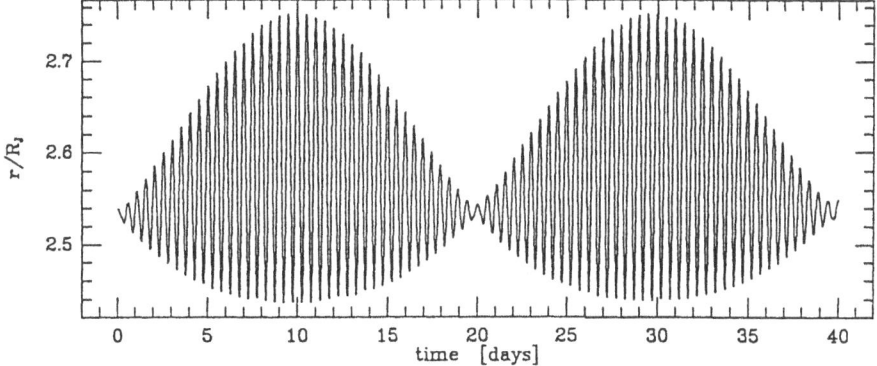

Figure 6. The evolution of a 1μm particle's orbit owing to *passage through Jupiter's shadow*, including solar radiation pressure. The latter force starts to drive an eccentricity oscillation as in Fig. 5. However shadow passage causes the charge to vary with the orbit period and thereby produces a force resonant with the orbit. The orbit precesses due to the planet's oblateness. The orbit's mean size is seen to vary with a twenty day period; the work done on the particle is non-zero over the orbit period because the charge on the grain when the particle is in the planet's shadow differs from that at similar orbital radii when in sunlight. The re-orientation ultimately changes the sign of the work done so that large orbital variations do not occur. (From Horanyi and Burns 1991).

5. Acknowledgments

Much of the work reviewed here was originally carried out in collaboration with Mark R. Showalter, Les Schaffer, Mihaly Horanyi or Douglas P. Hamilton, all of whom helped me understand the intricate physics of circumplanetary dust. NASA supported this research.

6. References

Burns, J. A. (1979) An elementary derivation of the perturbation equations of celestial mechanics. *Am. Jnl. Phys.* **44**, 944-949 (Erratum **45**, 1230).

Burns, J. A., and M. Horanyi (1990) Dynamics of the dust in Saturn's E ring. *Bull. Am. Astro. Soc.* **22** , 1042. Abstract in this volume.

Burns, J. A., and L. Schaffer (1989) Orbital evolution of circumplanetary dust by resonant charge variations. *Nature* **337**, 340-343.

Burns, J. A., P. L. Lamy and S. Soter (1979) Radiation forces on small particles in the solar system. *Icarus* **40**, 1-48.

Burns, J. A., M. R. Showalter, J. N. Cuzzi and J. B. Pollack (1980) Physical processes in Jupiter's ring: Clues to an origin by Jove! *Icarus* **44**. 339-360.

Burns, J. A., M. R. Showalter and G. E. Morfill (1984) The ethereal rings of Jupiter and Saturn, in R. J. Greenberg and A. Brahic (eds.), *Planetary Rings*, Univ. Arizona Press, Tucson, pp. 200-272.

Burns, J. A., L. Schaffer, R. J. Greenberg and M. R. Showalter (1985) Lorentz resonances and the structure of Jupiter's ring. *Nature* **316**, 115-119.

Burns, J. A., R. A. Kolvoord and D. P. Hamilton (1989) An assessment of potential hazards to the Cassini spacecraft from debris along satellite orbits. *JPL PD 699-11*, **13**, Section 6.

Cheng, A. F., P. K. Haff, R. E. Johnson and L. J. Lanzerotti (1986) Interactions of planetary magnetospheres with icy satellite surfaces, in J. A. Burns and M. S. Matthews (eds.), *Planetary Satellites*, Univ. Arizona Press, Tucson, pp. 403-430.

Colwell, J., and L. W. Esposito (1990) A numerical model of the Uranian dust ring. *Icarus* **86**, 530-560.

Cuzzi, J. N., and J. A. Burns (1988) Charged particle depletion surrounding Saturn's F ring: Evidence for a moonlet belt? *Icarus* **74**, 284-324.

Cuzzi, J. N., J. F. Cooper, L. L. Hood and M. R. Showalter (1989) Abundance and size distribution of ring material outside of the main rings of Saturn. *JPL PD 699-11*, **13**, Sec. 5.

Esposito, L. W., A. Brahic, J. A. Burns and E. A. Marouf (1991) Particle properties and processes in Uranus's rings, in J. T. Bergstralh, E. D. Miner and M. S. Matthews (eds.), *Uranus*, Univ. Arizona Press, Tucson, in press.

Goertz, C. K. (1989) Dusty plasmas in the solar system. *Rev. Geophys.* **27**, 271-292.

Grün, E., G. E. Morfill and D. A. Mendis (1984) Dust-magnetosphere interactions, in R. J. Greenberg and A. Brahic (eds.), *Planetary Rings* Univ. Arizona Press, Tucson, pp. 275-332.

Herbert, F. *et al.*(1987) The upper atmosphere of Uranus: EUV occultations observed by Voyager 2. *Jnl. Geophys. Res.* **92**, 15093-15109.

Horanyi, M., and J. A. Burns (1991) Orbital resonance due to planetary shadows. *Jnl. Geophys. Res.* in press. Abstract in this volume.

Horanyi, M., J. A. Burns, M. Tatrallyay, and J. G. Luhmann (1990) On the fate of dust lost from the Martian satellites. *Geophys. Res. Ltrs.* **17**, 853-856.

Johnson, R. E., L. A. Barton, J. W. Boring, W. A. Jesser, W. L. Brown and L. J. Lanzerotti (1985) Charged particle modification of ices in the Saturnian and Jovian systems, in J. Klinger, D. Benest, A. Dollfus and R. Smoluchowski (eds.), *Ices in the Solar System*, D. Reidel, Dordrecht, pp.301-317.

Mendis, D. A., J. R. Hill, W.-H. Ip, C. K. Goertz and E. Grün (1984) Electrodynamic processes in the ring system of Saturn, in T. Gehrels and M. S. Matthews (eds.), *Saturn*, Univ. Arizona Press, Tucson, pp. 546-589.

Mignard, F. (1984) Effects of radiation forces on dust particles in planetary rings, in R. J. Greenberg and A. Brahic (eds.), *Planetary Rings*, Univ. Arizona Press, Tucson, 333-366.

Northrop, T. G., D. A. Mendis and L. Schaffer (1989) Gyrophase drift and the orbital evolution of dust at Jupiter's gossamer ring. *Icarus* **79**, 101-115.

Ockert, M. E., J. N. Cuzzi, C. C. Porco and T. V. Johnson (1987) Uranian ring photometry: Results from Voyager 2. *Jnl. Geophys. Res.* **92**, 14969-14978.

Schaffer, L. E. (1989)The Dynamics of Dust in Planetary Magnetospheres. Ph. D. dissertation. Cornell Univ., Ithaca NY.

Schaffer, L., and J. A. Burns (1987) The dynamics of weakly charged dust: Motion through Jupiter's gravitational and magnetic fields. *Jnl. Geophys. Res.* **92**, 2264-2280.

Showalter, M. R. (1991) This volume.

Showalter, M. R., J. N. Cuzzi and S. M. Larson (1991) Structure and particle properties of Saturn's E ring. *Icarus* submitted.

Smith, B. A., and the Voyager imaging team (1989) Voyager 2 at Neptune: Imaging science results. *Science* **246**, 1422-1449.

Soter, S. L. (1971) The dust belts of Mars. *CRSR Rpt. 472*, Cornell Univ.

Stevenson, D. J. (1983) Planetary magnetic fields. *Rep. Prog. Phys.* **46**, 555-620.

Whipple, E. C. (1981) Potentials of surfaces in space. *Rep. Prog. Phys.* **44**, 1197-1250.

DUST IN PLANETARY RING SYSTEMS

Mark R. Showalter
Center for Radar Astronomy
Stanford University
Stanford, CA 94305
U.S.A.

ABSTRACT. Each of the outer gas giants Jupiter, Saturn, Uranus and Neptune is now known to be encircled by a system of rings. Some of these, such as the A, B, and C rings of Saturn and the nine narrow Uranian rings, are rather optically thick and are composed primarily of large bodies (1 cm to 10 m). However, every other system has been found to contain a large population of micron-sized dust. Such rings reveal the effects of a variety of physical processes that are also acting on interplanetary and interstellar grains. When such rings are examined as members of a general class, recurring patterns begin to emerge.

1. Introduction

Most of our knowledge about the diverse family of planetary rings comes from the recent reconnaissance of the outer planets by the Voyager spacecraft. This survey of the dusty rings will emphasize the current state of our knowledge about these rings' physical properties—what we know and how we know it. J. A. Burns presents in this volume a separate review of the dominant physical processes sculpting these rings.

For reasons that are not entirely clear, the known planetary rings show a nearly perfect anti-correlation between dust content and optical depth. This is undoubtedly related to the fact that dust tends to stick to larger particles, and so cannot survive for long in one of the denser rings. Hence, the study of dusty rings is synonymous with the study of the faintest planetary rings. Such rings are by their nature not always easy to detect, and so the data sets on some of these rings are quite limited in spite of the Voyager encounters. Indeed, some of the rings discussed below are only known because of a handful of Voyager observations, or in some cases just a single detection.

2. Physical Properties of the Dusty Rings

In this section I will present a very brief summary of the properties of the faint and dusty rings. Various general characteristics of these rings are also summarized in Table I.

A.C. Levasseur-Regourd and H. Hasegawa (eds.), Origin and Evolution of Interplanetary Dust, 349–356.
© 1991 *Kluwer Academic Publishers.*

2.1 JUPITER

Jupiter's ring system was discovered in a single image from the Voyager 1 flyby in 1979, and subsequently imaged in greater detail by Voyager 2. The most complete physical characterization of this ring system has been performed by Showalter *et al.* (1985, 1987). The ring appears to comprise three rather distinct components. The main ring is relatively thin and extends between orbital radii of 122,000 and 129,000 km, where the outer boundary is far more abrupt than the inner. Showalter *et al.* (1987) find some evidence for three brighter features, of which two appear to be associated with the embedded satellites Adrastea and Metis. Photometry indicates that the dust particle size distribution is compatible with a power law. It appears that a population of larger bodies is also present, which are seen in backscatter and presumably act as source bodies for the dust (Burns *et al.*, 1984).

Near the main ring's inner boundary arises the halo, a vertically extended cloud of material that seems to extend about halfway down to the top of Jupiter's atmosphere. This ring can be detected out to a full thickness of ~30,000 km. Its overall structure appears to be driven by resonant perturbations from Jupiter's inclined magnetic field (Burns *et al.*, 1985). No evidence for macroscopic bodies can be found in this region.

Finally, Showalter *et al.* (1985) were the first to note the presence of an even fainter ring extending outward from the main ring in a single Voyager image. This, dubbed the "gossamer" ring, appears to have a peak near the location of synchronous orbit at 160,000 km, which suggests a significant interaction with the local plasma. The ring does not have an abrupt outer boundary, but instead seems to fade out linearly, ending near the orbit of Thebe at 220,000 km. The lack of images at multiple phase angles precludes any possible determination of the particle sizes.

2.2 SATURN

The innermost, or D ring of Saturn seems to occupy most of the region between the C ring and the planet's cloudtops. In the few Voyager images where it can be seen, this ring shows a remarkable amount of structure. Two narrow ringlets are visible, with a number of fainter belts of material interspersed. This data set has never received the closer scrutiny it deserves, and as yet much remains to be learned about this ring. However, the distinct brightening of the ring at the highest phase angles clearly indicates diffraction by micron-sized dust. Recently, Marley and Porco (1990) have proposed that some of the D ring's structure may be associated with resonances with Saturnian f-mode oscillations; this possibility clearly warrants closer examination.

Travelling outward, the next ring to show a preponderance of dust is a narrow ringlet near the middle of the Encke gap in the A ring. This ring reveals a great deal of longitudinal variability, and also some clumps and kinks reminiscent of the "braided" F ring (Smith *et al.*, 1982). It is now known to share its orbit with the newly-discovered moon 1981S13 (Showalter, 1990). Once again, this ring has never received close scrutiny that it warrants, although an ample body of Voyager data exists. It is, however, clear that the ring brightens significantly at high phase angles, indicating a preponderance of dust.

The F ring itself was the first narrow and longitudinally variable ring observed (Smith *et al.*, 1981, 1982). It was, at the time at least, the prototypical "shepherded" ring, since it appears to be confined by the two nearby satellites Pandora and Prometheus. Cuzzi and Burns (1988) find evidence for an additional belt of smaller moonlets in the vicinity, based on charged particle absorption signatures detected by Pioneer 11. Photometry by Ockert *et al.* (1988) indicates that this ring has a narrow core of relatively high optical depth, along with a surrounding region composed primarily of dust. Furthermore, Burns *et al.* (1983) have found a very faint inward extension to this ring, which fills the region between the F and A rings.

Finally, Saturn has two very faint outer rings, designated G and E. Interest in these two rings has been rekindled recently because of the potential hazard they may pose to the Cassini orbiter. Hence, the most complete analyses of these rings was performed by a variety of investigators specifically for the Cassini Project (Cuzzi *et al.*, 1989). The isolated G ring was detected unambiguously in only two Voyager images, both at phase angles of 160°. It is roughly 7000 km wide, centered on a radius of 168,000 km, and is not associated with any known moon. Recently, however, I have found faint traces of the ring in a number of images taken at a slightly lower phase angle. Strangely, the ring appears to be far narrower in these views, implying a core/halo configuration. The ring also shows significant longitudinal variability.

On the other hand, the E ring of Saturn encompasses an area larger than that of all other planetary rings put together. It has a peak in density coincident with the orbit of the moon Enceladus, but encompasses the orbits of Mimas, Tethys and Dione as well. The ring is also quite thick, ranging from 10,000 to 50,000 km in vertical extent from its inner to its outer boundary. Showalter *et al.* (1991) combine a number of Earth- and Voyager-based observations to demonstrate that the ring contains a very narrow distribution of particle sizes, centered on roughly one micron. This narrow distribution precludes the possibility that the ring arises from collisional or disruptive processes, making it unlike any other known ring. Many investigators have proposed that the ring arises from "geysers" or "ice volcanos" directly from the surface of Enceladus.

2.3 URANUS

The nine "classical" Uranian rings have high optical depths and very little dust. One additional narrow ring discovered in the backscattered Voyager images (Smith *et al.*, 1986), 1986U1R (λ), has been found to be somewhat of an exception. In a single, very high phase Voyager image, this ring was found to be far brighter than any of its siblings, indicating that dust is a major constituent. Careful photometry by Ockert *et al.* (1987) has revealed this ring to have significant azimuthal variability.

The particular high-phase image noted above also reveals the presence of an extraordinary family of dust belts around and among the better known rings. Colwell and Esposito (1990) have developed models to describe how the neighboring rings can be both sources and sinks for this material.

Finally, one additional broad ring, designated 1986U2R, is visible in a single Voyager image at a 90° phase angle. This ring is interior to all of the structure discussed above, with a peak at a radius of 38,000 km and a radial width of ~5,000 km. Unfortunately, very little can be determined about the particle properties of this ring from a single view, although a predominance of dust is strongly suspected.

2.4 NEPTUNE

The Voyager cameras revealed a number of new dusty rings in orbit about Neptune, in addition to helping settle the mysteries of the ring "arcs" originally detected from the ground. Smith *et al.* (1989) have designated four major ring components. Two relatively prominent narrow rings, N53 and N63, have radii of 62,900 km and 53,200 km, respectively. Each of these lies roughly 1000 km beyond the orbit of a newly discovered satellite, and a dynamical association seems likely. It is the outer of these rings that harbors the three prominent arcs. In addition, a so-called "plateau" extends outward from N53 halfway to N63, with roughly constant opacity. Finally, N42 is a much fainter and broader ring at a radius of 41,900 km, which does not have abrupt boundaries like the others. Very preliminary photometric modeling by Smith *et al.* indicates that all of these rings contain a mixture of dust and larger bodies representing comparable optical depths.

3. Categorization of the Dusty Rings

Table 1 provides a quick summary of the properties of all the rings discussed above. It becomes apparent that the rings show recurring patterns, that are likely to be an outward manifestation of common underlying processes. To pursue this further, I have identified four general sets of traits that seem to accommodate all of the rings described. Each is discussed briefly in turn below.

3.1 CATEGORY 1: NARROW RINGS

The category of narrow rings includes the Encke ringlet, F ring and G ring core at Saturn, the Uranian λ ring and rings N53 and N63 of Neptune. Interestingly, all of these but N53 also show some evidence for longitudinal variability. The inverse observation, that all longitudinally variable rings are narrow, is not much of a surprise—a ring can only be variable if some force acts to counteract Kepler shear, and narrower rings have less shear to be counter-acted. Nevertheless, this observation seems to suggest that such counteracting forces are quite prevalent, so that rings will generally clump if Kepler shear is sufficiently small.

It is likely not a coincidence that all of these rings except the G ring have known moons nearby. The gravity of these moons probably provides the perturbations needed to confine clumps, although detailed mechanisms behind these effects are not always clear. These rings also tend to fall in the region around the planetary Roche limit, which is the place where mixtures of ring material and small moons are most common. The G ring is particularly puzzling if its azimuthal variability is confirmed, since Van Allen (1983) has placed very firm limits on the mass of any moons in the vicinity.

Lastly, these particular rings all contain populations of larger bodies mixed in with the dust. It may be that dust production is merely a byproduct of the processes required to keep the larger bodies in these rings confined.

3.2 CATEGORY 2: ONE-WAY EXTENSIONS

This classification describes rings that extend in one preferred direction from a region of higher optical depth. It includes the Jovian halo, the inward extension to Saturn's F ring, several of the individual Uranian dust belts, and Neptune's "plateau" outside of N53. Most of these rings are believed to comprise dust evolving radially away from the denser source region in one preferred direction. However, it remains puzzling why some of these rings have an opposite boundary that is also relatively abrupt.

With the exception of Neptune's "plateau", all of these rings are composed of 100% dust.

TABLE I. A general summary of the physical properties of dusty planetary rings.

	Optical depth	% dust	Thickness (km)	Width (km)	Ring Properties* a	b	c	d	e	f
JUPITER										
Halo	10^{-6}	100	~10,000	30,000	-	-	-	♦	-	2
Main Ring	3×10^{-6}	~50	≤300	7,000	-	♦	♦	-	♦	4
"Gossamer"	10^{-7}	?	≤1,000	80,000	-	♦	-	-	-	3
SATURN										
D Ring	$<10^{-3}$	50–100	≤10	100–10,000	-	-	♦	♦	♦	4
Encke Ringlet	0.1	100(?)	≤2	20	♦	♦	-	-	-	1
F Ring extension	$<10^{-2}$	100(?)	?	3,000	-	♦	-	♦	-	2
F Ring body	0.1–1	>50	≤10	1–100	♦	♦	-	-	-	1
G Ring body	10^{-6}	100	<500	7,000	-	-	♦	-	-	3
G Ring core	10^{-6}	>70	<50	<1,000	♦	-	-	-	-	1
E Ring	10^{-5}	100	~10,000	120,000	-	♦	-	-	-	3
URANUS										
1986U2R	10^{-4}–10^{-3}	100(?)	≤100	4,000	-	-	-	-	-	3
Dust belts	$≤10^{-5}$	100(?)	≤400	~10,000	-	♦	-	♦	♦	4
1986U1R (λ)	10^{-3}	>95	≤100	100	♦	♦	-	-	-	1
NEPTUNE										
N42	10^{-4}	40–70	?	12,000	-	-	-	-	-	3
N53	10^{-2}	40–70	?	~100	-	♦	-	-	-	1
"Plateau"	10^{-2}	20–50	?	5,000	-	-	-	♦	-	2
N63	10^{-2}–10^{-1}	20–50	?	20–100	♦	♦	♦	-	-	1

*Ring properties defined:
a: Azimuthal variations are observed.
b: Moons are nearby.
c: A denser core is observed.
d: Attached to a denser ring.
e: Internal structure is observed.
f: Assigned category number.

This seems consistent with the general impression that nongravitational drag forces are behind the observed configuration. To be complete, I must also mention that two of the "classical" Uranian rings, δ and η, also show faint radial extensions. However, these are generally not composed of dust, and may have an entirely different explanation.

It is perhaps interesting to note that the extension's direction is nearly as often radially outward as inward, suggesting that a variety of drag forces are responsible for the evolution (see Burns *et al.*, 1984). Also, one should recognize that the neighboring "source" region need only be denser in the relative sense—the Jovian halo is attached to a ring with $\tau \sim 10^{-6}$, whereas the Uranian dust belts are attached to rings with $\tau \geq 1$.

3.3 CATEGORY 3: "FUZZY" RINGS

This category includes the more diffuse rings, those lacking in any abrupt boundaries. The Jovian "gossamer" ring, the G and E rings of Saturn, 1986U2R at Uranus and N42 at Neptune all seem to fit into this category. Such rings are often regarded as showing dust as it spreads away from a source region, and in that sense might be better called "two-way extensions" by analogy to Category 2. Like the rings in Category 2, these also tend to be composed entirely of dust (although N42 is an exception). However, the mechanisms behind this dispersal are not clear, and in fact the source regions are also unknown in several instances (e.g. 1986U1R, N42). Perhaps, very low optical depth belts of moonlets are hidden away in these regions.

It should be noted that no single drag force can explain these rings, in which material seems to evolve in both directions simultaneously. Mechanisms yielding a "random walk" may be required. Burns and Horanyi (1990) have come the furthest in explaining the peculiar radial structure of Saturn's E ring, by noting how solar perturbations can pump up the eccentricities of dust grains. In this case, then, the radial profile does not represent orbital evolution; it is merely the superposition of many grains on different eccentric orbits. However, it is unclear how relevant this particular mechanism is to the other rings in this category.

3.4 CATEGORY 4: STRUCTURED RINGS

Finally, I take note of the remaining dusty rings, which generally show more complex kinds of structure. Included are the main Jovian ring, along with Saturn's D ring and the family of Uranian dust belts. These rings all show structures at a wider variety of radial scales. They may be nothing more than hybrid versions of the previous categories, in which the dust distribution is built upon a more complex "skeleton" of source bodies and rings.

3.5 OTHER RESULTS

Surprisingly, the categorization just accomplished also reveals that some of the most obvious ring traits do not correlate with ring type. First, there is no obvious connection between category and ring optical depth. Although low optical depth may be required for the dust to survive, the actual value of τ does not seem to correlate with the underlying processes. Second, ring types do not correlate with planet. In fact, it is interesting to note that every planet has a ring from at least three of the four categories (despite the fact that Jupiter and Uranus only have three rings tabulated!). Finally, ring type does not correlate with particle composition (which tends to be silicates at Jupiter, water ice at Saturn, and carbonaceous material at Uranus and Neptune).

The categorization I have outlined does have some predictive value, which can be tested as the rings described above are better understood. For example, the ratio of large bodies to dust in Neptune's N42 ring and plateau, based on the preliminary photometry of Smith *et al.* (1989), appears to be out of line with that expected for categories 2 and 3. It will be interesting to perform a more complete analysis of the data and see if this discrepancy persists. Nevertheless, we have found that the outward similarities between the faint and dusty rings do seem to imply some common underlying processes.

4. References

Burns, J. A., J. N. Cuzzi, and M. R. Showalter (1983). Discovery of gossamer rings. *Bull. Amer. Astron. Soc.* **15**, 1013–1014.

Burns, J. A., M. R. Showalter, and G. Morfill (1984). The ethereal rings of Jupiter and Saturn. In *Planetary Rings* (R. Greenberg and A. Brahic, Eds.), pp. 200–272. University of Arizona Press, Tucson.

Burns, J. A., L. E. Schaffer, R. J. Greenberg, and M. R. Showalter (1985). Lorentz resonances and the structure of the Jovian ring. *Nature* **316**, 115–119.

Burns, J. A., and M. Horanyi (1990). Dynamics of the dust in Saturn's E ring. *Bull. Amer. Astron. Soc.* **22**, 1042.

Colwell, J. E., and L. W. Esposito (1990). A numerical model of the Uranian dust rings. *Icarus,* in press.

Cuzzi, J. N., and J. A. Burns (1988). Charged particle depletion surrounding Saturn's F ring: Evidence for a moonlet belt? *Icarus* **74**, 284–324.

Cuzzi, J. N., J. F. Cooper, L. L. Hood, and M. R. Showalter (1989). Abundance and size distribution of ring material outside the main rings of Saturn. In *Cassini Mission Proposal Package.* JPL PD 699–11, **13**, (5-1)–(5-F-7).

Marley, M. S., and C. C. Porco (1990). D ring features and f-mode oscillations of Saturn. *Bull. Amer. Astron.Soc.* **22**, 1041.

Ockert, M. E., J. N. Cuzzi, C. C. Porco, and T. V. Johnson (1987). Uranian ring photometry: Results from Voyager 2. *J. Geophys. Res.* **92**, 14969–14978.

Ockert, M. E., J. B. Pollack, and M. R. Showalter (1988). Voyager photometry of Saturn's F ring II: the results. *Bull Amer. Astron. Soc.* **20**, 854.

Showalter, M. R., J. A. Burns, J. N. Cuzzi, and J. B. Pollack (1985). The discovery of Jupiter's "gossamer" ring. *Nature* **316**, 526–528.

Showalter, M. R., J. A. Burns, J. N. Cuzzi, and J. B. Pollack (1987). Jupiter's ring system:

New results on structure and particle properties. *Icarus* **69**, 458–498.

Showalter, M. R. (1990). Visual detection of 1981S13, the Encke gap moonlet. *Bull. Amer. Astron. Soc.* **22**, 1040–1041.

Showalter, M. R., J. N. Cuzzi, and S. M. Larson (1991). Structure and particle properties of Saturn's E ring. Submitted.

Smith, B. A., *et al.* (1981). Encounter with Saturn: Voyager 1 imaging science results. *Science* **212**, 163–191.

Smith, B. A., *et al.* (1982). A new look at the Saturn system: The Voyager 2 images. *Science* **215**, 504–537.

Smith, B. A., *et al.* (1986). Voyager 2 in the Uranian system: Imaging science results. *Science* **233**, 43–64.

Smith, B. A., *et al.* (1989). Voyager 2 at Neptune: Imaging science results. *Science* **246**, 1422–1449.

Van Allen, J. A. (1983). Absorption of energetic protons by Saturn's G ring. *J. Geophys. Res.* **88**, 6911–6918.

READING SATURN'S RING SPOKES

Laurance R. Doyle
SETI Institute, 245-7, NASA Ames Research Center, Moffett Field, CA 94035, U.S.A.

Eberhard Grün
Max-Planck-Institut für Kernphysik, Postfach 103980, D-6900 Heidelberg 1, F.R.G.

ABSTRACT. The micron-sized dust forming the radial spoke-like features in Saturn's rings are studied using radiative transfer analysis. Theories for their likely origin and evolution are discussed in light of these results, and future work is outlined.

1. Introduction.

Radial cloud-like features were first confirmed to exist in the outer B Ring of Saturn in October 1980 by the Voyager spacecraft. (A number of observers over the preceeding century had observed such features in the B and A Rings; see Doyle and Grün 1990 for a review.) These *spokes* were observed to be dark in backscatter, but brighter in forward scatter, indicative of micron-sized particles. Spokes are classified into three categories, *extended*, *narrow*, and *filamentary*, depending on shape, radial-azimuthal extent, location in the outer B Ring, and optical depth. Extended spokes were observed to form radially with the trailing edge maintaining a semi-corotating radial orientation and the leading edge taking on a Keplerian shear during the formation process, giving the spokes a triangular appearance. After complete formation, both edges of the spokes were observed to take on a Keplerian motion (Grün *et.al.* 1983).

Enhanced magnetic field activity (longitudinal regions of the rings where both the Saturn Electrostatic Discharge as well as the Saturn Kilometric Radiation originate) has also been correlated with periodically enhanced spoke optical depth, indicating that the spokes phenomena is closely tied to electromagnetic processes in the Saturn system (Porco and Danielson 1982, Porco 1988). One suggestion, for example, is that spokes may be related to turbulence in Saturn's magnetosphere (Burns *et.al.* 1983). However, the most popular theory for spoke formation is that they result from a plasma produced by the impact onto the rings of large *macrometeoritic* particles (meter-sized) (Goertz and Morfill 1983). This plasma, under Lorentz forces, propagates radially, electrostatically levitating regolithic dust off the surfaces of the larger B Ring boulders and producing the spokes. When the contribution from a secondary plasma (from the neutral hydrogen cloud) is taken into account, the effective size of the individual spoke particles is predicted to be around $0.6\mu m$ (Goertz 1984).

Several factors can be observed to differentiate between some of these theories and see what else the spokes can tell us about processes in the Saturn system. For example, there appears to be an asymmetry in the optical depth distribution of the spokes (they appear to be enhanced on the morning compared to the evening ansa). This can be understood in terms of the greater kinetic energy produced by the increased impact velocity of macrometeoritic material onto the midnight side compared to the noon side of the rings. This occurs because Saturn's orbital velocity vector happens to add to the rings rotational velocity vector on the midnight region of the rings, while it subtracts from the rotational velocity vector on the noon side of the rings (Durisen *et.al.* 1989). The individual size of the spoke particles themselves would also provide useful constraints on the plasma strength necessary to levitate these particles off the larger B Ring boulders. Such determinations can begin with radiative transfer modeling of the spoke and ring photometric data.

A.C. Levasseur-Regourd and H. Hasegawa (eds.), Origin and Evolution of Interplanetary Dust, 357–360.
© 1991 *Kluwer Academic Publishers.*

2. Phase Curve Analysis of Spokes.

The optical depth of the spokes can be determined through an analysis of the forward scattering portion of the spokes-rings phase curve - the low phase angle region being dominated by the backscattering regolith of the larger B Ring boulders. A typical spoke phase curve, where I is the reflected flux, and πF is the solar incident flux, is plotted in Figure 1. (We have shown the logarithm of the reflectivity to emphasize the high phase angle data, plotting the most accurately known data after Doyle et.al. 1989; jagged edges being due to varying geometry from image to image.) We modeled this phase curve with full multiple scattering considerations (including the effects of Saturnshine as well as the solar flux) varying the large and small particle albedos, phase functions, and ratios (Doyle et.al. 1989). Among other results we found that the B Ring is very backscattering, so that a Callisto large particle phase function fits best, while the small particle (dust) optical depth of the spokes was best fit at 2% (with the 0% model also shown), an amount significantly less than previous estimates. (This latter result significantly limits the effects of angular momentum transport by dust in the rings.)

We also found essentially *zero* free micron-sized dust outside the spoke regions in the outer B Ring, so that it can be expected that the larger B Ring boulders can sweep up free dust very rapidly (a more intuitive result when one considers the spokes' lifetimes of only a few hours). The large boulders' phase function also is seen to change slightly under the spokes, a strong indication that this regolith is indeed the source of the spoke particles. We found a rather high large particle albedo (about 0.56) indicating that the rings contain less than about 1% mass fraction of darker interplanetary *micro*meteoritic dust. Conservatively assuming that the micrometeoritic flux has been constant in time, and has the same density at 9 AU as at 1 AU, we find that the rings could not have been "catching" micrometeoritic dust for more than about 2×10^8 years, a time significantly younger than the age of the solar system.

3. Color Analysis of the Spokes.

We then examined identical spokes in various Voyager filter bandpasses. Accounting for possible changes in the large particle phase function under spokes (since the spoke particles appear to originate from the large particles' regolith) as well as the effects of multiple scattering and the large particle color albedo, the spoke particles nevertheless seem to increase in optical depth with increasing wavelength, as shown in Figure 2 (with a linear least-squares fit). This increase, we have found, is likely due solely to the single scattering efficiency term in the optical depth of the spoke particles. With this result we can scale a spoke's optical depth to extinction efficiency (Q_e), and the wavelengths of observations to size parameter, $x = 2\pi r/\lambda$, (where r is the particle effective radius and λ is the wavelength of observation), putting at least a lower bound on the spoke particle sizes. We have therefore scaled reflectivity to extinction efficiency, and wavelength to size parameter in order to use the spoke particles' sloping scattering properties (as indicated by their location on the extinction efficiency curve) as a template constraining the range of their effective radii. This is illustrated in Figure 3, where typical spoke photometric data from four separate filters is shown superimposed on the extinction efficiency curve (the solid curve for a single particle size and the dotted curve for a size distribution of about $r \pm \sqrt{0.1}r$, see Doyle et.al. 1989).

From this work we find that the effective spoke particle sizes themselves must be larger than about $0.4 - 0.5\mu m$ in radius and, in some cases, narrowly constrained around $0.6 \pm 0.2\mu m$. This result agrees surprisingly well with the effective spoke particle radius predicted by the theory for the spokes originating by a macrometeoroid-impact-produced plasma. Although such a narrow size distribution may be unusual in longer-lived phenomena, an individual spoke lasts only a few hours. The smaller constraint on size is likely due to the need for a large enough impact cross section to "catch" at least one electron from the propagating plasma, while the upper constraint is likely due to centrifugal disruption (Meyer-Vernet 1984). Our results here, then, appear to support again a macrometeoroid impact model for spoke formation.

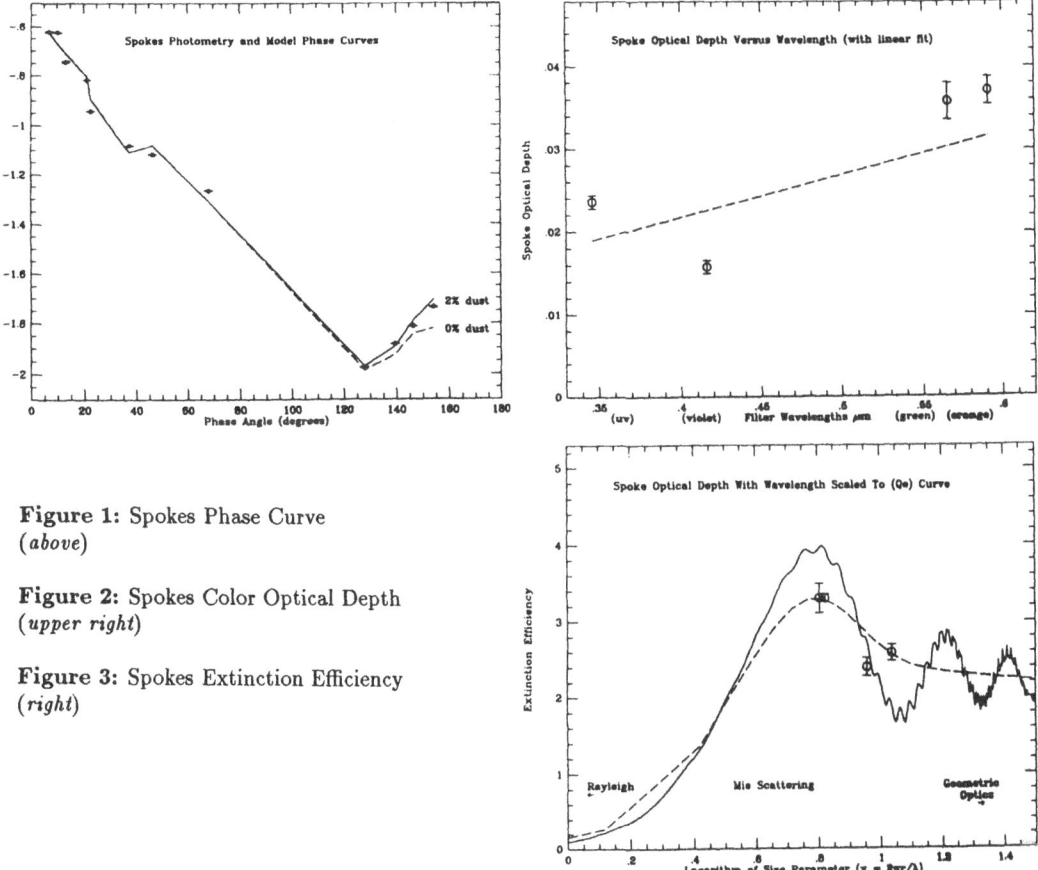

Figure 1: Spokes Phase Curve
(*above*)

Figure 2: Spokes Color Optical Depth
(*upper right*)

Figure 3: Spokes Extinction Efficiency
(*right*)

4. Conclusions.

Our study of the photometry and modeling of the spokes in Saturn's rings has lead to a number of interesting conclusions. First, we have determined that the spokes contain virtually all of the free dust in the outer B Ring (in the A Ring as well, from other work; Dones 1987). This result mitigates against theories of ring formation by loosely held particles. We have also determined that the spokes optical depth is only about 2%, allowing some relaxation on the total plasma strength required for their formation and propagation. We have also found that the spoke particle sizes we obtained support macrometeoroid impacts as a major source for the origin of spokes with effective spoke particle sizes, in general, narrowly constrained to a size distribution of around $0.6\mu m$. Finally, we have found that the general ring large particle albedo is too large to contain more than about 1% by mass of darker micrometeoritic dust, and so could not have been out in this interplanetary "rain" for more than a couple of hundred million years. The rings have about the same mass as Saturn's moon Mimas (3.65×10^{22} grams) which, as a sphere, has only about 10^{-5} the surface area of the rings. We therefore conclude that the ring material could have been in this form until disruptive capture by Saturn, and therefore that the rings of Saturn did not originate at the same time as Saturn formed, but are instead a fairly recent phenomena.

5. Future Work.

First: We would like to further deconvolve the processes by which spokes originate by examining the Fourier spectrum of spoke events. If a large contribution to spokes structure is due to Saturn ionospheric turbulent, then the spokes may occur with a turbulence or Kolmogoroff-like spectrum while, if they are due to macrometeoroid bombardment, a Fourier spectrum resembling, for example, a cratering distribution should be more evident.

Second: Polarization data on the spokes would also allow a better estimate of the spokes formational and evolutionary dependence on magnetic field effects. These could be obtained with the Hubble Space Telescope at low phase angles, but would have to await the Cassini Saturn Orbiter spacecraft for high phase angle coverage. The Cassini orbital insertion phase will also likely allow the only near-term opportunity to obtain wide phase angle coverage over a short enough time period (a few hours) to very accurately characterize the particle properties of individual spokes and determine how these characteristics (size distribution, for example) evolve over short time periods.

Third: We would like to model the spoke particles' formation trajectories and "sweep up" times onto the vertical cross-sections of the larger B Ring boulders. The B Ring contains fully 3/4 ths of the mass of Saturn's rings, and its optical depth has not yet been measured (only a lower limit has been determined by the saturation of the photopolarimeter occultation experiment aboard Voyager). With this approach we could estimate for the first time the vertical cross-sectional area of the optically thick outer B Ring and determine its optical depth as well as resolve if it is a many layered or mono-layered ring structure.

Fourth: Finally, if we can learn to read these "hieroglyphics" of the rings, we may likely be able to accurately determine the macrometeoroid flux in the outer solar system, a determination that would not only tell us more about conditions in the outer solar system and possibly formation processes in the outer part of the early protoplanetary disc, but also provide much better impact safety estimates for outer solar system bound orbiting spacecraft such as Cassini, as well.

6. References.

Burns, J.A., M.R. Showalter, J.N Cuzzi, and R.H. Durisen 1983, "Saturn's Electrostatic Discharges: Could Lightning Be the Cause?", *Icarus* 54, 280-295.

Dones, L. 1987, "Dynamical and Photometric Studies of Saturn's Rings", *Ph.D. Dissertation*, University of California, Berkeley.

Doyle, L.R., L. Dones, and J.N. Cuzzi 1989, "Radiative Transfer Modeling of Saturn's Outer B Ring", *Icarus* 80, 104-135.

Doyle, L.R., and E. Grün 1990, "Radiative Transfer Modeling Constraints on the Size of the Spoke Particles in Saturn's Rings", *Icarus* 85, 168-190.

Durisen, R.H., N.L. Cramer, B.W. Murphy, J.N. Cuzzi, T.L. Millikin, and S.E. Cederbloom, 1989, "Ballistic Transport in Planetary Ring Systems Due to Particle Erosion Mechanisms. I. Theory, Numerical Methods, and Illustrative Examples", *Icarus* 80, 136-166.

Goertz, C.K., and G.E. Morfill 1983, "A Model For the Formation of Spokes in Saturn's Rings", *Icarus* 53, 219-229.

Goertz, C.K. 1984, "Formation of Saturns's Spokes", *Adv. Space Res.* 4, 137-141.

Grün, E., G.E. Morfill, R.J. Terrile, T.V. Johnson, and G. Schwehm 1983, "The Evolution of Spokes in Saturn's B Ring", *Icarus* 54, 227-252.

Meyer-Vernet, N. 1984, "Some Constraints On Particles in Saturn's Spokes", *Icarus* 57, 422-431.

Porco, C.C., and G.E. Danielson 1982, "The Periodic Variation of Spokes in Saturn's Rings", *A.J.* 87, 826-833.

Porco, C.C. 1988, "Dual Periodicity in the Appearance of Spokes", *B.A.A.S.* 20, 852 (abstract).

CATASTROPHIC DISRUPTION OF SOLID BODIES BY COLLISION — EXPERIMENTAL APPROACH —

A. FUJIWARA
Department of Physics,
Kyoto University,
Sakyo-ku, Kyoto 606, Japan

ABSTRACT.   The results on mass, velocity, shape, and rotation of the fragments produced in the disruption experiments by impact are surveyed. Some future works are also suggested.

1. Introduction

Study of the outcomes of collisions between the solid bodies is crucial to understand the nature, origin, and evolution of interplanetary dust particles in many respects: (1) Some of the dust particles are undoubtedly supplied as the outcome of the asteroid-asteroid, or asteroid-dust collisions as found as IRAS dust bands, which are closely connected with some of asteroid families, and the structure and evolution of the dust bands are controled by the collisional processes (Sykes *et al.*, 1989). (2) Collisions among dust particles are believed to play an important role to supply the β-meteoroids, which are observed in the recent direct detection of dust particles by spacecrafts. (3) Dust particles are considered to be the end products of the collisional processes among the solid components of various size classes since the formation of the solar system.

To solve these collisional evolution and identify the sources of the interplanetary dust particles, the most fundamental quantities to be investigated on the collisional outcomes are: (1) mass (size) distribution of the fragments, and (2) velocity distribution of the fragments. Other quantities are also useful for some purposes; for examples, informations of the fragment shape is useful to identify whether dust particles are collisional origin or not, by the recovered samples or scattering properties of the dust particles, and to evaluate the Yarkovski effect working on the spinning dust particles, rotational frequencies of the collisional fragments must be known. Impact phenomena are classified into two classses: cratering and disruption. Concerning the size and velocity distribution of the fragments in cratering, Gault and Heitowit's (1963) classical work is still useful and valuable, although more systematic works are strongly desired. On the contrary, studies concerning the impact disruption have made great progress in the past decade. In this review, we mainly focus on the outcomes of the impact destruction found through experiments. Main

A.C. Levasseur-Regourd and H. Hasegawa (eds.), Origin and Evolution of Interplanetary Dust, 361–366.
© 1991 *Kluwer Academic Publishers.*

experimental results and scaling laws in the context of the catastrophic collision of asteroididal bodies until 1989 are summarized in Fujiwara *et al.*(1989). In this paper, some of the important points and recent topics are reviewed.

2. Experimental Methods

Most commonly made experiments are laboratory impact experiments using powder guns and sigle-stage or two-stage light-gas guns which can launch typically cm-sized projectiles up to several km/s. Shaped charge method is also used in higher velocity experiments (about 9km/s)(Capaccioni *et al.*,1986). Materials used for the projectiles and the targets are metals, rocks, and ices. Recently disruption experiments in the open air using explosives have been conducted by an Italian group (Martelli *et al.*, 1990). In these experiments aerodynamical drag for small fragments should be taken into account, while the experiments using guns are carried out in the vacuum state. For the dust particles, electrostatic dust accelerators are used, but the studies for the disruption are quite limited (McDonnell, 1976).

3. Fragment size

3.1. ENERGY FOR DISRUPTION AND THE LARGEST FRAGMENTS

Impact energy required to disrupt the targets consisting of various kind of materials and the degree of fragmentation were investigated by many authors (See Fujiwara *et al.*, 1989). Most of rocks start to disrupt at E/M-value (E; impact energy, M; target mass) of about several times 10^6 erg/g, while ices disrupt at one order of magnitude smaller value. For metals (*e.g.* iron) disruption occurs at around 10^8 erg/g. The mass fraction of the largest fragment to the target mass is roughly inversely proportional to the E/M-value for various materials.

3.2. SIZE DISTRIBUTION OF THE FRAGMENTS

It has been traditionally assumed in many papers that the size distribution is expressed by a power law;

$$n(m)dm = Cm^{-\alpha}dm,$$

where n(m)dm is the number of fragments of mass in the range of m to m+dm. The index has been conventionally assumed to be 0.8 in many references. However, the detailed studies show that is a function of fragment size and impact energy or more appropriate scaling parameters. Takagi *et al.*(1984) showed that the size spectrum consists of three segments and the power indices for these segments are expressed as a function of a scaling parameter P_I which is proposed as a good scaling parameter by Mizutani *et al.*(1990) and expressed as $P_I = P_0 V_{0p}/Y V_{0t}$ (V_{0p} ; initial volume of a projectile, V_{0t}; initial volume of a target, Y; yield strength of the target). The results show that the exponent changes with the parameter P_I. In the regime III (the small size region) most of data are slightly less than 0.8, while in the regime I (region of the largest fragment size), they take very large values and increase with the parameter P_I.

4. Velocity Distribution

Many investigators consider that velocity distribution of the fragments (relation between mass and velocity of individual fragments) is the most fundamental quantity to understand the collisional evolution of the solid particles in the solar system. However, the distribution in the whole size range has not yet been understood because of the experimental difficulties. Recently Nakamura and Fujiwara (1990)

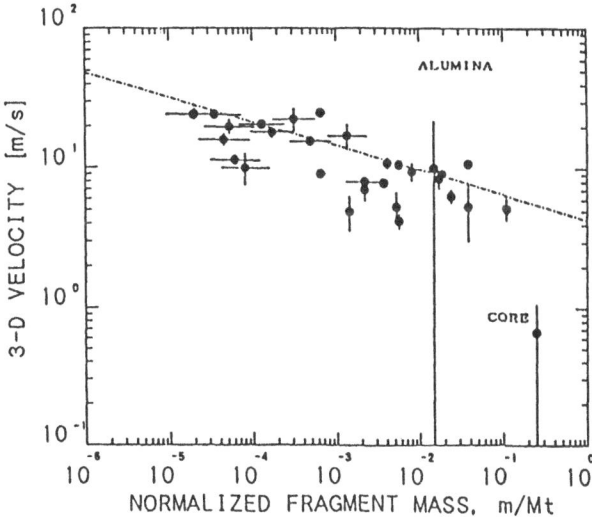

Figure 1. Velocity distribution of the alumina fragments. Velocity and mass of the projectile are 3.8km/s and 0.7g, respectively. Target diameter is 6cm. Nakamura and Fujiwara(1990).

took high-speed movie pictures of the fragmentation of an alumina ball and a basalt ball impacted by the nylon projectiles of 0.7 cm in diameter in the vacuum space. The 2-D velocity and 3-D velocities were determined for the fragment size down to a few mm. The example of the 3-D velocity distribution for the alumina target is shown in Fig.1. The best-fitted line is expressed by -1/6 th power of the fragment mass. These results are not inconsistent of the results by Davis and Ryan(1989). It is still not clear how this velocity distribution depends on the various impact parameters (masses and material properties of both the target and the projectile, and collisional velocity). However, the data on the velocity of the antipodal points of the target were obtained as a representative velocity (Fujiwara and Tsukamoto, 1980; Takagi *et al.*,1984), where the antipodal point is the opposite point on the target of the impact point. The data are shown in Fig.2. The velocity depends on the 0.76th power of E/M , and it is found from the figure that the data are well scaled by the Mizutani scaling parameter.

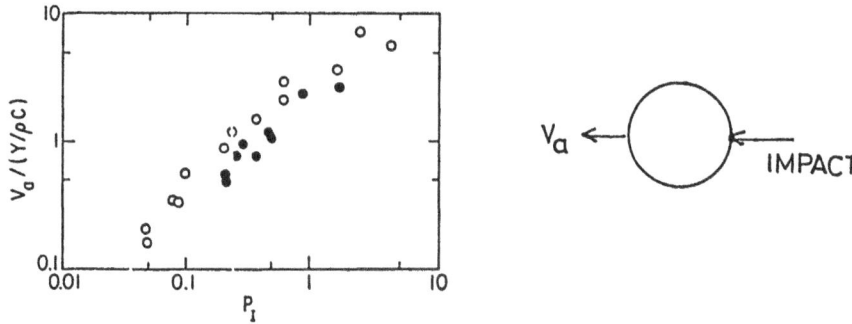

Figure 2. Velocity of fragments from the antipodal point V_a. PI is Mizutani parameter. C and ρ are sound velocity and density of the data (2.7km/s). Filled circle; the Takagi and Mizutani data (<1km/s). From Fujiwara et al.(1989).

5. Shape of the Fragments

Shape of the fragments is especially of interest for the study of dust particles, because it affects the scattering properties of the dust particles and it also becomes a clue to assign the origin of the dust particles. Many laboratory experiments show shapes of the fragments distributing around a very peculiar average shape(Fig.3); the average ratio for the longest, intermediate, and short axis is 2: $2^{1/2}$:1. Hence,

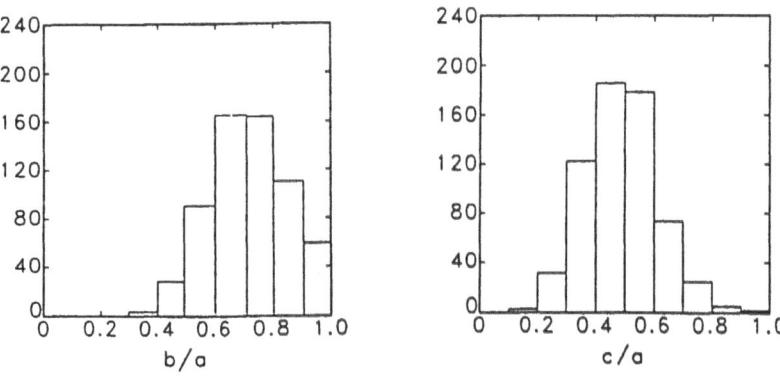

Figure 3. Histogram for b/a and c/a. Fujiwara et al.(1989).

if the dust particles are originated from collisional processes, the most probable shape of the dust particles will take this average value. There are two experiments which violate this very stable rule; one is Lange and Ahrens (1981) for ice targets, and another is the recent large scale experiments made by using explosives in the open air by Martelli et al. (1990). The reason of the deviation from the standard shape rule in these experiments is not clear, but the possibility that the original samples may have some anisotropy is not excluded.

6. Rotation of the Fragments

Rotational properties affect on the dust particles through Yarkovski effect. The data on the rotational frequency of the fragmented bodies are still limited to the one obtained by Fujiwara and Tsukamoto(1981). It is noted the fragments rotate at very high frequencies; for example, the fragments of several mm size rotate as high as a few hundreds rps. The maximum frequency rate is almost one order-of-magnitude less than the frequency for the rotational burst.

7. Future Works

There are many works to be done in the future on the impact phenomena connected with the interplanetary dust particles.

(1) Data of the impact experiments for wide varieties of physical parameters; varying density, size, strength of both targets and projectiles are required.

(2) Kinematical data of the fragments (velocity and spin distribution of the fragments, energy partition rate into various energy modes, angular momentum and momentum transfer rate) are still deficient.

(3) Impact experiments using various kind of icy and organic bodies are required.

(4) Collision between core-mantle bodies is of interest, because interstellar dust particles are considered to have such structure and also cometary dust particles may have similar structures.

(5) Most of experiments are aimed at the applications to the collisional evolution of asteroidal bodies. Semi-empirical laws or scaling laws proposed(Paolicchi et al., 1989; Housen and Holsapple, 1990; Mizutani et al., 1990) are intended to know how the laboratory scale experiments (10cm order bodies) can be extended to the 10^{6-7}cm-sized bodies. From the viewpoint of the interest in the dust particles, it is necessary to extend the scaling law to an order of 10^{-4}cm bodies. Especially interesting is to know how the disruption phenomena change for the object whose size is less than the fundamental size of the grains which constitute the body. McDonnell(1976) impacted 10^{-11}g particles (velocity 4km/s -15km/s) on a previously raised chip-essentially free standing- of a polished lunar sample. He suggested the catastrophic rupture limit of the body size is < 4 times the primary crater diameter. More experiments are required in this size range.

(6) Some of the Brownlee particles are aggregates of less smaller particles. Therefore collisions for these aggregates or rubble piles should be of great importance. Some experiments using rubble pile targets were made in the slow velocity range (Hartmann, 1980), and the results suggest the outcomes are strongly dependent on the initial size distribution of the rubbles. Moreover, recently, fractal structure is also proposed for the grain. To determine the cross sections for accretion or destruction using fractal grains are necessary in the discussion of growth rate of the grain in the space.

Acknowledgment

The author wishes to thank Dr. McDonnell for his valuable comments.

References

Capaccioni, F., P. Cerroni, M. Coradini, M. DiMartino, P. Farinella, E. Flamini, G. Martelli, P. Paolicchi, P. N. Smith, A. Woodward, and V. Zappala (1986) 'Asteroidal Catastrophic Collisions Simulated by Hypervelocity Impact Experiments' Icarus 66, 487-514.

Davis, D. and E. Ryan (1989) 'On Collisional Disruption: Experimental Results and Scaling Laws', Icarus 83, 156-182.

Fujiwara, A. and A. Tsukamoto (1981) 'Rotation of Fragments in Catastrophic Impact', Icarus 48, 329-334.

Fujiwara, A., P. Cerroni, D. R. Davis, E. Ryan, M. DiMartino, K. Holsapple, and K. Housen (1989) 'Experiments and Scaling Laws on Catastrophic Collisions' in R. P. Binzel, T. Gehrels, and M. S. Matthews (eds.), Asteroids II, Univ. Arizona Press, Tucson, pp. 240-265.

Gault, D. E. and E. D. Heitowit (1963) 'The Partition of Energy for Hypervelocity Impact Craters Formed in Rock', Proc. 6th Hypervelocity Impact Symp., 2 (Cleveland, OH, Firestone Rubber Co.), pp. 419-456.

Hartmann, W. K. (1980) 'Continued Low-velocity Impact Experiments at Ames Vertical Gun Facility: Miscellaneous Results' Lunar and Planet. Sci. Conf. XI Abstract 404-406.

Housen, K. and K. Holsapple (1990) 'On the Fragmentation of Asteroids and Planetary Satellites', Icarus 84, 226-253.

Lange, A., and T. J. Ahrens (1981) 'Fragmentation of Ice by Low-velocity Impact', Proc. Lunar Planet. Sci. Conf. 12, 1667-1687.

Martelli, J., P. Rothwell, P. N. Smith, I. Giblin, J. Martinsson, E. Ducrocq, M. Wettstein, M. DiMartino, P. Farinella (1990) 'Jets of Fragments from Catastrophic Break-up and Their Astrophysical Implications' presented in this colloquium.

McDonnell, J. A. M. (1976) 'Finite Target Hypervelocity Impact Measurements at Microscale Dimensions: Implications for Regolith Impacts', in Papers Presented to the Symposium on Cratering Mechanics. The Lunar Science Institute, Houston, pp. 73-75.

Mizutani, H., Y. Takagi, and S. Kawakami (1990) 'New Scaling Law on Impact Fragmentation', Icarus 87, 307-326.

Nakamura, A. and A. Fujiwara (1990) 'Velocity Distribution of Fragments Formed in Simulated Collisional Disruption' submitted to Icarus.

Paolicchi, P., A. Celino, P. Farinella, and V. Zappala (1989) 'A Semiempirical Model of Catastrophic Breakup Processes', Icarus 77, 187-212.

Sykes, M. V., R. Greenberg, S. F. Dermott, P. D. Nicholson, J. A. Burns, and T. N. Gautier, III (1989) 'Dust Bands in the Asteroid Belt', in R. Binzel, T. Gehrels, and M. Shapley (eds.) Asteroids II, Univ. of Arizona Press, Tucson, 336-367.

Takagi, Y., Mizutani, H., and S. Kawakami (1984) 'Impact Fragmentation Experiments of Basalts and Pyrophyllites', Icarus 59, 462-477.

METHODS, DIFFICULTIES, AND FIRST RESULTS IN LABORATORY SIMULATION OF COSMIC DUST ELECTRIC CHARGING

J. SVESTKA[1] and E. GRÜN[2]
[1]Prague Observatory, Petrin 205, 118 46 Prague 1, Czechoslovakia
[2]Max-Planck-Institut für Kernphysik, 6900 Heidelberg 1, Germany

ABSTRACT. Particles of radii 0.2 to 3 µm and of different materials were suspended in an electrodynamic quadrupole inside a vacuum chamber and exposed to beams of electrons and ions of energies up to 20 keV and 5 keV, respectively, with the aim to simulate electric charging of cosmic dust particles. It was found that the equilibrium surface electrostatic potential of glass particles of radii 0.2 to 2 µm charged by electrons of energies 1 to 20 keV is always positive. This can be explained by secondary electron emission at lower energies of electrons and by penetration of electrons through particles with subsequent secondary electron emission mainly from the exit side at higher energies. In case of charging by ions electrostatic potential of particles is generally much lower than expected values and interpretation of results of measurement is more complicated. The most promising way to eliminate instrumental influences disturbing processes of charging seems to be a construction of a smaller suspension system in which these influences would be negligible. Parameters of such a suspension system were derived from results of measurements.

1. Introduction

There are many phenomena connected to dust particles within the solar system which can be explained by their electric charging with subsequent interactions with electromagnetic fields and/or disruption by repulsive electrostatic forces - see e.g. Grün et al. (1984), Morfill et al. (1986), Boehnhardt and Fechtig (1987). Cosmic dust particles can be charged by a variety of mechanisms - see e.g. review of Whipple (1981). In case of solar system dust particles the most important charging processes are interactions with electrons and ions and photoemission by solar UV radiation. Theoretical calculations of electric charging by these processes are, however, especially in case of very small particles based on unreliable data extrapolated from results of measurements with plane surfaces (parameters of secondary electron emission and photoemission, capture probabilities of electrons and ions). Therefore we started experimental laboratory work on simulation of cosmic dust electric charging by electrons, ions and UV radiation - determination of equilibrium electrostatic surface potentials of particles of various sizes and materials exposed to beams of electrons and ions of different energies or illuminated by UV radiation of specific spectrum and intensity.

A.C. Levasseur-Regourd and H. Hasegawa (eds.), Origin and Evolution of Interplanetary Dust, 367-370.
© 1991 *Kluwer Academic Publishers.*

2. Methods

For charging of dust particles we used an electrodynamic quadrupole (minimum radius $r_0 = 25$ mm) inside a vacuum chamber (pressure down to 5.10^{-7} mbar) in which particles of radii 0.2 to 3 μm and of various materials (glass, carbon, tungsten etc.) were suspended and exposed to electron or ion (He, Ar, H_2) beams of energies up to 20 keV and 5 keV, respectively. The particles were illuminated by a HeNe laser and the light scattered by particles was observed by a telescope. A charge-to-mass ratio Q/M of a particle was determined from the amplitude of the quadrupole voltage V, applied frequency f, and oscillation frequency of a particle in vertical direction f_z (determined with help of a photomultiplier) by the formula

$$Q/M = const.r_0^2 f f_z / V \qquad (1)$$

The charge Q of the particle was found out by means of the charge induced on a metallic cylinder placed below the quadrupole through which the particle is forced to pass after completion of a charging process. From Q/M and Q the radius of the particle and its equilibrium surface electrostatic potential U was calculated. Schematic diagram of the vertical cross section of the suspension system is shown in Fig. 1 - for more details about the suspension system see Svestka et al. (1987) or Pinter et al. (1990) and references therein. A similar suspension system in which particles will be suspended and charged by photoemission due to UV radiation produced by a deuterium lamp with MgF_2 window is under construction.

The values of the equilibrium potential U can be influenced by a presence of a rest gas in a vacuum chamber and by the electric field inside the quadrupole. In order to quantitatively characterize these influences dependencies of the equilibrium electrostatic surface potential U on the vacuum pressure p and on the amplitude of the quadrupole voltage V for different energies of electrons and ions were studied.

3. Results

Test measurements were performed with ions because in absence of any disturbing instrumental influences equilibrium potential U (in volts) should be simply equal to the energy of ions E_i (in eV) so that it is easy to compare results of measurements with ideal values. It was found that U is increasing with decreasing V and it is practically constant if V is lower than approximately $E_i/2$. U is generally increasing with

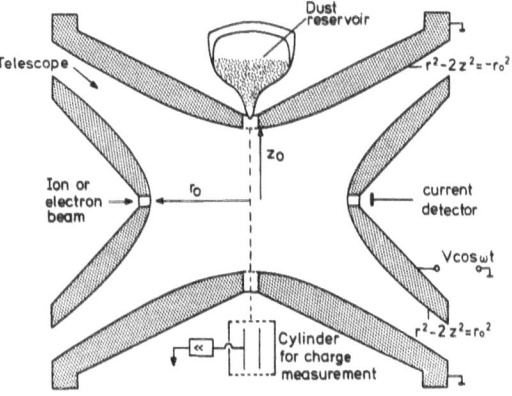

Figure 1. The vertical cross section of the particle suspension system.

decreasing p but within the energy range of charging ions it is always much lower than E_i even for the lowest vacuum pressure. From the analysis of dependencies $U = U(p)$ and charging curves (dependencies of U on time during charging) it was possible to derive an expression for the charging rate which contains the term corresponding to primary ions, the term due to electrons produced by collisional ionization of a rest gas which is proportional to vacuum pressure but also the third term which is independent of vacuum pressure and proportional only to ion current. This term is responsible for substantial lowering of U below the ideal values and it is probably caused either by secondary electron emission due to ion impacts from the inner surfaces of the quadrupole electrodes or by field ionization of water molecules adsorbed on the surface of particles (Pinter et al. 1990). Dependencies of U on E_i differ considerably from straight lines, at $E_i > 2$ keV the potential U is even decreasing with increasing E_i - see Fig. 2.

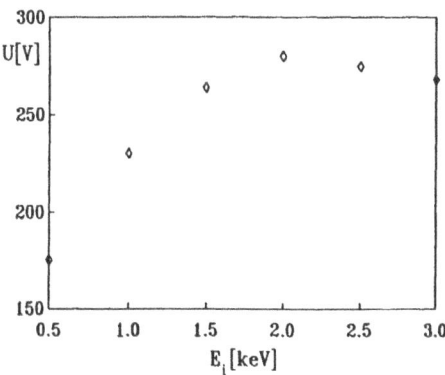

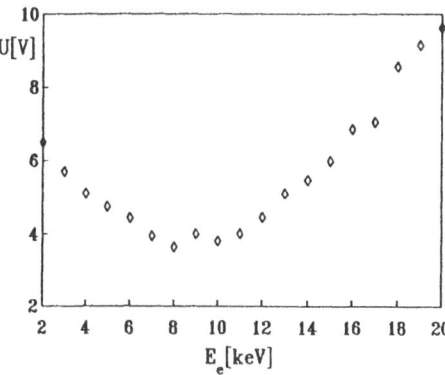

Figure 2. Typical dependence of particle equilibrium potential on energy of ions (glass particle of radius 0.47 μm, p = 10^{-6}mbar, He^+ ions).

Figure 3. Typical dependence of particle equilibrium potential on energy of electrons (glass particle of radius 1.12 μm, p = 10^{-6}mbar

Then we charged dust particles by electrons and studied analogously dependencies of U on V, p and E_e. With glass particles of radii up to 2 μm and electrons of energies 1 to 20 keV always positive particle potentials were found. U is decreasing with increasing E_e at lower E_e due to decreasing efficiency of secondary electron emission but then increasing due to electrons penetrating through particles and causing secondary electron emission from the exit side - see Fig. 3. From the analysis of dependencies U = U(p) and charging curves the expression for the charging rate was derived as in case of charging by ions. This expression contains again three terms corresponding respectively to primary electrons, secondary electrons (true secondaries and backscattered ones), and electrons produced by collisional ionization of a rest gas. The last term is responsible for measured potentials being lower compared to real "cosmic" ones and ensures that measured values give the lower limit to real potentials.

4. Discussion

From preliminary results of electric charging by electrons it follows that the equilibrium electrostatic surface potential of glass particles of radii smaller then 2 μm charged by electrons of energies 1 to 20 keV is always positive. One should get the same result in case of water ice particles because the secondary electron emission yield of glass and water ice is comparable - see Hashimov and Tarakanov (1982). This conclusion has already implications for solar system dust particles, e.g. for particles in planetary magnetospheres where electrons of energies 1 to 20 keV are present. It was assumed that dust particles exposed to electrons of these energies will be generally charged to high negative potentials equal approximately to energy of electrons (in eV) - see e.g. Grün et al. (1984). On the contrary to that it follows from our results that it could be that only the potential of larger particles is high and negative but the potential of particles of micron and submicron sizes is much lower and it is positive.

A quantitative determination of the terms lowering values of U is, however, rather difficult in cases of both electrons and ions. From equation (1) it follows that V is proportional to r_0^2 for constant Q/M. To eliminate the influence of the electric field inside a quadrupole it is necessary to reduce the dimensions of the present quadrupole by a factor of 5 to 10, i.e. to $r_0 = 2.5$ to 5 mm. From the analysis of dependencies U = U(p) and also from theoretical estimates it follows that the elimination of the influence of the rest gas requires $p \leq 10^{-10}$ mbar. The combination of low pressure with shorter path lengths of electrons and ions should make it possible to charge particles to equilibrium potentials without any collision of electrons or ions with an atom or molecule of a rest gas. For future work the construction of a smaller vacuum chamber, in which it will be possible to reach $p = 10^{-10}$ mbar and with a smaller quadrupole inside, seems to be the most promising possibility.

References

Boehnhardt, H. and Fechtig, H. (1987) 'Electrostatic charging and fragmentation of dust near P/Giacobini-Zinner and P/Halley', Astron. Astrophys. 187, 824-828.

Grün, E., Morfill, G. E. and Mendis, D. A. (1984) 'Dust-magnetosphere interactions', in R. Greenberg and A. Brahic (eds.), Planetary Rings, Univ. of Arizona Press, Tuscon, pp. 275-332.

Hashimov, N. M. and Tarakanov, V. L. (1982) 'Surface charge influence on sublimation rates of icy grains and cometary nuclei', Komety i Meteory 32, 3-9.(in Russian)

Morfill, G. E., Grün, E. and Leinert, C. (1986) 'The interaction of solid particles with the interplanetary medium', in R. G. Marsden (ed.), The Sun and the Heliosphere in Three Dimensions, D. Reidel Publishing Co., Dordrecht, pp. 455-474.

Pinter, S., Svestka, J. and Grün, E. (1990) 'Interaction of dust particles with electrons and ions', in E. Bussoletti and A. A. Vittone (eds.), Dusty Objects in the Universe, Kluwer Academic Publishers, Dordrecht, pp. 139-146.

Svestka, J., Grün, E., Pinter, S. and Schumacher, S. (1987) 'Laboratory charging of dust by electrons and ions', Publ. of Astron. Inst. of Czechoslovak Academy of Sci. 67, 277-280.

Whipple, E. C. (1981) 'Potentials of surfaces in space', Rev. Prog. Phys. 44, 1197-1250.

ELECTROSTATIC FRAGMENTATION OF IRREGULARLY SHAPED PARTICLES

TADASHI MUKAI
Dept. of Earth Sciences
Faculty of Science, Kobe University
Nada, Kobe 657, Japan

ABSTRACT. An enhancement of the electrostatic stress forces due to the charge concentration on a position with small radius of curvature on the surface of irregularly shaped particle causes the fragmentation of fluffy particle more than expected for a spherical particle. This mechanism may act to produce "dust clusters" as detected in comet P/Halley by Simpson *et al.*(1987) and also "dust swarms" near the Earth as reported in Fechtig(1982).

1. INTRODUCTION

There are many observed evidence to suggest the destruction of interplanetary dust particles. For example, Simpson *et al.*(1987) found higher counts of impact particles in smaller grains within short(1 or 2 seconds) time scale in the coma of comet P/Halley("dust clusters"). They have suggested that the fragmentation of parent dust particles produced these debris in the cometary coma. HEOS 2 experiments revealed an enhancement of dust impact events within a very short time interval. These groups of particles were also interpreted as the debris of large particles(see Fechtig 1982).

Many mechanisms to destroy the particles have been proposed in the previous papers, e.g. evaporation breakup(Mukai 1984), rotational bursting(Misconi 1976), collisional breakup with high energy particles(Mukai 1980) and electrostatic breakup(Rhee 1976, Boehnhardt and Fechtig 1987). In this short note, we will focus our study on the destruction of irregularly shaped particles caused by electrostatic forces.

2. MECHANISM

2.1. An equilibrium electric potential

A particle attains an equilibrium electric potential on its surface in the interplanetary environment. Several charging processes act on the grain(see, e.g. Mukai 1981, Boehhardt and Fechtig 1987). Photoelectric emission caused by sunlight mainly leads a positive state

371

A.C. Levasseur-Regourd and H. Hasegawa (eds.), Origin and Evolution of Interplanetary Dust, 371–374.
© 1991 *Kluwer Academic Publishers.*

of electric potential on the grain surface, and sticking of electrons leads its negative state. Field emission of electron prevents a grain from getting huge negative charges especially on small grains.

For example, the equilibrium potential for a graphite grain in the day side of the earth's magnetosphere is derived from a comparison of emission/sticking rates of charges as shown in figure 1. In this figure, a cross point between photoelectric emission rate (solid curve) and electron sticking rate(dashed curves) indicates the equilibrium potential U. A number density of electrons is assumed $0.1 cm^{-3}$. Other physical parameters of interest are referred in Mukai(1981). We can estimate from figure 1 that the electric potential U of graphite grain in the day side of the earth's magnetosphere becomes roughly $+(5 \sim 8)$ volts. By using a similar way, the values of U in the night side can be estimated, i.e. $U = (-12 \sim +4)$ volts.

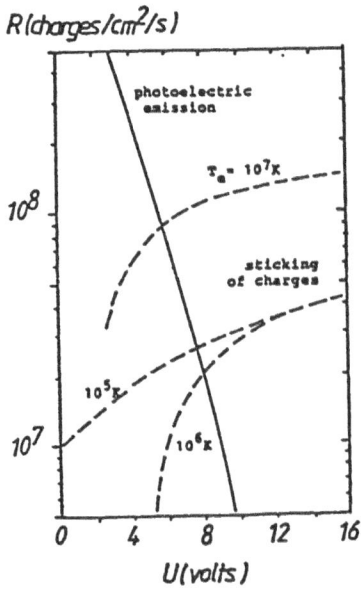

Figure 1. Emission/sticking rates (R) of charges on the surface of graphite grain in the day side of the Earth's magnetosphere. T_e denotes the electron temperature.

Boehnhardt and Fechtig(1987) derived the values of U in the coma of comet P/Halley as nearly $+(2 \sim 6)$ volts for carbon particle and $+(7 \sim 11)$ volts for silicate grain. It would be noted that extremely huge values of U are not attained in the interplanetary environment.

2.2. Fragmentation

The Maxwell's stress tensor force F is given by $F = \frac{1}{2}\varepsilon_o E^2$, where E is the strength of electric field on the surface of a grain, and ε_o denotes the dielectric constant of the vacuum.

For a sphere with a radius r and a surface electric potential U, the density of surface charges σ_o is expressed by $\sigma_o = Q/(4\pi r^2) = \varepsilon E = \varepsilon U/r$, where Q means the total charges on the surface of the grain and ε is the dielectric constant of the grain material.

It is well known that the charge distribution on the surface of an irregularly shaped particle is not uniform. We assume that the irregularity on the surface of a roughly spherical particle with a radius r is represented by a part of a sphere with a small radius of curvature ρ (see figure 2).

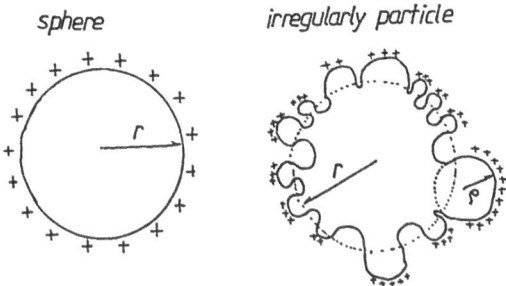

Figure 2. Charge distribution on the surface of a particle

The density of charges σ on a position with a radius of curvature ρ is given by

$$\sigma = \varepsilon \frac{U}{\rho} = \sigma_\circ \frac{r}{\rho} \tag{1}$$

Consequently, the stress tensor force F_i on this position becomes

$$F_i = F \left(\frac{r}{\rho}\right)^2. \tag{2}$$

This result implies that the tensor stress force F_i acted on the position with a small radius of curvature is significantly stronger than those F expected on the average surface of a sphere.

3. CONCLUSIONS

In figure 3, a dashed line shows the tensor stress force F caused by the surface charges of $U = 5$ volts as a function of a radius of a spherical particle r. Since the tensile strength of real grain material is rather high, we had to assume very fragile material to breakup the grains by the electrostatic fragmentation (see, e.g. Boehnhardt and Fechtig 1987).

When we consider the charge concentration on the surface of an irregularly particle, however, enough stress tensor forces to disrupt the particle can be expected, as shown in figure 3 by the solid curves. This suggests that the interplanetary particle with irregularly shape is broken up by electrostatic forces more than expected for a spherical grain. Until now, our knowledge of the irregularity of the interplanetary dust particles is very poor. Therefore, we need in situ measurements to know the shape of grains in the interplanetary space for studying not only their scattering/thermal emission properties, but their destruction mechanisms.

374

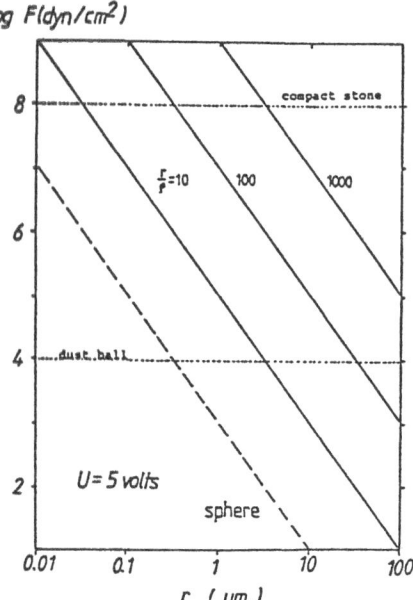

Figure 3. Stress force F vs. a grain radius r. A dashed line denotes the electrostatic forces on the surface of a sphere with a radius r. The solid lines show the forces on the position with small radius of curvature ρ.

Acknowledgement. This work was supported in part by the Scientific Research Fund of the Minister of Education, Science, and Culture(02640205)

References

Boehnhardt, H. and Fechtig, H.(1987) 'Electrostatic charging and fragmentation of dust near P/Giacobini-Zinner and P/Halley', *Astron.Astrophys.* **187**, 824-828.

Fechtig, H.(1982) 'Cometary dust in the solar system', *"Comets"(ed.L.L.Wilkening*, The Univ. of Arizona Press), 370-382.

Misconi, N.Y.(1976) 'On the rotational bursting of interplanetary dust particles', *Geophys. Res. Letters* **3**, 585-588.

Mukai, T.(1980) 'Grain disruption by collisions with solar energetic particles', *"Solid Particles in the Solar System" (eds. I.Halliday and B.A.McIntosh*, D.Reidel Publ. Co.) 385-389.

Mukai, T.(1981) 'On the charge distribution of interplanetary grains', *Astron. Astrophys.* **99**, 1-6.

Mukai, T.(1984) 'Heterogeneous grain destruction near the sun', *Earth, Moon, Planets* **30**, 99-103.

Rhee, J.W.(1976) 'Electrostatic disruption of lunar dust particles', *Lecture Notes in Phys.* **48**, 238-240.

Simpson, J.A., Rabinowitz, D., Tuzzolino, A.J., Ksanfomality, L.V. and Sagdeev, R.Z.(1987) 'The dust coma of comet P/Halley: measurements on the Vega-1 and Vega-2 spacecraft', *Astron. Astrophys.* **187**, 742-752.

PLASMA EMISSION FROM HIGH VELOCITY IMPACTS OF MICROPARTICLES ONTO WATER ICE

R. TIMMERMANN and E. GRÜN
Max-Planck-Institut für Kernphysik
6900 Heidelberg
Germany

ABSTRACT. Collisions of icy objects play a major role in the outer solar system. The purpose of this investigation is the experimental study of plasma production by dust impacts on icy surfaces. Impact speeds ranged from 3 to 60 km/s. It was found that the dominant ion species which were released are both positive and negative water clusters. The impact charge yield from icy surfaces is approximately a factor 100 below that from previously studied gold surfaces.

1. Introduction and Experimental Set-Up

In the outer solar system water ice exists in abundance. Many satellites of the giant planets consist of ice. At these distances from the sun the lifetime of water ice particles against sublimation is long enough (Mukai, 1986), that even meteoroids may contain large fractions of ice. In a model by Goertz and Morfill (1983) the formation of spokes on Saturn's B ring is explained by impacts of meter sized meteoroids onto icy ring particles. Plasma is produced which expands in Saturn's magnetic field. In this plasma submicron particles can be charged and lift-off the ring by electrostatic forces overcoming the gravitational force, thus giving rise to the spoke phenomenon. The purpose of this study is the investigation of plasma production by impacts on icy surfaces.

In our experiments water ice targets were exposed to high speed dust particles and the produced plasma was analysed. Charges were collected by electrodes at bias of potentials 1000 to 3000 V which were positioned in front of the target. In one set of experiments the ions were collected after they had passed a 0.8 m long field-free drift tube. This way mass spectra of the ions were obtained. The target temperature ranged from 80 to 170 K. Several hundred iron particles with masses ranging from 10^{-10} to 10^{-15} g were accelerated in the Heidelberg Van de Graaff to velocities ranging from 3 to 60 km/s, respectively. In addition, glass spheres in the mass range of 10^{-6} to 10^{-9} g were shot in the Munich plasma drag accelerator with speeds of up to 15 km/s.

A.C. Levasseur-Regourd and H. Hasegawa (eds.), Origin and Evolution of Interplanetary Dust, 375–378.

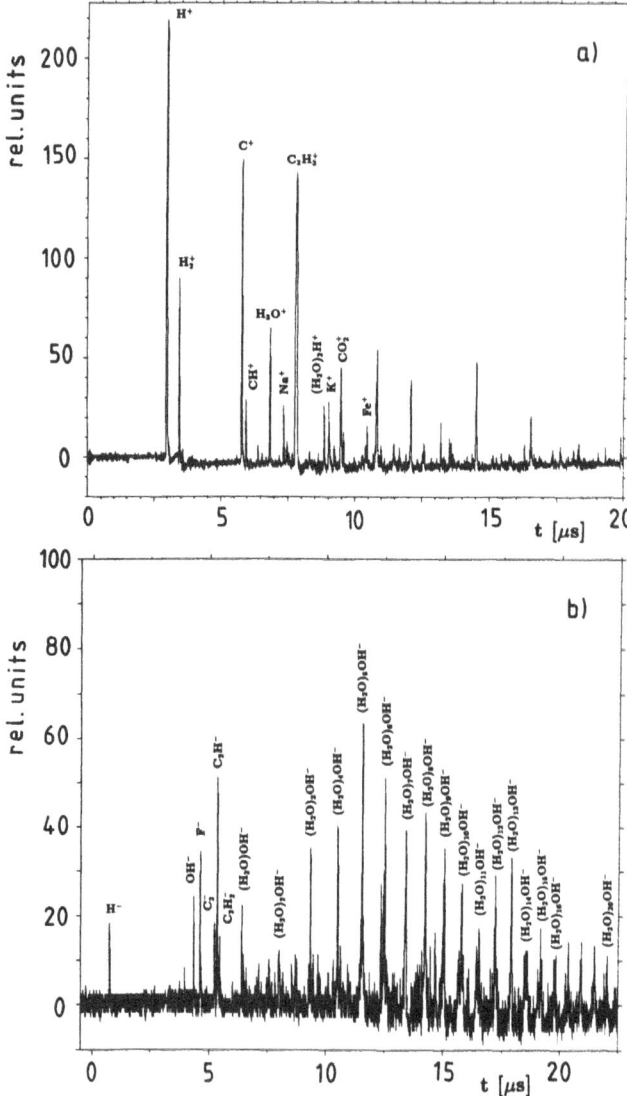

Figure 1. Time-of-flight mass spectra of ions released upon impacts of iron dust particles on ice.
a) Positive ion spectrum, impact speed 42.5 km/s.
b) Negative ion spectrum, impact speed 11.3 km/s.

2. Experimental Results

Positive and negative ions were mass analysed in a time-of-flight spectrometer. Figure 1 shows two examples of impact mass spectra obtained.

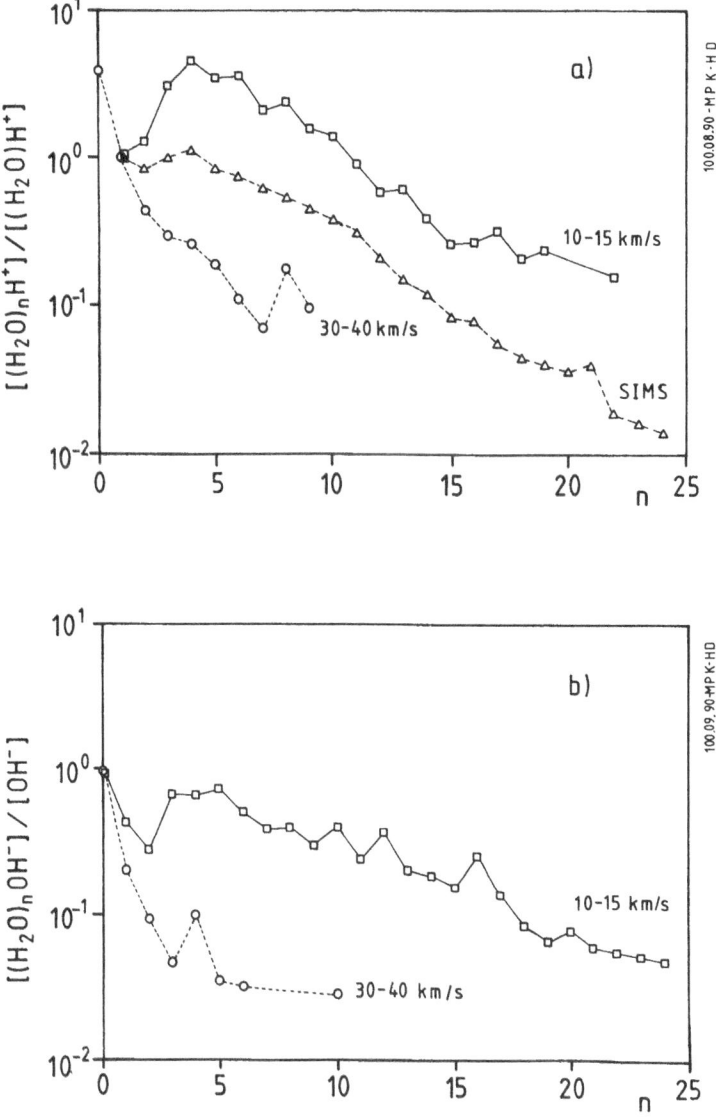

Figure 2. Cluster ion yield upon dust impacts on ice as a function of cluster size n for two different impact speed intervals.

a)   Positive ions, cluster ions produced by dust impacts and by SIMS (Estel et al., 1976) are compared

b)   Negative ions.

At low velocities (10-15 km/s) water clusters of the form $(H_2O)_nH$ and $(H_2O)_nOH$ could be detected where the number of water molecules n ranged up to 25 (cf. Figure 2). By increasing the velocity up to 40 km/s n decreased to small numbers; the ions released were composed mainly of H^+, H_3O^+, O^- and OH^-.

The total positive and negative charges released in these impact experiments fit to the empirical formula

$$Q^{\pm} = C\ m^{\alpha}\ v^{\beta}$$

with the following parameters $C = 1.4\ 10^{-7}$, $\alpha = 0.8$ and $\beta = 2{,}45$ where Q, m and v is given in C, g and km/s and when the particle velocity is higher than $v = 8$ km/s. Below this velocity the yield for negative charges decreases to a value of a factor 0.2 compared to the emitted positive charges. In comparison to gold as target material (Grün, 1984) the amount of plasma released is roughly a factor of 100 lower in the case of an ice target. A temperature effect of the plasma yield was not found.

The experimentally determined values for the charge yield and the knowledge on the impact plasma composition will improve our understanding of the Saturnian ring system. The plasma yield for ice targets will set boundaries for the projectile sizes which cause spokes on Saturns B-ring. Sofar only estimates for the yields have been used (Morfill and Goertz, 1983).

Ip (1983) proposed that the inner edge of the B-ring is close to the distance at which ions generated at the ring plane will syphon off the ring and will be lost to the ionosphere. Impact plasma is one source for such ions. With realistic impact charge yields better life times of particles in that region of the rings are obtained.

ACKNOWLEDGEMENTS. The authors are indebted to W. Frisch and E. Igenbergs for their support to the experiments at the Munich dust accelerator. The work was supported by the Deutsche Forschungsgemeinschaft DFG.

References

Estel, J., Hoinkes, H., Kaarmann, H., Nahr, H. and Wilsch, H. (1976) On the problem of water adsorption on alkali halide cleavage planes, investigated by secondary ion mass spectroscopy, Surface Sci., 54, 393-399

Goertz, C. and and Morfill, G. (1983) A model for the formation of spokes in Saturn's ring, Icarus, 53, 219-229

Grün, E. (1984) Impact ionization from gold, aluminum and PCB-Z, in Proc. of the Giotto Plasma Environment Working Group Meeting, ESA SP-224, 39-41

Ip, W.H. (1983) On plasma transport in the vicinity of the rings of Saturn: A siphon flow mechanism, J. Geophys. Res. 88, 819-822

Morfill, G.E. and Goertz C.K. (1983) Plasma clouds in Saturn's rings, Icarus, 55, 111-123

Mukai, T. (1986) Analysis of dirty water ice model for cometary dust, Astron. Astrophys., 164, 397-407

VELOCITY DISTRIBUTION OF FRAGMENTS IN COLLISIONAL BREAKUP

A. NAKAMURA and A. FUJIWARA
Dept. of Physics, Kyoto University
Kitashirakawa
Kyoto
606 Japan

ABSTRACT. One of the key outcomes of collisional disruptions to the dust system is the velocity distribution of fragments. A series of laboratory impact experiments were carried out to obtain the mass-velocity and the position-velocity relation of the fragments by taking movie films, and films for tow impacts were completely analyzed.

1. INTRODUCTION

Direct collisional process between solar system solid bodies ranging from dusts to planets is one of the major source of the interplanetary dust particles. The key outcomes of a collisional disruption to understand the evolution of the dust system is the velocity distribution as well as the size distribution of the fragments. The latter has been eagerly investigated experimentally and the sufficient amount of data afford to be compared with numerical simulations, while little experimental data about the motion of the individual fragments were obtained, although wide range of the fragment velocity was obtained from a cratering experiment against natural rock (Gault and Heitowit 1963).

Laboratory impact simulations were carried out to determine the velocity distribution of the fragments employing the popular high-speed photographic technique and an image processing system introduced in this field for the first time (Nakamura and Fujiwara 1990a).

2. METHODS

Here the experimental and the analytical procedures are outlined. The details are described in our other article (Nakamura and Fujiwara 1990a).

Spherical nylon projectiles of 0.70 cm in diameter were accelerated up to 3 ~ 4 km/sec by a two-stage light-gas gun horizontally

A.C. Levasseur-Regourd and H. Hasegawa (eds.), Origin and Evolution of Interplanetary Dust, 379–382.

aimed at basalt and alumina spheres of 6.0 cm in diameter. The incident angle of the projectile was changed in the plane including the trajectory and parallel to the floor. High-speed framing cameras were used at the framing rate of 6,000 frame/sec, and the pairs of film records taken from two orthogonal directions to the trajectory, parallel (side-view film) and perpendicular (top-view film) to the floor, respectively, were dubbed into a video tape frame by frame. These two dimensional (2-D) video pictures are digitalized by an image processor linked to a personal computer and analyzed to get the position and the size of the 2-D images of the fragments.

Table 1 shows the precise conditions of the experiments for which we got the final result of the velocity distribution. Hundreds of fragments were observed for each impact.

TABLE 1. Experiment conditions

Target diameter   = 6.0 (cm)
Projectile diameter = 0.70 (cm)
Projectile mass   = 0.21 (g)
Framing rate   = 6,000(frame/sec)

	Target mass(g)	Projectile velocity($\times 10^5$ cm/sec)	Incident angle to the surface
Alumina target	395.7	4.00	60°
Basalt target	303.2	3.20	60°

3. RESULTS AND DISCUSSION

The 2-D velocity and the initial position of each fragment were calculated by the least squares fit to its positions on the successive frames. The resulting 2-D velocity vectors are shown in Fig. 1. The most prominent difference between the alumina and the basalt targets are the velocity profile of the fragments from the antipodal site. In the case of the alumina target, the antipodal fragments locally become smaller than those of the neighborhood. This antipodal character, higher in velocity and smaller in size, is also observed at the other three normal shots to the similar alumina targets, while the basalt fragments at the antipodes are larger and slower regardless of the incident angle (90°~30° to the surface). Which physical property causes the difference between both targets has not been explained yet.

The general axial symmetry around the normal at the impact site is observed for both targets, which probably suggests that the fragments we observed play minor role as the carrier of the incident angular momentum. This speculation is reinforced by the following discussion about the energy partitioning.

Scores of recovered fragments were identified with the 2-D images and empirical power law relations between the image size and the mass for both alumina and basalt cases were derived. The mass of the rest of all fragments were estimated by using the relations. Some of the

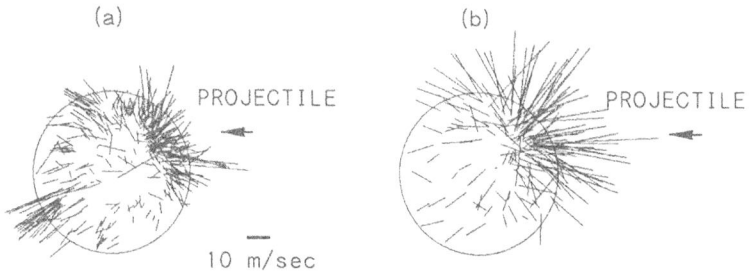

Figure 1. 2-D velocity vectors projected to the plane parallel to the floor of fragments from the alumina(a) target and the basalt(b) target listed in the Table 1. The arrowheads are omitted.

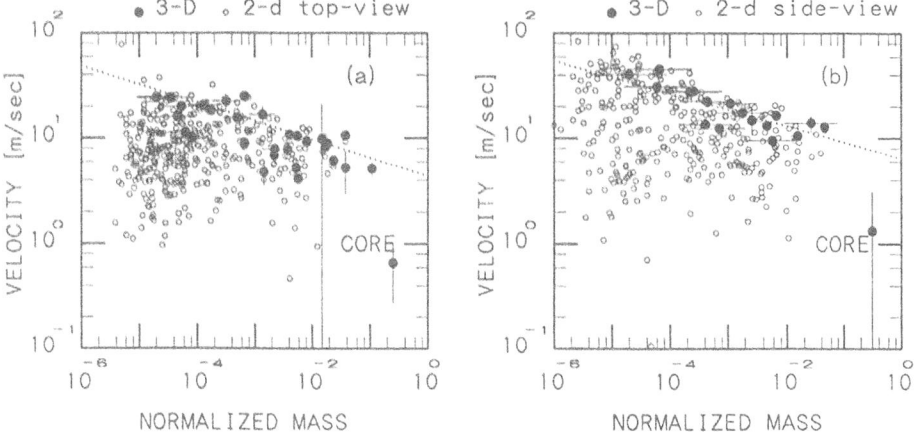

Figure 2. Mass-velocity distribution of fragments from the alumina target (a) and the basalt target (b) and the 3σ error bars associated with the least squares fits determining the mass from the size and the velocity from the position of the fragments. The dashed lines represent the least squares fits to the data of the 3-D velocity excluding the largest fragments. Roughly, the slopes are both -1/6.

fragments seen in the top-view film could be identified as the same fragments seen in the side-view. Figure 2 shows the 2-D and 3-D mass-velocity distribution of fragments. The slopes of the upper boundary of the 2-D velocity distribution agree reasonably with those of dashed lines obtained from least squares fits to the 3-D velocity data except those of largest fragments (called 'core'). The 3-D velocity of the largest fragments apparently deviate from the dashed lines. This deviation would be explained by the different sites they were created or different processes they were suffered. The largest fragments are spherical ones left at the center of the targets, while almost all the other fragments we got the 3-D velocity are from the surface of the targets created by the tensile stress generated at the surface. The slope of the fitted lines are both about -1/6 indicating the translational kinetic energy of individual fragments are proportional to the $(mass)^{2/3}$, i.e. the same dimension as the surface area of the fragments.

Total translational kinetic energy partitioned into fragments heavier than 10 mg was estimated from the above 3-D mass-velocity distributions and the mass distribution. Only one or two percent of the initial kinetic energy was partitioned to those larger fragments at both impacts. The rotational energy of the spinning motion of these heavier fragments could not be larger than the translational kinetic energy regarding, though insufficient, the experimental data(Fujiwara 1987; Nakamura and Fujiwara 1990b). Major fraction of the incident energy would be expended in the various energy mode, such as kinetic energy and comminution energy, of fine fragments from near the impact sites. The incident angular momentum is expected to be also carried away by these fine ejecta.

4. CONCLUSION

The employment of the image processor bring us a relatively systematic analysing method, however, there are much inconvenience which prevent us from a swift analysing and more improvements are desirable.

Although the present analysis is limited to the two impacts, the resulting velocity distributions are still fairly suggestive. More experimental data concerning to the motion of the fragments are required to confirm the -1/6 power law of the velocity distribution and the speculation that the kinetic energy is proportional to the surface area, to define the physical conditions where the -1/6 power law holds, or to find out that the results here are only the stochastic products, and eventually to understand the grinding process between solid bodies ranging from dusts to planets.

REFERENCES

Gault, D.E. and Heitowit, E.D.(1963) 'The partition of energy for hypervelocity impact craters formed in rock', Proc. 6th Hyper-velocity Impact Symp. Vol 2 (Cleveland. OH: Firestone Rubber Co.), pp 419-456.

Fujiwara, A.(1987) 'Energy partition into translational and rotational motion of fragments in catastrophic disruption by impact:An experiment and asteroid cases', Icarus 48, 329-334.

Nakamura, A. and Fujiwara, A.(1990a) 'Velocity distribution of frag-ments formed in simulated collisional disruption', Submitted to Icarus.

Nakamura, A. and Fujiwara, A.(1990b) 'Rotational motion of fragments from hyper-velocity impact experiments', Proc. 23th ISAS Lunar and Planet Symp.

JETS OF FRAGMENTS FROM CATASTROPHIC BREAK-UP AND THEIR ASTROPHYSICAL IMPLICATIONS

G. MARTELLI, P. ROTHWELL, P.N. SMITH, I. GIBLIN, J. MARTINSSON, E. DUCROCQ
M. WETTSTEIN
School of Mathematical and Physical Sciences, University of Sussex
Brighton BN1 9QH Sussex
U.K.

M. DI MARTINO
Osservatorio Astronomico di Torino
I-10025 Pino Torinese
Italy

P. FARINELLA
Dipartimento di Matematica, Universita' di Pisa
via Buonarroti, 2
I-56127 Pisa
Italy

ABSTRACT. We present some preliminary results of a series of catastrophic break-up experiments carried out in the open, against targets of natural and artificial rock, with and without a harder core. These experiments were aimed at investigating the outcomes of hypervelocity impact disruption phenomena, designed to understand the influence of large-scale collisions on the evolution of asteroids and other small solar system bodies. For the first time in this kind of experiments, evidence was found of collimated jets, i.e. the ejection of a statistically significant number of fragments all closely aligned about some preferential planes. Moreover, the presence of some groups of fragments lying close to each other on the ground was also detected.

1. EXPERIMENTAL TECHNIQUE

During the last few years the Space and Plasma Physics Group of the University of Sussex and the Planetology Group of the Turin Observatory have carried out a series of catastrophic fragmentation experiments in a travertine quarry (IMEG), located near Montemerano (Tuscany, Italy). The experiments were performed in the open, on a flat and relatively soft ground, in order to prevent secondary fragmentation of the primary fragments against the walls of a target chamber, and to record the uninterrupted ballistic trajectories of these fragments. We disrupted three targets, two of them being spheres of artificial rock (one with a core 10 cm in diameter, harder than the surrounding "mantle") fabricated using alumina cement, carborundum (SiC_2) and water in the correct stechiometric ratio, the other consisting of an irregular parallelepipedal block of limestone veined by calcite crystals. The targets were placed on a steel table, provided in the centre with a circular hole having a diameter of 5 cm, and were "impacted" from underneath (see Fig. 1). To simulate a high-velocity impact, we used a contact charge technique. This technique consists in filling a small cylindrical cavity in the target with plastic explosive and setting it off using a detonator. The momentum transferred by such an "equivalent projectile" is then measured using a ballistic pendulum. The pendulum consisted of a square block of steel weighing 4.7 kg suspended by 4 wires 6 m long. Upon detonation, the height gained by the pendulum was recorded by means of

383

A.C. Levasseur-Regourd and H. Hasegawa (eds.), Origin and Evolution of Interplanetary Dust, 383–386.

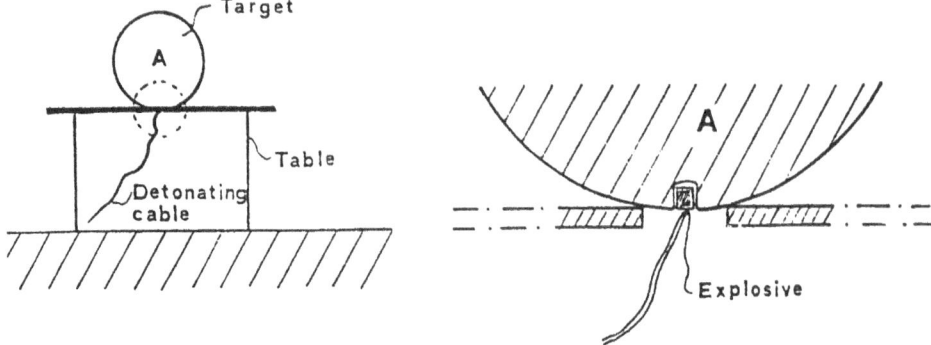

Figure 1. Experimental set-up.

fast-framing cameras. Using conservation of momentum, it was found that, within the errors, the equivalent mass of the "projectile", moving with the detonation velocity of the explosive (6.1 km/s, see later), was 1.80 g. The explosive we used in the present series of experiments was Gelatine 2b, which has a detonation velocity of 6.1 km/s, density 1.42 g/cm^3 and specific energy 4.44 kJ/g. The value of the equivalent projectile energy over target mass (E/M) ratio was of the order of some units x 10^7 erg/g, which is well known to be the critical specific energy required to achieve catastrophic fragmentation of hard rock. We believe that the contact charge technique provides a fairly realistic simulation of a rock-rock impact, because the acoustic impedance of the target made of artificial rock is well matched by that of the exploding charge located against the target itself ($\approx$7x10^5 g cm^{-2}s^{-1}). In order to record the target disruption we used two CCD TV cameras operating at 50 fields per second and provided with a mechanical shutter giving an exposure time of 1 ms. The CCD images have been analysed through a system which allowed two adjacent fields to be shown on a calibrated screen in rapid alternate motion, so that the displacement of the fragments from field to field could be measured accurately. Using a two-dimensional electronic cursor, it was also possible to plot the coordinates of any one fragment in a plane perpendicular to the axis of the cameras, and by simple software processing to obtain the component of the velocity of the fragments in this plane. The ground surrounding the table carrying the target was divided in 24 circular sectors of 15° each by a graduated string lying on the ground, which allowed accurate polar coordinate measurements of the fragments lying on the ground up to a 30 m distance from the original position of the target. The fragments were numbered and stowed away in polythene bags for subsequent measurements.

2. RESULTS

An interesting feature, which has been clearly detectable because the experiments were performed in the open, is the formation of collimated jets, i.e., the presence of statistically significant numbers of fragments all very closely aligned about some radial directions. This effect has been found both with targets made of natural rock and with targets made of artificial rock (cored and uncored). The natural rock target consisted of limestone with the intrusion of thin calcite crystals in the form of rather regular planes. Upon reassembling the target, it was noted that these planes coincided with some of the fracture planes, and that some of the jets appeared to be ejected from these planes. Since well defined jets have been produced also from the target made of homogeneous artificial rock, it would appear that jets are associated with fracture planes, wherever they are formed.

Figure 2 shows the frequency distribution of fragments as a function of the azimuthal angle for two targets, with and without the core. In order to enhance the presence of the jets and to show that a substantial fraction of the total mass can contribute to their formation, the numbers of fragments were multiplied by their respective mass. These plots exhibit peaks with high statistical significance. Since the experiments have been conducted in the open and in a relatively soft ground, it was also possible to observe some fragments grouped in "families" of up to some 10 members each. These fragments lied on the ground at small mutual distances (some centimeters) and many of their contours matched each other. Had the experiments been performed inside a target chamber, these fragments would have been scattered back at random into the chamber, and their common origin would most probably have been obscured.

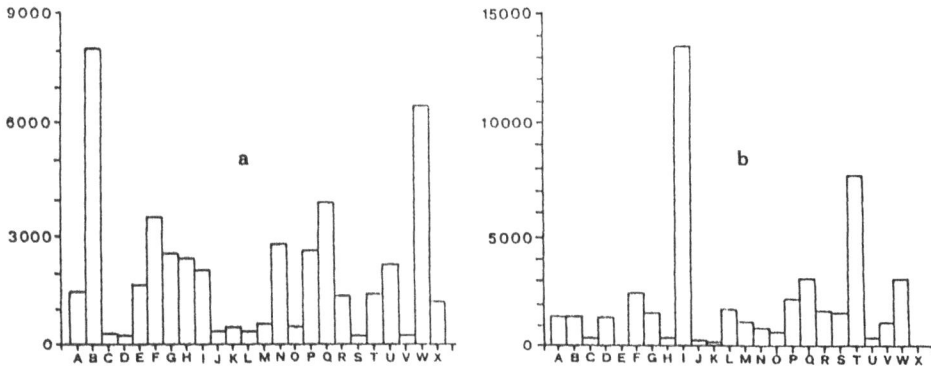

Figure 2. a) Fragment mass distribution from an artificial rock target (mass=8.724 kg, without core) versus the azimuthal angle. In abscissae the capital letters indicate circular sectors of 15° each, in ordinates the number of fragments multiplied by their mass in grams.

b) The same as in a) for an artificial rock target with core (mass=8.632 kg).

3. DISCUSSION

The existence of preferential planes and groupings among the ejected fragments is very interesting from the point of view of the formation of binary asteroids and of reaccumulated, self-gravitating asteroidal aggregates ("piles of rubble"), since these phenomena indicate that the fragment ejection velocity field after target break-up may be strongly non-isotropic. In other words, the relative velocity between two neighbouring fragments of mean radius r, ejected from a parent body of radius R, may be much less than the average value $\Delta v \cong (r/R)$ Vmax, where Vmax is the peak ejection velocity. On the other hand, if a significant fraction of the mass of a target asteroid is dispersed "to infinity" against its self-gravity, this means that Vmax is at least ~ 2Ve, Ve being the target escape velocity. But the mutual escape velocity of two adjacent fragments is about $(Gm/2r)^{1/2}$ (m being their mass), namely $(r/2R)$Ve $\cong (r/4R)$Vmax $\cong \Delta v/4$. Thus, fragments that escape from the parent body can keep gravitationally bound to each other, forming eventually binary or reaccumulated bodies, only if some jetting is present in the initial velocity field. From the observational point of view, the recent discovery by radar observations of a binary asteroid having small size (of the order of 1 km) and an Earth-approaching orbit (Ostro et al., 1990), suggests that the generation of gravitationally bound fragments during catastrophic collisions in the main asteroid belt may be fairly common. Further (indirect) evidence of the same process comes from the analysis of the spin rate distribution of small asteroids (Farinella et al., 1981; Binzel et al., 1989) and of the mass distribution of asteroid dynamical families (Zappala` et al., 1984; Chapman et al., 1989).

4. ACKNOWLEDGMENTS

This work was partially supported by NATO Grant 8605/43.

5. REFERENCES

Binzel, R.P., Farinella, P., Zappala`, V., and Cellino, A.   (1989)   Asteroids II University of Arizona Press, Tucson.
Chapman, C.R., Paolicchi, P., Zappala`, V., Binzel, R.P., and Bell, J.F.   (1989) Asteroids II, University of Arizona Press, Tucson.
Farinella, P., Paolicchi, P., and Zappala`, V.   (1981) 'Analysis of the spin rate distribution of asteroids', Astron. Astrophys. 104, 159–165.
Ostro, S., Chandler, J., Hine, A., Rosema, K., Shapiro, I., and Yeomans, D.   (1990) 'Radar images of asteroid 1989 PB', Science 248, 1523–1528.
Zappala`, V., Farinella, P., Knezevic, Z., and Paolicchi, P. (1984) 'Collisional origin of asteroid families: Mass and velocity distributions', Icarus 59, 261–285.

VII

ORIGIN OF INTERPLANETARY DUST

FROM COMETS AND ASTEROIDS, BACK TO INTERSTELLAR DUST

COMETARY AND ASTEROIDAL SOURCES
OF INTERPLANETARY DUST

MARK V. SYKES
Steward Observatory
University of Arizona
Tucson AZ 85721
USA

ABSTRACT. The Infrared Astronomical Satellite has provided extensive observations of the zodiacal cloud at high spatial resolution which will not be matched in the forseeable future. Within the zodiacal cloud, IRAS discovered extended dust structures providing the link between the interplanetary dust complex and the asteroids and comets which are its source. These are the asteroid dust bands and the cometary dust trails.

1. Introduction

The origin of the zodiacal cloud has long been thought to be primarily cometary [1,2,3]. Estimates of dust production by short-period comets at the current epoch, however, have fallen far short of that needed to maintain the cloud against losses by radiation forces (after collisional evolution) [4,5]. A cometary cloud would be replenished by the occasional capture of "new", highly active comets into short-period orbits. Whipple [2] has suggested P/Encke as having been such a source in the past.

Asteroid collisions also have been considered to be a source of interplanetary dust, but the lack of observational constraints on the population of boulder-sized precursors to dust, and uncertainties in the mechanics of collisional disruption of these and larger bodies, have made it difficult to estimate the asteroidal contribution. When the Pioneer spacecraft failed to detect an enhancement in the spatial density of interplanetary dust as they passed through the asteroid belt, the contribution of asteroid collisions to the zodiacal complex consequently was thought to be small compared to comets [6].

Our knowledge of the relationship between comets, asteroids, and interplanetary dust is now undergoing significant revision as a result of extensive observations made by the Infrared Astronomical Satellite (IRAS). The mission of this small orbiting telescope was to conduct the first sensitive survey of the entire sky at thermal

A.C. Levasseur-Regourd and H. Hasegawa (eds.), Origin and Evolution of Interplanetary Dust, 389–396.
© 1991 *Kluwer Academic Publishers.*

Figure 1. The sky at 25 μm as seen by IRAS. The field is 28° in width and is centered near 1^h RA and 0° DEC (1950). East is to the left and North is up. The bright band diagonally transecting the image is the central dust band complex, associated with dust generated in the Koronis and Themis asteroid families. The long narrow trail beneath the central dust band is the Tempel 2 dust trail. P/Tempel 2 is detected near the eastern edge of the field. The background cloudlike structures are the infrared cirrus.

wavelengths [7]. In the course of this successful survey IRAS detected previously unknown, extended dust structures: the missing links between the zodiacal cloud and its cometary and asteroidal sources. These were the cometary debris trails [8] and the asteroid dust bands [?] (Fig. 1).

2. A Cometary Source

Cometary dust or debris trails are long, narrow emission features which in the IRAS data appear very much like the airplane contrails from which their name is derived. They are associated with some short-period comets, generally extending over a small fraction of their orbits [8]. Millimeter and larger in size, trail particles are ejected from their parent comets at very low velocities ($\sim$ 1 m/s), and are relatively insensitive to radiation pressure, compared to comet *tail* particles. This is why they are seen very near the projected orbits of their parent comets. Trails are not rings. They do not extend completely around comet orbits. The timescale for such dispersion is large compared to the mean times between shifts in short-period

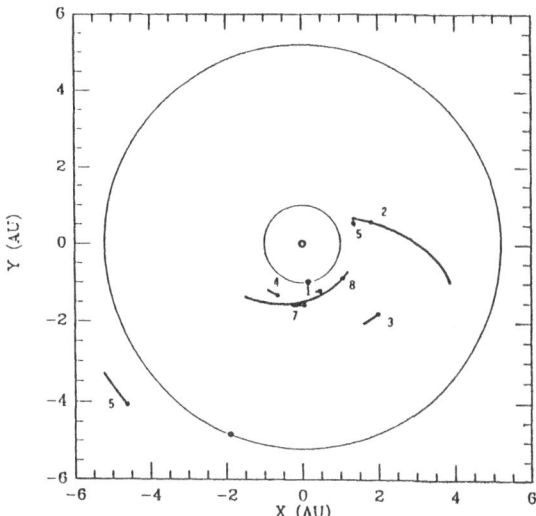

Figure 2. The eight trails associated with known short-period comets are projected onto the ecliptic plane. Positions were calculated for July 1, 1983. They are (1) Churyomov-Gerasimenko, (2) Encke, (3) Gunn, (4) Kopff, (5) Pons-Winnecke, (6) Schwassmann-Wachmann 1, (7) Tempel 1, and (8) Tempel 2. The orbits and positions of Earth and Jupiter are shown for scale.

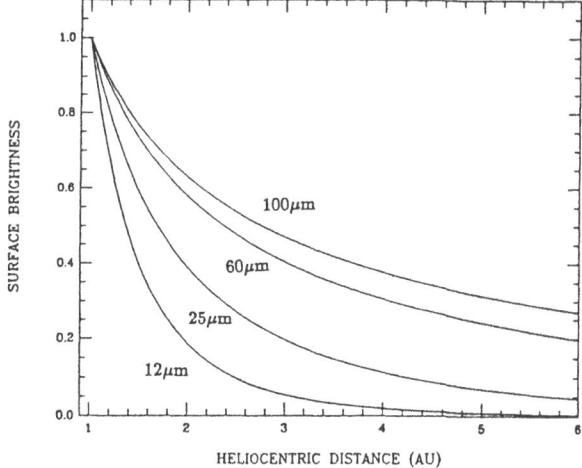

Figure 3. The surface brightness of a blackbody integrated over the IRAS passbands is shown as a function of heliocentric distance. Values are ratioed to 1 AU.

comet orbits due to gravitational perturbations by Jupiter [10].

A survey of all dust trails observed by IRAS has been completed [11]. A total of 8 trails were found in association with known short-period comets (Fig. 2). Nine fainter trails ("orphans") were also detected not associated with any known source, and it is presumed that they represent the first detections of 9 short-period comets. This is bolstered by the detection of cometary comas connected to two of the orphan trails. With the exception of Encke and Schwassmann-Wachmann 1, which are known to be anomalously active, trails tend to be observed near perihelion, and are associated with comets having the lowest perihelion distances. Trails were observed in orbital locations where they were hottest, and hence brightest at thermal wavelengths. Since short-period comets tend to have aphelion near Jupiter's orbit, and particles in elliptical orbits spend much more time near aphelion than perihelion, trail particles associated with other comets would tend to be much colder, hence fainter, in the IRAS passbands (Fig. 3). It is inferred that all short-period comets have trails, and that when the Infrared Space Observatory (ISO) is launched in 1993, it will see (with a few exceptions) a different ensemble of comet trails [11].

The generality of the dust trail phenomenon is important when considering the rate at which short-period comets supply mass to the interplanetary dust complex. A comparison with total mass-loss rates estimated from groundbased observations [17] indicates that at least 10 times more mass is lost in refractory trail particles [11]. Thus, comets are losing the great bulk of their mass in large particles, and at rates which may allow for the "steady-state" replenishment of the cloud. If the relative mass loss rates of refractory and volatile material inferred from trail studies represent the cosmogonic mass ratio, then our understanding the location and conditions of comet formation would have to be radically revised. However, it is more likely that short-period comets, having undergone significant thermal evolution, accumulate an increasingly refractory layer (mantle) over the nucleus surface as the available volatiles become depleted. An understanding of the physical properties of trail particles may then be extrapolated directly to the mantle.

Analysis of the thermal spectrum of the trail particles [11] indicates that they are very dark and porous or low-density. This is consistent with the Giotto and Vega observations of the Halley comet nucleus [12,13], and suggests that other short-period nuclei may be similar. Groundbased observations of comet nuclei have also suggested low-albedo surfaces [14,15,16].

In its observations of the zodiacal cloud, IRAS was most sensitive to particle radii of several microns at 12 and 25 μm [18]. Since trail particles are much larger, the cometary component of this cloud would arise from the collisional comminution of trail particles over time. However, IRAS also observed the cloud to have a high degree of azimuthal symmetry about the sun (to within a few percent of predicted surface brightness at 12 and 25 μm) [19]. As trail particles are comminuted into smaller sizes, they would be increasingly sensitive to radiation pressure and at some point might find themselves in Jupiter-crossing orbits. Numerical simulations of the dynamical evolution of such particles directly ejected from P/Encke, suggests that their orbital nodes would be distributed over all longitudes as a consequence

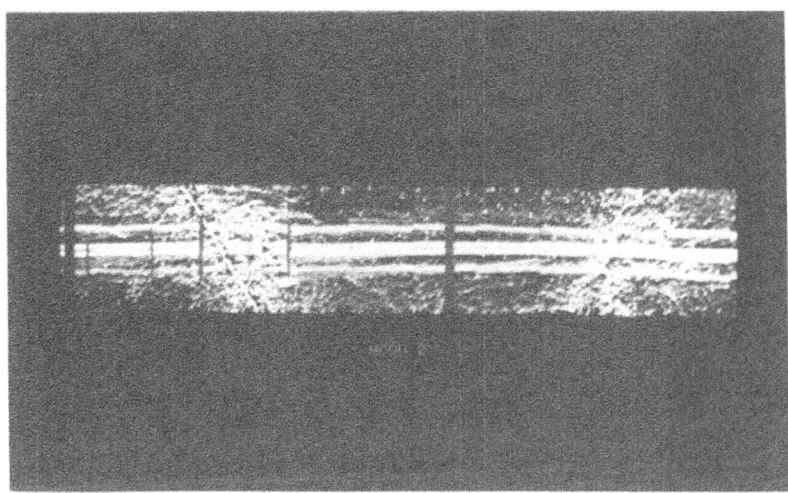

Figure 4. The ecliptic plane in cylindrical projection. IRAS scans have been high-pass filtered in ecliptic latitude to enhance the zodiacal dust band structures. Longitude increases to the left, beginning at 0°. Latitudes span ±30°. The principal bands are near 0° and ±9° latitude. Breaks and shifts in band positions are parallactic.

of gravitational interactions with Jupiter [20]. This could result in the cometary component of the zodiacal cloud having reasonable azimuthal symmetry about the sun. Otherwise, a cometary cloud would be very "bumpy".

3. Dust From The Asteroid Belt

Another source of interplanetary dust is asteroid collisions. One of the most significant discoveries made by IRAS was of regions of contemporary collisional dust production in the asteroid belt, evidenced by the zodiacal dust bands [9] (Fig. 4). The dust bands are tori of dust surrounding the inner solar system, formed when particles having similar orbital elements (a,e,i) have nodes distributed over all longitudes in the plane of symmetry of the torus. Orbit volume densities are maximum near the edges of the tori, and enhanced in the "corners" (Fig. 5) [21]. This gives rise to the appearance of parallel bands when viewed from the Earth.

The similarity in latitudes of the dust bands to the inclinations of the principal Hirayama asteroid families, in the outer main belt, led to the suggestion that the two were associated [22]. The outer pair were thought to arise from the Eos family, while the inner pair was associated primarily with the Themis family. Further analysis of the central band pair further resolved them into two components associated with

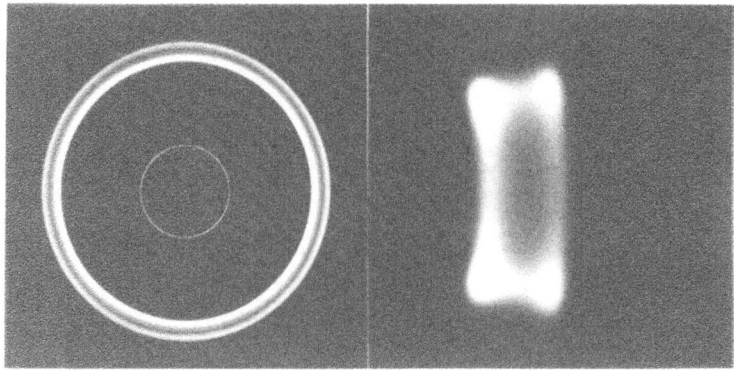

Figure 5. Model dust tori associated with the Eos asteroid family. On the left is a cut through the ecliptic plane, with the Earth's orbit shown for scale. On the right is the radial cross-section of the torus (the Sun is to the left). The mean orbital elements are the same as the known family members, and 2σ dispersions are assumed. The radial torus width is ~ 0.6 AU.

the Themis and Koronis families, respectively [21,23]. Several fainter bands also have been detected [23].

These asteroid families are the sites of the ancient catastrophic disruption of large asteroids. We are not seeing dust from that collisional event, however - such dust would have been lost long ago by Poynting-Robertson drag. Instead, we are witnessing the ongoing production of dust as the collisional fragments are themselves being broken up further into smaller and smaller sizes over time. Mass is being redistributed continually through collisions to smaller size debris until it is small enough to be removed from the families by radiation forces [24]. Calculations by Reach (this volume) indicate that material generated in the bands may be sufficient to supply the zodiacal cloud as it evolves towards the sun under Poynting-Robertson drag. In this case, the entire cloud observed by IRAS could be viewed as an extension of the major outer main belt asteroid families. This would satisfy the criterion of an immediate zodiacal cloud source that is distributed smoothly already in longitude. That asteroid dust production may be dominant in the outer main belt is also supported by the absence of significant dust bands associated with inner belt families [21].

Isolating the sources of dust production in the asteroid belt to a few specific and well-defined regions has consequences which may be testable at the Earth's orbit. Taking into consideration the dynamical evolution of torus particle orbits as they decay towards the Sun, dust collection experiments should evidence annual modulations of interplanetary dust components [25]. By determining the amplitude and node of these variations for a given particle size, the specific source region (e.g. asteroid family) may be identified. Thus, interplanetary dust studies may allow the minerologies associated with specific asteroid taxa to be determined [25].

5. Future Opportunities

Valuable new information in the study of specific sources of interplanetary dust will be gained with the upcoming Infrared Space Observatory (ISO) and the planned Space Infrared Telescope Facility (SIRTF). Archival analysis of data from the Cosmic Background Explorer (COBE) will add to our understanding of the relationship between the dust bands and the zodiacal complex, but is not expected to provide new detections of cometary dust trails (e.g. Tempel 2), because of its large field of view (and assuming pre-launch detector sensitivity requirements).

Since many of these sources were discovered after IRAS ceased functioning, ISO and SIRTF will offer the first opportunities to design specific observations of these phenomena to answer questions such as the existence of trails associated with other short-period comets and the nature and origin of the fainter dust bands and mysterious Type II dust trails [23]. Greater wavelength coverage and sensitivities will allow for a better understanding of the physical properties of dust as it is generated by comets and asteroids, as well as the underlying processes governing their release. By the next Colloquium on the origin and evolution of interplanetary dust, the global question contained in the title will have been answered in some detail.

This work was supported by NASA Grant NAG 5-1370.

[1] Whipple, F. (1955) 'A comet model. III. The zodiacal light', *Astroph. J.* **121**, 750-770.

[2] Whipple, F. (1967) 'On maintaining the meteoritic complex', in J.L. Weinberg (ed.), The Zodiacal Light and the Interplanetary Medium, pp. 409-426. NASA SP-150.

[3] Whipple, F. (1976) 'Sources of interplanetary dust', in H. Elsässer and H. Fechtig (eds.), *Lecture Notes in Physics No. 48, Interplanetary Dust and Zodiacal Light*, Springer-Verlag, New York, pp. 403-415.

[4] Delsemme, A. H. (1976) 'The production rate of dust by comets', in H. Elsässer and H. Fechtig (eds.), *Lecture Notes in Physics No. 48, Interplanetary Dust and Zodiacal Light*, Springer-Verlag, New York, pp. 314-318.

[5] Röser, S. (1976) 'Can short period comets maintain the zodiacal cloud?', in H. Elsässer and H. Fechtig (eds.), *Lecture Notes in Physics No. 48, Interplanetary Dust and Zodiacal Light*, Springer-Verlag, New York, pp. 319-322.

[6] Dohnanyi, J. S. (1976) 'Sources of interplanetary dust: Asteroids', in H. Elsässer and H. Fechtig (eds.), *Lecture Notes in Physics No. 48, Interplanetary Dust and Zodiacal Light*, Springer-Verlag, New York, pp. 187-205.

[7] Neugebauer, G. *et al.* (1984) 'The Infrared Astronomical Satellite (IRAS) mission', *Astrophys. J.* **278**, L1-L6.

[8] Sykes, M. (1986) 'The discovery of dust trails in the orbits of periodic comets' *Science* **232**, 1115-1117.

[9] Low, F. J. *et al.* (1984) 'Infrared cirrus: New components of the extended infrared emission', *Astrophys. J.* **278**, L19-L22.

[10] Bulyaev, N., Kresák, Ľ., Pittich, E., and Pushkarev, A. (1986) *Catalog of Short-Period Comets*, Astronomical Institute of the Slovak Academy of Sciences, Bratislava.

[11] Sykes, M. and Walker, R. (1991) 'Cometary Dust Trails. I. The IRAS Survey', Submitted to *Icarus*.

[12] Keller, H. *et al.* (1986) 'First Halley Multicolour Camera imaging results from Giotto', *Nature* **321**, 320-325.

[13] Sagdeev, R. *et al.* (1986) 'Television observations of comet Halley from Vega spacecraft', *Nature* **321**, 262-266.

[14] Campins, H., A'Hearn, M., and McFadden, L.-A. (1987) 'The bare nucleus of Comet Neujmin 1', *Astrophys. J.* **316**, 847-857.

[15] Millis, R., A'Hearn, M., and Campins, H. (1988) 'An investigation of the nucleus and coma of P/Arend-Rigaux', *Astrophys. J.* **324**, 1194-1209.

[16] A'Hearn, M., Campins, H., Schleicher, D., and Millis, R. (1989) 'The nucleus of comet P/Tempel 2', *Astrophys. J.* **347**, 1155-1166.

[17] Kresák, Ľ. and Kresáková, M. (1987) in E. J. Rofe and B. Battrick, *Symposium on the Diversity and Similarity of Comets*, ESA SP-278, pp. 729-744.

[18] Reach, W. (1988) 'Zodiacal emission. I. Dust near the Earth's orbit', *Astrophys. J.* **335**, 468-485.

[19] Good, J. (1991) 'Zodiacal dust cloud modelling using IRAS data', submitted to *Astrophys. J.*

[20] Gustafson, B., Misconi, N. and Rusk, E. (1987) 'Interplanetary dust dynamics. II. Poynting-Robertson drag and planetary perturbations on cometary dust', *Icarus* **72**, 568-581.

[21] Sykes, M. (1990) 'Zodiacal dust bands: Their relation to asteroid families', *Icarus* **85**, 267-289.

[22] Dermott, S., Nicholson, P., Burns, J., and Houck, J. (1984) 'Origin of the solar system dust bands discovered by IRAS', *Nature* **312**, 505-509.

[23] Sykes, M. (1988) 'IRAS observations of extended zodiacal structures', *Astrophys. J.* **334**, L55-L58.

[24] Sykes, M. and Greenberg, R. (1986) 'The formation and origin of the IRAS zodiacal dust bands as a consequence of single collisions between asteroids', *Icarus* **65**, 51-69.

[25] Sykes, M. (1989) 'Asteroidal sources of dust at the Earth's orbit', *Meteoritics* **24**, 330.

CONSTRAINTS ON THE PARENT BODIES OF COLLECTED INTERPLANETARY
DUST PARTICLES

S. A. SANDFORD
NASA/Ames Research Center
Mail Stop 245-6
Moffett Field, CA 94035 USA

ABSTRACT. Samples of interplanetary dust particles (IDPs) have now been collected
from the stratosphere, from the Earth's ocean beds, and from the ice caps of Greenland and
Antarctica. The most likely candidates for the sources of these particles are comets and
asteroids. Comparison of the infrared spectra, elemental compositions, and mineralogy of
the collected dust with atmospheric entry models and data obtained from cometary probes
and telescopic observations has provided important constraints on the possible sources of
the various types of collected dust. These constraints lead to the following conclusions.
First, most of the deep sea, Greenland, and Antarctic spherules larger than 100 µm are
derived from asteroids. Second, the stratospheric IDPs dominated by hydrated layer-lattice
silicate minerals are also most likely derived from asteroids. Finally, the stratospheric IDPs
dominated by the anhydrous minerals olivine and pyroxene are most likely from comets.
The consequences of these parent body assignments are discussed.

1. Introduction

Sites from which samples of interplanetary dust have been collected include the Earth's
stratosphere and ocean beds, and the Greenland and Antarctic ice caps. The collected
interplanetary dust particles (IDPs) generally have diameters less than 200 µm and survive
atmospheric entry because their small masses allow them to be decelerated high in the
Earth's atmosphere. The thinness of the upper atmosphere allows the particles to decelerate
over large distances, and many are not heated to sufficiently high temperatures to be
vaporized, although the larger particles may be melted. It is these melted particles that make
up the majority of the collected deep sea spherules. The IDPs collected in the stratosphere
are typically smaller than the spherules (d < 50 µm) and consequently have usually survived
unmelted. Because of their more pristine nature, this paper will concentrate largely on the
stratospheric IDPs. The properties of the collected IDPs will not be presented here in great
detail but will only be discussed in-so-far as they provide information about their parent
bodies. More detailed reviews of the properties of IDPs and the techniques used for their
collection can be found in the literature (Sandford 1987; Bradley et al. 1988).

Most of the IDPs collected in the stratosphere can be placed into one of three classes
defined by their characteristic infrared spectra (Sandford and Walker 1985) (Figure 1a).
The IDPs in these classes are dominated by olivine, pyroxene, and layer-lattice silicate
minerals, respectively. The first two categories are often referred to collectively as the
'anhydrous' IDPs and the last category is referred to as the 'hydrous' IDPs. The existence
of several distinct IDP types strongly suggests that multiple sources are responsible for the

397

A.C. Levasseur-Regourd and H. Hasegawa (eds.), Origin and Evolution of Interplanetary Dust, 397–404.
© *1991 Kluwer Academic Publishers.*

398

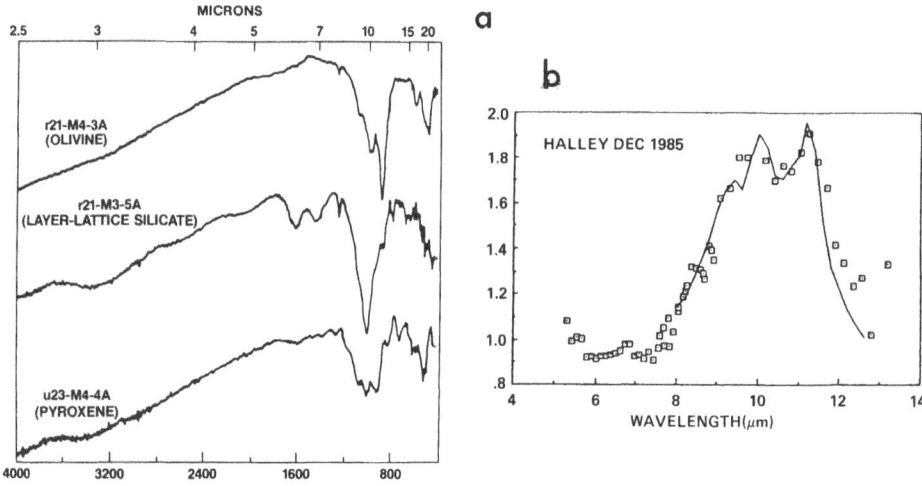

Figure 1 - (a) representative infrared spectra of the three main IDP types: pyroxene, layer-lattice silicate, and olivine, and (b) a comparison between the spectrum of Comet Halley (points) and an IDP composite (see text).

interplanetary dust complex. If the various collected dust types can be matched with sources, it then becomes possible to infer some of the sources' properties and evolution histories from the measured properties of the collected dust. Until recently, IDP-source relationships could only be suggested by comparing the collected IDPs with our preconceptions of what materials from comets and asteroids ought to look like. In the past few years, however, with the aid of good infrared cometary spectra, data returned from the Halley spaceprobes, and additional modelling of the atmospheric capture of the dust, it has become possible to constrain the possible sources of the dust more quantitatively. This paper reviews the present state of our knowledge about the sources of the collected IDPs and discusses some of the implications of the suggested IDP-source relationships.

2. Sources of the Collected Dust

The three techniques which have proven to be the most useful in constraining the sources of the collected dust take advantage of comparisons between: (i) the infrared spectra of the IDPs and comets, (ii) the elemental compositions of the IDPs and Halley dust (as measured by spacecraft), and (iii) the maximum temperatures experienced by the IDPs during atmospheric entry and those expected for different kinds of Earth-encounter orbits. Each of these three techniques will be discussed separately below. Note that the terms 'asteroid' and 'comet' will be used in a genetic sense throughout the discussion in this paper. Thus, dormant or defunct comets are still assumed to be comets, not asteroids.

2.1 CONSTRAINTS FROM INFRARED SPECTRA

As already mentioned, the majority of the stratospheric IDPs fall into one of three classes defined by their infrared spectra (Sandford and Walker 1985) (Figure 1a). The spectra from each class show distinctive silicate absorption features diagnostic of the major mineral present (olivines, pyroxenes, or layer-lattice silicates). The advent of improved telescopic infrared spectrometers now make it possible to obtain good quality, moderate resolution spectra of comets in the infrared spectral region. Thus, direct comparisons can be made between the infrared spectra of comets and the various collected IDP types.

Figure 1b shows a comparison of the spectrum of Comet Halley with a 'best fit' composite spectrum containing 55%, 35%, and 10% contributions from the olivine, pyroxene, and layer-lattice silicate IDP classes, respectively (Bregman et al. 1987). Several points are immediately apparent. The presence of substructure within the overall '10 µm' silicate feature clearly demonstrates that crystalline silicates are present, i.e. *cometary dust is not entirely amorphous.* (It will be argued later that cometary dust is probably dominated by crystalline material). Comparisons with the IDP spectra demonstrate that the presence of olivine is required to fit the strong sub-peak near 11.3 µm and that substantial amounts of pyroxene are required to match the width of the Halley feature. The fit does not suffer, however, if the 10% contribution from layer-lattice silicates is removed. Thus, there is no convincing spectral evidence that comet Halley contains layer-lattice silicates. The 11.3 µm sub-peak has since been detected in several other comets (Lynch et al.1989; Hanner et al. 1990). Comparisons of the IDP spectra with the spectra of these other comets yield matches similar to that shown in Figure 1b. Thus, the presence of crystalline olivines and pyroxenes may be a general property of comets.

In summary, comparison of the infrared spectra of comets and the collected stratospheric IDPs suggests that *comets are a viable source for the collected anhydrous IDPs dominated by the minerals olivine and pyroxene.* By process of elimination, this suggests the IDPs dominated by layer-lattice silicates may have an asteroidal origin.

2.2 CONSTRAINTS FROM DUST COMPOSITIONS

Detailed transmission electron microscope (TEM) studies demonstrate that the different IDP types exhibit distinctly different chemical and mineralogical properties. The anhydrous IDPs generally consist of a porous structure of crystalline materials that are usually not in

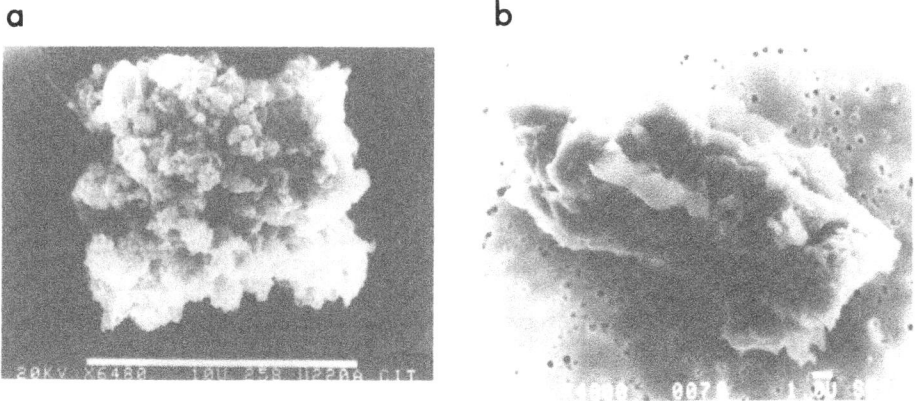

a   b

Figure 2 - TEM photographs of (a) a representative anhydrous IDP, and (b) a representative hydrous IDP (photos courtesy of J. Bradley, McCrone Associates).

chemical or mineralogical equilibrium with each other (Bradley 1988) (Figure 2a). This suggests that the various subcomponents within the particles were formed in very different environments and subsequently mixed together. The lack of equilibration between the minerals indicates that very little alteration occurred to these particles after they were incorporated into their parent body(s). Such a composition is consistent with that expected for particles from comets if comets are true repositories of pristine, primitive materials (a common belief, but one that still has not been unequivocally demonstrated). In contrast, the IDPs dominated by layer-lattice silicates generally show much lower porosities and a much higher degree of chemical and mineralogical equilibration (Figure 2b). The minerals and their structures are characteristic of hydrothermal alteration processes similar to those thought to have occurred to several types of meteorites (although the majority of the layer-lattice silicate IDPs differ in detail from the known meteorites) (Tomeoka and Buseck 1985). These similarities suggest an asteroidal origin for the hydrous IDPs. These arguments are not particularly rigorous, however, since they depend to some extent on our preconceptions of what cometary and asteroidal grains should look like.

Lawler et al. (1989) have compiled the relative abundances of Fe, Mg, and Si from the best data sets returned by the Vega-1 and Giotto spacecraft and compared them with data from the CI meteorite Orgueil and several IDPs. They found that Halley dust spanned a large compositional range, suggesting the presence of unequilibrated phases. The compositional ranges spanned by both the CI meteorite and the IDPs dominated by layer-lattice silicates were found to be much smaller than that observed from Halley dust. The compositional scatter of an anhydrous IDP, however, was found to provide a good match to the Halley data.

Thus, the spacecraft data appear to be largely consistent with our general preconceptions, namely that *the anhydrous IDPs are more likely to be derived from cometary sources and the hydrous IDPS are more likely to come from asteroids.*

2.3 CONSTRAINTS ON THE PRE-CAPTURE ORBITS OF THE COLLECTED DUST

Determination of the pre-atmospheric orbits of collected IDPs could provide extremely useful information about the dust sources. Unfortunately, it is not possible to determine the exact orbital parameters of the collected dust since this information is lost upon atmospheric entry. However, determination of the peak temperature the IDPs experienced during atmospheric entry and the density of solar flare particle tracks in their constituent minerals provide information about the general characteristics of the particles' orbits.

Before continuing, it is appropriate to briefly review the orbital evolution of dust particles in the size range of the collected IDPs as they are influenced by Poynting-Robertson (PR) radiation drag effects (Poynting 1903; Robertson 1937). Particles evolving under the PR effect have orbits whose semi-major axes decreases with time. Aphelia distances decay much more rapidly than perihelia distances and elliptical orbits are rapidly circularized before the perihelion distance decreases appreciably. Thus, particles injected into the interplanetary medium in circular orbits outside 1 AU spiral towards the sun and arrive at 1 AU in nearly circular orbits. Particles injected into the interplanetary medium in elliptical trajectories have their orbits circularized before their perihelia distance is significantly decreased after which time they spiral sunward in relatively circular orbits.

There are several cases to consider if we are to model the Earth encounter orbits expected for particles from comets and asteroids (Sandford 1986). First, particles from the main belt asteroids should arrive at 1 AU in nearly circular, prograde orbits having low inclinations. Such particles will have low Earth encounter velocities and a narrow (mass scaled) range of non-zero exposure ages in space (about 10^4 years for a 10 μm particle). In contrast, particles from highly eccentric Earth-crossing comets will have a range of inclinations up to and including retrograde motion. Depending on the nodes of the orbit, the particle could be captured by the Earth immediately after injection into the interplanetary medium or as late as the time at which the aphelion distance falls below 1 AU. Thus, dust from Earth-crossing comets will have higher Earth encounter velocities and show a much larger range in space exposure ages ($0-10^4$ yrs for a 10 μm particle). Comets with modest inclinations and perihelia outside ~1.2 AU provide a special intermediate case. Here, the particle orbits will circularize before their perihelia fall to 1 AU. As a result, particles from these sources will reach Earth in orbits similar to those of particles from asteroids. These particles would have space exposure ages similar to those of asteroidal particles but could have slightly higher Earth encounter velocities if their inclinations were significant. While this last case covers a restricted orbital phase space, it applies to many short period comets which may be major contributors to the overall zodiacal dust complex. Thus, with the possible exception of confusion between dust from asteroids and low-inclination prograde comets with perihelia outside 1 AU, the IDPs from comets and asteroids might be separated by determining either their PR infall times or their Earth encounter velocities.

The PR infall times of the collected stratospheric IDPs, as determined by the density of solar flare tracks within their constituent minerals, could be used to constrain the sources of the different classes of dust (Sandford 1986). Solar flare particle tracks are produced within the olivine and pyroxene minerals of IDPs by the passage of solar flare iron nuclei and have been detected in many IDPs using TEM techniques (Bradley et al. 1984). Since these energetic nuclei only have a range of ~100 μm in silicates, the track density in an IDP is a measure of the time it was exposed to space as a small particle. The populations of asteroidal and cometary IDPs may then be distinguishable by their different track density distributions. Unfortunately, it has so far proven difficult to carry out such a study due to complications associated with limitations of the TEM technique and to uncertainties associated with the possibility of track annealing during atmospheric entry heating.

Better progress towards defining the source of the dust has been made by comparing expected Earth encounter velocities with those inferred from the collected IDPs. Several investigators have modelled the entry of small dust particles into the Earth's atmosphere and computed the maximum heating expected as a function of particle velocity, entry angle, mass, and density (cf. Fraundorf 1980). Recently, Sandford and Bradley (1989) derived rough 'thermometers' that could be applied to the collected IDPs to determine upper limits for the heating experienced by individual grains during atmospheric entry. It was found that the hydrous IDPs generally experience minor atmospheric heating consistent with encounter velocities of about 12 km/sec, i.e. near Earth's escape velocity. This indicates that the hydrous IDPs arrive at 1 AU in nearly circular, prograde, low-inclination orbits and *suggests that the IDPs dominated by layer-lattice silicate minerals have an asteroidal origin.* In contrast, the olivine-rich IDPs, as a class, have been heated to higher temperatures than expected for a distribution of orbital velocities consistent with sporadic meteors, implying more eccentric, Earth-crossing orbits. *This suggests that the IDPs dominated by olivine minerals have a cometary origin* . As a class, the pyroxene-rich IDPs seem to have experienced heatings intermediate to those of the layer-lattice silicate and olivine IDPs. This suggests orbits which are only moderately elliptical and/or have modest inclinations. This suggests the pyroxene IDP class is also derived from comets, although the case is less conclusive than that for the olivine-rich IDPs. To summarize, the Earth encounter orbits inferred for the IDPs are consistent with previous conclusions, namely, the anhydrous IDPs are most likely from comets and the hydrous IDPs are most likely from asteroids.

At this point it is appropriate to make several comments about the source(s) of the larger particles (mostly spherules) collected from Greenland, Antarctica, and the ocean bottoms. These particles are probably derived predominantly from asteroids. Particles of this size (diameters on the order of 100 μm) are expected to be very strongly heated upon atmospheric entry (Love and Brownlee 1990). Calculations show that particles of this size arriving in Earth-crossing, cometary orbits will usually be heated to sufficiently high temperatures to be vaporized. In contrast, particles derived from asteroids could have low enough encounter velocities to survive, although a large fraction of the particles would be expected to be melted (as is observed). Also, studies of ^{26}Al and ^{10}Be in deep sea spherules show that the particles have cosmic ray exposure ages up to 10^7 years (Nishiizumi et al. 1990). Such long time scales are more consistent with exposure in an asteroidal regolith than cometary surfaces.

3. Implications

Three independent means of constraining the sources of the various classes of stratospheric IDPs have all yielded similar results: the anhydrous, olivine- and pyroxene-rich IDPs are most likely from comets, and the hydrous IDPs dominated by layer-lattice silicates are probably from asteroids. The majority of the particles collected from Greenland, Antarctica, and deep sea sediments are probably from asteroids. In this section, some of the implications of these suggested assignments are considered.

In many respects the properties of the stratospheric IDPs are consistent with general expectations of the properties of their source assignments . For example, the anhydrous IDPs generally contain many void spaces that may have once held volatile ices which were lost after ejection from a comet parent body. In addition, the anhydrous IDPs show larger chemical and mineralogical diversity than the hydrous IDPs. This suggests that the anhydrous IDPs are the more 'primitive' particles, at least in the sense that they have undergone the least alteration since they were incorporated into their parent body. Since comets are often assumed to have preserved their original starting materials better than asteroids, the suggested source identifications are consistent with general preconceptions.

The identification of the collected IDPs with different source types leads to several unanticipated conclusions, however. First, the assignment of the anhydrous IDPs to cometary parent bodies implies that most of the silicates in comets are crystalline, not amorphous. This is inconsistent with models that assume the cometary silicate grains consist of 'amorphous' olivine containing small amounts (5-15%) of crystalline olivine. Many IDPs contain some glassy materials, but these materials never dominate the mass. We know that most of the crystalline minerals in the particles are original to the particle, i.e. they are not formed by the heating of previously amorphous materials during atmospheric entry, because these mineral grains contain solar flare particle tracks. If the anhydrous IDPs really are from comets, models that assume cometary grains are made of mixtures of crystalline olivine and pyroxene are the most appropriate.

Raman spectra from the various IDP classes show that all three types contain C-rich materials (Allamandola et al. 1987). The exact state of the carbon is not well-defined but it is known that the material contains aromatic chemical units smaller than 25 Å. While not yet fully demonstrated, the C-rich material in IDPs may be similar to the polymeric kerogens found in meteorites. There is little or no graphite in IDPs and this is probably not an appropriate analog material for modelling of the carbonaceous component of comets and asteroids. The Raman spectra of IDPs rarely show silicate bands, even when the infrared spectra of the particles are dominated by silicates. This suggests that, while the silicates dominate the particle masses, the C-rich material mediates the interaction of the particles with visible light. Thus, models containing 'bare' silicate particles may not be appropriate. Finally, the IDP parent body assignments discussed here raise several interesting issues related to the isotopic compositions of the dust. Ion probe measurements of IDPs have demonstrated that many of the particles contain large deuterium excesses (McKeegan et al. 1985). These excesses are sufficiently large that they cannot be explained by simple Solar System processes and have been taken as an indication that the particles contain interstellar materials. Whether the D-rich materials exist in IDPs in discrete grains, or only represent a molecular 'memory' of altered interstellar materials, is not presently clear. In any event, the presence of isotopic anomalies can be considered to be an indication of 'primitiveness.' It is interesting to note, therefore, that D enrichments have so far only been detected in IDPs in the pyroxene and layer-lattice silicate classes, but not in the olivine-rich IDPs (McKeegan 1987). Given that comets are generally assumed to be more 'primitive' than asteroids, and that our best assignments connect the olivine-rich IDPs with comets and the hydrous IDPs with asteroids, this might be considered somewhat of a surprise. Why do the hydrous IDPs, which show properties characteristic of secondary alteration and elemental redistribution, contain 'primitive' isotopic anomalies, while the apparently cometary olivine-rich particles do not? Clearly, the presence of one 'primitive' characteristic within a material does not imply other primitive characteristics can be assumed. In the future it will be necessary to carefully define what criteria are being used when materials are referred to as 'primitive'.

4. Conclusions

Three independent means of constraining the sources of the different classes of stratospheric IDPs all yield similar conclusions, namely that the anhydrous, olivine- and pyroxene-rich IDPs are probably from comets and the hydrous IDPs dominated by layer-lattice silicates are probably from asteroids. It is likely that most of the particles collected from Greenland, Antarctica, and deep sea sediments are from asteroids. In many respects the observed properties of the stratospheric IDPs are consistent with general preconceptions of what dust from the assigned source types should look like. For example, the IDPs thought to be from comets show higher porosities and greater chemical and mineralogical diversity than those thought to be from asteroids. If the anhydrous IDPs do indeed represent cometary materials, they can be used as a source of ground truth for cometary models. It is then most appropriate to assume that the C-rich component of cometary grains contains little or no graphite, but is instead similar to meteoritic kerogens. The carbonaceous material is expected to mediate the interaction of the grains with visible light even when silicates are the dominant phase present, i.e. 'bare' silicates are not appropriate. Finally, the grains most likely consist largely of crystalline olivines and pyroxenes.

The detection of deuterium enrichments in particles from both the hydrous and anhydrous IDP classes demonstrates that both types of particles are 'primitive', at least in the sense that both preserve material that has a molecular memory of a pre-solar environment. Clearly, the assumption that comets are in all respects more 'primitive' than asteroids is called into question.

The information gleaned from the collected stratospheric IDPs in the past 15 years has resulted in significant advancements in our understanding of the interplanetary dust population and its sources. However, it is clear from observations like those of the D/H ratios in IDPs that there is still much to be learned. The interplanetary dust community can expect to see significant new results from the study of the collected IDPs in the future.

References:

Allamandola, L.J., Sandford, S.A., and Wopenka, B. (1987) 'Interstellar polycyclic aromatic hydrocarbons and carbon in interplanetary dust particles and meteorites', Science 237, 56-59.

Bradley, J.P. (1988) 'Analysis of chondritic interplanetary dust thin-sections', Geochim. Cosmochim. Acta 52, 889-900.

Bradley, J.P., Brownlee, D.E., and Fraundorf, P. (1984) 'Discovery of nuclear tracks in interplanetary dust', Science 226, 1432-1434.

Bradley, J.P., Sandford, S.A., and Walker, R.M. (1988) 'Interplanetary dust particles', in J. Kerridge and M. Matthews (eds.), Meteorites in the Early Solar System, Univ. Arizona Press, Tucson, pp. 861-895.

Bregman, J.D., Campins, H., Witteborn, F.C., Wooden, D.H., Rank, D.M., Allamandola, L.J., Cohen, M., and Tielens, A.G.G.M. (1987) 'Airborne- and ground-based spectrophotometry of comet P/Halley from 5-13 micrometers', Astron. Astrophys. 187, 616-620.

Fraundorf, P. (1980) 'Distribution of temperature maxima for micrometeorites decelerated in the Earth's atmosphere without melting', Geophys. Res. Lett. 10, 765-768.

Hanner, M.S., Newburn, R.L., Gehrz, R.D., Harrison, T., Ney, E.P., and Hayward, T.L. (1990) 'The infrared spectrum of comet Bradfield (1987s) and the silicate emission feature', Ap. J. 348, 312-321.

Lawler, M.E., Brownlee, D.E., Temple, S., and Wheelock, M.M. (1989) 'Iron, magnesium, and silicon in dust from comet Halley', Icarus 80, 225-242.

Love, S.G. and Brownlee, D.E. (1990) 'Heating and thermal transformation of micrometeoroids entering the Earth's atmosphere', Icarus, in press.

Lynch, D.K., Russell, R.W., Campins, H., Witteborn, F.C., Bregman, J.D., Rank, D.M., and Cohen, M. (1989) '5- to 13-µm airborne observations of comet Wilson 1986I', Icarus 82, 379-388.

McKeegan, K.D. (1987) 'Ion microprobe measurements of H, C, O, Mg, and Si isotopic abundances in individual interplanetary dust particles', Ph.D. thesis, Washington University in St. Louis.

McKeegan, K.D., Walker, R.M., and Zinner, E. (1985) 'Ion microprobe isotopic measurements of individual interplanetary dust particles', Geochim. Cosmochim. Acta 49, 1971-1987.

Nishiizumi, K., Arnold, J.R., Fink, D., Klein, J., Middleton, R., Brownlee, D.E., and Maurette, M. (1990) 'Exposure history of individual cosmic particles', Earth Planet. Sci. Lett., submitted.

Poynting, J.H. (1903) 'Radiation in the Solar System: Its effect on temperature and its pressure on small bodies', Philos. Trans. R. Soc. London Ser. A 202, 525-552.

Robertson, H.P. (1937) 'Dynamical effects of radiation in the Solar System', Mon. Not. R. Astron. Soc. 97, 423-438.

Sandford, S.A. (1986) 'Solar flare track densities in interplanetary dust particles: The determination of an asteroidal versus cometary source of the zodiacal dust cloud', Icarus 68, 377-394.

Sandford, S.A. (1987) 'The collection and analysis of extraterrestrial dust particles', Fund. Cosmic Phys. 12, 1-73.

Sandford, S.A. and Walker, R.M. (1985) 'Laboratory IR transmission spectra of individual interplanetary dust particles from 2.5 to 25 microns', Ap. J. 291, 838-851.

Sandford, S.A. and Bradley, J.P. (1989) 'Interplanetary dust particles collected in the stratosphere: Observations of atmospheric heating and constraints on their interrelationships and sources', Icarus 82, 146-166.

Tomeoka, K. and Buseck, P.R. (1985) 'Hydrated interplanetary dust particle linked with carbonaceous chondrites', Nature (London) 314, 338-340.

CHARACTERISTICS OF INTERSTELLAR AND CIRCUMSTELLAR DUST

A.G.G.M. TIELENS
NASA Ames Research Center,
MS 245-3, Moffett Field, CA 94035.

ABSTRACT: This paper reviews our current knowledge of interstellar and circumstellar dust from an observational point of view. It is concluded that interstellar dust is highly heterogenous. It consist of a large number of different materials formed in a great variety of stellar birthsites under varying conditions and each with a highly isotopically anomalous composition in its main elements.

1. The Carbonaceous Components of Interstellar Dust

Various dust components have been identified in the interstellar medium (ISM). Interstellar silicates make up about 30% of the total dust volume. The remainder has to be carbonaceous in nature and graphite, amorphous carbon, Polycyclic Aromatic Hydrocarbons (PAHs), and organic grain mantles have been suggested[1]. Although notably uncertain, the total identified dust volume seems to fall short of the total dust volume required to explain the interstellar extinction curve and a yet unidentified dust component might be important in the ISM (ie., diamonds[2]).

1.1 *Organic grain mantles and the 3.4μm absorption feature.* A weak absorption feature at ≈3.4μm, observed in the interstellar medium (ISM), is very characteristic for saturated aliphatic hydrocarbons (fig. 1), but its detailed substructure is poorly fit by normal parafins (eg., $CH_3(CH_2)_nCH_3$) and instead indicates the presence of nearby (but not adjacent) electronegative groups. A particular good (but non-unique) fit in peak position and width is obtained with the spectra of alcohols[3] (fig. 1). Clearly, the resolution and quality of this data is so high that the presence of a feature near 3.4μm in laboratory spectra cannot be claimed as an identification anymore ! Thus, for example, hydrogenated amorphous

405

A.C. Levasseur-Regourd and H. Hasegawa (eds.), Origin and Evolution of Interplanetary Dust, 405–412.
© 1991 *Kluwer Academic Publishers.*

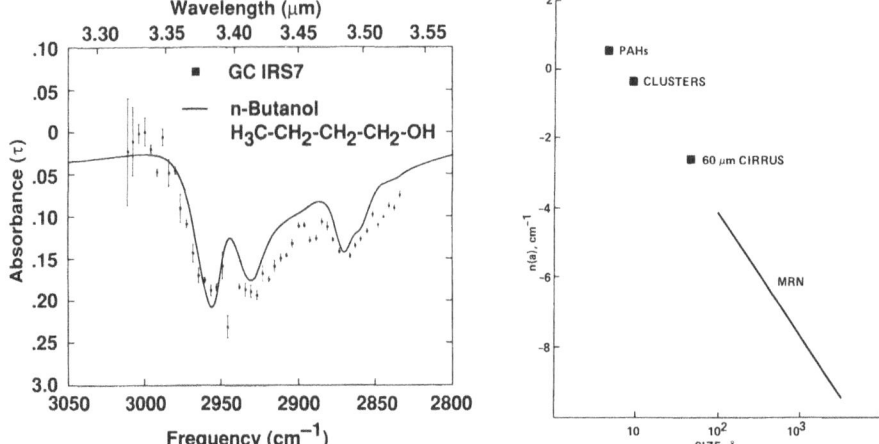

Fig. 1 (Left): The 3.4μm feature in the ISM is fit remarkably well by the CH_2 & CH_3 stretching modes in butanol [3].
Fig. 2 (Right): The grain size distribution derived from observations [16].

carbon, a-C:H or HAC[4], can be ruled out as its carrier. High resolution 5-8μm studies will be important for further characterization of this dust component. Although uncertain, the derived total volume for these organic grain mantles is only about half that of silicates and it does not appear to dominate interstellar dust[1]. Further observations along many more lines of sight will be very important to study its galactic distribution.

1.2 *Graphite and the 2200Å bump*. The 2200Å bump in the interstellar extinction curve has been traditionally identified with graphite[5] and most interstellar dust models are based upon that premise[6,7]. Small graphite spheres (a≈200Å) or "flakes" (axial ratio≈1.6 and a<50Å) provide an excelent fit to the observed profile[8]. Moreover, graphite grains - with an anomalous isotopic composition betraying a stardust origin - have been isolated from carbonaceous meteorites[9], providing some support for this assignment. However, the calculated 2200Å profile (ie., peak position and width) for graphite grains depends strongly on the grain characteristics (ie., shape and size) and thus should vary widely in the galaxy. Indeed, the width is observed to vary by 25%, but the peak position is constant to within 0.1%[10]. This invariance in the peak position actually forms a problem for any material (ie., amorphous carbon, silicates). Given the abundance constraints, the 2200Å bump has to be due to an

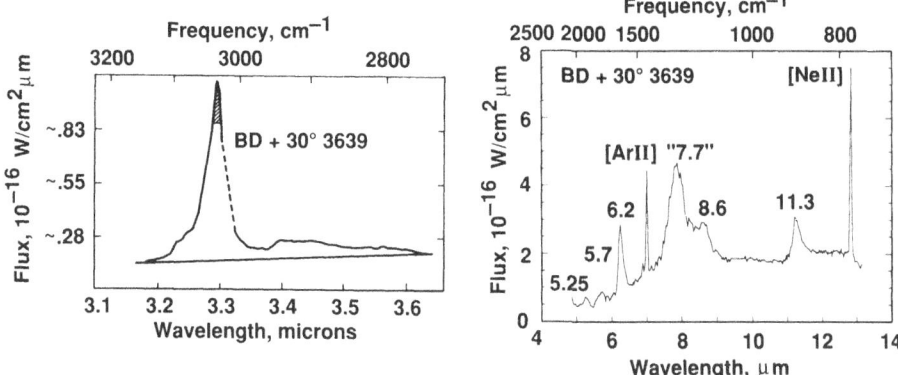

Fig. 3: The planetary nebula BD +30°3639 shows a rich spectrum of IR emission features due to PAH molecules [12].

intrinsically very strong transition[9]. Therefore, its peak position should be very sensitive to surface modes effects and thus to particle shape[11]. This implied shape selectivity in the ISM suggest we may be missing an important point (eg., C_{60} ?).

1.3 *PAHs and the IR emission features.* IR spectra of many interstellar and circumstellar sources are dominated by emission features at 3.3, 6.2, 7.7 and 11.3µm, which are very characteristic for aromatic hydrocarbon materials[12] (fig 3). Since they are also observed in regions far from the illuminating star, the equilibrium dust temperature cannot possibly be high enough to cause emission around 3µm. Consequently, the emission process has to be non-thermal in origin; ie., absorption of a single UV photon, highly excites the carrier which then relaxes rapidly through the emission of IR vibrational photons (the emission features). This constraints the carrier to small sizes ($\approx$50 C-atoms); ie., PAH molecules consisting of a planar, hexagonal C lattice with peripheral H atoms[12]. There is very good spectroscopic agreement between observed interstellar spectra and laboratory spectra of a collection of PAHs[12]. Indeed, based upon the PAH hypothesis, weak overtone and combination bands were predicted in the 5-6µm range and these have subsequently been observed[13]. Nevertheless, it should be emphasized that spectroscopic analysis of the observed features mainly implies an aromatic hydrocarbon structure and the molecular character of the carrier is inferred indirectly from an analysis of the energetics. In some sources, the observed spectra also show broad plateaus underlying these narrow emission bands[12]. Again based upon energetics, these are attributed to somewhat larger aromatic hydrocarbon units (ie., PAH clusters) containing ≈500 C atoms (≈10Å). The spectra of some sources are dominated by these broad plateaus[14] and their study may help further elucidating their carrier. Likewise, IRAS observations of the ISM reveal a contribution from a dust component with a fluctuating temperature[15] at 60μm (ie., a size of ≈50Å) and the interstellar grain size distribution extends smoothly into the molecular domain (fig. 2)[16].

2.   The Composition of Stardust

There is a bewildering zoo of objects known to form dust in their outflows or ejecta, including M-, S-, and C-giants, supergiants, planetary nebulae, novae, RCBr stars, WC stars, and type I and II supernovae (SN)[17,18]. Among the identified stardust components are silicates, amorphous carbon, PAHs, and oxides. Although these dust components are generally identified by a generic name (ie., silicates), they are formed under distinctly different conditions in the various objects and thus will differ in their detailed composition and structure (ie., mineralogy). For example, C-dust formation in red giant outflows will resemble soot formation in laboratory hydrocarbon flames, resulting in highly hydrogenated C-dust. In contrast, H is absent in the WC star outflows and their dust is probably similar to that formed in laboratory laser vaporization experiments on graphite[18]. Likewise, the Fe/Mg ratio is ≈1 in M giants, but <<1 in Si-rich zones of SN and this will be reflected in the silicate mineralogy. This may actually have some bearing on silicate compositions measured in Halley[19]. There is also some observational support for such variations. Thus, while most M-giants show a broad and structureless 10μm band (ie., amorphous silicates, but see below), some show evidence for crystaline olivine or aluminous oxides/aluminates[20,21]. Further spectral and compositional characterization of stardust in its various birthsites is an important goal of the next decade.

Each of these stardust birth sites has an anomalous isotopic composition in the main dust forming elements as well as many others[17]. For example, M-giants are enriched in ^{17}O (and ^{13}C and ^{14}N) by factors up to 10 due to dredge up from the H-burning shell (CNO cycles)[22] and this will have been preserved in the condensing silicate stardust. Likewise, C-giants (which evolve from M-giants

Table 1: The stardust budget of the galaxy

Source	contribution (10^{-6} $M_\odot$ kpc^{-2} yr^{-1})		
	carbon-dust	silicates	SiC
C-giants	2	- -	0.07
M-giant	- -	3	- -
novae	0.3	0.03	[0.007]
planetary nebulae	0.04	- -	- -
Red Supergiants	- -	0.2	- -
WC stars	0.06	- -	- -
type II supernovae	[2]	[12]	- -
type Ia supernovae	[0.3]	[2]	- -

Ref.: [17], [18]

through the dredge up of freshly synthesized ^{12}C) will show a highly variable $^{12}C/^{13}C$ ratio. Typically, these enrichments are a factor of a few, but much larger variations are possible. For example, stardust formed from the ashes of He-burning (ie., in type II SN or WC stars) consist exclusively of ^{12}C. Contrariwise, C-soot formed in nova-ejecta is expected to have $^{13}C/^{12}C > 1$. Indeed, such isotopic variations have been the key in identifying interstellar dust components preserved in carbonaceous meteorites[2].

While a large variety of objects contribute, the stardust budget is dominated by only a few (table 1)[17,18]. C-giants dominate the C-stardust injection. The He-burning shell in type II SN, containing mainly C and He, might contribute equally, provided mixing in the ejecta is unimportant. Otherwise, C/O<1 and no C-soot condensation is expected. The injection of elemental Si into the ISM is dominated by SN with some ($\approx$20%) contribution by M-giants. However, although dust condensation has been observed in SN 1987a, no 10μm silicate feature has been observed[23] (but large (>1μm) silicates are not excluded). Its dust condensation efficiency is also poorly constrained (>0.01%)[24]. Nevertheless, based upon the observed high depletion of Si from the gas phase, it is generally assumed that most of the Si enters the ISM in solid form and thus that SN dominate the silicate stardust budget. SiC stardust has only been observed around C-giants, but studies of meteorites suggest that novae might contribute as well[2]. In any case, the former is expected to dominate (table 1).

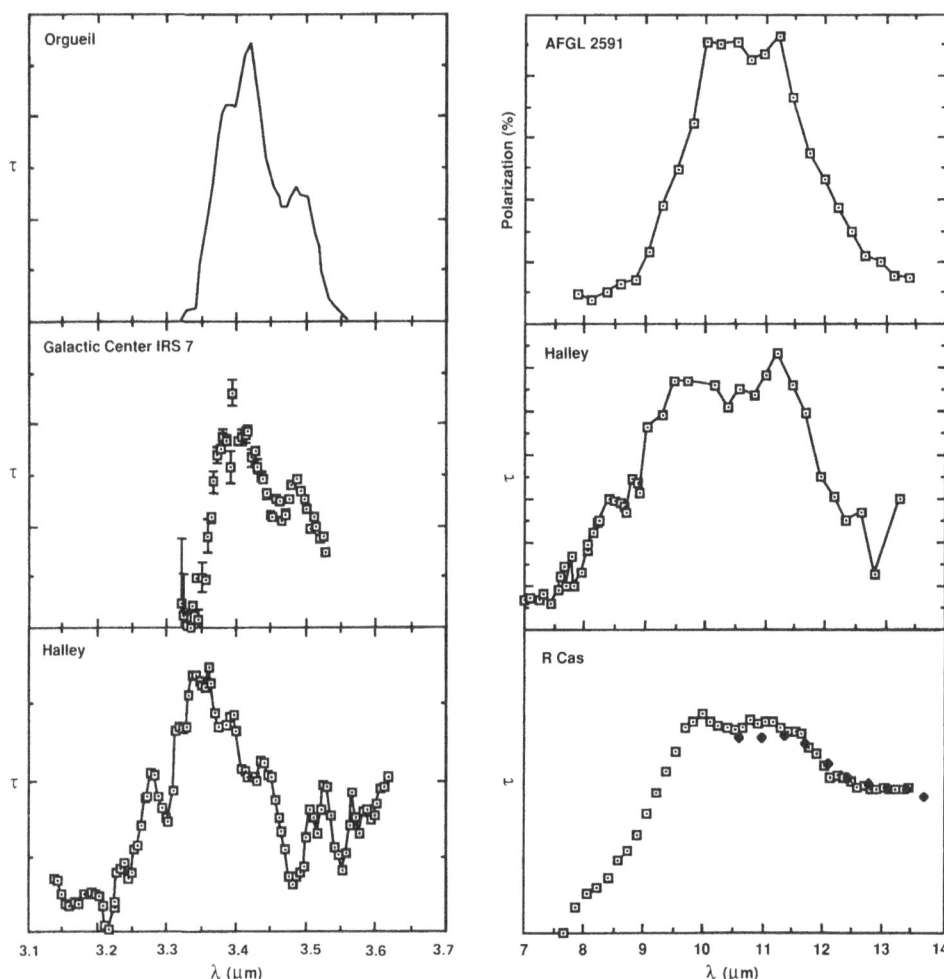

Fig. 4 (Left): The 3.4μm features observed in the acid-insoluable residue of the carbonaceous meteorite Orgueil, the galactic center source IRS 7, and comet Halley [3,24,30].

Fig. 5 (Right): The structure in the 10μm features observed in the protostar AFGL 2591, comet Halley, and the late type giant R Cas [20, 28, 29].

3   Interstellar Dust and Solar System Materials

It is of some interest to compare the observed properties of interstellar and circumstellar dust with those of solar system materials. A 3.4μm feature, characteristic for CH_2 & CH_3 stretching modes, has been detected in the ISM, in the acid-insoluable part of carbonaceous meteorites, and in comet Halley (fig. 4). While the interstellar and meteoritic spectra show similar structure, the increased 3.38/3.42 ratio shows that the meteoritic hydrocarbons have ≈20% larger CH_2/CH_3 ratio than in the ISM. The comet feature is distinctly different from the others, consisting of three components, at 3.28, 3.35, and 3.52μm[24]. These positions imply the presence of electronegative groups adjacent to the CH_2 & CH_3 groups. The first two features might be due to aromatic H and CH_2 & CH_3 groups on PAHs[25], but HAC[4] is excluded. Their relative intensities imply then that CH_2 & CH_3 sidegroups are much more abundant in comets than on interstellar PAHs (cf., fig. 3). In this model, the 3.52μm feature might originate in CC overtones of PAHs[26]. Alternatively, all three bands might have an origin in CH_2 & CH_3 groups connected to various unsaturated aliphatic (rather than aromatic) hydrocarbons[27]. In particular, the dominant 3.35μm feature is very characteristic for such groups adjacent to C=O groups (ie, ketones and esters but not aldehydes). If these cometary hydrocarbon grains are related to the CHON particles detected by Giotto[19], then the latter identification is perhaps more likely. In any case, it is discomforting that the meteoritic hydrocarbon component, which has certainly been further processed on planetary bodies, resembles the interstellar 3.4μm feature so well while the "pristine" cometary component is distinctly different.

While 10μm cometary spectra show evidence for crystalline silicates[28], the broad and structureless interstellar 10 μm feature is generally taken to imply an amorphous carrier. However, this might also be caused by a mixture of crystalline materials. Supporting this, some M-giants show a weak 11.2μm feature (Fig. 5), probably due to crystalline olivine. Likewise, a 10μm polarization study of a protostar reveals the presence of crystalline olivine in the ISM, although there is hardly a trace of it in absorption (fig 5)[29].

4. References

[1]Tielens, A.G.G.M. and Allamandola, L.J., 1987, in *Interstellar Processes*, eds. D. Hollenbach and H. Thronson, (Reidel, Dordrecht), p.397.

[2]Anders, E., Lewis, R., Tang, M., and Zinner, E., 1989, in *Interstellar Dust*, eds. L.J. Allamandola and A.G.G.M. Tielens, (Kluwer, Dordrecht), p.389.

[3]Sandford, S., et al., 1991, *Ap. J.*, in press.

[4]Borghesi, A., Bussoletti, E., and Colangeli, L., 1987, *Ap. J.*, **314**, 422.

[5]Stecher, T.P. and Donn, B., 1965, *Ap. J.*, **142**, 1681.

[6]Greenberg,J.M., and Hong, S.S., 1974 , in *Galactic Radio Astronomy*, eds. F. Kerr, and S. Simonson, (Reidel, Dordrecht), p.155.

[7]Mathis, J.S., Rumpl, W., and Nordsieck, K.H., 1977, *Ap. J.*, **217**, 425.

[8]Draine, B.T., 1989, in *Interstellar Dust*, eds. L.J. Allamandola and A.G.G.M. Tielens, (Kluwer, Dordrecht), p.313.

[9]Amari, S., Anders, E., Virag, A., and Zinner, E., 1990, *Nature*, **345**, 238.

[10]Massa, D., and Savage, B.D., 1989, in *Interstellar Dust*, eds. L.J. Allamandola and A.G.G.M. Tielens, (Kluwer, Dordrecht), p.3.

[11]Bohren, C.F., and Huffman, D.R., 1983, *Absorption and Scattering of Light by Small Particles*, (Wiley, New York).

[12]Allamandola, L.J., Tielens, A.G.G.M., and Barker, J.R., 1989, *Ap. J. Suppl. Ser.*, **71**, 733.

[13]Allamandola L.J., et al., 1989, *Ap. J. Letters*, **345**, L59.

[14]Buss, R.H., et al., 1990, *Ap. J. Letters*, in press.

[15]Ryter, C., Puget, J.L., and Perault, M., 1987, *Astr. Ap.*, **186**, 312.

[16]Tielens, A.G.G.M., in *Submillimetre Astronomy*, eds., G.D. Watt and A.S. Webster, (Kluwer Dordrecht),p.13.

[17]Tielens, A.G.G.M., 1990, in *Analysis of Returned Comet Nucleus Samples*, ed. S. Chang, NASA CP, in press.

[18]Tielens, A.G.G.M., in *Carbon in the Galaxy*, eds., S. Chang, J. Tarter, and D. deFrees, NASA CP-3061, p.59.

[19]Jessberger, E.K., Christoforidis, A., and Kissel, J., 1988, *Nature*, **332**, 691.

[20]Tielens, A.G.G.M, 1990, in *From Miras to Planetary Nebulae*, eds. M.O. Mennessier and A. Omont, (Editions Frontiers, Montpellier), p.186.

[21]Onaka, T. deJong, T., and Willems, F., 1989, *Astr. Ap.*, **218**, 169.

[22]Lambert, D.L., 1989, in *Evolution of Peculiar Red Giant Stars*, eds., H. Johnson and B. Zuckerman, (Cambride Univ. Press, Cambridge), p.101.

[23]Wooden, D., 1989, Ph.D. Thesis, University of California, Santa Cruz.

[24]Baas, F., Geballe, T.R., and Walther, D.M., 1986, *Ap. J. Letters*, **311**, L97.

[25]Knacke, R.F., Brooke, T.Y., and Joyce, R.R., Astr. Ap., 187, 625.

[26]Schutte, W., Tielens, A.G.G.M, Allamandola, L.J., Cohen, M., and Wooden, D., 1990, *Ap. J.*, **360**, 577.

[27]see also Sakata et al. elsewhere in this volume

[28]Bregman, J.D., et al. 1987, *Astr. Ap.*, **187**, 616; See also Sandford in this volume.

[29]Aitken, D.K., Roche, P., Smith, C.H., James, S.D., and Hough, J.H., 1988, *M.N.R.A.S.*, **230** 629.

[30]Wdowiak, T.J., Flickinger, G.C., and Cronin, J.R., 1988, *Ap. J. Letters*, **328**, L75.

CHEMICAL COMPOSITION OF DUST EXPECTED FROM CONDENSATION MODELS

TETSUO YAMAMOTO
Institute of Space and Astronautical Science
Yoshinodai 3-1-1, Sagamihara
Kanagawa 229
Japan

ABSTRACT. This review examines to what degrees the present chemical equilibrium condensation models are effective in predicting chemical composition of grains observed in a variety of cosmic environments. The composition expected from the equilibrium calculations is reviewed separately for refractory (rocky and metallic) and volatile (icy) components. Comments are given on the limitation of the equilibrium calculations in predicting the grain composition. By taking cometary ice as a typical cosmic volatile condensate, it is pointed out that its composition is far from that expected from the equilibrium models. Theories on the formation of cometary volatiles are reviewed, and an observational clue helpful to testing the theories is pointed out. Discussion is given on the advantage for formation of organic materials from volatile solids.

1. Introduction

Condensation is an elementary process for formation of dust grains. Models of condensation of gases having various elemental composition provide a basis of a study of the chemical composition of grains formed in a variety of cosmic environments. This review concentrates on the composition as viewed from the condensation theory. Spectroscopic studies of the chemical composition are addressed to a review by Tielens in this volume.

The condensates are classified into refractories and volatiles. The refractory solids include rocky and metallic materials, whereas the volatiles are ices of various composition composed mainly of H, C, N, O and S. Organic materials are another important component of grains, which have similar elemental composition as volatiles, but are more refractory. Table 1 gives a rough idea on the elements relevant to the above classification of the grain materials. Condensation (or sublimation) temperatures mainly depend on the strength of the bond; volatiles have the weakest bond such as van der Waals or hydrogen bonds, whereas refractories have strong bonds such as valence or metallic bonds. Condensation temperatures of volatiles are lower than 100 K in general, whereas those of rocky and metallic grains are higher than 1000 K under typical pressures relevant to cosmic condensation. Table 1 also indicates that the abundances of C, N, and O, the elements composing volatiles

413

A.C. Levasseur-Regourd and H. Hasegawa (eds.), Origin and Evolution of Interplanetary Dust, 413–420.
© 1991 *Kluwer Academic Publishers.*

and organics, are about ten times larger than those of Si, Mg, Fe, and Ni, the elements composing rocks and metals. This implies that ice and organics have potential to be as abundant as rocky and metallic solids or more.

TABLE 1. Properties of abundant elements
$(T = 10 \sim 1000\,\mathrm{K},\ P \sim 10^{-4}\,\mathrm{atm})$

	volatility	phase	bond	abundance
H He	very volatile	gas		10^{10}
C N O	volatile refractory	ice organics	van der Waals H-bond partially valence	10^7
Mg Si	refractory	rock	valence	10^6
Fe Ni	refractory	metal (+oxide, sulfide)	metallic	10^6

2. Refractory Composition

Chemical equilibrium condensation calculations have been established for rocky and metallic grains. The equilibrium calculations of a gas of solar elemental composition were intensively done in connection with the explanation of the composition of minerals in meteorites (see review by Grossman and Larimer, 1974). Major condensates that the equilibrium theory predicts are Mg-silicate such as pyroxene and olivine, and Fe-Ni alloy. Silicates are observed in many interstellar and solar system objects in both amorphous and crystalline forms (*e.g.* Tielens, in this volume). Whether amorphous or crystalline solid is formed depends on the kinetics of condensation such as a crystalline nucleation rate and a cooling rate of the grains in the environment where the grains condense (Seki and Hasegawa, 1981). On the other hand, iron grains have not been observed in interstellar space, but the presence of iron or magnetite is suggested by depletion of Fe in the gas of interstellar space (*e.g.* Whittet, 1984)

Equilibrium calculations for condensation of a gas of elemental composition other than solar have also been done (Gilman, 1969; see also Woolf, 1975). The composition of condensates depends crucially on the C/O ratio of the vapor. For $C/O < 1$ as in the case of the solar composition, C atoms are locked in CO gas, which is stable at high temperatures and low pressures, so no free carbon is available for condensation. For $C/O > 1$, on the other hand, free carbon is available for condensation. In this case, major condensates are C and SiC. The presence of carbonaceous grains are suggested in circumstellar and interstellar space (*e.g.* a review by Duley, 1988), and also presolar graphite grains are identified, though very minor, in meteorites (Amari *et al.*, 1990). SiC is observed in circumstellar envelopes of carbon stars and planetary nebulae (*e.g.* Bode, 1988), and pre-solar SiC is detected in meteorites (Bernatowicz *et al.*, 1987).

The chemical equilibrium theory has limitation for predicting solid composition formed through non-equilibrium processes. For carbonaceous grains, PAH (Léger and Puget, 1984) and QCC (Sakata *et al.*, 1984), which would be non-equilibrium products, have been proposed from a comparison of the laboratory spectra with the interstellar and circumstellar spectra. The possibility of formation of hydrocarbon grains in the case of $C/O > 1$ is suggested from thermodynamic calculations, because of the presence of abundant hydrogen (see review by Salpeter, 1977). For siliceous grains, Wada *et al.* (in this volume) propose non-stoichiometric SiO solid trapping H_2O molecules in it, which is formed by rapid cooling of SiO gas, as a candidate dust composition around protostars. It must be pointed out that the elementary processes for the formation of those non-equilibrium condensates have not yet been clarified.

3. Volatile Composition

Volatiles have not been detected in interplanetary dust, but have been observed in the dust in interstellar clouds, and in comets. Volatiles are an excellent probe for studying the origin and evolution of dust including refractories, since volatiles are sensitive to temperature and radiation in the environment where they condensed and have been placed. In the study of the interrelation between interplanetary and interstellar dust, a comet is a key object (*e.g.* Greenberg, 1982; 1988), since a comet is one of the most volatile-rich and pristine objects in the solar system probably preserving interstellar dust at the time of the formation of the solar system, as well as one of the main sources of present interplanetary dust. Volatile composition of comets is inferred from the gaseous composition observed in the coma.

Table 2 summarizes candidate molecules composing cometary volatiles and their abundances observed in the comae of comets (Weaver, 1989). Although the abundance values are heavily weighted towards the results of comet Halley and have uncertainties, it is clear that both oxidized (CO, CO_2) and reduced (CH_4, NH_3, HCN) species constitute cometary volatiles. Namely, it is a remarkable characteristic that the cometary volatiles are a mixture of reduced and oxidized compounds. It must be noted that the chemical condensation theory (*e.g.* Lewis, 1974) predicts reduced compounds. Cometary volatiles are a typical sample of non-equilibrium condensates. This implies that the equilibrium theory breaks down at low temperatures, and non-equilibrium consideration is required for volatiles, as expected from the fact that the time required to achieve chemical equilibrium increases exponentially as the temperature is lowered.

TABLE 2. Relative abundances of molecules in the comae of comets (from Weaver, 1989)

H_2O	CO	CH_4	CO_2	H_2CO	NH_3	HCN	N_2
1	0.02-0.07	0.01-0.05	~ 0.03	≤ 0.05	0.003-0.02	~ 0.001	≤ 0.001

Two types of the theories have been proposed for the formation of cometary volatiles. The theory of the first type assumes that cometary volatiles are a mixture of materials condensed in the solar nebula and in the Jovian subnebulae from a gas whose composition was quenched at a high temperature (Prinn and Fegley, 1989; Fegley and Prinn, 1989).

Another theory of this type assumes that cometary volatiles are clathrate hydrates formed in the solar nebula. (Lunine, 1989; Engel *et al.*, 1990). On the other hand, the theory of the second type asserts that cometary volatiles had originally condensed in the interstellar cloud, and lost very volatile species in the solar nebula (Yamamoto *et al.*, 1983, Yamamoto, 1985). Namely cometary volatiles are regarded as a sublimation residue of the interstellar ice. These two types of the theories are distinguished by whether cometary volatiles originates as the condensates in the solar nebula including the subnebulae of the Jovian planets, or as sublimation residues of interstellar ice. In the following, I will briefly describe the mixture model and the interstellar-ice residue model. More detailed discussion is given by recent reviews by Yamamoto (1990a, b).

The *'mixture model'* is based on a chemical equilibrium calculation of the molecular composition in a gas of solar composition, with taking account of quenching. In this model, the solar nebula is assumed to be initially hot (> 1000 K). In the course of the nebular cooling, the gaseous composition is quenched at a certain temperature called a quenching temperature, since the time scale for achieving the chemical equilibrium becomes much longer than the nebular dynamical time scale at low temperatures. Condensation of volatiles occurs in the quenched gas. For carbon compounds, for example, a ratio of CO/CH_4 is a measure of the redox state. In a chemical equilibrium scheme, CO is a dominant carbon species of the gas at high temperatures and low pressures, and CH_4 is the dominant species at low temperatures and high pressures. Actually, however, the gaseous composition is fixed in the cooling at a certain temperature called a quenching temperature. In consequence, CO dominates CH_4 even at low temperatures at the pressures of the solar nebula that Fegley and Prinn (1989) suppose. On the other hand, CH_4 dominates CO at the pressures of the Jovian subnebulae, since the total pressure of their hypothetical Jovian subnebulae is much higher than that of the solar nebula. Similar situation holds for N_2-to-NH_3 conversion: the abundances are $N_2 \gg NH_3$ in the solar nebula, whereas $NH_3 \sim N_2$ in the Jovian subnebulae. In this model, cometary volatiles are assumed to be a mixture of condensates formed under both conditions.

It must be pointed out that H_2O ice condenses at ~ 150 K under the solar nebula conditions of the pressure of $\sim 10^{-3}$ to 10^{-5} bar (Lewis and Prinn, 1980) assumed in this model, and at a higher temperature under the subnebula conditions.

In the *'interstellar-ice residue model'*, the formation process of cometary volatiles is divided roughly into two stages. The first is the interstellar cloud stage, with its fragment being the parent cloud of the solar nebula. At this stage, gaseous molecules in the cloud condense onto grain surfaces to form volatile mantles on them. Since the grain temperature in dense regions of the cloud is as low as ~ 10 K, the grains can be coated with volatile mantles composed of molecules up to very volatile ones except H_2 and He. The volatile mantles are composed of a mixture of reduced and oxidized species as characterized by the Greenberg particle (Greenberg, 1982). Note that the chemical composition of the gas and volatile mantles is far from that expected from the thermochemical equilibrium, since the chemistry prevailing in interstellar clouds is not thermal chemistry, but is based on ion-molecule reactions at low temperatures and densities. The second is the primordial solar nebula stage. In the outer solar nebula where cometary nucleus formed, the grains in the parent interstellar cloud would have survived, but their mantles lost highly volatile species from their composition, since the temperature of the solar nebula would have been higher than that of the interstellar cloud. According to this model, cometary nuclei are regarded as planetesimals formed from grains coated with volatile mantles of the sublimation residues.

This model assumes that each of the volatile species composing the ice sublimes independently, which is rather too simplified in view of recent sublimation experiments of ice mixtures (see Yamamoto (1990a, b) for the implications of these experiments). Another assumption concerns the volatile mantle composition in the interstellar cloud. Namely, this model adopts the interstellar molecule (*gaseous*) composition. The composition of the volatile mantle (*i.e. solid* composition) is not necessarily the same as the gaseous composition (*e.g.* d'Hendecourt *et al.*, 1985). It should be pointed out that the study of the ice in interstellar clouds is closely connected with the study of the cometary volatiles and of the origin of interplanetary dust.

Under the assumptions stated above, this model places a constraint upon the temperature that cometary volatiles experienced as: $T_{subl}(N_2) \sim 20 \,K < T < T_{subl}(CO_2) \sim 70 \,K$ from a comparison of the abundances of cometary and interstellar molecules, where T_{subl} is the sublimation temperature. Note that cometary volatiles have not experienced the temperatures higher than $\sim 70 \,K$ according to this model.

Crystalline structure of the ice is one of the keys to testing the two types of the theories described above. Laboratory experiments (*e.g.* Klinger (1990) for a review) show that the crystalline structure of ices depends on the temperature at which they condense. At low pressures relevant to cosmic ices, the structure is amorphous when condensed at low temperatures, whereas it is crystalline when condensed at high temperatures. For H_2O ice, the transition temperature is 130 to 140 K. As stated previously, the mixture model predicts that the H_2O ice condenses at about 150 K in the solar nebula, and at a higher temperature in the Jovian subnebulae. According to the interstellar-ice residue model, on the other hand, the ice have not experienced temperatures higher than $\sim 70 \,K$. Thus, the mixture model predicts crystalline ice, whereas the interstellar-ice residue model predicts amorphous ice, though there is a possibility of alteration of the crystalline structure after condensation, for example, by UV irradiation (Kouchi and Kuroda, 1990). The crystalline structure of fresh ice in the interior of a cometary nucleus is a key to determining which theory is plausible. Direct observations of the ice in the interior of the nucleus may be difficult at present, and we may have to wait for a mission to the nucleus. At present, the spin temperature derived from the observations of the ortho/para abundance ratio of H_2O molecules observed in comets Halley and Wilson (Mumma *et al.*, 1987; Larson *et al.*, 1988) will provide a clue to infer the formation temperature (see Yamamoto, 1990a).

4. Formation Conditions of Organic Matter

Organic materials are present ubiquitously in space. In interstellar space, the presence of organic refractory grains was proposed by Greenberg (1971) as resulting from photoprocessing of ice, and is confirmed from IR observations (Tielens, in this volume). Furthermore molecules of organic composition occupy a fraction of the observed interstellar molecular species in a gas phase. In the solar system, organic materials are found in carbonaceous chondrites, interplanetary dust as tar balls (Bradley and Brownlee, 1986), and cometary dust as CHON particles (Kissel and Krüger, 1987), and are suggested to be present on the

418

surfaces of icy satellites of the Jovian planets. In a gas phase, atmospheres of the Jovian planets and their satellites contain organic species.

In the formation of organic materials, a redox state is one of the important factors. Many experiments suggest that a reduced state provides more favorable conditions for forming organic materials than an oxidized state. It is interesting to compare the redox state of the gas and solid phases in space by noting molecular species composed of carbon, nitrogen, or both. The gas phase carbon and nitrogen chemistry indicates that C is mainly in the form of CO, and N in the form of N_2 in interstellar clouds (*e.g.* Irvine and Hjarmarson, 1983) and in the solar nebula (Prinn and Fegley, 1989; Fegley and Prinn, 1989). Namely, C and N are mainly in the oxidized state in the gas phase. The reduced state is realized only locally such as in the atmospheres of the Jovian planets, and in the Jovian subnebulae. On the other hand, volatile solid composition is not known well at present compared with the gaseous composition. For the volatile solid composition in solar nebula, we can refer to the composition of cometary volatiles. If cometary volatiles represent primitive solar nebula volatiles, their composition indicates that the solar nebula volatile solids are more reduced than the nebular gas as stated in §3. In interstellar clouds, on the other hand, recent progress in infrared observations of dense clouds reveals the presence of H_2O, CO ices, and suggests CH_3OH, and NH_3 as the ice composition (Tielens, in this volume). Figure 1 shows equilibrium temperatures of ices which condense from a gas of the interstellar molecule composition (Yamamoto *et al.*, 1983), and is calculated based on the simplified assumption as stated in §3. It is clearly seen that the reduced compounds such as hydrocarbons, nitriles and ammonia condense more easily than the oxidized compounds such as CO and N_2. These lines of results indicate that reduced compounds tend to be concentrated more in the volatile solids, and suggest that the solid phase provides more favorable conditions for the formation of precursors of organic materials than the gaseous phase.

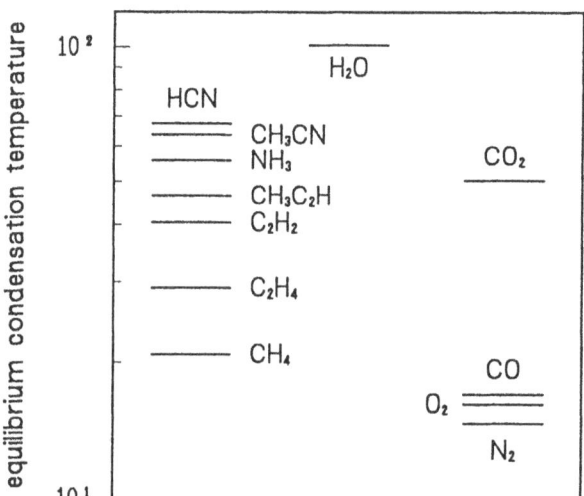

Figure 1. Equilibrium condensation temperatures (in K) of a gas of the interstellar molecule composition for the gas (mainly of H_2) density of 10^5 cm^{-3}. The species of the left side are reduced ones, and those in the right side are oxidized ones.

Acknowledgments

This work was partly supported by the Special Research Project on Evolution of Matter, University of Tsukuba.

References

Amari, S., Anders, E., Virag, A, and Zinner, E. (1990): 'Interstellar graphite in meteorites', *Nature* **345**, 238-240.

Bernatowicz, T., Fraundorf, G., Ming, T., Anders, E., Wopenka, B., Zinner, E., and Fraundorf, P. (1987): 'Evidence for interstellar SiC in the Murray carbonaceous meteorites', *Nature* **330**, 728-723.

Bradley, J.P. and Brownlee, D.E. (1986): 'Cometary particles: Thin sectioning and electron beam analysis', *Science* **231**, 1542-1544.

Bode, M.F. (1988): 'Observations and modelling of circumstellar dust', *Dust in the Universe*, eds. M.E. Bailey and D.A. Williams, Cambridge University Press, Cambridge, pp. 73-102.

d'Hendecourt, L.B., Allamandola, L.J., and Greenberg, J.M. (1985): 'Time dependent chemistry in dense molecular clouds I. Grain surface reactions, gas/grain interactions and infrared spectroscopy', *Astron. Astrophys.* **152**, 130-150.

Duley, W.W. (1988): 'Models of interstellar grains', *Dust in the Universe*, eds. M.E. Bailey and D.A. Williams, Cambridge University Press, Cambridge, pp. 209-218.

Engel, S., Lunine, J.I., and Lewis, J.S. (1990): 'Solar nebula origin for volatiles in Halley's comet', *Icarus* **85**, 380-393.

Fegley, B., Jr., and Prinn, R.G. (1989): 'Solar nebula chemistry: Implications for volatiles in the solar nebula', *The Formation and Evolution of Planetary Systems*, eds. H.A. Weaver, and L. Danley, Cambridge Univ. Press., Cambridge, pp. 171-211.

Gilman, R.C. (1969): 'On the composition of circumstellar grains', *Astrophys. J* **155**, L185-L187.

Grossman, L. and Larimer, J.W. (1974): 'Early chemical history of the solar system', *Rev. Geophys. Space Phys.* **42**, 71-101.

Greenberg, J.M. (1971): 'The chemical and physical properties of interstellar dust', *Molecules in the Galactic Environment*, eds. M.A. Gordon and L.E. Snyder, John Wiley & Sons, New York, pp. 94-124.

Greenberg, J.M. (1982): 'What are comets made of? A model based on interstellar dust', *Comets*, ed. L.L. Wilkening, Univ. Arizona Press, Tucson, pp. 131-163.

Greenberg, J.M. (1988): 'The interstellar dust model of comets: post Halley', *Dust in the Universe*, eds. M.E. Bailey and D.A. Williams, Cambridge University Press, Cambridge, pp. 121-143.

Irvine, W.M. and Hjarmarson, Å. (1983): 'Comets, interstellar molecules, and the origin of life', *Cosmochemistry and the Origin of Life*, eds. C. Ponnamperuma, D. Reidel, Dordrecht, pp. 113-142.

Klinger, J. (1990), 'Physical properties of frozen volatiles - Their relevance to the study of comet nuclei', *'Comets in the Post-Halley Era'*, eds. R. Newburn and J. Rahe, Kluwer Academic Publishers, in press.

Kissel, J and Krüger, F.R. (1987): 'The organic component in dust from Comet Halley as measured by the PUMA mass spectrometer on board Vega' *Nature* **326**, 755-760.

Kouchi, A. and Kuroda, T. (1990): 'Amorphization of cubic ice by ultraviolet radiation', *Nature* **344**, 134-135.

Larson, H.P., Weaver, H.A., Mumma, M.J., and Drapatz, S. (1988): 'Airborne infrared spectroscopy of Comet Wilson (1986l) and comparison with Comet Halley', *Astrophys. J.* **338**, 1106-1114.

Lewis, J.S. (1974): 'The temperature gradient in the solar nebula', *Science* **186**, 440-443.

Lewis, J.S., and Prinn, R.G. (1980): 'Kinetic inhibition of CO and N_2 reduction in the solar nebula', *Astrophys. J.* **238**, 357-364.

Léger, A. and Puget, J.L. (1984): 'Identification of the 'unidentified' IR emission features of interstellar dust?', *Astron. Astrophys.* **137**, L5 - L8.

Lunine, J.I. (1989): 'Primitive bodies: Molecular abundances in Comet Halley as probes of cometary formation environment', *The Formation and Evolution of Planetary Systems*, eds. H.A. Weaver, L. Danley, and F. Paresce, Cambridge Univ. Press., Cambridge, pp. 213-242.

Mumma, M.J., Weaver, H.A., and Larson, H.P. (1987): 'The ortho-para ratio of water vapor in comet P/Halley', *Astron. Astrophys.* **187**, 419-424.

Prinn, R.G., and Fegley, B., Jr. (1989): 'Solar nebula chemistry: Origin of planetary, satellite, and cometary volatiles', *Origin and Evolution of Planetary and Satellite Atmospheres*, eds. S. Atrea, J. Pollack, and M. Matthews, Univ. Arizona Press, Tucson, pp. 78-136.

Sakata, A., Wada, S., Tanabé, T., and Onaka, T. (1984): 'Infrared spectrum of the laboratory-synthesized quenched carbonaceous composite (QCC): Comparison with the infrared unidentified emission bands', *Astrophysical J. Lett.* **287**, L51 - L54.

Salpeter, E.E. (1977): 'Formation and destruction of dust grains', *Ann. Rev. Astron. Astrophys.* **15**, 267-293.

Seki, J. and Hasegawa, H. (1981): 'Origin of amorphous interstellar ice grains', *Prog. Theor. Phys.* **66**, 903-912.

Weaver, H.A. (1989): 'The volatile composition of comets', *Highlights Astron.* **8**, 387-393.

Whittet, D.C.B. (1984): 'Interstellar grain composition: A model based on elemental depletions', *Mon. Not. R. Astron. Soc.* **210**, 479-487.

Woolf, N.J. (1975): 'Circumstellar dust', *Dusty Universe*, eds. G.B. Field and A.G.W. Cameron, Neale Watson Academic Publications, Inc., New York, pp. 59-87.

Yamamoto, T. (1985): 'Formation environment of cometary nuclei in the primordial solar nebula', *Astron. Astrophys.* **142**, 31-36.

Yamamoto, T. (1990a): 'Chemical theories on the origin of comets', *'Comets in the Post-Halley Era'*, eds. R. Newburn and J. Rahe, Kluwer Academic Publishers, in press.

Yamamoto, (1990b): 'The origin of comets as viewed from the gaseous composition', *Primitive Solar Nebula and Origin of the Planets*, eds. H. Oya, K. Nakazawa, and H. Mizutani, Terra Publishing Company, Tokyo, in press.

Yamamoto, T., Nakagawa, N., and Fukui, Y. (1983): 'The chemical composition and thermal history of the ice of a cometary nucleus', *Astron. Astrophys.* **122**, 171-176.

DISTRIBUTION OF DUST IN THE DISK AROUND BETA PICTORIS

Takenori Nakano
Nobeyama Radio Observatory, National Astronomical Observatory
Nobeyama, Minamisaku, Nagano 384-13, Japan

1. Introduction

Large far-infrared excesses in some nearby main-sequence stars, revealed by the *Infrared Astronomical Satellite* (*IRAS*), have been interpreted as being due to thermal radiation from dust orbiting the stars, heated to about 100 K by the stellar radiation (Aumann *et al.* 1984; Aumann 1985; Sadakane and Nishida 1986). The existence of solid circumstellar material is commonly interpreted in the context of planet formation, and the dust has been suggested to be formed by collisions of planetesimals (Nakano 1987, 1988).

The excesses in β Pic are outstanding among these stars. Using a CCD camera supplied with a coronagraph, Smith and Terrile (1984) obtained at the I band centered at wavelength $\lambda \approx 0.89\mu$m a high resolution image of a thin disk around β Pic which is seen nearly edge-on, and found that the surface brightness along the central line of the image can be approximated by a power law

$$I(\epsilon) \propto \epsilon^{-\mu} \tag{1}$$

with $\mu \approx 4.3$ for separation angle ϵ from the star between $6''$ and $25''$, or separation distance between 100 and 400 AU. Paresce and Burrows (1987) obtained CCD images of the β Pic disk at the B, V, R, and I_c bands. The surface brightness at the I_c band centered at 0.79μm is approximated by equation (1) with $\mu \approx 3.6$ between 100 and 200 AU (Artymowicz, Burrows, and Paresce 1989).

There has been some controversy on the density distribution of dust, $n(r)$, at distance r from the star. Assuming that dust is a nearly isotropic scatterer and has the same size distribution everywhere and that the disk has a finite radius of 500 AU, Smith and Terrile (1984) found that the observed surface brightness distribution can be reproduced if $n(r) \propto r^{-3}$. Using the inversion equation Buitrago and Mediavilla (1986) reached a conclusion that although a slowly changing distribution like $n \propto r^{-1}$ is acceptable, the distributions steeper than $n \propto r^{-1.5}$ should be rejected because these distributions can be consistent with the observed surface brightness distribution only with a physically unacceptable scattering function. If dust is replenished at a constant rate from sources which are distributed only outside a circle of radius, say r_s, and dust is in equilibrium under the Poynting-Robertson drag, we have $n \propto r^{-1}$ at $r < r_s$ (Leinert, Röser, and Buitrago 1983). This fact was sometimes regarded as a support to Buitrago and Mediavilla's result (*e.g.*, Backman and Gillett 1987).

A.C. Levasseur-Regourd and H. Hasegawa (eds.), Origin and Evolution of Interplanetary Dust, 421–424.
© 1991 Kluwer Academic Publishers.

Recently I investigated the dust density distribution using the inversion equation (Nakano 1990). This is a short report of this investigation with some addition.

2. The Intensity of the Scattered Light

A distant observer seeing a dust disk edge-on receives the scattered light of intensity

$$I(\epsilon) = \int n(r)\sigma(r,\theta)F_0\left(\frac{r_0}{r}\right)^2 dx, \tag{2}$$

where $\sigma(r,\theta)$ is the differential scattering cross section of a dust particle at scattering angle θ, r_0 is the outer radius of the disk, F_0 is the stellar radiation flux at $r = r_0$, and x is the distance from the observer along the line of sight. We have assumed that the disk is transparent to both stellar and scattered radiation as is the case for the β Pic disk at least at $r \gtrsim 100$ AU (Smith and Terrile 1984). When dust has some size distribution, $n(r)\sigma(r,\theta)$ should be regarded as the sum over the size distribution. Using relations $r/\sin\epsilon = x/\sin(\theta-\epsilon) = D/\sin\theta$, D being the distance from the observer to the star, we can rewrite equation (2) as

$$I(\epsilon) = \frac{F_0 r_0^2}{D\sin\epsilon} \int_{\theta_0(\epsilon)}^{\pi-\theta_0(\epsilon)} n(r)\sigma(r,\theta)d\theta, \tag{3}$$

where $\theta_0(\epsilon) \equiv \arcsin[(D/r_0)\sin\epsilon]$ is the scattering angle at the near end of the disk along the line of sight.

We assume that the θ-dependence of $\sigma(r,\theta)$ is the same everywhere and adopt a power-law distribution

$$n(r)\sigma(r,\theta) = n_0\sigma_0\left(\frac{r}{r_0}\right)^{-\nu} f(\theta), \tag{4}$$

where n_0 and σ_0 are the number density of dust particles and the total scattering cross section of a dust particle, respectively, at the outer boundary $r = r_0$. From equations (3) and (4) we have

$$I(\epsilon) = \frac{F_0 n_0\sigma_0 r_0^{\nu+2}}{D^{\nu+1} \sin^{\nu+1}\epsilon} \int_{\theta_0(\epsilon)}^{\pi-\theta_0(\epsilon)} f(\theta)\sin^\nu\theta d\theta. \tag{5}$$

Although the β Pic disk has been imaged only inside 400 AU, it must have somewhat larger extent. We shall consider the following two cases on the distribution of the integrand $f(\theta)\sin^\nu\theta$.

a) *The case where most of the scattered light comes from the part of the line of sight nearest to the star.*
In this case we can take r_0 sufficiently large and then $\theta_0(\epsilon) \approx 0$, and we have from equation (5)

$$I(\epsilon) \propto \epsilon^{-(\nu+1)}, \tag{6}$$

because $\epsilon \ll 1$. Thus the ϵ-dependence of the surface brightness is solely determined by the distribution of particles, and the observed surface brightness of the β Pic disk can be reproduced with $\nu \approx 2.6 - 3.3$ in agreement with Smith and Terrile (1984).

b) The case where most of the scattered light comes from the outer part of the line of sight.

Differentiation of equation (5) with ϵ and some manipulation lead to the so-called inversion equation (Buitrago and Mediavilla 1986)

$$f(\theta_0) + f(\pi - \theta_0) = -\frac{I(\epsilon)\cos\theta_0}{F_0 n_0 \sigma_0 r_0}\left(\nu + 1 + \frac{d\log I}{d\log\epsilon}\right). \tag{7}$$

If the scattered light comes mostly from the outer part of the disk because of the steep forward scattering by dust and/or slowly changing density distribution, $F_0 n_0 \sigma_0 r_0 f(\theta_0)$ must be greater than $I(\epsilon)$. In this situation the quantity in parentheses in equation (7) must be at least of the order of unity. For $\nu \approx 1$ and the observed $I(\epsilon)$, it is definitely non-zero and is almost independent of ϵ. Hence from equation (7) the ϵ-dependence of the surface brightness must be determined by the scattering function $f(\theta)$ independent of the density distribution. For the observed surface brightness given by equation (1) with $\mu \approx 3.6 - 4.3$, the particles must be markedly forward-scattering. The backward scattering may not be so conspicuous as the forward scattering as inferred form the scattering function of the interplanetary dust which is responsible for the zodiacal light (Leinert *et al.* 1976). Therefore equations (1) and (7) require that the scattering function must satisfy

$$f(\theta) \propto (\sin\theta)^{-\mu}\cos\theta \tag{8}$$

at small θ. The scattering function for the interplanetary dust (Leinert *et al.* 1976) is pretty well reproduced by equation (8) with $\mu \approx 2$ at $2.5° \lesssim \theta \lesssim 30°$. The scattering function given by equation (8) with the observed values $\mu \approx 3.6 - 4.3$ is quite different from that for the interplanetary dust. Thus if the β Pic disk has a slowly changing density distribution with $\nu \lesssim -(d\log I/d\log\epsilon) - 2 = \mu - 2$, which makes the quantity in parentheses of equation (7) definitely negative, the particles must have the special scattering property given by equation (8), and they must be much more steeply forward-scattering than the interplanetary dust.

3. Discussion

The inversion equation (7) must be satisfied even for case (a) where the scattered light comes mainly from the part of the line of sight nearest to the star. In this case we have $I(\epsilon) \gg F_0 n_0 \sigma_0 r_0 f(\theta_0)$. Hence the left-hand side of equation (7) is a minor term and the two terms in parentheses, $\nu + 1$ and $d\log I/d\log\epsilon$, must nearly cancel each other, or $\nu + 1 + d\log I/d\log\epsilon \approx 0$. From this we again obtain equation (6).

Buitrago and Mediavilla (1986) considered that the density distribution with $\nu \approx 3$ was unacceptable because equation (7) with the observationally determined $d\log I/d\log\epsilon$ gives a physically unreasonable scattering function, *e.g.*, $f(\theta)$ taking a negative value at some ranges of θ.

However, the observed surface brightness must include some uncertainty. The derivative $d\log I/d\log\epsilon$ calculated from the observed $I(\epsilon)$ may have even larger uncertainty. Thus the quantity in parentheses of equation (7) with $\nu \approx \mu - 1$ must fluctuate around zero. A model fitting with a single power-law distribution given by equation (4) may also give rise to a fluctuation around zero. For instance, if the actual distribution of $n\sigma$ has an index ν slightly larger in the outer region than in the inner region, $|d\log I/d\log\epsilon|$ must be somewhat larger in the outer region than in the inner region. If we approximate such a distribution of $n\sigma$ with a single power law, the right-hand side of equation (7) would be negative in the inner region and positive in the outer region. Thus the behavior of the scattering function obtained from equation (7) cannot be used as a check of simplified models when the terms in the parentheses nearly cancel each other. A small fluctuation in $d \log I / d \log \epsilon$ should be attributed to the deviation of $n\sigma$ from a simple power law as well as to uncertainties in the observations.

If we approximate the scattering function as $f(\theta) \propto \theta^{-\lambda}$ at some range of small θ, the contribution of the outer region to the scattered light is small when $\nu > \lambda - 1$ as seen from equation (5). Thus the assumption in case (a) in §II is consistent with the result for, e.g., $\lambda \approx 2$, a value suggested from the observations of the zodiacal light.

For case (a) in §II we have obtained $\nu \approx \mu - 1$ by assuming that the disk has a size much larger than the observed size. If the size is not so large, ν must deviate somewhat from $\mu - 1$. In reality Smith and Terrile (1984) obtained $\nu \approx 3.1$ instead of 3.3 for $\mu \approx 4.3$ by taking $r_0 \approx 500$AU, only 25 % larger than the radius of the region observed by them.

So far we have assumed that the apparent thickness of the β Pic disk is nearly equal to the real thickness and the line of sight can be taken nearly on the midplane of the disk. If the apparent thickness is due mainly to tilting of the disk to the observer and the real thickness is much smaller, we can find immediately that the surface brightness is given by equation (6). Thus the index to $n\sigma$ must again be $\nu \approx 2.6 - 3.3$.

The distribution $n \propto r^{-1}$ has found some acceptance because this distribution is realized when dust sources are distributed only outside the observed region and dust is in equilibrium under the Poynting-Robertson drag as mentioned in §1. However, the dust distribution can be steeper than r^{-1} when dust sources are distributed in the observed region, and can be consistent with $n\sigma \propto r^{-(\mu-1)}$.

I would like to thank Dr. T. Mukai for valuable discussion. This work was supported partly by the Grants-in-Aid for General Scientific Research (63540193) and for Scientific Research on Priority Areas (Origin of the Solar System 01611006) of the Ministry of Education, Science, and Culture, Japan.

References

Artymowicz, P., Burrows, C., and Paresce, F., 1989, *Astrophys. J.*, **337**, 494.

Aumann, H. H. 1985, *Publ. Astr. Soc. Pacific*, **97**, 885.

Aumann, H. H., *et al.* 1984. *Astrophys. J. (Letters)*, **278**, L23.

Backman, D. E., and Gillett, F. C. 1987, *Cool Stars, Stellar Systems, and the Sun*, ed. J. L. Linsky and R. E. Stencel (Berlin: Springer), p. 340.

Buitrago, J., and Mediavilla, E. 1986. *Astr. Astrophys.*, **162**, 95.

Leinert, C., Link, H., Pitz, E., and Giese, R. H. 1976, *Astr. Astrophys.*, **47**, 221.

Leinert, C., Röser, S., and Buitrago, J. 1983, *Astr. Astrophys.*, **118**, 345.

Nakano, T. 1987, in *IAU Symposium 115, Star Forming Regions*, ed. M. Peimbert and J. Jugaku (Dordrecht: Reidel), p. 301.

Nakano, T. 1988, *Mon. Not. Roy. Astr. Soc.*, **230**, 551.

Nakano, T. 1990, *Astrophys. J. (Letters)*, **355**, L43.

Paresce, F., and Burrows, C. 1987, *Astrophys. J. (Letters)*, **319**, L23.

Sadakane, K., and Nishida, M. 1986, *Publ. Astr. Soc. Pacific*, **98**, 685.

Smith, B. A., and Terrile, R. J. 1984, *Science*, **226**, 1421.

OFF-DISK IMPLANTATION OF EARLY SOLAR WIND INTO A PLANETESIMAL-DUST CLOUD

SHO SASAKI
Institute of Geology and Mineralogy
Faculty of Science
Hiroshima University
Hiroshima 730
Japan

ABSTRACT. Off-disk implantation of ancient solar wind into a protoplanetary dust cloud can explain the present amounts of solar-type noble gases in gas-rich meteorites and Venus, even if the dust cloud is very opaque along its midplane.

1. Introduction

The high abundances of solar-type noble gas in lunar soils and some air-borne particles are considered to come from the implantation of present solar wind. In the present solar system, noble gases with solar-type elemental and isotopic abundances are also found in gas-rich meteorites (Goswani et al., 1984) and Venus' atmosphere (Pollack and Black, 1982). Furthermore, He and Ne in terrestrial mantle show solar-type isotopic signatures. The implantation of *ancient* solar wind into a protoplanetary dust cloud is a possible mechanism to explain these ubiquitous solar-type noble gases (Wetherill, 1981; McElroy and Prather, 1981).

Although the remnant grains of initial condensates dissipated along with the solar nebula, dust grains were produced by planetesimal collisions (Nakano, 1988). The opacity of the dust swarm is proportional to dust number density and inversely proportional to dust size. If the dust cloud was opaque and optical thickness along the disk midplane was high, solar wind species would be trapped only at the inner edge (0.05 – 0.1AU) of the disk (Wetherill, 1981; Prinn and Fegley, 1989) (see Fig. 1(a)). If the dust cloud was transparent, the solar wind would penetrate but no significant implantation would take place. Then the implantation of a large amount of the ancient solar wind flux did not occur ubiquitously in the solar system?

The dust swarm should have a vertical distribution, reflecting orbital inclinations of dust-forming planetesimals. Even if optical thickness along the disk midplane is

A.C. Levasseur-Regourd and H. Hasegawa (eds.), Origin and Evolution of Interplanetary Dust, 425–428.
© *1991 Kluwer Academic Publishers.*

very large, the solar wind as well as radiation can penetrate through the off-disk region of the disk (Sasaki, 1989; 1990) (see Fig. 1 (b)) . In the present report, we estimate trapped amounts of solar-wind noble gases, taking into account a vertical distribution of dust grains.

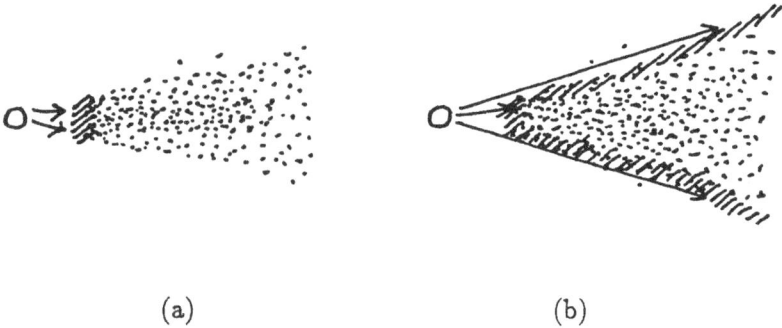

(a)   (b)

Fig. 1 Schematic picture of solar wind implantation: (a) Inner boundary implantation (b) Off-disk implantation

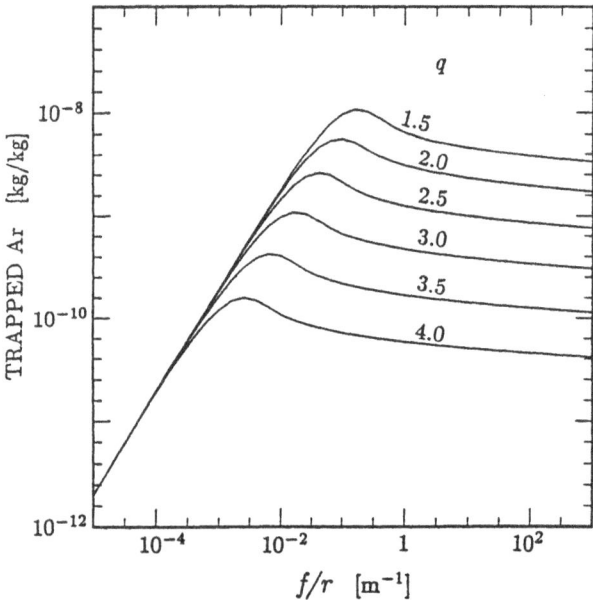

Fig. 2   Total trapped amount of ^{36}Ar between 3 and 3.5 AU. The values are given by kg Ar per kg total solid mass in the region. The horizontal axis is $f/r[\text{m}^{-1}]$. Solar wind intensity is chosen to be 100 times the present value and implantation duration is assumed to be 10^7yr.

2. Dust Distribution

The dust cloud is assumed to have a power-law density distribution radially and a gaussian distribution vertically. We assume that the disk is truncated at $a_0 = 0.1[\text{AU}]$ by dust evaporation. We have mass distribution of dust

$$\rho_{dust}(a, z) = \rho_{de} f \left(\frac{a}{1\text{AU}}\right)^{-q} \exp\left\{-\left(\frac{z}{h}\right)^2\right\},$$

where f is mass fraction of dust relative to the total solid mass and $\rho_{de} f$ expresses the dust mass abundance at the midplane of the terrestrial region ($a = 1[\text{AU}]$ and $z = 0$). Using the average orbital inclination of dust i, the vertical scale height h is written by $h = ia$. We have $i = 0.01$ from random velocities of planetesimals in the late stage of planetary formation. This inclination gives $\rho_{de} = 2.7 \times 10^{-8}[\text{kg/m}^3]$ with the present smeared solid mass distribution (Hayashi, et al., 1985). Since the opacity determines the solar wind implantation, the ratio of dust mass fraction to dust size f/r is an important parameter controlling our results.

3. Results

We obtain integrated amounts of solar wind ^{36}Ar in a vertical column. Figure 2 shows the trapped amount of solar wind ^{36}Ar between 3.0 and 3.5 AU. Solar wind intensity is assumed 100 times larger than the present value, and implantation duration is assumed to be 10^7yr. When f/r is small, disk is so transparent that efficient wind capture does not occur. When f/r is large ($> 10^{-1}[\text{m}^{-1}]$), the solar wind implantation takes place in the off-disk region. Because larger f/r simply moves trapping region vertically outward, the total trapped amount does not decrease largely (Sasaki, 1990). Gas-rich meteorites with high noble gas concentration have ^{36}Ar abundances around $10^{-9}[\text{kg/kg}]$ (Sasaki, 1990). Since a plausible value of q is 2.5–2.75 from the present mass distribution, Ar abundance in gas-rich meteorites is explained when f/r is larger than $10^{-2}[\text{m}^{-1}]$.

4. Discussions

In the later stage of solar system formation, gravitational scattering by massive Jupiter should enhance the inclination of planetesimals in the asteroidal region. If the vertical height of the disk there is larger than that of the inner regions, the trapped amount of noble gas should increase greatly. Therefore even if solar wind duration was shorter or intensity is weaker, a large amount of solar-type gas could be supplied to the dust (or larger bodies) in the asteroidal region.

In Venus, ratio of ^{36}Ar mass to the total planetary mass is $2.4 \times 10^{-9}[\text{kg/kg}]$, which is comparable to values of gas-rich meteorites with high noble gas abundances. The off-disk implantation can also supply Venus' solar-type noble gas if the dust with implanted gases accumulates onto the planet. During or after the implantation, both He and Ne should escape from grains because temperature there is higher than that in the asteroidal region. This may explain low Ne/Ar of Venus.

Acknowledgments: The author thanks M. Ozima and T. Futagami for discussions. He is grateful to R. O. Pepin for valuable comments. Most parts of this work were done while the author stayed at University of Arizona. He thanks D. M. Hunten for supports and encouragements.

References

Goswani, J. N., Lal, D. and Wilkening, L. L. (1984) 'Gas-rich meteorites: probes for particle environment and dynamical processes in the inner solar system', *Space Sci. Rev.* **37**, 111-159.

Hayashi, C., Nakazawa, K., and Nakagawa, Y. (1985) 'Formation of the solar system', in D. C. Black and M. S. Matthews (eds.), *Protostar and Planets II*, The University of Arizona Press, Tucson, pp.1100-1153.

McElroy, M. B. and Prather, M. J. (1981) 'Noble gases in the terrestrial planets', *Nature* **15**, 535-539.

Nakano, T. (1988) 'Formation of planets around stars of various masses – II. Stars of two and three solar masses and the origin and evolution of circumstellar dust cloud', *Mon. Not. R. astr. Soc.* **230**, 551-571.

Pollack, J. B. and Black, D. C. (1982) 'Noble gases in planetary atmospheres: implications for the origin and evolution of atmospheres', *Icarus* **51**, 169-198.

Prinn, R. G. and Fegley, B. Jr. (1989) 'Solar nebula chemistry: origin of planetary, satellite and cometary volatiles', in S. K. Atreya, J. B. Pollack, and M. S. Matthews (eds.) *Origin and Evolution of Planetary and Satellite Atmospheres.*, The University of Arizona Press, Tucson, pp.78-136.

Sasaki, S. (1989) 'Penetration of the solar wind after dissipation of the solar nebula: origin of Venusian Ar by off-disk implantation of the solar wind', *Proc. NIPR Symp. Antarct. Meteorites* **2**, 326-334.

Sasaki, S. (1991) 'Off-disk penetration of ancient solar wind', *Icarus* **90**, in press.

Wetherill, G. W. (1981) 'Solar wind origin of ^{36}Ar on Venus', *Icarus* **46**, 70-80.

COMPARISON OF 3 MICRON FEATURES OF TRAPPED H2O AND H2O FROST IN SiO CONDENSATE WITH OBSERVED DUST FEATURES

S.WADA[1], A.SAKATA[2], and A.T.TOKUNAGA[3]

1. Dept. of Chemistry, University of Electro-Communications, Chofugaoka, Chofu, Tokyo 182, Japan
2. Dept. of Applied Phys. and Chem., University of Electro-Communications, Chofugaoka, Chofu, Tokyo 182, Japan
3. Institute for Astronomy, University of Hawaii, 2680 Woodlawn Dr., Honolulu, HI 96822 U.S.A.

ABSTRACT   We synthesized a SiO condensate trapping H_2O and H_2O ice deposited on it. An IR spectrum of the condensate and that of a protostar NGC 7538/IRS 9 were compared. The spectrum of the condensate agreed well with the protostellar spectrum.

1. Introduction

In spectra of protostellar objects absorption features are observed at 3.07 μm and near 10 μm (Willner et al., 1982). They are attributed to H_2O ice and amorphous silicate, respectively. Day and Donn (1978) and Nuth and Donn (1982) synthesized grains from SiO and Mg gas and showed that their IR spectrum exhibited a broad feature peaked near 10 μm. We synthesized a condensate from SiO gas. The condensate trapped H_2O gas into its structure. This H_2O causes a broad 3 μm feature peaked at 2.94–2.99 μm (Wada et al., 1990). We deposited H_2O ice on the SiO condensate. It was examined whether the protostellar feature can be accounted for the SiO condensate frosted with ice on it.

2. Experiments and Results

Powder of SiO_2 and Si mixed with equal mole amounts was heated in a tantalum boat to 1,300 °C, and SiO vapor was produced. The SiO vapor was condensed onto a KBr or KRS-5 crystal substrate, and a copper substrate, both of which were cooled by liquid nitrogen. When the condensate was formed, H_2O was trapped into the SiO condensate.

We made a cell specially designed for IR measurement at low temperature. Crystalline H_2O ice was deposited onto the condensate (referred to "frosted SiO") at about –50 °C. Then the deposited ice was defrosted by evacuation in the cell (this material is referred to "defrosted SiO"). A JASCO-810 IR spectrophotometer was used to obtain IR spectra of the condensates.

The 3 μm spectrum of the SiO condensate trapping H_2O was compared to the observed spectrum toward the Galactic Center source IRS 7 (Butchart et al., 1986, Fig.1). IR spectra

429

A.C. Levasseur-Regourd and H. Hasegawa (eds.), Origin and Evolution of Interplanetary Dust, 429–432.

of the "frosted SiO" in the cell were compared to the observed spectrum of protostellar dust (Willner et al., 1982) in Fig.2.

3. Discussion

The 3 μm feature caused by trapped H_2O in SiO condensate agreed well with the feature toward the Galactic Center source IRS 7, in which the peak is at 2.95–3.00 μm (Fig. 1). It is uncertain as yet whether the dust which causes the 3 μm feature exists in interstellar space or near the Galactic Center. Tanaka et al. (1990) found a similar broad absorption feature peaked at 2.95 μm around M type stars. This 3 μm feature is observed in hot circumstellar space or diffuse cloud regions. In this "dry" condition, molecular H_2O ice frost or dirty ice cannot exist. Trapped H_2O into Si-O structure may survive the condition.

There is a clear difference between the IR feature toward the Galactic Center sources and protostellar dust features. The peak is at 2.95–3.00 μm for the feature toward the Galactic Center sources, including IRS 7 (McFadzean et al., 1989), and at 3.07 μm for the protostellar feature. The difference in the 3 μm features is caused by different chemical and physical conditions of H_2O. In a dense cloud H_2O is deposited onto the surface of "dry" core dust grain containing trapped H_2O. The deposited H_2O ice adds a peak at 3.07 μm on the broad 2.95–3.00 μm feature. Therefore, the 3 μm feature of protostellar dust can be attributed mainly to a mixed feature of the two H_2O components.

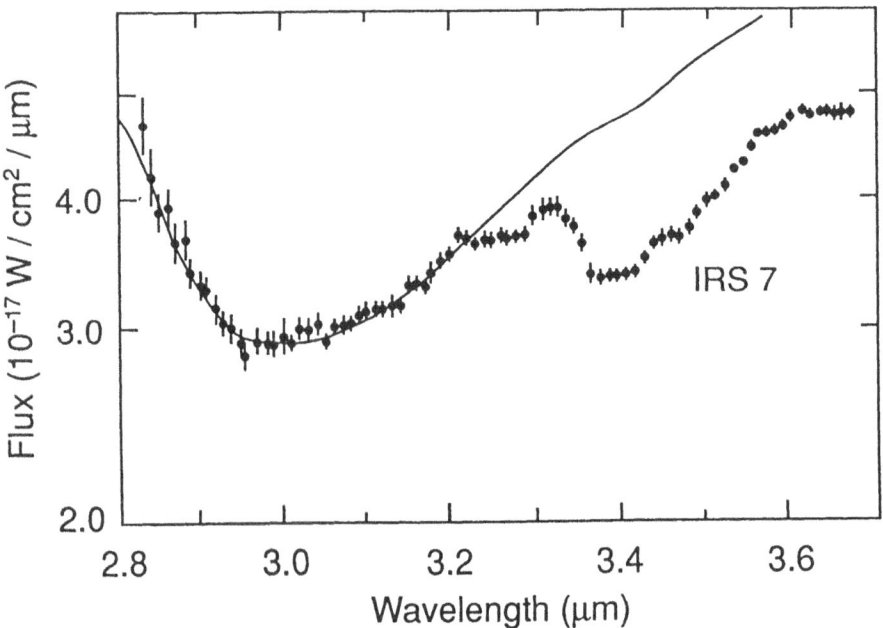

Fig.1. Comparison of a 3 μm spectrum of trapped H_2O in SiO condensate to that toward the Galactic Center source IRS 7.

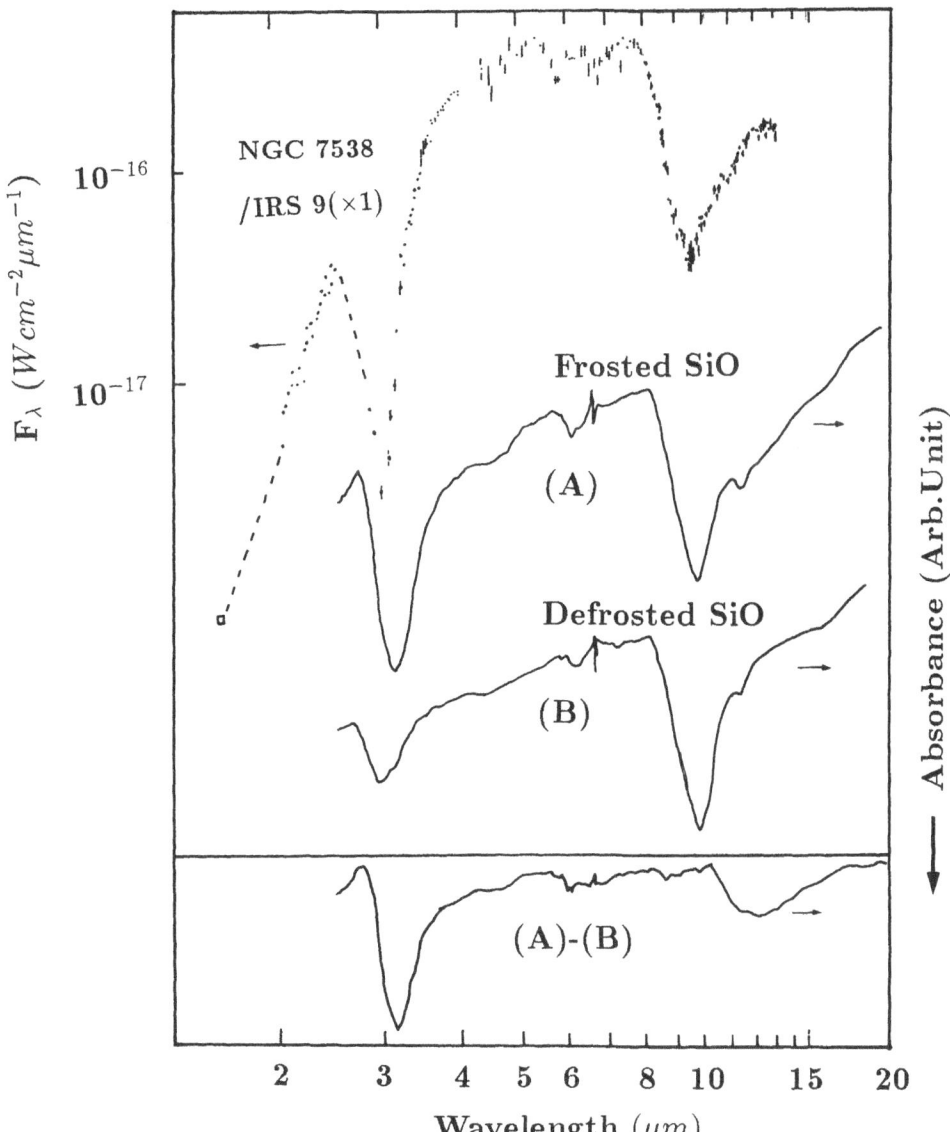

Fig.2. Comparison of IR spectra of "frosted SiO condensate" and "defrosted SiO condensate" to a spectrum of protostar NGC 7538/IRS 9 (after Willner et al., 1982). (A)–(B) shows a difference spectrum of (A) and (B).

In the 10 μm wavelength region, there is another difference between the dust feature toward the Galactic Center source IRS 7 and protostellar dust features. The 10 μm absorption feature toward the Galactic Center sources is narrower than that of the protostellar feature (Aitken et al., 1986). Pure H_2O ice causes a broad feature peaked near 12 μm (Léger et al., 1979). In our experiment, "frosted SiO" did not show a hump at 12 μm, instead resulted in broadening the 10 μm feature in long wavelength region. This "frosted SiO" exhibits a feature similar to the NGC 7538/IRS 9.

In a course of star formation, protostellar dust which is coated with H_2O ice would be dried. "Defrosted SiO" may be an analog to this kind of dust grains. Its IR spectrum shows a peak at 2.94 μm and 10 μm (Fig.2 (B)).

4. Conclusion

1. Deposition of small amount of crystalline H_2O ice on SiO condensate containing trapped H_2O shows a similar feature to that of a protostar NGC 7538/IRS 9.

2. The 3 μm feature of protostellar dust can be attributed mainly to a mixed feature of the two H_2O components, trapped H_2O into Si-O structure and H_2O ice.

3. H_2O ice deposited on SiO condensate resulted in broadening the 10 μm feature in long wavelength region.

4. In a 3 μm spectrum of defrosted SiO condensate the peak is located at 2.94 μm.

References

Aitken, D. K., Roche, P. F., Bailey, J. A., Briggs, G. P., Hough, J. H., and Thomas, J. A. (1986) 'Infrared spectropolarimetry of the Galactic Centre: magnetic alignment in the discrete sources' Mon. Not. R. astr. Soc. **218**, 363–384.

Butchart, I., McFadzean, A. D., Whittet, D. C. B., Geballe, T. R., and Greenberg J. M. (1986) 'Three micron spectroscopy of the Galactic Center Source IRS 7.' Astron. Astrophys. **154**, L5–L7.

Day, K. L., and Donn, B. (1978) 'An Experimental Investigation of the Condensation of Silicate Grains.' Ap. J. **222**, L45–L48.

Léger, A., Klein, J., de Cheveigne, S., Guinet, C., Defourneau, D., and Belin, M. (1979) 'The 3.1 μm absorption in molecular clouds is probably due to amorphous H_2O ice' Astron. Astrophys. **79**, 256–259.

McFadzean, A. D., Whittet, D. C. B., Longmore, A. J., Bode, M. F., and Adamson, A. J., (1989) 'Infrared studies of dust and gas towards the Galactic Centre: 3–5 μm spectroscopy' Mon. Not. R. astr. Soc. **241**, 873–882.

Nuth, J. A., and Donn, B. (1982) 'Laboratory Measurements of Amorphous Silicate Smokes and the Infrared Spectra of Oxygen-rich Star.' Ap. J. **257**, L103–L105.

Tanaka, M., Sato, S., Nagata, T., and Yamamoto, T. (1990) 'Three micron ice-band features in the ρ Ophuchi source' Ap. J. **352**, 724–730.

Wada, S., Sakata, A., and Tokunaga, A. T. (1990) 'Trapped H_2O in SiO condensate: an explanation for the 3 μm band observed toward the Galactic Center' Ap. J. submitted.

Willner, S. P. Gillett, F. C., Herter, T. L., Jones, B., Krassner, J., Merrill, K. M., Pipher, J. L., Puetter, R. C., Rudy, R. J., Russell, R. W., and Soifer, B. T. (1982) 'Infrared Spectra of Protostars: Composition of the Dust Shells' Ap. J. **253**, 174–187.

THE SOURCE COMPOSITION OF GALACTIC COSMIC RAYS AS POSSIBLY ORIGINATED FROM THE DUST IN THE CIRCUMSTELLAR AND INTERSTELLAR SPACE

Kunitomo SAKURAI
Institute of Physics, Kanagawa University
Rokkakubashi, Yokohama 221, Japan

ABSTRACT. The chemical composition of galactic cosmic rays in their sources is similar to that of interstellar clouds or grains which are relatively enriched in refractory and siderophile elements as compared with the chemical composition of the solar atmosphere. Taking into account this fact, it is shown that the cosmic ray source matter can be identified as the dust or grains observed in the envelopes of red supergiant stars or the matter originally ejected from supernova explosions.

Key words: Galactic cosmic rays, Cosmic ray source matter

1. INTRODUCTION

At present, it is thought that the cosmic ray source matter is mainly supplied from gases ejected from supernova explosions (e.g., Oda et al., 1988). However, we not known even now on how the chemical composition of this matter is formed by using the matter ejected from supernovae. As have been shown by many people, both refractory and siderophile elements are enriched by a few times more in the chemical composition of this matter as compared with that of the solar atmosphere, which is now thought of as the standard for the chemical composition of the local galactic matter (Sakurai, 1989). Since the first ionization potentials for the most of these elements are relatively lower than those of volatile elements as C, N, O, He, Ne and others, it has been pointed out that the most elements contained in the cosmic ray source matter may be only partially ionized before and during their acceleration somewhere in the interstellar space. In fact, the ionization states of the elements play some important role in the efficiency of their

A.C. Levasseur-Regourd and H. Hasegawa (eds.), Origin and Evolution of Interplanetary Dust, 433–436.

acceleration as pointed out by Cassé and Goret (1978) for the first time.

After ejected in association with supernova explosions, the matter embedded in the expanding outer envelopes of parent stars begins to cool down. While drifting in the nearby space surrounding suprenovae, this matter further cools down into lower temperature state. Therefore, it seems possible that, after cooled down into a state as low as 1000K in temperature, the matter initially ejected from supernova explosions becomes the source matter of galactic cosmic rays. In this paper, we will examine if this possibility is valid and acceptable, and then discuss that the chemical composition of the cosmic ray source matter is formed in association with the condensation process of the elements in such environments as seen in the circumstellar gases of red supergiant stars or in the gases ejected from supernova explosions.

2. THE SOURCE COMPOSITION OF GALACTIC COSMIC RAYS

Both refractory and siderophile elements are relatively enriched in the chemical composition of the cosmic ray source matter as compared with that of the solar atmosphere. In this case, these two compositions are normalized by the Si abundances in them. It is noted that the condensation temperature for each of these elements is usually higher than 1000K, though being dependent on the ambient temperature (Wasson, 1985). This means that, as compared with volatile elements, they are relatively easily condensed into dusts or grains in the process of their formation while gases containing those elements are being cooled down in the interstellar space.

It thus seems that the condensation process as related to the formation of dusts or grains plays an important role in making the observed chemical composition of the cosmic ray source matter. By examining the possible contribution of this process, we have obtained such a relation of the relative enrichment of various elements to their condensation temperatures as shown in Fig. 1. It follows from this figure that, for the source composition of galactic cosmic rays, both refractory and siderophile elements are mostly enriched by a few times more in comparison with the chemical composition of the solar atmosphere (Sakurai, 1989). Following the procedure as ever made by many people, these two compositions are normalized by the abundances of the element Si in them. All volatile elements with the condensation temperature less than 500K are well underabundant in the chemical composition of the cosmic ray source matter as compared with that of the solar atmosphere. These

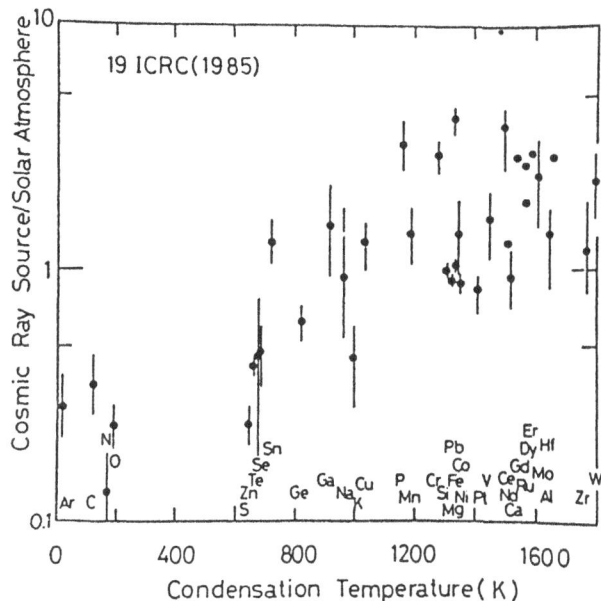

Fig. 1   The abundance ratios of the elements
of the chemical composition of the cosmic ray
source matter to that of the solar atmosphere
as a function of the elemental condensation
temperatures, normalized by the element Si.

results suggest that the elements with the condensation
temperature higher than about 1000K are relatively enhanced
in the cosmic ray source matter by some process as the
condensation of gases into dusts or grains in the interstellar space.

3.   A POSSIBLE PROCESS AS RELATED TO THE FORMATION OF COSMIC RAY SOURCE MATTER

As shown in Fig. 1, the chemical composition of the cosmic
ray source matter is similar to that of dusts or grains in
the circumstellar gases and in the dense giant gas clouds
in the interstellar space.   Furthermore, these dusts or
grains there seem to have been formed as the result of the
efficient condensation of non-volatile elements associated
with the cooling of these gases.   Therefore, the chemical
composition of dusts or grains thus formed may be nearly
the same as that of carbonaceous chondrites classified as
C2 (or CM).   As well known, both refractory and siderophile

elements are generally more enriched in these chondrites as compared with that of Cl chondrites (Wasson, 1985).

The results summarized above suggest that the cosmic ray source matter is formed through the condensation process of gases which takes place in the circumstellar gases or in gases ejected from supernova explosions. Since this process works efficiently within those gases with temperature less than about 1000K, the cosmic ray source matter must have passed through such a low temperature state when it is formed somewhere in the interstellar space. The most probable places for this condensation to occur seem to be identified as the deep inside of the circumstellar gases of red supergiant stars or of the low-temperature dense gases surrounding supernova remnants.

4. CONCLUDING REMARKS

Even now,we have not found as yet any real mechanism for cosmic rays to be accelerated from the matter enriched by refractory and siderophile elements in the galactic space.

According to our scenario (Sakurai, 1990), the cosmic ray source matter in a low-temperature state would become heated and ionized partially by shock waves associated with stellar winds and/or supernova explosions. The elements, being partially ionized, would be accelerated to cosmic ray energy due to their interaction with shock waves just mentioned.

REFERENCES

Cassé, M. and Goret, P. (1978) Ionization models of cosmic ray sources, Astrophys. J. **221**, 703-712.

Oda, M., Nishimura, J. and Sakurai, K. (1988) *Cosmic Ray Astrophysics*, Terra Sci. Pub., Tokyo.

Sakurai, K. (1989) The origin of cosmic rays as viewed from their source composition, Adv. Space Res. **12**, 149-152.

Sakurai, K. (1990) Proc. NATO Summer Inst., Erice(in press).

Wasson, J.L. (1985) *Meteorites: Their Record of Early Solar System History*, W.H. Freeman, New York.

LABORATORY STUDIES OF GRAIN MANTLES IN INTER-STELLAR SPACE

C.X. MENDOZA-GOMEZ and J.M. GREENBERG
Laboratory Astrophysics, University of Leiden, Postbus 9504
2300 RA Leiden, The Netherlands

ABSTRACT. At Laboratory Astrophysics we are simulating the most relevant conditions in interstellar space in order to follow the chemical and physical evolution of, among others, interstellar organic grain mantles.

1. Introduction

In the atmospheres of evolved stars, small silicate condensates are formed. When they are injected into the interstellar space, they are cooled down to temperatures of about 10 K. It is on these sub- micron particles that atoms and molecules in the molecular clouds freeze down to form icy mantles (Greenberg and Yencha, 1972, Greenberg, 1979). At the same time that these icy condensates form, they are being irradiated by ultraviolet light which comes from stars and is generated by cosmic ray particles. Some of the molecules in the mantles are broken down, free radicals are created and new molecules are formed. The ice mantles evolve thermally by impulsive heating triggered by grain-grain collisions (Greenberg 1979, Greenberg, 1982, Schutte, 1988 and Grim, 1988), by nearby protostellar objects (Lacy et al, 1984), by energetic particles (Léger et al, 1985), or by local heating in accretion disks (Cohen 1983 and van de Bult et al, 1985).

2. Laboratory Studies

In order to study the physical and chemical processes taking place in the interstellar medium in general, and in particular on these interstellar mantles, the most relevant conditions in interstellar space are reproduced at Laboratory Astrophysics (see e.g. Hagen et al, 1979 and Greenberg, 1986). We deposit gas mixtures (consisting mainly of H_2O, CO, NH_3, CH_4, etc in various ratios) on an aluminum block (representing the silicate core of interstellar grains) cooled down to interstellar temperatures (10 K) and we irradiate them with UV light (simulating the radiation field in interstellar space). These irradiated ice mixtures are warmed-up to room temperature and what is left over on the block (the so-called organic residue) is being analysed by several methods.

3. Results

The more soluble part is being analysed by the group of Prof. J. Ferris at the R.P.I. (Troy, N.Y.) by GCMS (Gas Chromatography - Mass Spectrometry) and HPLC (High Pressure Liquid Chromatography).

A.C. Levasseur-Regourd and H. Hasegawa (eds.), Origin and Evolution of Interplanetary Dust, 437–440.

The results of these analyses are reported elsewhere (see e.g. Agarwal *et al*, 1986 and Schutte, 1988). The less soluble part is analysed by pyrolysis chemical ionization mass spectrometry (PYCI- MS) at the FOM-Institute in Amsterdam (see e.g. Westmore and Alauddin, 1986 and Pouwels *et al*, 1989). All the samples consist of compounds of very high molecular weight - many (especially those with initial composition containing CH_4 and C_2H_2) with mass peaks of more than 500 AMU (see figures) well above the background noise level. In order to check whether these big molecules do not come from contamination (e.g. pump oil in the system), we labelled some of the samples and compared their spectra with the ones of unlabelled samples. It was found that the peaks shifted, and thus the peaks are real and do not come from contamination. We also checked whether the peaks were reproducible or not, and we found they were, i.e. two different samples with the same parent gas mixture and irradiation time, gave the same main peaks.

Although not yet completely characterized, we can say that the pyrolysate is a very complex mixture of hydrocarbon chains since a significant number of main peaks are separated by mass number 14 (CH_2 groups) and, following a main peak, decrease in intensity with increasing fragment weight. The complexity of the spectra would arise from there being many different isomers within homologous series from C_5 onto beyond C_{30}. One very interesting thing is that our spectra looks very much like the one found for the aliphatic hydrocarbons of the Murchison Meteorite (Cronin, 1990), since its most prominent fragment ions belong to a C_nH_{2n-1} series (m/z = 69, 83, 97, 111, ...) that declines in intensity with increasing m/z.

In order to better characterize the composition of the laboratory photoproduced organic residues, they are also being analyzed by other methods. Residues examined for us by the group of Dr. Colin Pillinger using stepwise heating and oxidation, for isotopic analysis, indicated an evaporation sequence similar to that of meteoritic organics, i.e., the presence of aromatic kerogens (see Franchi *et al*, 1989). This is yet to be confirmed, although the preliminary mass spectra obtained by Dr. Krueger, similar to the one performed for Comet Halley (see Kissel and Krueger, 1987) also gave such indications.

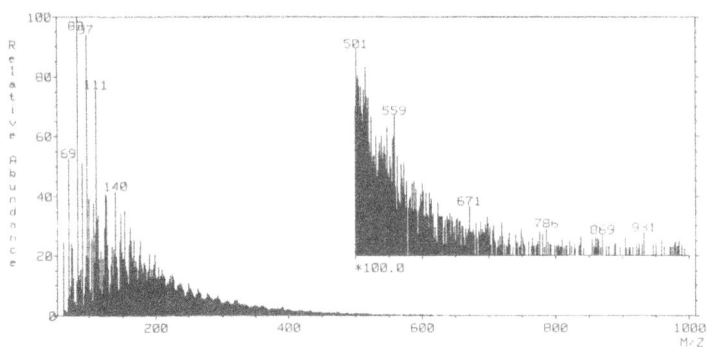

Figure 1. PYCI-mass spectra of irradiated H_2O: CO: C_2H_2: NH_3 = 5:5:2:2

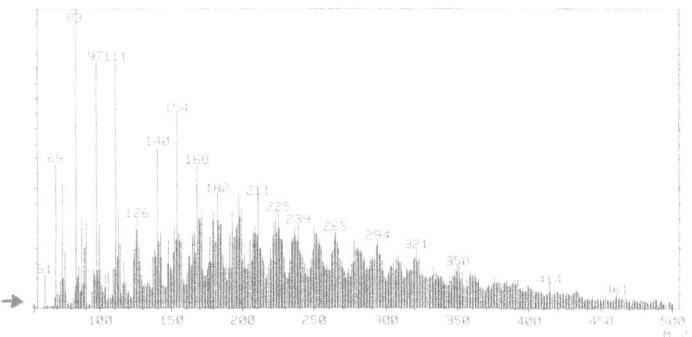

Figure 2. Amplification of the first region of figure 1.
The arrow indicates the background noise level.

4. Conclusions

What has been found up to now on the analysis of our residues would agree well with what has been found for the hydrocarbons of the Murchison Meteorite (Cronin, 1990) and for other meteoritic organics (Franchi *et al*, 1989). The confirmation of this will establish a direct connection between meteorites and interstellar dust via the organics, as well as the connection through mineral compositions. Both carbonates and hydrated silicates may be used to show connections with interstellar dust via aggregates in which the ice of the dust is melted.

Acknowledgements

This work is supported by NASA-Grant nr. NGR 33-018-148, and one of us (C.X. Mendoza-Gómez) acknowledges a grant from DGAPA, University of Mexico (UNAM). The authors are deeply grateful to the useful comments and suggestions of Dr. John Cronin, to the work being done by the group of Dr. C. Pillinger and Dr. F. Krueger, and would like to also thank J. Pureveen, G.B. Eijkel and J.J.Boon for their help and use of the mass spectrometer at the FOM-Institute in Amsterdam.

References

Agarwal, V.K., Schutte, W., Greenberg, J.M., Ferris, J.P., Briggs, R., Connor, S., van de Bult, C.P.E.M. and Baas, F., 1985, in: Origins of Life 16, D. Reidel Publ. Co., 21-40.
Cohen, M., 1983, in: Ap. J. Letters, 270, L69.
Cronin, J., and Pizarrello, S., 1990, in: Geochim. Cosmochim. Acta, Vol. 54, 2859-2868.
Franchi, I.A., Alexander, C.M.O., Pillinger, C.T., Mendoza-Gómez, C.X. and Greenberg, J.M. 1989, in: Physics and Mechanics of Cometary Materials (ESA SP-302), 89-92.
Greenberg, J.M. and Yencha, A.J., 1972, in: Interstellar dust and related topics, eds. J.M. Greenberg and H.C. van de Hulst. D. Reidel Publ. Co., Dordrecht, Holland. p. 3.

Greenberg, J.M., 1979, in: Stars and Star Systems, ed. Westerlund, B.E., Dordrecht, Reidel, 1973.

Greenberg, J.M., 1982, in: Submillimetre Wave Astronomy, eds. Beckman, J.E. and Phillips, J.P., Cambridge University Press.

Greenberg, J.M. 1986, in: Astrochemistry, eds. M.S. Vardya and S.P. Tarafdar. 501-523.

Grim, R., 1988. Ph.D. Thesis. University of Leiden, The Netherlands.

Hagen, W., Allamandola, L.J. and Greenberg, J.M., 1979, in: Astrophys. Sp. Sc. 65, 215-240.

Kissel, J., and Krueger, F.R., 1987, in: Nature 326, 755.

Lacy, J.H., Baas, F., Allamandolla, L.J., Persson, S.E., Mc. Gregor, P.J., Lonsdale, C.J., Geballe, T.R., and van de Bult, C.E.P.M., 1984, in: Ap. J., 276, 533.

Léger, A., Jura, M., and Omont, A., 1985, in: Astron. Astr., 144, 147.

Oró, J., 1961, in: Nature 190, 384-390.

Pouwels, A.D., Eijkel, G.B., Arisz, P.W. and Boon, J.J., 1989, in: J. Anal. Appl. Pyrol. 15, 71-84.

Schutte, W.A., 1988. Ph.D. Thesis, Univ. of Leiden, The Netherlands.

van de Bult, C.E.P.M., Greenberg, J.M., Whittet, D.C.B., 1985, in: M.N.R.A.S., 214, 289.

Westmore, J. and Alauddin, M., 1986, in: Mass Spectrom. Rev., 5, 381.

COLLOQUIUM SUMMARY

THE INTERPLANETARY MEDIUM IS THRIVING

J. MAYO GREENBERG
Laboratory Astrophysics
Huygens Laboratorium
P.O.Box 9504
2300 RA Leiden
The Netherlands

ABSTRACT. Usually in presenting a summary report of a meeting one tries to pick out the highlights. In the present instance I am tempted to say that the meeting was its own highlight because there were so many fine presentations covering such a wide range of interrelated topics. The organizers are clearly to be congratulated on their timing.

1. MULTIFACETED APPROACH

The first impression one gets is the multifaceted approach, involving new as well as extensions of old techniques brought to bear on the subject of the 'Origin and Evolution of Interplanetary Dust'.

1.1. Spectral range

The classical visual and ultraviolet observations have been extended into the infrared, the far infrared and, most recently, into the submm region. The contributions of IRAS have been outstanding in the infrared while the longward wavelength extension by COBE and the James Clerk Maxwell Telescope (JCMT) are just coming into play.

1.2. Space sensing

The delayed return of the Long Duration Exposure Facility (LDEF) is already providing important but still preliminary new data on space particles while the Soviet Space Station MIR will also be used for such analyses. In both of these, there is the opportunity to study returned samples chemically as well as morphologically and, from tracks etc., dynamically as well (mass, velocity, orbits density). The Galileo spacecraft will enormously extend the data return from the three previous detectors aboard the Pioneer 8 and 9 and the Helios spacecraft which recorded only a few hundred impacts. Further in situ measurements by space vehicles will be coming from the Munich Dust Counter (MDC) aboard the MUSES-A mission of Japan. In the future (according to current plans) the

A.C. Levasseur-Regourd and H. Hasegawa (eds.), Origin and Evolution of Interplanetary Dust, 443–451.
© 1991 *Kluwer Academic Publishers.*

CASSINI/NASA mission will return data on the interplanetary particles. The NASA solar probe mission will partially probe dust near the sun.

1.3. Collection

Particles are collected in the atmosphere by high flying airplanes.Many of these are positively identified as being of extraterrestrial origin - so-called interplanetary dust particles (IDP's). Even though many of these are of such low density that their entry is well cushioned by the atmosphere they are somewhat heated and modified in their chemical and morphological structure. Nevertheless they are probably closer to their original form than such other collected particles as meteorites and spherules.

1.4. Chemical and physical analysis

Microprobes for the study of laboratory available samples - whether created in the laboratory or returned with space probes or collected in the earth's atmosphere or at the surface(ground, ocean, antarctic ice) are being exploited for both chemical and morphological data.

1.5. Laboratory simulation

Laboratory simulation of cometary, interstellar, meteoritic materials are providing basic comparisons of the use of techniques for studying the actual materials.

1.6.Theoretical developments

New theories of scattering by fluffy and irregular particles are being applied to both interplanetary and cometary dust particles.

2. MAJOR PROBLEMS.

The major problems which were the focus of the meeting were:

2.1. Sources and sinks of interplanetary particles

2.2. Interrelation between interstellar dust, meteors, meteorites asteroids, comets, stratospheric particles, spherules

2.3. Observations and theories of chemical, and morphological evolution of dust particles, their lifetime and orbits.

3. SPACE DATA.

The accounts by McDonnell, Mandeville, Hüdepohl, and Grün of the past, present and future space efforts give great promise for separating the sources of interplanetary dust - asteroids vs comets vs terrestrial (in the case of LDEF and MIR). We should have a great deal more definitive data on fluxes, orbits, masses and chemical analyses on residual material. Already the Pioneer data revealed the previously unexpected result that the interplanetary particle density certainly has no cut-off beyond 2 AU and falls off very slowly all the way out to 16 AU. But confirmation of the result remains a job for the Galileo detector. In the meanwhile the question of why the zodiacal light (Z.L.) cut-off exists even with the extended dust presence is receiving attention with the answer already being strongly suggested by the IRAS infrared emission data.

4. ZL CLOUD MODELS.

The classical ideas for various spatial models for the visual scattering properties of the interplanetary particles were summarized by Kneissel and some further suggestions were made by Hovenier. An interesting new conceptual approach using the gegenschein as a means of choosing among several models for the spatial distribution was reported by Hong with the proviso of the need for much more precise data. Some observed ultraviolet zodiacal light scattering properties were used by Lillie to indicate the possible presence of submicron particles resembling the cores of the core-mantle interstellar particles. The infrared properties of the dust were shown by Levasseur-Regourd to require models of the ZL particles in which both size distribution and chemical or morphological structure depend on distance from the sun and distribution out of the ecliptic. Such ideas, involving a radial dependence on material density and, by inference, albedo, have been either deduced of were deducible, from the earlier Helios data and the material density falloff of meteors whose orbits carry them further from the sun. Now the infrared data on albedo may be used to deduce the same property which is being called degree of fluffiness. As the theory of Greenberg and Hage on comet dust implies, the increase of dust fluffiness easily explains the slower temperature decrease with distance than $r^{-0.5}$ for $r \rangle 0.5$ AU, perhaps as slow as $r^{-0.3}$.

5. WHAT ARE THE SOURCES OF THE INTERPLANETARY DUST PARTICLES?

Is it possible to explain the Zodiacal Light, and the chemical and morphological structure of collected stratospheric particles and the meteorites in a coherent way, and what is the link to interstellar and circumstellar dust? This question deservedly received the most attention and was viewed from a variety of approaches.

5.1. Mineralogical and chemical analysis and laboratory simulations.

This approach alone was the subject of 20% of the oral presentations and the posters. Results on stratospheric particles, spherules, meteorites, and comet dust were presented along with laboratory simulations on relevant ices and minerals. The first talk on this by Bradley laid the foundation for much of the later discussion. The micron and submicron grain dimensions in the fine grained morphology of interplanetary dust particles requires the use of microprobes. The mineralogical composition of the particles generally falls into two categories -anhydrous and hydrated (layer) silicates.The existence of purely anhydrous phases in some particle precludes the earlier presence of liquid water and seems to demonstrate that these are direct descendants of cometary debris. The particles which have both hydrated and anhydrous phases would appear to be similar to CI and CM (the most 'primitive') meteorites. A further distinction is in the porosities of these two particle types with the former being much fluffier - closer to cometary as shown by the Greenberg-Hage model.

However, along with the positive distinctions there are enough particles of mixed characteristics to raise the question of whether comets and asteroids form totally discontinuous population types or whether perhaps there are at least some intermediate type bodies. We still do not know that comets are all born at the same distance from the sun and, in fact, there are some who suggest that there is evidence of different birthplaces from their orbits. In any case both meteorites and comet debris show mineralogical signatures of interstellar dust with the former sometimes indicating an earlier state (asteroidal) involving liquid water. The range of studies using laboratory and theoretical simulations of formation and evolution of silicates and other material constituents presumed relevant to asteroidal and cometary formation was impressive and too numerous to mention individually. The importance of such studies can not be overemphasized. Analysis of laboratory created samples under controlled conditions and development of standards for various types of detectors will provide a reliability basis for the analysis of solar system particles. One example was the chemical identification of comet infrared emission by laboratory created organics, another was the infrared study of SiO/water mixtures by Sakata's group. Another was the considera-tion of silicate-water interaction measured over time depending on temperature with relevance to the creation of hydrated silicates found in asteroidal meteorites. The Bussoletti group presented an example of spectra of laboratory produced glassy and crystalline silicates for comparison with cometary spectra.

5.2. Asteroid and cometary debris trails and tails.

Asteroidal and cometary debris trails should be the most direct evidence for the source of debris in the solar system other than the comet dust observation. From these one may attempt to make estimates of the dust mass input. Only preliminary results are yet available from IRAS studies of comet trails which provide data on the very low velocity component of short period comet dust debris.The fact that the brightest IRAS bands associated with the principal Hirayama asteroid families led Sykes to conclude that they result from collisions. However he also pointed out that, in general, the dust production rates are not correlated with the local asteroidal concentration. Although the detection of zodiacal dust bands and cometary debris trails by IRAS may not yet be sufficient to explain all that is required to give the observed zodiacal light, the application of the new capabilities for detecting the largest masses in dust streams using submillimeter observations, may provide the required deficit. These large particles, relatively invisible in the infrared, could erode or break up to supply the needed dust seen in the visible. Long period comets do not have trails locally so that one must provide a firm theoretical foundation to calculate what happens to the dust which leaves the comet at all velocities and masses. Apparently, according to the latest such calculations on orbits and ejection times presented by Fulle, the mass input by long period comets is quite similar to that provided by short period comets.

5.3. Orbital distribution of meteoroids.

The source of interplanetary particles can be deduced from their orbits in the neighborhood of the earth if these orbits are significantly different for particles coming from comets or asteroids. Remote detection is limited to photographic, radar, and television observations. The most up-to-date survey of these data presented by Steel seems to show that about equal contributions to the interplanetary particle complexes are made by asteroids and comets. But, just as in the data from tails and trails the uncertainties are such that both the absolute amounts and the relative amounts of mass *and* input require much more investigation along these directions to be totally convincing.

5.4. Optical scattering and IR emission by the interplanetary dust cloud.

The simultaneous interpretation of the visible scattered light and the infrared emission by the zodiacal dust cloud has required an expanded approach to the nature of the responsible particles. It is no longer possible to consider simple solid material particles in the theories. The direct evidence for much more complex particles in the form of IDP's and indirect but convincing evidence from comet dust studied by Greenberg and deduced from meteor densities, require new theoretical scattering methods. Not yet fully appreciated is the concomitant need for reinvestigation of the dynamical properties and orbital evolution of complex particles in the solar system. Fluffy aggregated comet dust particles should hardly be visible as zodiacal light particles of *equal mass*. Even IDP's are relatively poor scatterers of visible radiation. If the zodiacal light comes from such low albedo particles we would require a much higher mass. There must exist a hierarchy of fluffiness or, inversely, solidity. This is already evident in the differences between the radial distribution of the effectively scattering and the effectively infrared emitting particles, with the latter extending in their distribution to larger distances from the sun than the former. Could asteroid debris be relatively more responsible for the visibly scattering (more solid) particles and cometary debris for the less scattering - higher infrared emitting - particles? If so, this should show up in the relative degree of concentration of these two different types in and out of the ecliptic; asteroid debris being expected to be more concentrated to the ecliptic. A very important deduction from the existence of low albedo particles is that the most likely cause of such a low albedo is not alone the fluffiness but the presence of material more absorbing than rocky (silicate) material. The inference that this absorbing component is organic - as now proven to exist in cometary dust as well as IDP's - and as shown to exist in meteorites is a critical connecting link between all dust components and leads to a further implication that the ultimate source of solid material in the interplanetary system is interstellar dust. A very novel and potentially important theoretical study of the physical processing of micrometeoroids impacting on the upper atmosphere by Kamijo may provide a connection between the interplanetary particles collected in the upper atmosphere (IDP's) and spherules.

The fact that short period comets and asteroids both are the prime sources of particles in the ecliptic while long period comets are the prime sources of particles out of the ecliptic with probably a smaller contribution from periodic comets would imply that out-of-ecliptic dust is fluffier on the average than dust in the ecliptic. The optical manifestations observed in and out of the ecliptic were shown by Levasseur Regourd to be indeed different in their polarizing properties but whether the corresponding albedos can be the total cause is yet to be determined. In any case we seem to be seeing at least 2 populations of particles.

6. ASTEROID AND COMET ORIGINS.

After having examined and, hopefully, having answered the question of the sources of interplanetary particles, the final stage is to study the material source of the asteroids and comets. The basic question here is to what extent these objects can trace their composition *directly* to the collapsing interstellar cloud out which the solar system was born. Interstellar clouds are well known to contain a wide range of molecular constituents in both the gas phase and solid state -interstellar dust. During the process of collapse to the disk form of the protosolar nebula and subsequent to the turn-on of the sun these constituents must certainly undergo some processing as a result of heating, turbulence and radiation. One class of theories is that all of the solid material was evaporated before being reconstituted into the present solar system bodies, including asteroids and even comets. The current understanding is that at least in comets and, to a lesser extent, in asteroids the interstellar dust grains preserved many of their pre-solar system properties. This theme emerged in a large number of papers presented at the symposium.

6.1. Interstellar dust

Currently we have knowledge of a wide variety of material constituents of the dust. Some of these are volatiles in the form of frozen ices, the others being relatively non-volatile or refractory. However even the refractories have varying degrees of volatility. The major refractory components which have been observed may be characterized as silicates and organics. The former category is probably dominated by amorphous magnesium/iron silicates, produced by stars. The latter consist of (1) complex organic molecular mixtures created as residues from ultraviolet processing of 'ices' in space; (2) large aromatic molecules called PAH's (polycyclic aromatic hydrocarbons) and; (3) other yet uncertain carbon or carbonaceous molecules. The connection of these constituents with what is found in interplanetary particles involves not only *what* they are but *how* they are configured. The most abundant of the carbon materials appear as organic mantles on the silicates. Another property of great importance in establishing the *cosmic dust connection* is the interstellar dust size, or size distribution. Theoretical modelling of interstellar polarization and extinction show that the bulk (by mass) of the interstellar particles consist of tenth micron silicate cores with organic mantles whose mean, or characteristic, volume is about twice that of the silicates. in pre-solar interstellar space the dust grains have substantial outer mantles of frozen molecules generally dominated by H_2O and CO but containing many other species such as CO_2, CH_3OH, H_2CO with a total of 10-15 well identified by solid state infrared spectroscopy. What role do these 'ices' play in the pre and early solar system chemistry and are these completely or partially evaporated before the formation of asteroids and comets? In the same way that one may question where in the pre-solar nebula, the silicates are preserved or evaporated one may also question where and to what extent the ices and the organics may have been preserved before becoming a part of the larger bodies.

6.2   Interstellar dust-comets

The invited review by Yamamoto was a careful consideration of the former question and he presented persuasive arguments to show why, in the regions where comets are formed, the H_2O ice and some of the molecules trapped in it are maintained in place on the aggregating grains. However he suggested that under some conditions the more volatile constituents like CO are expected to partially or fully evaporate so that one may expect to observe some variability in the CO abundances from one comet to the other depending on the initial formation region.   Accordingly, in all cases the organics and obviously the silicates should be preserved. In the paper by Hage some of the consequences of considering unaltered interstellar dust to be the ingredients of comets were taken up in some detail. Of major interest was the conclusion that the observational properties of comet can best (probably *only*) be satisfied if comets are very fluffy aggregates of interstellar dust as modelled by Greenberg; that is, not only the chemical composition but also the size and configuration is relevant. Thus the source of low albedo interplanetary particles is naturally explained along with low density meteors. Although the interplanetary particles known as chondritic IDP's are also fluffy, their degree of fluffiness is considerably less (by a factor of about 5). Nevertheless they bear sufficient resemblances in morphological structure to aggregated interstellar dust(but with reduced organics) to have evolved to their present state by natural processes occurring during their up to 10,000 year sojourn in the interplanetary space (heating, solar wind, photolysis) and their atmospheric entry. And, furthermore, such physical changes are just what are needed to explain the greater amount of particles with decreasing albedos further removed from the sun. Sandford drew the connection between interstellar and comet dust and IDP's via their correlated mineralogical properties and (inferred) orbits of the latter. Those particles with higher-than-average earth encounter velocities and therefore relatively eccentric orbits contain predominantly anhydrous silicates as observed in comet dust.

6.3   Interstellar dust-asteroids

Establishing asteroidal connections with interstellar dust has to be done via meteorites and other debris. Since already interstellar matter must have undergone substantial metamorphosis not only during but after parent body formation the problem is much more difficult than for comets. Among other things water can not have survived in solid form. But there is evidence that in liquid form it had effects which can be deduced from meteorites and IDP's. For example, Sandford suggests an asteroidal source for those IDP's which have predominantly layer-lattice silicates which result from hydration in the presence of liquid water. His reason is that these IDP's indicate very low encounter velocities implying low inclination prograde orbits implying asteroidal debris. If liquid water survives in asteroid formation then trapped CO_2 may also have survived. This may perhaps be the source of carbonates in meteorites. It would interesting to determine whether there exists a correlation between the presence of carbonates and hydrated silicates because both indicate the presence of liquid water at some stage of formation of the parent body.

7.   CONCLUSIONS

The study of interplanetary dust has been shown at this meeting to provide a rich field for providing a clearer understanding of the origin and formation of the solar system.

A schematic summary of the meeting is shown in Figure 1.

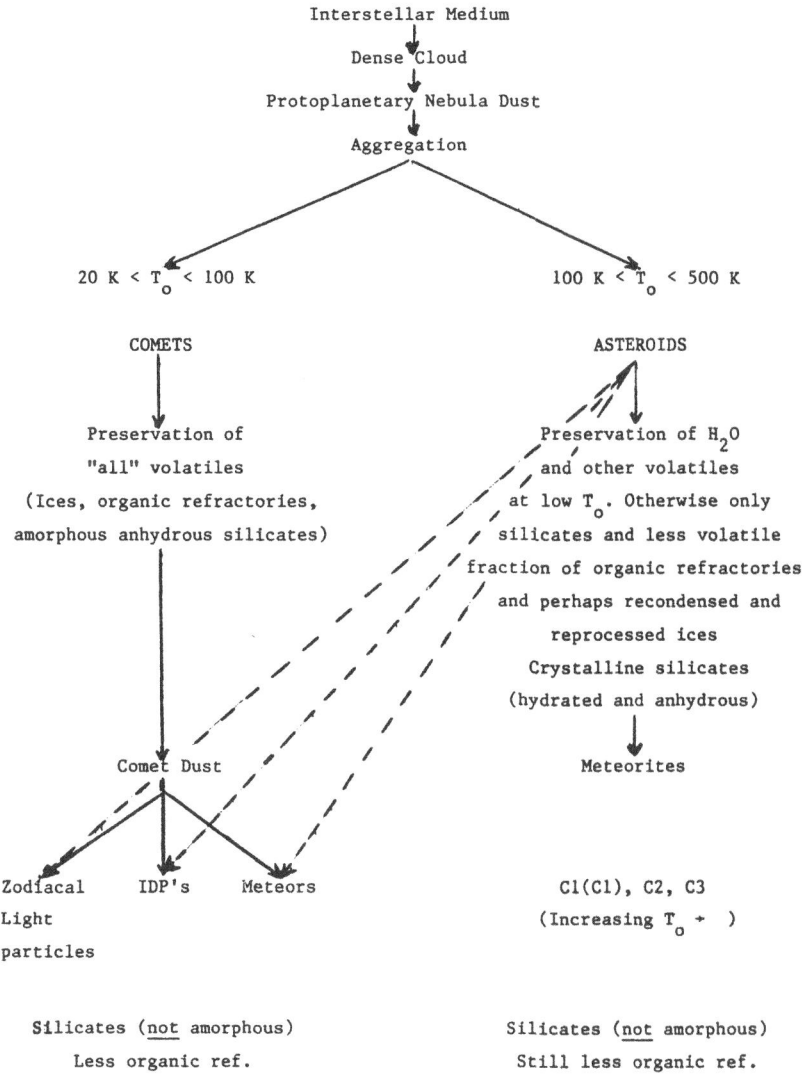

Figure 1. Some suggested pathways from the interstellar
     to the interplanetary medium.

AUTHOR INDEX

Akisawa H.	269
Arakawa E.T.	99, 102
Araki H.	15, 45
Asher D.	327
Blanco A.	125
Boice D.C.	265
Bosma P.B.	155
Bradley J.P.	63
Brandon E.	277
Burns J.A.	341
Bussoletti E.	91, 125
Clairemidi J.	217, 277
Clube V.	327
Colangeli L.	91, 125
Cremonese G.	225
Deshpande M.R.	281
Di Martino M.	383
Doyle L.R.	357
Ducrocq E.	383
Dumont R.	131, 199
Eklund P.C.	102
Fang Y.	195
Farinella P	383
Fechtig H.	21
Fonti S.	125
Fujii N.	281
Fujimura A.	15, 45
Fujiwara A.	281, 361, 379
Fulle M.	225
Giblin I.	383
Gilmour I.	99
Giovane F.	191
Glasmachers A.	15, 45
Green S.F.	37
Greenberg J.M.	261, 437, 443
Grün E.	15, 21, 45, 257, 357, 367, 375
Gucun W.	33
Gyssens M.	335
Hage J.I.	261
Hajduk A.	323, 331
Hajdukova M.	323
Hanchang P.	57
Hanner M.S.	21, 171
Hayashi T.	15, 45
Hong S.S.	147, 179, 183
Hovenier J.W.	155
Hoyle F.	235
Hüdepohl A.	15, 45
Huebner W.F.	221, 265
Igenberg E.	15, 45
Iglseder H.	15, 45
Ishibashi T.	281
Ishii N.	15, 45
Jianguo M.	53, 105
Joshi U.C.	281
Kadono T.	281
Kaito C.	117
Kamijo F.	303
Kawamura K.	113
Keller H.U.	229
Khare B.N.	99, 102
Kikuchi S.	249
Kissel J.	21
Kneissel B.	139
Knöfel A.	335
Kohl H.	257
Koike C.	95, 121
Koller G.	15, 45
Konno I.	221
Kouchi A.	87
Koutchmy S.	191
Kuroda T.	87
Kwon S.M.	147, 179, 183
Lamy P.	163, 191, 195
Levasseur-Regourd A.C.	131, 199
Lillie C.F.	151
Lindblad B.A.	21, 299, 311
Linkert D.	21
Llebaria A.	191, 195
Lumme K.	159
Mandeville J.C.	11
Mann I.	139, 187
Martelli G.	383
Martinsson A.	383
Matsumura M.	203
Maucherat A.	191
McDonnell J.A.M.	3, 37
Meglick	245
Meisse C.	99
Mendoza-Gomez C.X.	437
Mennella V.	91
Misawa K.	105
Misconi N.Y.	183
Mizutani H.	15, 45
Moreels G.	217, 277
Morfill G.	21
Muinonen K.	159
Mukai S.	249
Mukai T.	249, 371

Naegeli D.W.	265
Nagasawa K.	307
Nakada Y.	49
Nakamura A.	281, 379
Nakamura N.	79
Nakano T.	421
Niblett D.H.	3, 37
Nishioka K.	253
Nogami K.	15, 45, 53, 105, 109
Ohtsuka K.	315
Oka T.	269
Omori R.	53, 105
Onaka T.	49
Orofino V.	125
Perrin J.M.	163
Randolph J.E.	29
Reach W.T.	211
Renard J.B.	131, 199
Rendtel J.	335
Richter K.	229
Rietmeijer F.J.M.	49
Roggemans P.	335
Rothwell P.	383
Rousselot P.	217, 277
Sagan C.	99, 102
Saito Y.	117
Sakata A.	241, 429
Sakurai K.	433
Sandford S.A.	397
Sasaki S.	425
Schwehm G.	15, 45
Seki M.	203
Sen A.K.	285
Shibai H.	121
Shijie Z.	57
Shimaoka T.	79
Showalter M.R.	349
Shulan M.	53
Smith P.N.	383
Stammes P.	207
Steel D.	41, 291, 327
Stephens J.R.	49, 125
Stevenson T.J.	3, 37
Sugawara K.	269, 273
Sullivan K.	3, 37
Suzuki K.	319
Svedhem H.	15, 45
Svestka J.	367
Sykes M.	389
Thompson W.R.	99, 102
Tielens A.G.G.M.	405
Timmermann R.	375

Tokunaga A.T.	241, 429
Tomeoka K.	71
Tsuchiyama A.	83, 95, 113
Tsurutani B.T.	29
Uesugi K.T.	15, 45
Visvanathan N.	245
Wada S.	241, 429
Watanabe J.	253, 273
Weinberg J.L.	179, 183
Wettstein M.	383
Wickramasinghe D.T.	245
Wickramasinghe N.C.	235
Xiguang W	33
Xueying M.	57
Yamakoshi K.	15, 45, 53, 105, 109
Yamamoto T.	15, 45, 413
Yiwen X.	33
Young D.W.	102
Zhang J.M.	102
Zhenkun L.	57
Zhifang C.	57
Ziyuan O.	33
Zook H.A.	21

The manufacturer's authorised representative in the EU is Springer
Nature Customer Service Centre GmbH, Europaplatz 3, 69115 Heidelberg,
Germany. If you have any concerns regarding our products, please
contact ProductSafety@springernature.com

Printed and bound by CPI Group (UK) Ltd, Croydon, CR0 4YY
23/04/2026
02095624-0008